Parks and Protected Areas in Canada

Planning and Management

Third Edition

Parks and Protected Areas in Canada

Planning and Management

Third Edition

Edited by
Philip Dearden and Rick Rollins

OXFORD
UNIVERSITY PRESS

OXFORD
UNIVERSITY PRESS

70 Wynford Drive, Don Mills, Ontario M3C 1J9
www.oupcanada.com

Oxford University Press is a department of the University of Oxford.
It furthers the University's objective of excellence in research, scholarship,
and education by publishing worldwide in

Oxford New York
Auckland Cape Town Dar es Salaam Hong Kong Karachi
Kuala Lumpur Madrid Melbourne Mexico City Nairobi
New Delhi Shanghai Taipei Toronto

With offices in
Argentina Austria Brazil Chile Czech Republic France Greece
Guatemala Hungary Italy Japan Poland Portugal Singapore
South Korea Switzerland Thailand Turkey Ukraine Vietnam

Oxford is a trade mark of Oxford University Press
in the UK and in certain other countries

Published in Canada by Oxford University Press

First published 2009

Library and Archives Canada Cataloguing in Publication

Parks and protected areas in Canada : planning and management /
edited by Philip Dearden and Rick Rollins. — 3rd ed.

ISBN 978-0-19-542734-9

1.National parks and reserves—Canada—Management.
2. Natural areas—Canada—Management. 3. Wilderness areas—
Canada—Management. I. Dearden, Philip II. Rollins, Rick
QH77.C3P37 2008 333.78'30971 C2008-902359-5

1 2 3 4 – 11 10 09 08

Cover Image: Chris Parker/Getty Images

This book is printed on permanent (acid-free) paper ♾.
Printed in Canada

Contents

List of Figures

List of Tables

List of Boxes

Contributors

Jim Butler is Professor Emeritus of the Faculty of Renewable Resources at the University of Alberta, where he was a founder of the program in interpretation studies at the University of Alberta. He has had a long career in the field of interpretation, including with the National Audubon Society, as Chief Naturalist for the state of Kentucky, and as Head of Interpretation and Education for the provincial parks of Alberta. Since his retirement, he is a popular lecturer worldwide for Holland America Cruise Ships.

Rosaline Canessa is Assistant Professor in Geography at the University of Victoria. Her research interests focus on integrated coastal management, marine protected areas, and the development and use of Geographic Information Systems to support planning and management, particularly in collaborative settings.
E-MAIL: Rosaline@uvic.ca

Philip Dearden is Professor of Geography at the University of Victoria. He is a member of the World Commission on Protected Areas of IUCN and has been active in the planning and management of protected areas in many different countries, especially in Asia. He is the Chair of Canada's Working Group on Marine Protected Areas under the Ocean Management Research Network and Co-chair of Parks Canada's NMCA Marine Science Network. He is particularly interested in incentive-based conservation in marine environments and zoning and has an active research program on the topic in Southeast Asia.
E-MAIL: pdearden@office.geog.uvic.ca

Jessica Dempsey is a Trudeau Scholar at the University of British Columbia, currently studying market-based conservation mechanisms en route to a Ph.D. in Geography. She is also actively involved in lobbying and organizing around the UN Convention on Biological Diversity.
E-MAIL: jdempsey@interchange.ubc.ca

Paul Eagles is Professor in the Department of Recreation and Leisure Studies at the University of Waterloo. He has extensive experience in many aspects of park planning and management gained through work in more than 20 countries. Currently he is Chair of the Tourism Task Force for the World Commission on Protected Areas, a commission of the World Conservation Union (IUCN).
E-MAIL: eagles@healthy.uwaterloo.ca

Wolfgang Haider is Associate Professor in the School of Resource and Environmental Management at Simon Fraser University. His research and teaching interests include protected areas management and planning, outdoor recreation and tourism behaviour, and landscape perception. His work frequently involves choice modelling to determine preferences in a trade-off context.
E-MAIL: whaider@sfu.ca

Glen Hvenegaard is Associate Professor of Geography and Environmental Studies at the University of Alberta's Augustana Campus in Camrose. He teaches courses on parks, environmental studies, and physical geography, and conducts research on conservation issues related to ecotourism, birds, parks, and environmental education. He has published widely and received national and international awards for his work. He is a member of the World Commission on Protected Areas and its Task Force on Tourism, and is a Fellow with Leadership for Environment and Development.
E-MAIL: hveng@marcello.augustana.ab.ca

Margaret Johnston is Professor in the School of Outdoor Recreation, Parks and Tourism at Lakehead University. Her research and teaching focus on polar tourism, recreation behaviour in natural areas, and community outcomes of sporting events.
E-MAIL: mejohnst@lakeheadu.ca

Steve Langdon is the Field Unit Superintendent for Coastal British Columbia, responsible for Parks Canada interests on Vancouver Island and the Lower Mainland. He has had a long career with Parks Canada, holding a variety of positions including: Director, NMCA Establishment—West Coast; Director, Aboriginal Affairs Secretariat in the National Office; Superintendent, Gwaii Haanas National Park Reserve; Superintendent, Nahanni National Park Reserve; Superintendent, Ft St James National Historic Site; and Chief, Visitor Services, Riding Mountain National Park.
E-MAIL: steve.langdon@pc.gc.ca

Harvey Lemelin is Assistant Professor with the School of Outdoor Recreation, Parks and Tourism at Lakehead University. His research and teaching interests involve sustainable tourism, northern communities and protected areas, and the human dimension of wildlife–human interactions.
E-MAIL: harvey.lemelin@lakeheadu.ca.

Chris Malcolm is Associate Professor in the Department of Geography at Brandon University. He spent many summers growing up camping, canoeing, and fishing in Ontario provincial parks such as Samuel de Champlain, Esker Lakes, and Lady Evelyn. These experiences have led to an academic career teaching Wildlife Management, Parks and Protected Areas, and Biogeography. His research projects include habitat selection of northern pike in southern Manitoba, the ecological role of marbled murrelets in Clayoquot Sound, BC, and the management of beluga watching in Churchill, Manitoba.
E-MAIL: malcolmc@brandonu.ca

Kevin McNamee is Director of the Park Establishment Branch with Parks Canada Agency, where he is responsible for overseeing the federal government's efforts to create and expand national parks and to establish new national marine conservation areas. He was one of the 21 original signatories of the Canadian Wilderness Charter in 1989, and served as the federal Endangered Spaces co-ordinator with the Canadian Nature Federation from 1989 to 2000 with a focus on lobbying the federal government to complete the national park system. He was the Executive Director, then Conservation Director, of the Canadian Parks and Wilderness Society from 1983 to 1989, and taught wilderness and recreation courses at Trent University in Peterborough.
E-MAIL: Kevin.McNamee@pc.gc.ca

Mark D. Needham is Assistant Professor in the Recreation Resource Management Program at Oregon State University. He received his BA and MA in Geography and Environmental Studies at the University of Victoria and his Ph.D. in Human Dimensions of Natural Resources at Colorado State University. His special interests and research are in recreation resource management, human dimensions of wildlife, nature-based tourism, and survey methodology and analysis.
E-MAIL: Mark.Needham@oregonstate.edu

Joe Pavelka teaches in the Applied Ecotourism and Outdoor Leadership program at Mount Royal College in Calgary and in the Haskayne School of Business at the University of Calgary. He is also completing his doctorate in the Department of Geography at the University of Calgary, examining amenity migration and its impacts on nature-based tourism communities. Joe has provided market research and strategic support to tourism and recreation organizations throughout Canada for over 15 years.
E-MAIL: jpavelka@mtroyal.ca

Bob Payne is Professor in the Department of Outdoor Recreation and Tourism at Lakehead University, with special interests in human use management in protected areas.
E-MAIL: rjpayne@flash.lakeheadu.ca

Rick Rollins is a faculty member in the Department of Recreation and Tourism at Malaspina University College and an adjunct faculty member in the Department of Geography at the University of Victoria. His teaching and research deal with recreation behaviour and management in natural settings.
E-MAIL: rollins@mala.bc.ca

John Shultis is Associate Professor in the Outdoor Recreation and Tourism Management Program at the University of Northern British Columbia. He has conducted research, published, and taught in the areas of protected areas planning management and environmental interpretation for over 15 years. He is a member of the IUCN World Commission on Protected Areas and an executive editor of the *International Journal of Wilderness.*
E-MAIL: Shultis@unbc.ca

Scott Slocombe is Professor of Geography and Environmental Studies at Wilfrid Laurier University. His research and teaching focus is on regional environmental planning and sustainability, including protected areas, regional planning, environmental assessment, and systems approaches.
E-MAIL: sslocomb@wlu.ca

Jeannette Theberge is a wildlife and monitoring ecologist in Mount Revelstoke and Glacier national parks. She has conducted research in a cross-section of parks in Canada and the US, and, with her co-author/father, she has visited many of the world's park systems. She has a Ph.D. from the University of Calgary in conservation ecology of grizzly bears.
E-MAIL: jen.theberge@pc.gc.ca

John Theberge is a wildlife ecologist specializing in wolves and predator–prey relationships. He taught at the University of Waterloo for 30 years and has been active in the establishment and ecologically based management of wilderness parks in Canada. He received the Harkin Award from the Canadian Parks and Wilderness Society for his contribution to protected areas in Canada. He currently lives in the Okanagan Valley of British Columbia and is active in research, writing, and promoting the establishment of a national park.
E-MAIL: johnmarythe@xplorenet.com

Stephen Woodley is an ecologist, and Chief Ecosystem Scientist with Parks Canada. He works on a number of issues related to protected areas, including developing techniques for monitoring and assessing ecological integrity, ecological restoration, and sustainable forestry.
E-MAIL: Stephen_Woodley@pc.gc.ca

Pamela Wright has an interdisciplinary natural resource background specializing in both social and ecological sciences with a focus on parks and protected areas and related issues in forest sustainability and monitoring. She served as vice-chair of the Panel on the Ecological Integrity of Canada's National Parks. She is currently Associate Professor in the Outdoor Recreation and Tourism Management program at the University of Northern British Columbia and she serves as the founding chair of the BC Protected Areas Research Forum.
E-MAIL: pwright@unbc.ca

Foreword to the Third Edition

The world is a very different place from what it was when I wrote the Foreword to the first edition of this book in 1991. Canada was a country that was heavily influenced by its historic relations with England and France and its modern relations with the United States. We received many of our ideas and evaluated our progress against those cultures, adapted to our circumstances and preferences. Today we live in a country increasingly charting its own course in the world and we think we have less and less to learn from our previous role models. There are no models for a nation composed of people from all over the planet that seeks to embrace its founding Aboriginal peoples and move forward as a citizen of the world without seeking to bend others to our will. We are a work in progress that, surprisingly, is becoming a global role model.

One thing continues to stand at the centre of Canadian experience. All Canadians, whether newly arrived or with deep roots, are shaped by the single common fact that they are aware they live in a country of enormous space, cold climate, and wild landscapes that is, paradoxically, one of the most intensely urbanized and prosperous societies on earth. At international parks gatherings it becomes quickly obvious that Canada, despite the challenges thoughtfully analyzed in this volume, is a global leader in terrestrial protected areas (we are, to date, catastrophically bad at marine conservation).

Several challenges stand out for Canadian parks and protected areas in the twenty-first century. Two are carried forward from the past. We need to act with urgency to create more interconnected protected areas while we still can, because the pace of industrial exploitation and its malignant offspring, climate change, are radically transforming our natural heritage and depleting wildlife populations. The second challenge is the eternal curse of existing parks. It was best stated by the great J.B. Harkin, the first Commissioner of National Parks in Canada: 'the battle to keep them inviolate (after establishment) is never won. Claims for the violation of their sanctity are always being put forward under the plausible plea of national or local needs' (Harkin, 1957: 15). These two challenges must always be addressed in all thinking and actions directed towards parks and protected areas.

But in 2008, these traditional challenges are joined by a new one that is less visible but perhaps more important. The decline of our former role models, combined with the general anxiety young people feel for the state of the world and the contempt they feel for those who got us into this situation, has led to widespread cynicism and a post-modern mistrust of any broad social narratives. National parks and protected areas are creatures of a narrative that values wild nature as necessary and beautiful, and that is, in and of itself, worthy of protection. If we do not believe in that narrative and spread it, parks and protected areas will cease to exist.

Canada will be increasingly called on to chart its own social course in what unfortunately promises to be a very turbulent twenty-first century. We will be required to develop and act on our own narrative if we are to flourish. This presents both an opportunity and a challenge to those of us who care about the health of the world and our own legacy. We must ensure that the new Canadian narrative has the preservation of big, wild nature at its centre and a vast, well-managed, and interconnected system of parks and protected areas on land and water as a key measure of social progress.

Harvey Locke
Senior Advisor, Conservation, Canadian Parks and Wilderness Society
Member, World Commission on Protected Areas
Montreal, Quebec

REFERENCE

Harkin, J.B. 1957. *The History and Meaning of the National Parks of Canada.* Saskatoon: H.R. Larson.

Acknowledgement

We would like to acknowledge Ole Heggen of the Department of Geography at the University of Victoria for his work on the maps and diagrams.

PART I

Overview

> National Parks are maintained for all the people—for the ill that they may be restored, for the well that they may be fortified and inspired by the sunshine, the fresh air, the beauty, and all the other healing, ennobling, and inspiring agencies of Nature. National Parks exist in order that every citizen of Canada may satisfy a craving for Nature and Nature's beauty; that we may absorb the poise and restfulness of the forests; that we may steep our souls in the brilliance of wild flowers and the sublimity of the mountain peaks; that we may develop in ourselves the buoyancy, the joy, and the activity we see in the wild animals; that we may stock our minds with the raw materials of intelligent optimism, great thoughts, noble ideas; that we may be made better; happier and healthier.
>
> *James B. Harkin, first Commissioner of Canadian National Parks*

Part I, Overview, sets the stage for issues pursued in subsequent chapters. We begin by asking, what are protected areas and what are the different types of protected area? Why do we have parks and protected areas? What scientific, ethical, aesthetic, spiritual, or other reasons have been used to rationalize our passionate feeling about these special places? James B. Harkin, Canada's first Commissioner of National Parks, provided his thoughts most eloquently on the topic over 75 years ago, as noted above. Chapter 1, by Philip Dearden and Rick Rollins, reveals many reasons for protecting natural areas and the many benefits realized from existing parks, and discusses how the relative importance among these reasons can change over time. We also provide an overview of how Canada is doing in terms of establishing and managing protected areas and some of the key issues being faced, such as declining visitation and the challenges of planning for global climate change. The chapter finishes with an overview of the book contents.

From a historical perspective the Canadian vision for parks has changed over time, as Kevin McNamee explains in Chapter 2. In the early days, parks were viewed mainly as playgrounds or tourist attractions set in exotic natural surroundings. Over the past century, our concept of the purpose and value of parks has shifted, to a large extent because the amount of wilderness or undeveloped landscape in Canada has been drastically reduced. As a consequence of this historical evolution of thinking about why we need parks, the style or approach to management has varied considerably. Hence, in the past 20 years we have seen revisions to parks legislation, park policies, and park regulations. In short, we have been agonizing over, 'How should parks be managed?' The present view places a much higher emphasis on ecological integrity. However, the legacy of past thinking remains. We still have townsites in Banff, Jasper, and Riding Mountain national parks. We condone alpine skiing, golfing, hunting, and other questionable recreation activities in some parks. Logging occurs in the heart of Algonquin Provincial Park in Ontario, as does mining in the heart of Strathcona Provincial Park in BC. These park management decisions were made in the past when we viewed parks differently. To make sense of these apparent contradictions in the context of thinking about parks, one needs an appreciation for the history of parks, as provided in Chapter 2.

The book acknowledges the diversity of different types of protected areas but concentrates mainly on large natural areas managed primarily for the conservation of biodiversity and recreational use. Next to the national parks, then, the provincial parks are most significant in Canada in this regard. In Chapter 3, Chris Malcolm outlines some of the differences between national and provincial parks, provides an overview of the status of provincial parks in Canada, and discusses some of the main management issues being faced.

CHAPTER 1

Parks and Protected Areas in Canada

Philip Dearden and Rick Rollins

INTRODUCTION

The World Conservation Union (IUCN, 1994) defines a protected area as 'an area of land and/or sea especially dedicated to the protection and maintenance of biological diversity, and of natural and associated cultural resources, and managed through legal or other effective means'. Although societies throughout the world for generations have set aside areas of special value for protection, the modern manifestation of protected areas was really ushered in with the advent of Yellowstone National Park in the United States in 1872. Unlike many earlier protected areas, Yellowstone was large and set aside mainly for conservation reasons and dedicated to all peoples of the US as a common good, rather than to a privileged class (Nash, 2001).

The idea caught on, and Canada was one of the first countries to adopt a similar strategy, although with a much stronger emphasis on income generation through tourism, as described in Chapter 2. Since that time protected area (PA) systems have spread throughout the world, as described in Chapter 17, and now cover some 12 per cent of the global land surface. Not only have PAs grown in coverage, they have also been broadened in terms of their management objectives. National parks are the best-known type of PA, but there are many other types of designation that also confer protection. The IUCN recognizes this diversity through having different categories of protected areas (Table 1.1).

Different types of protected areas give priority to different values (Table 1.2). Thus, Category I protected areas exhibit the strongest restrictions on use to match their prime function of biodiversity protection and include designations such as ecological reserves and wilderness lands (Box 1.1). Category II lands, usually administered as national parks, have somewhat less stringent protective regulations and a mandate for human use and enjoyment, as well as biodiversity protection. The relaxation of protective standards continues through to Category VI, which, although ostensibly managed primarily for biodiversity protection, also allows sustainable use; such use may include forestry and other extractive activities. In principle, the idea of different categories of PA is of value because it recognizes the plurality of approaches necessary to provide a comprehensive approach to landscape conservation. However, Locke and Dearden (2005) have identified several problems with the implementation of the IUCN system. These are discussed in Chapter 17, and the IUCN is now revising the categories and their implementation.

TABLE 1.1 IUCN Categories of Protected Areas

Category	Description
I Strict protection	Strict Nature Reserve(a)/ Wilderness Area(b)
II Ecosystem conservation and recreation	National Park
III Conservation of natural features	Natural Monument
IV Conservation through active management	Habitat/Species Management Area
V Landscape/seascape conservation and recreation	Protected Landscape/Seascape
VI Sustainable use of natural ecosystems	Managed Resource Protected Area

Source: IUCN (1994).

TABLE 1.2 Matrix of Management Objectives and IUCN Protected Area Management Categories

Management Objective	Ia	Ib	II	III	IV	V	VI
Scientific research	1	3	2	2	2	2	3
Wilderness protection	2	1	2	3	3	—	2
Preservation of species and genetic diversity	1	2	1	1	1	2	1
Maintenance of environmental services	2	1	1	—	1	2	1
Protection of specific natural/cultural features	—	—	2	1	3	1	3
Tourism and recreation	—	2	1	1	3	1	3
Education	—	—	2	2	2	2	3
Sustainable use of resources from natural ecosystems	—	3	3	—	2	2	1
Maintenance of cultural/traditional attributes	—	—	—	—	—	1	2

Key: 1, Primary objective; 2, Secondary objective; 3, Potentially applicable objective; —, Not applicable
Source: IUCN (1994).

BOX 1.1 Wilderness in Canada and the World

The term 'wilderness' is difficult to define and can mean many things to different people in different times. From Biblical times, the conception of wilderness was of wastelands (i.e., deserts) unsuitable for human habitation, and this association remained dominant until the rise of the Industrial Revolution. Leaders of the Romantic movement, which arose as a result of the degraded social and environmental conditions resulting from the Industrial Revolution, were the first to start associating wilderness with positive attributes. In the United States, transcendentalists such as Ralph Waldo Emerson, Henry David Thoreau, and John Muir accentuated the positive attributes of wilderness among the intelligentsia, suggesting that wild areas provided psychological and physical solace to those affected by the vicissitudes of urban life. Fast forward to the present, where the term 'wilderness' is commonly used to define natural areas that appear to be unaffected by human activities and where humans do not normally reside. Indeed, the term 'wilderness' is often equated with national or provincial parks by the Canadian public.

There are two distinct ways that 'wilderness' can be defined. The first meaning is noted above: a highly personal definition that every person holds but that generally refers to natural areas outside of urban centres largely unaffected by human presence or activities. In this definition, 'wilderness' is whatever a person thinks it is: one person's wilderness is another person's urban park. The second meaning refers to specific legislated areas termed 'wilderness' or 'wild lands'. That is, many governments around the world—particularly in Western nations—have passed legislation that specifically defines wilderness areas. The US Wilderness Act (1964) is the best-known and by far most influential example.

In Canada, the first legislation related to wilderness was the 1959 Wilderness Area Act in Ontario. The 1988 amendments to the Canadian National Parks Act also provided for the designation of wilderness areas within national parks. However, both these pieces of legislation were largely ignored by successive governments. In Ontario, the Wilderness Area Act has recently been identified as being largely redundant due to subsequent changes in provincial park legislation and policies: 10 areas were designated under this Act. In Canada, only three designated wilderness areas have been created since 1988.

Provinces and territories in Canada have widely varying policies regarding wilderness. The most common mention of 'wilderness' is in park zoning systems throughout Canada. For example, Parks Canada policy identifies Zone II as the wilderness zone; although the term 'wilderness' is not specifically defined, this policy notes that motorized forms of recreation are not normally permitted and that large wilderness areas should provide opportunities for remoteness and solitude (echoing the US definition). Approximately 90 per cent of the total area of Canada's national parks is classified as wilderness zone. Most provinces have zoning systems that either use the term 'wilderness' or provide equivalent terms. Wilderness areas can also be identified as a specific *category* of provincial park (e.g., Alberta).

At the global scale, the IUCN Protected Areas Management Category Ib is termed a 'wilderness area', defined as a 'large area of unmodified or slightly modified land, and/or sea, retaining its natural character and influence, without permanent or significant habitation, which is protected and managed so as to preserve its natural condition'. Many of the large Canadian national and provincial parks are categorized as wilderness areas under the IUCN model. However, there has been significant debate over the appropriateness of the existing categories, including the wilderness category, both as a result of the duality of the definition of 'wilderness' noted above (individual differences in defining 'wilderness' are even less significant than cultural differences) and the suggestion that the term 'wilderness' mistakenly erases the long history and contemporary presence of Aboriginal peoples in supposedly 'wild' (i.e., uninhabited) places.

John Shultis, University of Northern British Columbia

This book recognizes the diversity of protected areas but concentrates mainly on those in the higher categories, particularly national parks in Canada. However, some chapters focus specifically on provincial parks (Chapter 3) and on other landscape stewardship approaches (Chapter 16), and a wide variety of protective designations are considered in other chapters (e.g., Chapter 11).

PROTECTED AREA VALUES

We expect protected areas to play a variety of roles in the landscape. These roles represent the values that we wish to see maintained within the landscape that, without protection, would not be able to withstand market forces. That is why they are protected. These values are outlined below and summarized in Table 1.3, which depicts the values as different locations and institutions in a city. The purpose of the latter is to illustrate that PAs are not 'single-use' areas, any more than all the functions of specific urban locations could be united into one, and to illustrate the diversity of values represented in PAs.

- *Art gallery*: Just as people visit an art gallery for aesthetic reasons, many parks were designated for their scenic beauty. When Banff was first established, Canadian Pacific Railway president William Van Horne declared that 'since we can't export the scenery we shall have to import the tourists', and appreciation of scenery is still a major reason why people visit parks.
- *Zoo*: Parks are usually easy places to watch wildlife in relatively natural surroundings. Park wildlife, at least in the national parks, is protected from hunting and, therefore, not as shy of humans as wildlife outside parks.
- *Playground*: Parks provide excellent recreational settings for many outdoor pursuits and provision of recreational opportunities has been one of the main functions of parks over the years. However, in keeping with the remarks of the first Commissioner

TABLE 1.3 Protected Area Values with Suggested Allegories

Value	Allegory
aesthetic	art gallery
wildlife viewing	zoo
historical	museum
spiritual	cathedral
recreation	playground
tourism	factory
education	schoolroom
science	laboratory
the 'extra'ordinary	movie theatre
ecological capital	bank
ecological processes	hospital
ecological benchmarks	museum

Source: Dearden (1995).

of National Parks, J.B. Harkin, park recreation should be thought of in a more profound sense of 're-creation' of self, rather than merely playing games.

- *Movie theatre*: Like a movie, parks are able to transport us into a different setting from our everyday existence. We go there to do different things than we would normally do. This is one of the arguments against having golf courses in our parks. Golf might be a fine recreational pursuit but it does not require a park to play it and, furthermore, it introduces everyday activities into the park environment at the cost of discovering new activities that are dependent on a park environment. An excellent treatise on this topic is *Mountains without Handrails* (Sax, 1980).
- *Cathedral*: Many people derive spiritual fulfillment from nature, just as others go to human-built structures, such as churches, temples, and mosques. Such sites, irrespective of denomination, help us appreciate the existence of forces more powerful than ourselves and remind us that humility is a virtue.
- *Factory*: The first national parks in Canada were designated with the idea of generating income through tourism. Since these early beginnings, the economic role of parks has been recognized, although it is a controversial one due to the potential conflict with other roles. A study by Eagles et al. (2000) on national parks, wildlife areas, and provincial parks in Canada calculated a total daily economic impact of $120.47 to $187.69 per park visitor. Overall, the economic impact was between $13.9 billion and $21.6 billion for the year. Although these totals are high, of perhaps greater significance is the distribution of the benefits. As Walton and Simon (2003) point out, in poor, northern, Aboriginal communities, income potential is extremely limited and the factory role of parks takes on great significance, perhaps requiring a different perspective on park policies from those extant in the south.
- *Museum*: Parks protect the landscape as it might have been when European colonists first arrived in North America. As such, parks act as museums to remind us of these conditions. Nash (1967), in his well-known history of the wilderness movement, describes this as a main reason behind the early growth of the parks movement in the US, although it is certainly a less prominent motivation now. These museums also serve a valuable ecological function as they provide important areas against which to measure ecological change in the rest of the landscape, often known as the *benchmark* role. Davis et al. (2006) provide a review of this role, and a Canadian example of a comparative monitoring system, Environment Canada's Ecological Monitoring and Assessment Network (EMAN), is described in more detail by Craig et al. (2003).
- *Bank*: Parks are places in which we store and protect our ecological capital, including threatened and endangered species (see Chapter 5 and Dearden, 2001, for more details). From these accounts we can use the interest to repopulate areas with species that have disappeared. Examples include the muskoxen from the Thelon Game Sanctuary that have recolonized terrain outside the sanctuary, and the elk and bison from Elk Island National Park outside Edmonton that have been used to start herds elsewhere.
- *Hospital*: Ecosystems are not static and isolated phenomena but are linked all over the planet. Protected areas constitute some of the few places where these processes still operate in a relatively natural manner. As such, they may be considered ecosystem

'hospitals' that help to maintain processes that may be damaged and not working effectively elsewhere. Much attention, for example, is now focused on the carbon cycle as imbalances caused by the burning of fossil fuels contribute to global warming. Forests are major carbon 'sinks' where carbon dioxide is taken in from the atmosphere and stored in organic form. Thus, forests that are protected play a major role in maintaining some balance in the carbon cycle. Kulshreshtha and Johnston (2003), for example, calculate that Canada's national parks have sequestered 4.43 gigatonnes of carbon that would cost society $72–$78 billion to replace. In a more modest but also interesting example of the *hospital role*, Knowler et al. (2003) calculate the value to salmon of protecting freshwater habitat in protected areas on Canada's west coast. Although the hospital function is a commonly overlooked role of protected areas, it is becoming one of increasing interest to scientists.
- *Laboratory*: As relatively natural landscapes, parks provide outside laboratories for scientists to unravel the mysteries of nature. For example, Killarney Provincial Park in Ontario provided an important laboratory for early research on acid precipitation in Canada.
- *Schoolroom*: Parks can play a major role in education as outdoor classrooms. Direct physical contact with the complexities of the natural environment help inspire awe, humility, and respect. This function can also play a key role in helping influence visitors to change behaviours to more environmentally benign practices in their everyday lives.

It is useful to bear this full range of values in mind as the emphasis among them can change over time. As already mentioned, when Canada's national parks were inaugurated there was an overwhelming emphasis on the playground and factory roles. As the rest of the landscape became dominated by market forces for forestry, agriculture, and urban uses, the ecological roles of parks became increasingly recognized. For many years, Parks Canada legislation and policies laboured under what was known as the 'dual mandate', which required making full use of the parks for the enjoyment and benefit of the people while ensuring that they were unimpaired for future generations (see Chapter 2). Various administrators interpreted this balance in differing ways, but finally, in 1988, the National Park Act was amended to give clear priority to the protection of ecological integrity over human use, and this has been further emphasized in more recent policy and legislation (see Chapter 9 and Dearden and Dempsey, 2004, for a more extensive review of these changes).

Similar changes are evident at provincial levels as the need for more protection becomes evident. For example, the oldest and one of the most significant provincial parks in the country, Algonquin, some 260 km northeast from Toronto, has been subject to extensive logging activity over the years and contains over 8,000 km of logging roads. The Ontario Parks Board recommended in 2007 that the area protected from logging needed to be raised from 22 per cent to 54 per cent of the park in order to increase protection for park ecosystems.

There are also further nuances regarding changes in role priority. For example, historically, within the ecological context, most emphasis was placed on the *bank role*. Recently, increasing interest has been shown in the *hospital role* and the value of ecosystem services provided by PAs. This interest reflects greater scientific understanding of ecosystem linkages and heightened public and political awareness of environmental degradation in general (Box 1.2).

There is now also a growing appreciation of the link between visitation to parks and maintaining ecological integrity of the landscape overall. Historically, park visitation and ecological integrity were seen as mutually exclusive, and to protect ecological integrity visitation had to be limited. It is now recognized, however, that PAs are part of the larger landscape and that visitation can actually be used to enhance rather than detract from environmental quality if visitation is managed appropriately.

We can think of landscape as a continuum of values bounded on one side by market values and on the other by protected area values (Figure 1.1). How much of the landscape should be devoted to providing these different kinds of values for society? The target of 12 per cent has been used by many jurisdictions both in Canada and abroad (WCED,

BOX 1.2 Something Happened in 2006 . . .

In the fall of 2006 there was an unprecedented rise in public awareness in Canada of environmental degradation. Much of this awareness could be attributed to the increased publicity being given to global climatic change. Al Gore released his film *An Inconvenient Truth* as part of his personal battle to make the public more aware of global warming. The UN's Intergovernmental Panel on Climate Change, a consensus of the world's top 2,000 climate experts, also released a very strongly worded report that confirmed human agency as being the main cause of global change and warning of some of its consequences. The report was probably instrumental in persuading previous skeptics, like Canada's Prime Minister, Stephen Harper, of the need to introduce serious measures to address global change.

By late 2006 the environment ranked second to health as the major concern of the Canadian public (www.cbc.ca/canada/story/2006/11/08/environment-poll.html). An IPSO-REID poll released on 4 July 2007 found that 91 per cent of Canadians felt they should do their bit to fight global warming, even if they had to pay for it (www.cbc.ca/canada/toronto/story/2007/07/04/enviro-survey.html) and an earlier poll found that 62 per cent of Canadians would be willing to have the economy grow at a 'significantly slower rate' to reduce global warming (www.theglobeandmail.com/servlet/story/RTGAM.20070129.wclimatepoll29/BNStory/ClimateChange).

Although increases in environmental awareness and motivation to do something to contribute are good trends, they do not always translate into the most effective environmental measures. For example, Canada's need to pass endangered species legislation, the Species at Risk Act, to comply with the UN Convention on Biological Diversity may, in fact, have siphoned funds from protected area programs, that, in the long run, may have contributed more effectively to endangered species protection. Similarly, with politicians now rushing to be seen as sympathetic to the changes necessary to mitigate global change, this may detract from funding for improving protected area systems.

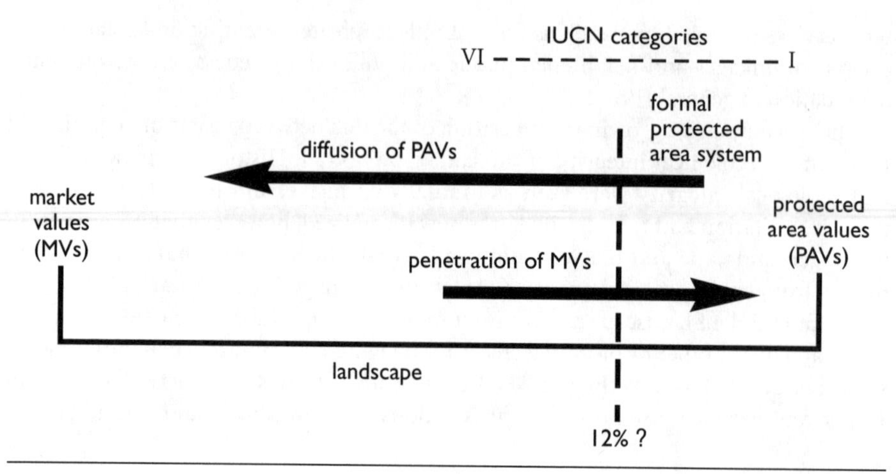

FIGURE 1.1 The landscape as 'valuescape', showing unwanted penetration of market values into protected areas and the use of IUCN protected area categories and an ecosystem-based approach to diffuse protected area values throughout the landscape. A major question is how much land should fall under protective designations, with 12 per cent providing an average international guideline.

1987). However, there is a danger in adopting such a figure that the landscape becomes divided into two dichotomous types along the 12 per cent line, either protected or non-protected, with the assumption that in the latter we do not need to pay any attention to the protected area values outlined in Table 1.1. This produces not only everyday landscapes of destruction, but also will lead to the eventual choking of the protected areas, as they become islands of extinction cut off from other PAs (Chapter 4).

A more ecosystem-based approach to incorporation of protected area values (PAVs) needs to be taken (Chapter 13). This involves making greater use of the less restrictive categories of PA (Categories IV–VI) to diffuse PAVs into a higher proportion of the landscape, as shown in Figure 1.1. It also involves placing a greater emphasis on a broader range of stewardship approaches to landscape planning that complement, but are not substitutes for, more centralized, government initiatives such as park systems (see Chapter 16). And it also involves a full recognition of the inspirational value of parks to engage visitors in understanding current environmental challenges and ways that they can become involved in addressing those challenges. In other words, visitation becomes a main means of spreading PAVs throughout the landscape through personal action. Park visitors also can become strong advocates for parks, and as park creation is part of the political process, then such advocates are essential for future park establishment and resourcing.

The realization of the complementary nature of visitation and maintenance of park ecological integrity and landscape quality represents another subtle shift in the balance among the various park roles (Table 1.3). Parks Canada has recognized the importance of this connection by creating a new External Relations and Visitor Experience

Directorate (ERVE) with a specific mandate to augment visitor experiences. This is a particularly interesting development since only 10 years ago the agency was withdrawing support for interpretive services as being not cost-effective. Interpretation, as discussed in Chapter 8, is now seen as being key to creating memorable experiences for visitors that will turn them into 'ambassadors of Parks Canada's natural and historic sites' and 'lifelong stewards' of the environment (Parks Canada, 2007).

Parks provide an ideal environment for encouraging personal changes in our everyday lives. People visit parks voluntarily to do something different (the *movie theatre role* in Table 1.3). They are not constrained by the habits of their everyday existence. This is a perfect opportunity to introduce them to new perspectives that may well change many of their environmentally damaging day-to-day habits when they return to their homes. After visiting the magnificent wilderness of the Columbia Icefields in Jasper National Park and finding out that these apparently pristine glaciers are polluted by activities that many people engage in around their homes, particularly the use of chemicals, then people will gain a more engaged perspective on their role in creating environmental damage, and hopefully, will change some of their habits. Designing appropriate visitor experiences to create these wider changes is one of the main challenges of our park systems.

So the balance among the various roles we expect our protected areas to play has fluctuated over time. In general, a shift has occurred from roles focused more on the playground mandate to those emphasizing ecological protection (Figure 1.2). At the same time, the nature of threats to parks has changed. In the past these threats were mainly from inside the parks as a result of excess visitation. However, management of these visitor pressures has improved considerably and visitor numbers have fallen (Box 1.3).

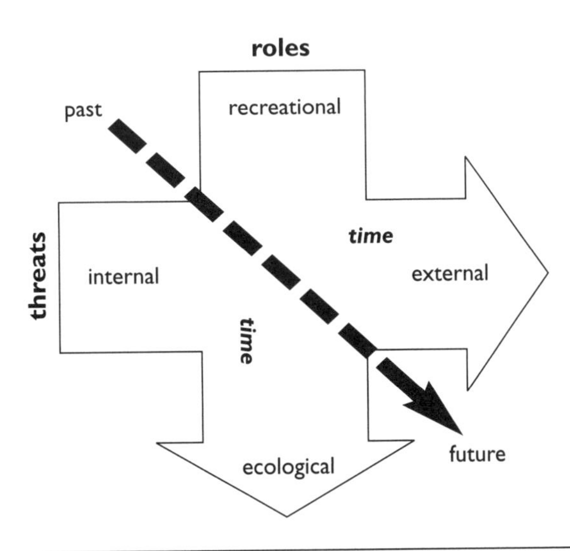

FIGURE 1.2 The changing emphasis in park roles over time.

Yet, the parks have become increasingly isolated from natural surroundings as resource extraction industries have flourished, and they are also as vulnerable as the rest of the landscape to the effects of global climate change, acid rain, and other widespread phenomena. Consequently, most threats to park ecological integrity are now seen as coming from outside rather than inside the parks. Formerly, management attention was directed largely at managing the recreational role of parks and its attendant threats to park environments; now the emphasis is more on the ecological roles and mitigating the external threats (Figure 1.2).

BOX 1.3 Where Have All the Visitors Gone?

Visitation to national parks has declined by roughly 20 per cent since 1995. Even if changes, implemented in 2000, in how visitation is counted are taken into account and the mountain parks excluded, visitation still shows a 9 per cent decline. Visits to national parks declined in every province except BC, Alberta, and Newfoundland and Labrador over the last five years.

These figures reflect provincial and international trends. In the US, overnight stays fell by 20 per cent between 1995 and 2005, and tent camping and backcountry camping each decreased 24 per cent during the same period. In Yosemite National Park, famous for its high summer visitation, there were 20 per cent fewer visitors in July 2005 than in July 1995. Three main factors seem to be at work:

- The baby boomers who were active in parks throughout their lives are getting older and not as physically capable of strenuous exercise. Even those who still engage in outdoor activities on a regular basis now prefer a gourmet dinner and a soft bed afterwards rather than sleeping in the pup tent they did when younger. Surveys conducted for Parks Canada show that 63 per cent of Canadians over the age of 65 said they used to visit parks but no longer do so.
- The other main age group found to be missing in surveys is people in the 18–24 age category, historically one of the main age-group users of parks. Similar results have been found in the US. This drop in visitation by younger people has been attributed, at least in part, to the so-called 'nature deficit disorder' (Louv, 2005). Young people in this generation have grown up in a more urbanized world, cut off from everyday ties to the natural environment and with a strong predilection for electronic gadgetry. They understand and feel more comfortable with their electronic world than with the challenges posed by the outdoor world. Given a free day, they are more inclined to spend it on Facebook than hiking or canoeing. This trend, apparent in many countries throughout the world, affects not only park visitation but also the entire gamut of society's interaction with the environment. What happens when the nature-deficit generation of today become the decision-makers of tomorrow?
- Surveys have also found that minorities and new immigrants are unlikely to be park visitors and supporters. Immigrants born outside of Canada make up close to 25 per cent of the population (and over 50 per cent in Toronto) but only 10

per cent of visitors to national parks, for example. This is understandable, for many immigrants, especially if from developing countries, are likely to be from large cities rather than poor rural farmers. They are used to an urban lifestyle, a large majority settle in Canada's three largest cities, and they need to establish themselves in their new country and may not have the necessary free time or money to get out to the parks. Many are also unaware of the vastness of Canada and what there is to see in the parks.

The main questions are: Should these visitor declines generate concern? And, if so, what can be done about it? On the one hand, some have suggested that the parks should be marketed to provide more of what the people want, including more commercialized activities and businesses and more motorized recreation. Others, however, caution that park policies cannot be driven simply by changing public tastes; otherwise, they will become enslaved to recreational fashion, with cellphone towers, touchscreen computers, and jet skis replacing the nature the parks were created to protect.

Sources: Surveys conducted for Parks Canada by Environics Research in 2002 and 2005; US data from Cart (2006), at: <seattletimes.nwsource.com/text/2003453655_webparks29.html>.

BIOREGIONAL CONSERVATION APPROACHES

One way to facilitate ecological connections and buffer external threats is through the development of large-scale, bioregional conservation landscape models. Such approaches have evolved from concentration on nodes of conservation, such as national parks, to ways of connecting networks of different types of protected areas throughout the landscape. One of the best-known initiatives is the Yellowstone to Yukon (Y2Y) Conservation Initiative started in 1993 by The Wildlands Project, the Canadian Parks and Wilderness Society (CPAWS), and others (Locke, 1993) and now supported by over 180 Canadian and US conservation groups, First Nations, industries, and governments. Prompted by science, which showed that certain species migrated all along the Rockies from the US up to Alaska, the initiative attempts to see how different protected areas along the way can be linked together (Figure 1.3). There are 11 national parks and dozens of state and provincial and territorial parks along this corridor. However, there are still many weaknesses and missing links in connectivity.

Two designations by the BC government have been key in realizing the vision. The Muskwa–Kechika Management Area (MKMA) covers some 4.4 million ha and is managed primarily for conservation, including 1.1 million ha in parks. Any development in the areas outside of the parks must, by law, allow for environment as a prime consideration. Recently, after seven years of roundtable discussions, 475,000 ha of new parks have been added to the existing 116,000 ha, with an additional 410,000 ha of special management zones and 900,000 ha of no-logging adjacent to the MKMA. Together with the MKMA, these new areas create a conservation matrix of 6.3 million ha connecting the ecosystems of the Rocky and Cassiar mountains with those of the Pacific coast.

This area provides a home to the greatest combined abundance and diversity of large mammals in North America, among them thousands of moose, elk, and caribou and the

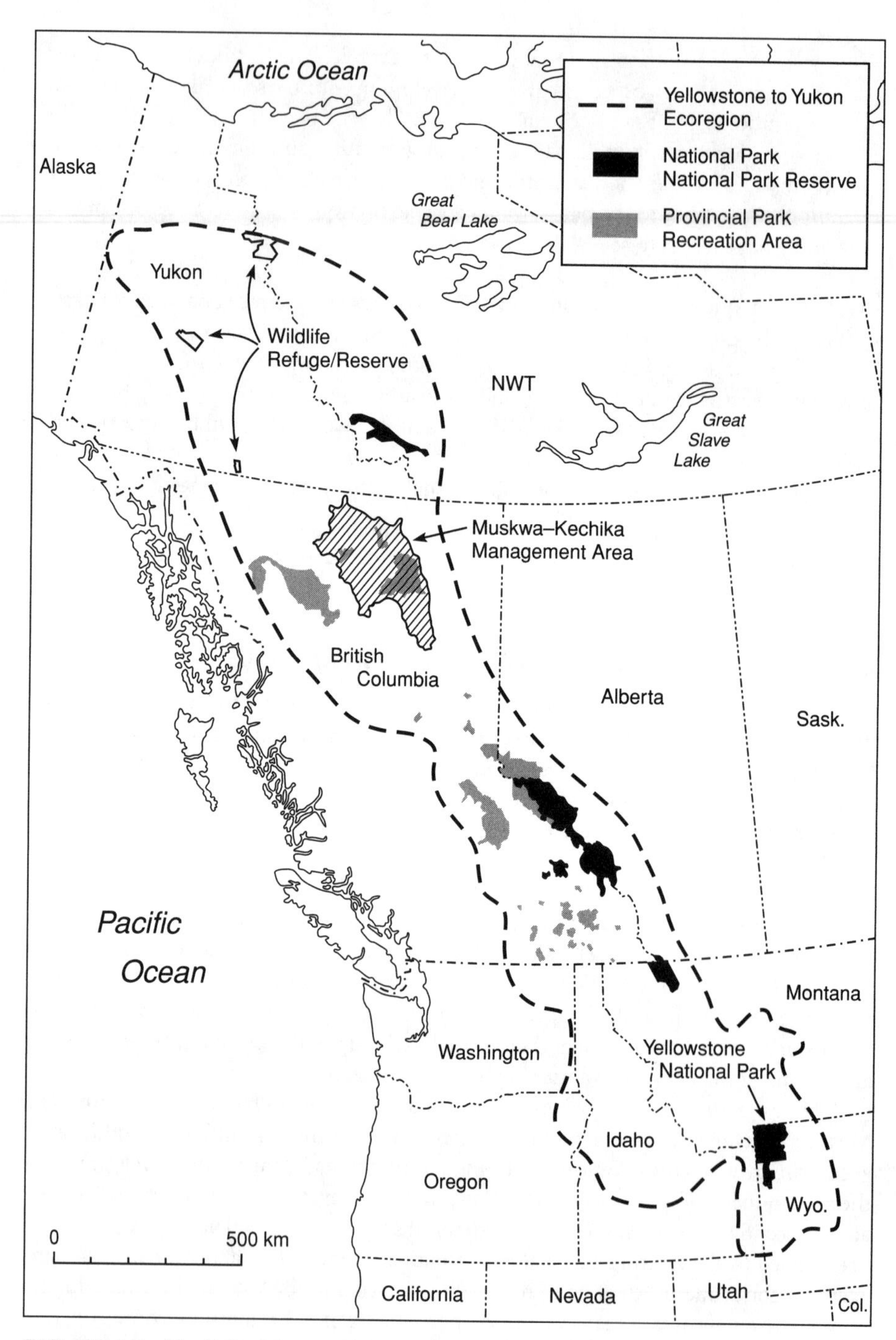

FIGURE 1.3 The Yellowstone to Yukon conservation initiative.

continent's largest concentration of Stone sheep, all accompanied by a healthy population of predators. In fact, wolves from this area and Alberta were used in the reintroductions to Yellowstone. As emphasized by Rivard et al. (2000), large size is still one of the most important criteria for an effective protected area, and will be even more important as the full effects of global change become more apparent (Box 1.4).

THE STATUS OF CANADA'S PROTECTED AREAS

Canada was the third country in the world to proclaim a national park, back in 1885, and also has the oldest national park service in the world. We have a long history in protected area designation and management, but how are we doing in terms of meeting international and national objectives?

BOX 1.4 Where Have All the Critters Gone?

Global climate change is the most severe problem faced by our planet. Although scientists identified human industrial activity as a possible source of climate change over a hundred years ago, and the science behind this has been part of many university courses for at least 30 years, only during the last few years has the severity of the challenge become accepted by politicians. The potential impacts of global change on natural ecosystems are huge, and already there are many documented examples of changes in species range in response to climate change (Parmesan, 2006).

The climate change scenarios for Canada, because of our high latitudes, show that we will be one of the most affected countries in the world. Already the northern regions are showing large-scale reductions in snow and ice cover, reductions in permafrost, coastal inundation, and stressed populations of northern species such as polar bears. Scientists predict that each 1°C rise in temperature will cause biomes to migrate northward some 300 km. Given the predicted minimum increase of 2–5°C in 70–100 years, this will translate to 600–1500 m in elevation and 300–750 km in distance. Species must either be able to migrate fast enough to keep up with these changes, evolve to deal with them, or become extinct. Certain biomes, such as arctic-alpine and the boreal forest, will be very vulnerable to these changes, as will others (see Chapter 4).

For protected areas, climate change obviously has very serious implications. On the one hand, PAs will play a huge *hospital role* (Table 1.3) in helping sequester carbon from the atmosphere. On the other hand, the *bank role*, providing refuge for natural populations, will be very vulnerable to the changes described above. PA networks must be made as resilient as possible against these changes. One of the main mechanisms for doing this is through the kind of large-scale bioregional planning illustrated by Y2Y that emphasizes connectivity, especially north–south connectivity among PAs. New PAs will be required that help facilitate migration, provide source populations, and provide suitable habitat for incoming populations. Private lands will also be important (see Chapter 16). Whitelaw and Eagles (2007) provide an example from Ontario that illustrates planning on private lands for the kind of long and wide conservation corridors that will be required in the future. Biodiversity implications are likely to be especially severe in the oceans, yet Canada has created virtually no marine protected areas, let alone functioning networks of MPAs (see Chapter 15; *MPA News* 8, 6 [special issue on climate change]).

Managers will have to rethink their conservation strategies to take into account predicted changes in species requirements. Some of the necessary steps are discussed in more detail by Hannah (2003). One particularly troubling aspect raised by Hannah is the current gap between what needs to be done in PA management and what is being done. Given the discussion in the next section on the limited management capacity of park systems in Canada, particularly at the provincial level, this gap will become even wider with the new and severe challenges raised by global warming. A survey of PA jurisdictions in Canada found that 80 per cent had not completed a comprehensive assessment of the potential impacts and implications of climate change on policy and management and have no adaptation strategy or action plan. Furthermore, 86 per cent of managers surveyed felt that they did not have the capacity necessary to deal with climate change issues (Lemieux et al., 2007).

Hannah (2003) suggests that one of the most important steps to deal with global change in PAs is closing this management gap as soon as possible before it gets too wide to bridge. Unfortunately, in almost all provincial jurisdictions in Canada, politicians have been driving things the other way by consistently cutting the funding available to park agencies.

One often overlooked value of PAs in addressing climate change is the *schoolroom role*, as discussed earlier, where visitors can be made more aware of the challenges of global change, and also of things that they can do to help. Peart et al. (2007) suggest park messaging that emphasizes:

- Large ecosystem conservation initiatives are a positive approach to preparing for climate change.
- It is important to maintain as much of the present wildlife, fisheries, and coastal habitat as possible to help minimize the effects of climate change and to help as many species as possible survive. These conservation efforts will act as insurance against the negative effects of high-risk transformations.
- With good public response, especially related to land and water allocation, it is possible to minimize some climate-related changes so that the consequences to wildlife will be less significant.
- Maintaining and bolstering the resilience of native ecosystems will minimize the opportunity for the invasion and establishment of harmful exotic species and the outbreak of native species (e.g., mountain pine beetle).
- Wildlife must be able to migrate (emphasizing the importance of cores and connectivity) and climate change will make that more difficult for many species. Messages should include the importance of community planning based on 'backcasting' from the future, not on forecasting from the current situation.
- Mitigation alone is not the solution.

In 1992 Canada's federal, provincial, and territorial Ministers of Environment, Parks, and Wildlife signed *A Statement of Commitment to Complete Canada's Network of Protected Areas.* Terrestrial systems were to be completed by 2000 and marine designation was to be 'accelerated'. Despite impressive growth, with the area protected up 19 per cent between 2000 and 2005, Canada is still far from meeting these commitments; indeed, four jurisdictions—Yukon, Nunavut, Newfoundland and Labrador, and Environment Canada—still do not have the frameworks in place to attempt completion (Environment Canada, 2006). Even Parks Canada has only 72 per cent of its system plan complete (see Chapter 9). In terms of overall protection of Canada's ecoregions, 29 per cent are provided a high level of protection (i.e., over 12 per cent of their area), 12.4 per cent have moderate protection (6–12 per cent), 41.9 per cent low protection (<6 per cent), and 16.6 per cent have no protected areas (ibid.). The situation is even worse on the marine front, with under 0.5 per cent set aside in protective designation and Canada ranking seventieth globally in terms of the percentage of oceans protected.

Canada obviously has a long way to go in terms of meeting our own expectations, let alone those of the international community. Canada is signatory to the UN Convention on Biological Diversity, which calls for 'the establishment and maintenance by 2010 for terrestrial and by 2012 for marine areas of comprehensive, effectively managed, and ecologically representative national and regional systems of protected areas'. Although progress is being made, Canada will not be able to meet these international commitments until quite some time after the target dates. Some of the difficulties faced are discussed in more detail in Chapters 3, 9, 14, and 15. The main challenge, however, is getting the political will to make good on the commitments that governments have made to the people of Canada and the international community. Political will is driven by voter interest, which again indicates the importance of park visitation emphasized in the previous section.

To date, about 10 per cent of Canada's terrestrial area has been awarded protective designation, short of the international goal of 12 per cent first suggested by the World Commission on Environment and Development in 1987 (WCED, 1987) and well short of the average of 14.6 per cent protected by OECD countries (Environment Canada, 2006). Canada manages 5.1 per cent of the world's terrestrial protected area estate. However, 95 per cent of Canada's terrestrial protected areas fall within IUCN categories I–IV and hence have a strong protective mandate. Among OECD countries, Canada ranks sixteenth out of 30 in terms of the proportion of land protected (the US, for example, protects almost 25 per cent compared with our 10 per cent), yet ranks fourth in terms of proportion of land with strong protection (IUCN Categories I–IV). Furthermore, Canada has some two-thirds of its protected area within a small number of sites that are in excess of 300,000 ha in size. Few countries have the ability to preserve such large intact landscapes.

Despite the large size of some areas and high levels of protection under which they are designated, there is some doubt as to the abilities of management agencies to provide adequate management. In a national survey the majority of park agencies 'reported significant deficiencies in their ability to manage or monitor their protected area networks' (ibid., 28). For example, only about a quarter of Canada's terrestrial PAs have management plans in place, and where plans are in place, with the notable exception of Parks Canada, few agencies are implementing them.

In sum, Canada is making progress in terms of increasing the amount of designated protected area, but progress is slow and falls short of both national and international commitments. Protection of water, both freshwater and marine systems, is particularly poor. Management is also variable. Parks Canada is deservedly among the most respected park management agencies in the world, yet most other agencies at both the federal and provincial levels have few resources to devote to PA management and doubt their own abilities to maintain ecological integrity in these areas. Improvements in management that strengthen protection are often driven by non-governmental organizations (NGOs), such as CPAWS, and other outside bodies.

The status of Canada's PAs changes all the time in terms of both size and management. One useful way to keep updated on these changes is to join an NGO such as CPAWS and receive their newsletters. Another important source of up-to-date technical information on parks and park management is the Science and Management of Protected Areas Association. SAMPAA holds regular conferences attended by many park scientists, academics, and NGOs. The proceedings of these conferences are available either in print or directly off the web and contain state-of-the-art papers on park management in Canada. Regional parks conferences are also held regularly in Ontario, Manitoba, and BC, and these are also excellent ways to keep informed.

THEMES

Six sections provide the overall framework for this book. The first section, 'Overview', continues after this chapter, with a historical review of the development of the protected area system in Canada. Focusing mainly on the national park system, Kevin McNamee in Chapter 2 traces the growth of the system, the main management issues, and the critical role that politicians and conservationists have played as the system has evolved. Chapter 3, by Chris Malcolm, examines the status of provincial parks in Canada and some of the main challenges being faced.

Part II, 'Conservation Theory and Application', recognizes the key role that understanding and application of ecological principles play in park planning. In Chapter 4, Jeannette and John Theberge provide an ecological primer on some of the principal concepts underlying the designation and management of protected areas. Chapter 5, by Stephen Woodley from Parks Canada, outlines some of the ways in which the agency is approaching management for ecological integrity, with a particular emphasis on active management. Several examples, which help to ground the theories of the previous chapter in actual management practices, are discussed.

The third section, 'Social Science Theory and Application', complements the previous section by providing an overview of some of the main social science concepts that provide the basis for park planning. In Chapter 6, Mark Needham and Rick Rollins describe the social science approaches to looking at protected area management. Most park management is people management, yet, as emphasized by the Panel on Ecological Integrity (Parks Canada, 2000), this area is often overlooked and understaffed. Needham and Rollins outline various approaches that give insight into why and how people use protected areas, crowding, conflict management, and some of the social and economic impacts of parks. Wolfgang Haider and Bob Payne, in Chapter 7, further develop these concepts and relate them specifically to different approaches towards visitor management frameworks, such as the Visitor Activity Management Process (VAMP) of Parks Canada. They provide many examples to ground the theory into actual management approaches in various parks. Chapter 8, by Glen Hvenegaard, John Shultis, and Jim Butler, looks at one critical aspect of human use management in parks—interpretation. Perhaps no aspect of park services suffered so badly during the economic climate of the last decade. Interpretation infrastructure was allowed to fall into disrepair, new initiatives were put on hold, and many existing services either were cut completely as 'uneconomic' or were made into 'pay-for-service' features, with costs that often discouraged all but the most ardent park lover. Since interpretation is a prime way in which park agencies can contact their public and in which environmental awareness can be raised and management problems eased, this was definitely a retrograde step. However, the pendulum has swung back, and particularly within Parks Canada there is renewed interest in interpretation.

Part IV, 'Putting It Together', presents different aspects of park management from a more holistic perspective. In Chapter 9, Pam Wright and Rick Rollins provide a comprehensive overview of national park management, outlining the main changes in legislation and policy, park management planning and administration, and the current status of the system plan. Joe Pavelka and Rick Rollins, in Chapter 10, concentrate on just one park, but deal with many different dimensions. The site is Canada's oldest and best-known park, Banff, and the case study embodies many of the challenges felt at other parks in the system.

One distinctive factor about PAs in Canada is the large areas protected in high latitudes and the special relationship that remains between Aboriginal people and the land in the North. Harvey Lemelin and Margaret Johnston, in Chapter 11, discuss these aspects in more detail in regard to protected areas in the North. Unlike most of the other chapters, Chapter 11 considers many different types of PA in addition to national parks. Promoting and sustaining appropriate tourism in PAs is always a challenge, and in Chapter 12 Rick Rollins, Paul Eagles, and Philip Dearden examine some of the issues and approaches involved. The final chapter in this section, by Scott Slocombe and Philip Dearden, looks at the need to place PAs within an ecosystem-based context for their planning and management. A synthesis is provided of different approaches and tools, but the authors conclude that there are still few working examples of ecosystem-based approaches to the management of protected areas.

Part V (Chapters 14–17) examines various thematic issues. Chapter 14 looks at the impact of an Aboriginal presence on protected areas. In 1993, Dearden and Berg suggested that parks have witnessed penetration of different interests in their designation and management over time (Figure 1.4). Although entrepreneurs were dominant for many years, the power of environmental interests steadily grew to become a dominant force in the 1970s and 1980s until these interests, too, were eclipsed by a more pervading influence—First Nations. In parts of Canada—especially the territorial North, British Columbia, Labrador, and northern Quebec—no treaties were entered into with the people already living on the land when white people colonized. The European settlers simply took the land, and treaties that were negotiated in the nineteenth and early twentieth centuries often were not adhered to by governments. This has now been recognized as not only unethical, but also illegal. As a result, treaty negotiations are underway in

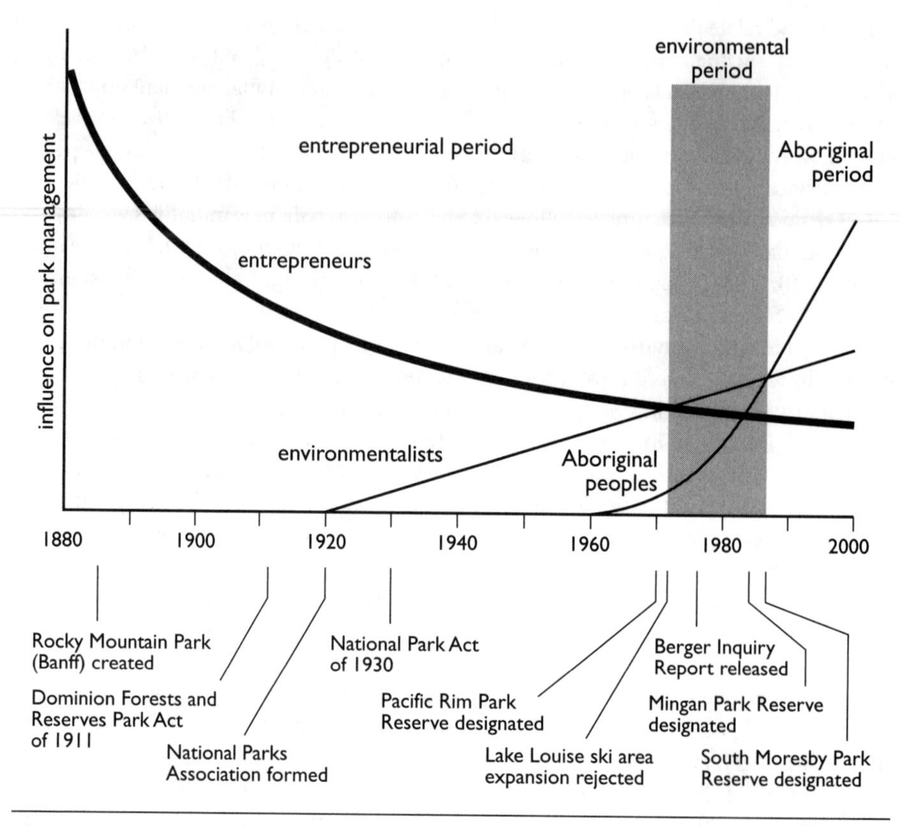

FIGURE 1.4 Suggested influences of various external groups on park management over time. *Source*: Dearden and Berg (1993).

much of Canada, many years after the land was taken. In a number of instances this has been a very positive move for protected areas, as First Nations have made conservation one of their top priorities in treaty negotiations. However, there are also challenges. Philip Dearden and Steve Langdon in Chapter 14 provide some historical perspective on this situation before discussing some of the implications at different parks across the country and the progress being made in incorporating Aboriginal peoples into protected area designation and management.

As noted earlier, Canada has been particularly slow in developing protection of marine habitats. Since the last edition of this book was published new legislation has been passed, the National Marine Conservation Areas Act of 2002, but there are still no areas actually designated under this legislation; the hope is that this might change in 2008. Nonetheless, Canada will fall far short of its international commitment to create a network of MPAs by 2012. In Chapter 15, Dearden and Rosaline Canessa discuss the current status of MPA establishment and outline the main challenges being faced.

This book has a stated focus on the establishment and management of large, natural parks established by governments. However, there has been a blossoming of other approaches to the diffusion of protected area values throughout the landscape, and many of these are for smaller pieces of land and do not rely solely on government coffers. Yet, these acquisitions can be critically important in linking larger protected areas together as well as in protecting smaller units for their intrinsic value. In Chapter 16 Jessica Dempsey and Dearden describe some of the initiatives now being made outside the main park systems, with special emphasis on private land stewardships and the emerging role of nature conservancies in the purchase and protection of conservation lands. They raise some interesting issues not only about the benefits of these initiatives, but also in regard to some of the potential problems. The final chapter in Part V, by Philip Dearden, responds to requests from book users for a chapter dealing with international perspectives on protected areas. Dearden has been engaged in research on tropical PAs for many years and brings this expertise to bear to provide some insight on the international scene and, particularly, to address the challenge of conflicts between local peoples and parks, which arise mainly out of poverty.

Part VI, 'Concluding Perspectives', has one chapter, by the editors, that draws the material together and summarizes some of the key issues for the future.

REFERENCES

Craig, B., H. Vaughan, and M. Doyle. 2003. 'EMAN: Delivering information to improve resource decisions in communities, parks and protected areas', in Munro et al. (2003).

Davis, G.E., D.M. Graber, and S.A. Acker. 2006. 'National parks as scientific standards for the biosphere; or, how are you going to tell how it used to be, when there's nothing left to see?', *George Wright Forum* 23: 34–42.

Dearden, P. 1995. 'Park literacy and conservation', *Conservation Biology* 9: 1–3.

———. 2001. 'Endangered species and terrestrial protected areas', in K. Beazley and R. Boardman, eds, *Politics of the Wild: Canada and Endangered Species.* Toronto: Oxford University Press, 75–93.

——— and L. Berg. 1993. 'Canada's national parks: A model of administrative penetration', *Canadian Geographer* 37: 194–211.

——— and J. Dempsey, J. 2004. 'Protected areas in Canada: Decade of change', *Canadian Geographer* 48: 225–39.

Eagles, P.F.J., D. McLean, and M.J. Stabler. 2000. 'Estimating the tourism volume and value in parks and protected areas in Canada and the USA', *George Wright Forum* 17, 3: 62–82.

Environment Canada. 2006. *Canadian Protected Areas Status Report 2000–2005.* Ottawa.

Hannah, L. 2003. 'Protected areas management in a changing climate', in Munro et al. (2003).

IUCN (World Conservation Union). 1994. *Guidelines for Protected Area Management Categories.* Gland, Switzerland: CNPPA/IUCN.

Knowler, D., B. Macgregor, M. Bradford, and R. Peterman. 2003. 'Valuing freshwater salmon habitat as a benefit of protected areas on the west coast of Canada', in Munro et al. (2003).

Kulshreshtha, S., and M. Johnston. 2003. 'Economic value of stored carbon in protected areas: A case study of Canada's national parks', in Munro et al. (2003).

Lemieux, C., T. Beechey, and D. Scott. 2007. 'A survey on protected areas and climate change (PACC) in Canada: Survey update', *ECO* (Canadian Council on Ecological Areas) 16: 2–3.

Locke, H. 1997. 'The role of Banff National Park as a protected area in the Yellowstone to Yukon Mountain Corridor of western North America', in J.G. Nelson and R. Serafin, eds, *National*

Parks and Protected Areas: Keystones to Conservation and Sustainable Development. Berlin: Springer, 117–24.

——— and P. Dearden. 2005. 'Rethinking protected area categories and the "new paradigm"', *Environmental Conservation* 32: 1–10.

Louv, R. 2005. *Last Child in the Woods: Saving Our Children from Nature-Deficit-Disorder.* Chapel Hill, NC: Algonquin.

Munro, N.W.P., P. Dearden, T.B. Herman, K. Beazley, and S. Bondrup-Nielsen, eds. 2003. *Making Ecosystem-Based Management Work: Connecting Managers and Researchers.* Proceedings of the Fifth International Conference on Science and Management of Protected Areas, Victoria, BC, 11–16 May. Wolfville, NS: Science and Management of Protected Areas Association. At: <www.sampaa.org/publications.htm>.

Nabham, G.P., and S. Trimble. 1994. *The Geography of Children: Why Children Need Wild Places.* Boston: Beacon Press.

Nash, R. 2001. *Wilderness and the American Mind,* 4th edn. New Haven: Yale University Press.

Parks Canada. 2000. *Unimpaired for Future Generations? Conserving Ecological Integrity with Canada's National Parks—Volume 2: Setting a New Direction for Canada's National Parks.* Report of the Panel on the Ecological Integrity of Canada's National Parks. Ottawa.

———. 2007. 'Visitors learn, grow, discover', *Experiences* 1, 2: 1.

Parmesan, C. 2006. 'Ecological and evolutionary responses to recent climate change', *Annual Review of Ecology and Systematics* 37: 637–69.

Peart, B., S. Patton, and E. Riccius. 2007. *Climate Change, Biodiversity and the Benefit of Healthy Ecosystems.* Vancouver: CPAWS-BC.

Rivard, D.H., J. Poitevin, D. Plasse, M. Carleton, and D.J. Currie. 2000. 'Changing species richness and composition in Canadian national parks', *Conservation Biology* 14: 1099–1109.

Sax, J. 1980. *Mountains without Handrails.* Ann Arbor: University of Michigan Press.

Walton, M., and J. Simon. 2003. 'The industry of economic development and protected areas', in Munro et al. (2003).

Whitelaw, G.S., and P.F.J. Eagles. 2007. 'Planning for long, wide conservation corridors on private lands in the Oak Ridges Moraine, Ontario, Canada', *Conservation Biology* 21: 675–83.

World Commission on Environment and Development. 1987. *Our Common Future.* Oxford: Oxford University Press.

KEY WORDS/CONCEPTS

bioregional conservation approaches
Convention on Biological Diversity (CBD)
ecological integrity
ecological stress
decreasing visitation
global change
international commitments
marine protected areas (MPAs)
nature deficit disorder
protected area categories
protected area values (PAVs)
status of PAs in Canada
Y2Y

STUDY QUESTIONS

1. Why are parks important to Canadians? Provide a list of benefits or reasons for having parks.
2. What forces in society seem to be opposed to parks? What individuals, organizations, or sectors of the economy can you identify that are opposed to parks or would benefit if parks were reduced or eliminated? Develop a list of disadvantages or reasons that could be advanced to reject the idea of parks.
3. Discuss and evaluate the validity of each list (advantages and disadvantages of parks).
4. Six types of protected areas have been identified by the IUCN. Briefly describe each of these types of parks, and identify examples of protected areas in Canada that approximate each of these IUCN designations.
5. What is happening to PA visitation in many parts of the developed world? What are the reasons behind these trends? Are they of concern? What might be done about them?
6. Why are bioregional approaches to conservation planning going to become even more important in the future?
7. Outline the main changes that have occurred relating to protected areas in Canada over the last decade. What do you think are the main forces behind these changes?
8. What is your personal definition of 'wilderness': that is, what are the characteristics of your conception of wilderness? How do you think you developed this conception of wilderness? For example, what is the role of your cultural or ethnic background or previous experience in creating your personal wilderness ideal?
9. How do the spatial, temporal, and cultural differences in defining wilderness make its global management more difficult?

CHAPTER 2

From Wild Places to Endangered Spaces: A History of Canada's National Parks

Kevin McNamee

INTRODUCTION

Canada's national parks system plays a critical role in conserving biological diversity and wilderness landscapes. At the beginning of the twenty-first century, almost 276,250 km^2, or practically 2.8 per cent of Canada, is protected from industrial development in 42 national parks and reserves. These parks constitute over 30 per cent of all lands conserved from industrial development within Canada's network of protected areas, making the federal government the largest custodian of protected areas within the nation. Figure 2.1 shows the current distribution of national parks.

In its early years, the evolution of the national park idea in Canada was influenced more by the nation's focus on economic development and prevailing social values, and less on the need to preserve wilderness. Historically, government, industry, and local communities emphasized the economic value of national parks as places of recreation and as tourism destinations. However, as Canada's wilderness continues to dwindle, the essential role of national parks and protected areas in preserving important natural ecosystems is more broadly acknowledged. Parks are now viewed as agents of conservation rather than just for recreation, particularly in that the preservation of natural areas from industrial development is increasingly acknowledged as a critical step in conserving the world's biological diversity. And a growing element in promoting parks and protected areas as agents of conservation is their critical role in helping visitors to develop a deeper appreciation and connection to the land, to understand the need to protect the land and its inherent diversity of wildlife, and to recognize the role they can play in conserving nature.

This chapter explores the evolution of the national parks system and some of the landmark events that have shaped it over the last century—particularly the expansion of the national parks system—and the range of management issues that have preoccupied governments and conservation groups for most of its history. Also featured is the critical role that politicians, Aboriginal people, and conservationists have played in shaping the parks system.

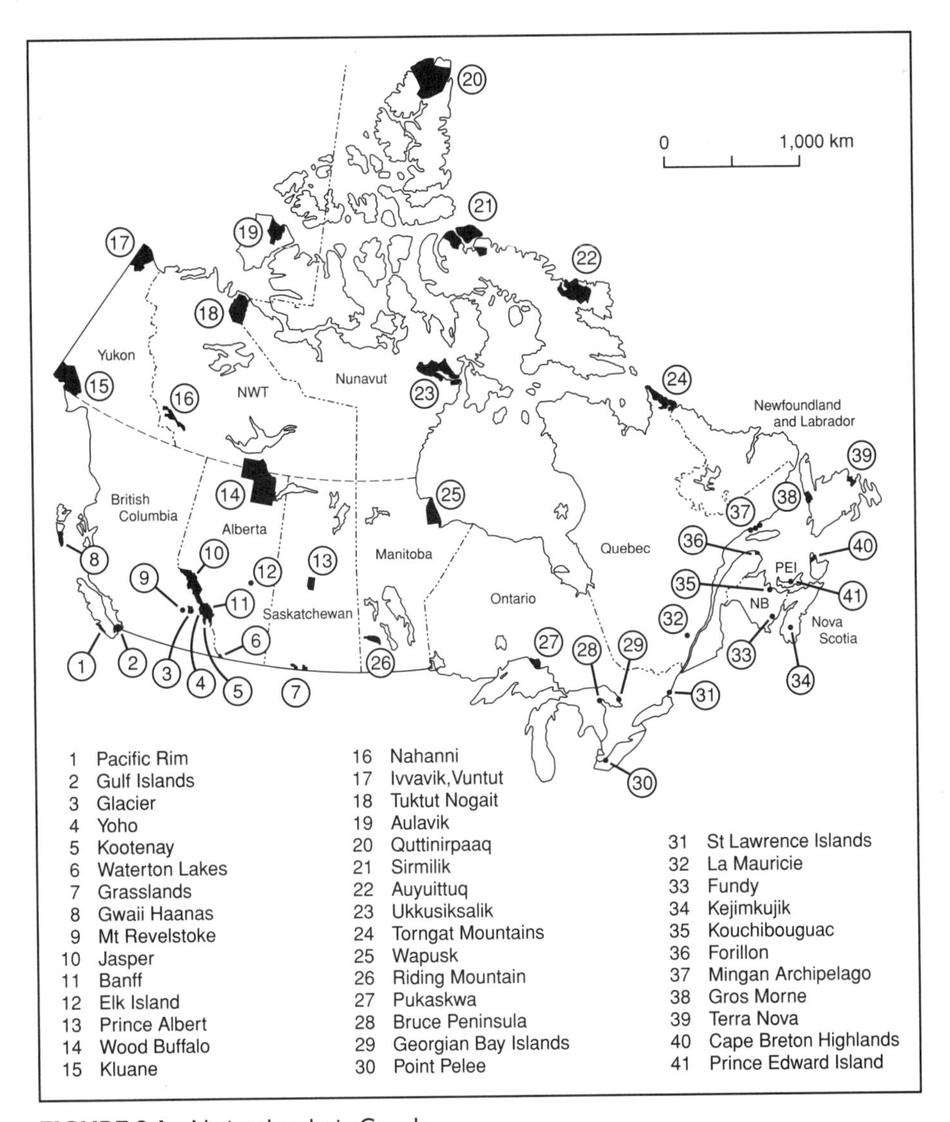

FIGURE 2.1 National parks in Canada.

BIRTH OF THE CANADIAN NATIONAL PARKS NETWORK (1885–1911)

The impetus for Canada's first national park, Banff, was the discovery of the Cave and Basin mineral hot springs by three employees of the Canadian Pacific Railway (CPR). They sought to establish a claim so that they personally could profit from the commercial development of the hot springs. The federal government denied the claim. Instead, in November 1885, it established a 26 km^2 (10 sq. mi.) reservation around the

Banff hot springs on the slopes of Sulphur Mountain. The hot springs were now protected in the public interest and no longer available for 'sale, or settlement, or squatting'. The government, in partnership with the CPR, sought to exploit the economic benefits of the hot springs (Figure 2.2). The wording of the Order-in-Council that established the reserve reflected the value of the hot springs to the government: 'there have been discovered several hot mineral springs which promise to be of great sanitary advantage to the public' (Lothian, 1976: I, 20).

BOX 2.1 Key Dates in the Evolution of National Park Management

1885 Canada's first national park (Banff) established.
1911 Dominion Parks Branch established as world's first national park service.
1911 Dominion Forest Reserves and Parks Act establishes Canada's national park system.
1930 Canada's first National Parks Act passed by Parliament.
1930 Transfer of resources agreement confirms a large number of national parks as federal reserves.
1937 First federal–provincial agreement to establish a national park (Cape Breton Highlands).
1964 First comprehensive statement of national parks policy.
1971 First National Park System plan approved.
1976 Canada signs World Heritage Convention; Nahanni designated the world's first natural World Heritage Site by UNESCO.
1976 First northern national park reserves established.
1979 Revised National Parks Policy gives first priority to ecological integrity.
1984 Canadian Rocky Mountain Parks World Heritage Site established.
1984 Ivvavik National Park: first national park established by a land claim agreement.
1986 Approval of first national policy and system plan for National Marine Conservation Areas.
1988 Amendment to the National Parks Act legally formalizes the principle of ecological integrity.
1990 *Canada's Green Plan* commits to complete the national park system by the year 2000.
1991 Tabling of first *State of the Parks Report* in Parliament.
1992 Canada signs the UN Convention on Biological Diversity.
1993 Gwaii Haanas Agreement establishes unprecedented co-management terms between the federal government and the Haida Nation.
1994 Revised Guiding Principles and Operational Policies tabled in Parliament.
1996 Banff–Bow Valley Study sets a new benchmark for ecological management in Canada.
1998 Moratorium announced on commercial development outside of park communities within national parks.

1998	Parks Canada becomes an operating agency through the proclamation of the Parks Canada Agency Act and government declares the parks will not be commercialized or privatized.
1998	Parliament votes to establish Tuktut Nogait National Park and to reject proposals to carve from it land for mining.
1999	Three Arctic national parks are established through the first Inuit Impact and Benefits Agreement under the Nunavut Land Claim Agreement.
1999	Series of decisions and guidelines on commercial development in the Rocky Mountain Parks are released.
2000	Panel on the Ecological Integrity of Canada's National Parks releases report.
2000	Revised Canada National Parks Bill (Bill C-27) given royal ascent.
	Government announces an action plan to establish ten new national parks and five new national marine conservation areas, to expand three existing national parks, and to implement an ecological integrity program.
2002	Canada National Marine Conservation Areas Act given royal ascent.
2003	Federal government announces $219 million allocation to create new national parks and national marine conservation areas, and to protect the ecological integrity of the national park system.
	Agreements signed to create the first two national parks of the twenty-first century—Gulf Islands in British Columbia and Ukkusiksalik in Nunavut.
2005	Torngat Mountains National Park Reserve established as first national park reserve in Labrador.
2007	Prime Minister announces significant expansion of Nahanni National Park Reserve that will eventually see the park expanded to six or seven times its current size.

Wilderness preservation had little to do with the establishment of the Banff Hot Springs reserve and other national parks around the turn of the century. The Banff hot springs were to become, as the Deputy Minister of the Interior stated in 1886, 'the greatest and most successful health resort on the continent' (ibid., 23). To achieve this, he called for a plan 'to commence the construction of roads and bridges and other operations necessary to make of the reserve a creditable national park' (ibid.). The hot springs, the clean air, and the mountain scenery would attract tourists to Banff on the newly constructed railway. They would stay at new hotels, such as the Banff Springs Hotel, that were constructed within the new parks. Thus, both the government and the CPR, which constructed the hotels, would profit.

Creating the First National Parks

The federal government moved quickly both to expand the Banff Hot Springs reserve and to establish other parks. A Dominion Land Surveyor hired to complete a legal survey of the Banff reserve drew the government's attention to 'a large tract of country lying outside of the original reservation' with 'features of the greatest beauty'. The surveyor noted that these lands 'were admirably adapted for a national park' (ibid.).

Acting on the surveyor's find, Parliament passed the Rocky Mountain Park Act in June 1887 to establish the boundaries for a more extensive park of 673 km^2 (260 sq. mi.), which would later be called Banff National Park. The area was to be 'a public park and pleasure ground for the benefit, advantage and enjoyment of the people of Canada'. Under the legislation, the government could make rules for 'the protection and preservation of game, fish (and) wild birds', and to preserve some of the park's natural features and to control the cutting of timber (ibid., 11). This was one of the first times the need to conserve park resources was acknowledged.

The creation of Rocky Mountain Park was partly modelled on earlier actions in the United States. Prime Minister Sir John A. Macdonald was advised to protect the commercial value of the Banff hot springs by establishing a reserve similar to the Arkansas Hot Springs Reserve, which was created by Congress in 1832. The Rocky Mountain Park Act used similar language to that contained in the legislation establishing in 1872 Yellowstone in Wyoming as the world's first national park. The Yellowstone Park Act set land aside 'as a public park or pleasuring ground for the benefit and enjoyment of the people' (ibid., 24).

Here the similarity to Yellowstone ends. While Yellowstone sat unattended for almost two decades, the Canadian federal government made sure that Rocky Mountain Park was made useful and contributed to the national economy (Figure 2.3). The Prime Minister confirmed this policy: 'the Government thought it was of great importance that all this section of country should be brought at once into usefulness' (Craig-Brown, 1969: 49). Timber cutting, mineral development, and grazing were allowed. Mineral claims were worked in Banff for almost half a century.

There was virtually no opposition to resource development activities within the park. The National Policy of the Macdonald government in the 1880s stressed the need to develop and exploit natural resources as the means for developing a national economy. National parks that produced profits from tourism and resource development were viewed simply as an extension of that policy. The underlying assumption of the National Policy was that there were plenty of natural resources to exploit, and that government and industry had a shared responsibility to develop those resources. The first parks were manifestations of that policy (ibid.).

During the parliamentary debate, at least one MP gave voice to the more contemporary view of national parks. Samuel Burdette, a Liberal MP, warned the House of Commons: 'Allow coal miners and hunters and lumberers to frequent and work it and it ceases to be a national park and they would certainly destroy the game, and the fish, and the scenery, and all the beauties we have heard so much about' (Foster, 1978: 25). Even Prime Minister Macdonald acknowledged that 'as much attention as possible should be paid to the protection of the timber in the general line of the park' (MacEachern, 2001: 18). Nevertheless, the extent to which the government developed and manipulated park resources in the decades that followed suggest that while some noble words were spoken, the park was made 'useful'.

In addition to Rocky Mountain Park, several other parks were created under the National Policy. In 1886, A.W. Ross, a member of Parliament from Manitoba, suggested

FIGURE 2.2 The first bathing establishment constructed at the Cave and Basin in Rocky Mountain Park (c.1887–8). The architecture for the two bath houses is fashioned after Swiss-style buildings. In 1887, 3,000 availed themselves of the park's mineral hot springs. Cascade Mountain rises in the background.

FIGURE 2.3 The sleepy frontier town of Banff is pictured here in 1887, with Cascade Mountain rising over what is now Banff Avenue. Regulations passed under the National Parks Act in 1890 prohibited 'furious riding and driving on public roads' in the town. Cows, however, were permitted to graze on the main street.

the federal government examine the potential for more parks along the railway. As a result, Glacier and Yoho parks were established in 1888 to make the British Columbia and the mountain sections of the CPR as 'popular as possible'. And to ensure their popularity, the CPR constructed the Glacier Park Lodge and Mount Stephen House hotels within the two parks. The parks were also created 'to preserve the timber and natural beauty of the district' (Foster, 1978: 31). Policies promoting both the exploitation and conservation of park resources were consistent, because preserving natural scenery was central to retaining the first parks as tourist attractions.

Other parks were established outside the Banff area. F.W. Godsal, a local rancher from Cowley, Alberta, spearheaded efforts to establish the Waterton Lakes National Park despite the opposition of the federal bureaucracy. Godsal urged the government to protect the Crowsnest Pass and Waterton Lakes in a national park, 'otherwise a comparatively small number of settlers can control and spoil these public resorts' (Rodney, 1969: 172). The bureaucracy, however, felt the government should focus on the three existing parks and manage them properly. One civil servant warned the Deputy Minister of Interior: 'Don't you think it possible to overdo this park reservation business?' The comment was ignored. T. Mayne Daly, Minister of the Interior, established the Waterton Lakes Forest Park in 1895 (Figure 2.4), observing that 'Posterity will bless us' (Lothian, 1976: I, 32). His observation was quite prescient given that a century later, the United Nations jointly declared Waterton Lakes National Park, along with its neighbour to the south, Montana's Glacier National Park, a World Heritage Site, ranking it with some of the planet's most astonishing natural wonders.

By 1911, the federal government had protected a number of areas for posterity: Rocky Mountain Park, the Yoho and Glacier Park Reserves, and the Waterton Lakes and Jasper Forest Parks. These areas were multiple-use parks, inspired by a profit motive, and were not founded in any environmental ethic. Yet, their creation was prompted by a need to protect the spectacular scenery for its tourism value, and proved critical to today's legacy of wilderness reserves. Through the first federal parks, the federal government acknowledged its responsibility to hold lands in trust for the public benefit, that there was a need to conserve natural resources, and that the creation and maintenance of parks was a government responsibility. Canadians continue to benefit from the far-sighted decisions of Canada's early political leaders, no matter what the early rationale for creating the nation's first national parks.

This period of park development, from 1885 to 1911, requires more research. As MacEachern (2001) points out in his history of national parks in Atlantic Canada, the 'doctrine of usefulness thesis has been accepted rather uncritically as the perspective

FIGURE 2.4 Visitors to Waterton Lakes Park pause beside their car at the park entrance west of Pincher Creek, Alberta, c.1930. The Pincher Creek Automobile Club began construction on the Pincher Creek–Waterton Road in 1911. Only 64 visitors arrived in Waterton that year. In 1938, Ottawa imposed fees on motor vehicles entering Waterton Lakes Park: 25 cents for a single trip and one dollar for the season.

from which to view Canadian park history' (ibid., 16). He cites several 'preservation impulses' in the early political debates and actions of civil servants, suggesting that some also saw these early national parks as places of preservation.

BRINGING THE NATIONAL PARKS TO CANADIANS (1911–1957)

This period is characterized by expansion of the national parks system beyond its initial Rocky Mountain focus to include natural areas in central and eastern Canada. In order to secure the political support for an enlarged network of national parks, federal civil servants promoted the recreational and tourism benefits of the parks. During this period, also, some of the first examples of public advocacy for park values were exhibited.

In 1911, there was no system of national parks: instead, there was one legislated park and four created by Order-in-Council; there were parks, park reserves, and forest reserves. Each was run with no real policy direction by a superintendent under the Minister of the Interior. It became clear to the government that there was a need for a separate branch of government to administer the parks and to bring them under one system.

The parks were growing in popularity and proving to be a national asset. The success of the first parks convinced the government of the need to protect their scenery. Frank Oliver, the Minister of Interior, introduced legislation into the House of Commons in 1911 that shifted parks policy from promoting parks as 'primarily places of business' to places where 'there will be no business except such as is absolutely necessary for the recreation of the people' (Foster, 1978: 75).

To this end, Parliament passed the Dominion Forest Reserves and Parks Act in 1911 to accomplish several things: it created two categories of conservation lands—forest reserves and dominion parks; it reduced the level of development permitted in the parks; and it placed the dominion parks under the administration of the world's first national parks branch, known variously over the years as the Dominion Parks Branch, the National Parks Branch, Parks Canada, Canadian Parks Service, and now the Parks Canada Agency. Through this Act, Parliament also provided the first overall statement of purpose for the dominion parks: 'they shall be maintained and made use of as public parks and pleasure grounds for the benefit, advantage and enjoyment of the people of Canada' (Lothian, 1976: II, 12–13).

The legislation was a disappointment to the new parks branch because it dramatically reduced the size of the Rocky Mountain, Jasper, and Waterton Lakes Dominion Parks. Rocky Mountain, for example, was reduced by more than half. The reductions occurred because the government concluded that large parks were not required for recreation and for providing people access to natural areas. Thus, land was withdrawn from the parks system and placed into the forest reserves, which concentrated on protecting wildlife.

The government's action to reduce the parks was not a popular decision with James B. Harkin, the first commissioner of the Dominion Parks Branch. The chief civil servant in charge of the dominion parks from 1911 to 1936, Harkin left an indelible mark on Canada's national parks system. He brought a philosophy to the position that was a

FIGURE 2.5 The bath house at the Upper Hot Springs in Banff National Park, c. 1935. William McCardell claims to have discovered these springs on New Year's Day in 1884. The first road for carriages to the spring, which was then privately operated, was completed in 1886. The bath house pictured here was constructed in 1932 by unemployed local men.

mixture of reverence for the power of nature and a pragmatic view of the economic value of nature and the parks to society.

Harkin was heavily influenced by the writings of John Muir, one of America's foremost naturalists and national park advocates. Harkin believed that the national parks 'exist in order that every citizen of Canada may satisfy his soul-craving for Nature' (ibid., 81). He believed the parks provided people with a chance for 'wholesome recreation that would physically and spiritually rejuvenate them' (ibid.). In parks, people could experience nature and beauty, and absorb the peace of the forests.

Harkin believed that Canadians had a responsibility to safeguard Canada's wildlands by establishing more parks. Harkin also believed that anything that impaired the natural beauty of the park, or its peaceful tranquillity, had to be excluded. Under Harkin the national parks system expanded from its western base to eastern Canada, the number of parks increased from 5 to 16, and resource extraction activities were prohibited.

To obtain the political support and government finances to accomplish this, Harkin promoted the economic value of the national parks. He was particularly impressed by the tourism value of the parks (Figures 2.5, 2.6). In one annual report Harkin wrote that the 'National Parks provide the chief means of bringing to Canada a stream of tourists and a stream of tourist gold' (Marty, 1984: 98). The CPR agreed with Harkin, because their calculations demonstrated that the Rocky Mountains generated $50 million a year in tourism revenue (Foster, 1978). Harkin calculated the value of scenic lands to be $13.88 an acre, while wheat land was worth $4.91 (ibid.). Armed with Harkin's statistics, politicians rose in the House of Commons time and again to defend government expenditures on the dominion parks.

Harkin's promotion of the tourism value of parks produced improved visitor accommodation, the provision of minor attractions to supplement natural features, and the construction of roads, such as the Banff–Jasper Highway, and trails so that natural attractions could be reached in safety and comfort. The national park regulations were changed

FIGURE 2.6 Mountain guide Rudolph Aemmer is pictured here on the summit of Mount Victoria near Lake Louise in Banff National Park in 1931. In July 1954 four Mexican women and their male guide were killed after falling 600 m in Abbot Pass. Three other women were rescued by a Swiss guide and Canadian Pacific Railway staff. *Photo: W.J. Oliver.*

in 1911 to allow the first automobiles into the parks. Several decades later, Harkin's work in developing the recreational potential of national parks would begin to cause their deterioration through overuse. His intention, though, was always one of trying to facilitate access by visitors, but not at the cost of what made the national parks special:

> while we were forced in the beginning to stress the economic value of national parks we realized that there were other values far more important which would be recognized in time. . . . The day will come when the population of Canada will be ten times as great as it is now but the National Parks ensure that every Canadian . . . will still have free access to vast areas possessing some of the finest scenery in Canada, in which the beauty of the landscape is protected from profanation, the natural wild animals, plants and forests preserved, and the peace and solitude of primeval nature retained. (Harkin, 1957: 9)

The Birth of a Public Constituency for Parks

Public protests against the reduction in size of some of the dominion parks in 1911 provide some of the first examples of how public pressure shaped the national parks system. Organizations such as the Alberta Game and Fish Protective Association, the Camp Fire Club of America, and the Canadian Northern and Grand Trunk Pacific Railway lobbied for a return of the parks to their larger sizes. Backed by the Camp Fire Club of America and other civil servants, Harkin was able to get the government to expand the Waterton Lakes park from 34 to 1,095 km^2 (13 to 423 sq. mi.) so that it would form a natural continuation of Glacier National Park in Montana. While the park was again reduced by half in 1921, a core wilderness area had been protected because of public pressure. Harkin was also successful in using public pressure to expand Rocky Mountain Park to protect park wildlife.

The first Canadian organization to promote the value of national parks was formed in 1923 to oppose the Calgary Power Corporation's plan to dam the Spray River near Canmore, inside Rocky Mountain Park. The National Parks Association of Canada was formed to promote the conservation of national parks for 'scientific, recreational and scenic purposes, and their protection from exploitation for commercial purposes' (Bella, 1987: 51). Arthur Wheeler, who led the fight against the dam, encouraged the Association to 'protest against any actions that will create a precedent for commercial encroachment upon the integrity of the Canadian National Parks' (Johnston, 1985: 9). Thus ensued a debate over the need to protect national park lands versus the desire to develop their natural resources.

The Spray Lakes debate resembled a similar issue in the United States. A 1913 decision to flood the Hetch Hetchy Valley in Yosemite National Park, California, was preceded by a tremendous national debate on the value of national parks and the need to preserve wilderness. While the valley eventually was lost, the fight, led by John Muir, who had been instrumental in the establishment of the park, promoted a broad swell of public support for the concept of national parks and wilderness. In Canada, where the Spray River cause also was lost, no large base of public support for the Canadian national parks emerged and the National Parks Association faded from view.

The Spray Lakes fight and other issues reinforced Harkin's efforts to further protect the dominion parks. Harkin was especially concerned that a 1926 agreement between Canada and Alberta to transfer control of natural resources to the province would impact on the parks. In 1927 he convinced the Minister of the Interior to introduce legislation in the House of Commons that would establish the principle of the absolute sanctity of the national parks.

Premier Brownlee of Alberta opposed the legislation because he did not want to lose control over the natural resources contained in the parks, citing the water power resources of the Upper Spray Lakes and the coal deposits in Rocky Mountain and Jasper parks as examples. Brownlee's opposition prompted a survey of the parks to identify and remove areas that had important industrial potential. This was an extension of the 'parks must be useful' philosophy: if resource development is to be prohibited in the national parks, then resources capable of being developed should not be included within the boundaries of the parks. In 1928, the survey recommended that the Kananaskis and Spray Lakes watersheds with their potential for hydro power, as well as other areas suitable for grazing, or for coal and timber extraction, be withdrawn from the parks.

In 1930, Parliament passed legislation transferring control over natural resources to the governments of Alberta, Saskatchewan, and Manitoba. As part of the deal, the boundaries of the Rocky Mountain Park were changed to delete the Kananaskis and Spray Lakes and areas now known as Canmore and Exshaw. However, in the process, it was agreed that the control of the national parks, such as the Rocky Mountain parks, Prince Albert National Park in Saskatchewan, and Riding Mountain National Park in Manitoba, would rest solely with the federal government.

Parliament Passes the First National Parks Act

Parliament also passed the National Parks Act in 1930, providing a sweeping statement on parks that reflected Harkin's philosophy. Neither could new parks be established nor existing parks be eliminated, nor their boundaries changed, without Parliament's approval. Mineral exploration and development were prohibited and only limited use of green timber for essential park management purposes was allowed. The parks also were confirmed as absolute sanctuaries for game. At the same time, the dominion parks were renamed national parks, and Rocky Mountain Park became Banff National Park.

Parliament did exercise its right to eliminate several national parks that had been established as wildlife sanctuaries for species threatened with extinction. For example, Buffalo National Park at Wainwright, Alberta, was eliminated in 1947 because the buffalo had been saved with the establishment of Wood Buffalo and Elk Island national parks in the 1920s. The Nemiskan National Park in Saskatchewan, established to protect the pronghorn antelope, also was abolished in 1947 because of the growth in antelope populations and because farmers and ranchers wanted to use the land for cattle grazing.

Harkin's efforts to promote the value of national parks resulted in more parks being established under the new National Parks Act. For example, Nova Scotia was the first province to agree to transfer provincial land to the federal Crown to create Cape

Breton Highlands National Park in 1936. Prior to that, the national parks were established from lands owned or administered by the federal government or from lands that were purchased.

During Harkin's tenure, political support for national parks was high: Cape Breton Highlands was established through the support of the Yarmouth Fish and Game Protective Association and the Premier of Nova Scotia; a member of Parliament and the mayor of Dauphin, Manitoba, lobbied for Riding Mountain Park; and another member of Parliament wanted a park in Prince Edward Island. While there was a decidedly commercial basis to this support, it clearly resulted in the preservation of large tracts of natural areas.

Sometimes national parks were the product of more intricate political dealing. The Prince Albert Liberal Riding Association presented a list of demands to Mackenzie King in 1926 before agreeing to nominate him as their candidate to the House of Commons. The list included a request for a national park. King defeated a young lawyer named John Diefenbaker in 1926, and as Prime Minister presided over the opening of Prince Albert National Park in 1928. Diefenbaker, a future Prime Minister, later referred to Prince Albert National Park as 'that mosquito park offered to Prince Albert as a reward for the election of Mackenzie King' (Waiser, 1989: 43).

In expanding the national park system into eastern Canada, the federal government had to deal with local populations that made a living from the lands that they sought to protect in new national parks. As MacEachern (2001: 19) notes, the process was simple—the 'Parks Branch chose land it thought appropriate for a park, the provinces expropriated the land, and the landowners settled' (ibid.). Families were expropriated to create Cape Breton Highlands (1936), Prince Edward Island (1937), Fundy (1948), and Terra Nova (1957) national parks. Feeling they had no choice, landowners accepted the government's offer and moved into communities next to the parks. As MacEachern details in his history, the bad feelings generated by these actions lingered for years. While this period—1911 to 1957—was characterized by a tremendous growth in the number of national parks, in eastern Canada it was at the expense of local landowners. It would not be until the 1980s that the federal government would deal with communities and landowners as part of the process, rather than as an impediment to the creation of national parks.

One year before he died at the age of 80 in 1958, Harkin lamented the lack of a public constituency in Canada for the protection of wilderness (Nash, 1969). Perhaps, if Harkin had lived into the next decade, he would have been pleased at the sudden growth in citizens' organizations dedicated to promoting the value of national parks to all Canadians.

GROWING SUPPORT, DWINDLING WILDERNESS, NEW PARKS (1958–1984)

The 1960s saw dramatic growth in public concern for the environment. Rapid industrial and urban development, air and water pollution, threats to northern wilderness areas in Alaska and Canada, and the publication of Rachel Carson's *Silent Spring* (1962) energized citizens to demand government action to protect the environment. Part of their agenda included a demand for more parks and for less industrial and recreational development within the boundaries of existing parks.

The minister in charge of national parks generally has to make decisions on the establishment of new national parks and on the commercial uses to be permitted within them. Such decisions usually were ad hoc, and while the parks were managed under the National Parks Act for the 'benefit, education and enjoyment' of Canadians, the public had little or no influence on those decisions.

Several parks ministers decried the lack of an organized constituency for national parks. In 1960, the minister in charge of national parks, Alvin Hamilton, made an impassioned plea in the House of Commons for help in defending national park values: 'How can a minister stand up against the pressures of commercial interests who want to use the parks for mining, forestry, for every kind of honky-tonk device known to man, unless the people who love these parks are prepared to band together and support the minister by getting the facts out across the country' (Henderson, 1969: 331).

Participants at the 1961 Resources for Tomorrow Conference in Montreal agreed with Hamilton. They concluded that there was a need for a non-government organization 'to perform a watchdog role over those areas now reserved for park purpose' (ibid., 332). The National and Provincial Parks Association of Canada (now the Canadian Parks and Wilderness Society) was formed in 1963 to perform this watchdog role by promoting the value of parks and advocating the expansion of park networks.

The National and Provincial Parks Association of Canada (NPPAC) subsequently lobbied successfully for the creation of Kluane and Nahanni National Park Reserves. It helped defeat proposals to hold the 1972 Winter Olympics in Banff National Park and to construct a multi-million dollar resort complex in Lake Louise. These were major battles that focused public and political attention on the ecological value of national parks. It also marked the beginning of a policy shift away from the recreational value of national park lands to their ecological value. The NPPAC, along with other organizations such as the Alberta Wilderness Association and the Canadian Nature Federation, were instrumental in developing environmental standards for national parks to ensure that they were 'maintained and made use of so as to leave them unimpaired' as required by the National Parks Act.

Adopting the First National Parks Policy

Among the NPPAC's most important achievements was its successful advocacy for the first comprehensive national parks policy. In 1958 the government completed a broad policy statement to guide the use, development, and protection of the national parks. However, it sat on the shelf for years. In response to the NPPAC's lobbying, the federal cabinet adopted and implemented the policy in September 1964.

Until the adoption of the 1964 policy, the national parks were administered piecemeal. The National Parks Act, with its dual mandate for use and protection, left ample room for varying opinions as to what parliamentarians meant exactly when they passed the legislation in 1930. Parliament dedicated the national parks 'to the people of Canada for their benefit, education and enjoyment', and they were to be 'maintained and made use of so as to leave them unimpaired'. What kind of benefits were the parks to provide to Canadians? What did 'unimpaired' mean? Where do the limits to recreational development end and requirements for conservation begin?

Each successive government, and the ministers in charge of the national parks, had different interpretations of these requirements. The National Parks Branch wanted a policy statement that would provide continuity for the management of the parks extending beyond the terms of office for any particular government, and that was not in danger of being changed on a political whim (Canada, 1969).

The 1964 policy established the preservation of significant natural features in national parks as its 'most fundamental and important obligation'. Other provisions were to guard against private exploitation, overuse, improper use, and inappropriate development of parklands. It drew a distinction between urban types of recreation, which were to be discouraged, and recreation that involved the use and conservation of natural areas within the parks.

Since then, the federal government has approved two substantive revisions to the policy, placing a progressively stronger emphasis on the preservation of ecological values over public use. In 1979, cabinet put to rest the debate over the dual mandate contained in the National Parks Act by establishing the maintenance of the ecological integrity of parklands as a prerequisite to use. It committed the government to setting legislative limits to the size of downhill ski areas and the size of the towns of Banff and Jasper, both of which are inside national park boundaries. Tourism facilities and overnight accommodation were to be located outside the parks wherever possible. The public was to be consulted on the development of park management plans, changes to the park zoning systems, and revisions to the policy. Reflecting strong public concerns over the loss of parklands to development, the policy revision stated that the 'majority of National Park lands and their living resources are protected in a wilderness state with a minimum of man-made facilities' (Canada, 1982: 40).

In 1995, further revisions to the Parks Canada policy defined how the national parks would be managed to maintain their ecological integrity. There is now a strong emphasis on working with other jurisdictions to co-ordinate the management of lands both inside and outside the national parks so that the greater ecosystems, of which the national parks are part, sustain the overall ecological processes and components that the parks are trying to represent and protect. While the 1979 policy also stressed the need for co-operation between Parks Canada and its neighbours, it was couched in the context of economic development rather than the current ecological context.

A Systematic Approach to Creating New National Parks

Progress in expanding the national parks system came to a halt after Harkin's resignation in 1936. Only two national parks were created between 1936 and 1968—Fundy in 1948 and Terra Nova in 1957. The federal government's 1930 goal of establishing a national park in each province had not been attained because there were still no parks in Quebec. And there were no plans to expand the system.

In 1962, the Canadian Audubon Society, now Nature Canada, sought to reverse this situation by challenging the federal and provincial governments to expand the national park system to mark the nation's 1967 Centennial year. Observing that Canada lacked 'a representative west coast national park . . . , no true example of the once-vast prairie grassland, and no truly good example of Great Lakes shoreline' (Anon., 1962: 74), the Society recommended the creation of at least 12 new national parks that are 'representative of the tremendously varied scenic, topographic and cover features of the nation' (ibid.). The idea did not spur much action because of the lack of co-operation of the provinces, who owned most of the land suitable for the new parks.

However, political impetus was given to expanding the national parks system when Jean Chrétien took charge of the national parks portfolio in 1968, just after the Centennial year. Chrétien immediately declared that to achieve an adequate representation of Canada's heritage, 40 to 60 new national parks would be required to complete the national park system by 1985. He saw a need for urgent action: costs for new parkland were reaching 'prohibitive' levels; and potential national parklands could be quickly spoiled 'by different forms of economic and social development' (Chrétien, 1969: 10).

Chrétien's statement was significant for acknowledging the need for some criteria to guide the location of new national parks. The park establishment process was still largely ad hoc, with parks being created wherever there was sufficient political support and interest. In 1971, Parks Canada adopted a natural regions system plan to guide park expansion activities. The government's goal is to represent the characteristic physical, biological, and geographic features of each of 39 natural regions within the national parks system. While Chrétien had suggested completing the national parks system by 1985, the centennial year of Banff's establishment, no target date would be confirmed until two decades later.

Chrétien did meet one target. The president of the NPPAC wagered the minister five dollars he could not establish nine parks over a five-year period (Anon., 1972). Chrétien won the bet handily, establishing 10 new national parks totalling 52,870 km^2 (18,500 sq. mi.). His successes included: the first national parks in Quebec, including La Mauricie, located in Chrétien's riding of Shawinigan; the first new national park in British Columbia in almost four decades, Pacific Rim; and the first national parks in northern Canada—Kluane, Nahanni, and Auyuittuq.

In the 1960s political and local response to new national parks was different from what it had been during Harkin's era, particularly because the federal government, and not local communities and politicians, was initiating many of the new park proposals. Chrétien encountered strong opposition from local communities and Aboriginal people to new park proposals. The Association for the Preservation of the Eastern Shore successfully opposed the creation of the proposed Ship Harbour National Park in Nova Scotia. Intense opposition from the Inuit and Innu in Labrador forced the postponement in 1979 of federal–provincial negotiations for two new national parks—the Torngat Mountains and the Mealy Mountains—in Labrador (Bill, 1982).

While this opposition stopped the creation of several national parks, it resulted in a park establishment process vastly more sensitive to the social and economic concerns of local residents. For example, this new approach helped revive the two Labrador proposals. In 2005, the governments of Canada and Newfoundland and Labrador, along with the Labrador Inuit Association, signed agreements that resulted in the establishment of the Torngat Mountains National Park Reserve as part of the Labrador Inuit land claim agreement. And in March 2001, both the Labrador Inuit and the Innu Nation joined with Parks Canada and the government of Newfoundland and Labrador in launching a study to determine the feasibility of a national park in the Mealy Mountains of southern Labrador. Progress on either project would have been out of the question had Parks Canada not adopted a more co-operative approach to working with local communities and with Aboriginal people.

Local Communities Force Changes to Parks Policy

Government policy until the 1970s was to expropriate and remove local communities located within the boundaries of proposed national parks. More than 200 families were expropriated to create Forillon National Park in Quebec (Figure 2.7). Some 1,200 residents in 228 households were also removed from their land and communities to complete the land acquisition process for Kouchibouguac National Park in New Brunswick.

Prior to the 1960s, there was little civic resistance to expropriations for national parks. However, society began to reassess its relationship with authority (La Forest and Roy, 1981). Residents affected by government plans to establish Kouchibouguac and Gros Morne national parks demonstrated their opposition in a massive resistance.

This resistance resulted in a number of actions. In 1980, Canada and New Brunswick commissioned a special inquiry into the violence and public controversy that surrounded the establishment of Kouchibouguac National Park. The inquiry condemned

FIGURE 2.7 Old fishing village along the Bay of Gaspé in Forillon National Park, Quebec, east of Grande-Grave, c. 1968–70. Park visitors can now bike or hike along the old road to the most easterly point in the park, Cap Gaspé. Most of the houses pictured here were removed after the Quebec government expropriated local residents to make way for the new federal park. In an ironic twist, the central theme of the park's interpretive program is 'Harmony between man, the land, and the sea'.

the government's policy of requiring mass expropriations of lands required for national park purposes. And while it urged the government to proceed with developing the park, it recommended that affected residents be allowed to continue commercial fishing and clam-digging activities within the park, that Parks Canada emphasize bilingual staffing because the park lay in a predominantly French-speaking area, and that the history of the Acadian community be stressed in the park's communication programs.

The government began to change its approach to dealing with local people starting in Gros Morne National Park, Newfoundland. Under the Family Homes Expropriation Act, passed by the Newfoundland legislature in 1970, the 125 families affected by the park were not forced to move. Few accepted the offer to leave their outport communities. Today, seven communities are located within several park enclaves where the park boundary simply is drawn around them. The National Parks Act also was amended in 1988 to allow local residents to continue cutting firewood and snaring rabbits within the national park, thereby redressing two of their main grievances against the park.

The government amended its policy in 1979 and again in the Canada National Parks Act in 2000, to prohibit the expropriation of private landowners in areas where it wants to establish or enlarge national parks. Private land can now be acquired for parks purposes only if the owner is willing to sell the land to the government. Cases in point are the current land acquisition programs for the Grasslands and Bruce Peninsula national parks. Land is being acquired only from those landowners willing to sell their land.

Finally, the government now must ensure that there is local support for new national parks before proceeding with park establishment. While this has lengthened the time it takes to successfully negotiate a national park, it involves local communities in the negotiations for new parks. The government is trying to ensure that new national parks are supported by local communities and that they make a positive contribution to the community's way of life.

Aboriginal Land Claims and New National Parks

In 1962 the federal government began to examine potential national park sites in the Yukon and Northwest Territories. Plans to mine the Kluane Game Sanctuary in the Yukon and to dam the Nahanni River in the Northwest Territories prompted public campaigns for their protection. Jean Chrétien, the parks minister in the cabinet of Prime Minister Pierre Elliott Trudeau, wrote that 'when I saw the Nahanni River in the Northwest Territories and the Kluane Range in the Yukon, I wanted to protect them forever and eventually did' (Chrétien, 1985: 68). Trudeau's government announced plans to turn both areas into national parks in 1972. Chrétien was similarly moved to create Auyuittuq National Park after flying over Baffin Island. 'I was so excited that I said to my wife, "Aline, I will make this a national park for you"' (ibid.).

Legislation to create the three parks was opposed, however, by several Native organizations representing Aboriginal people who lived in the territories. For example, the Inuit Tapirisat of Canada (now Inuit Tapiriit Kanatami), the national Inuit organization, contended that the government was acting unilaterally and taking Inuit land in the eastern Arctic to establish the parks. The Inuit charged that the government was, in effect, expropriating land from the Aboriginal people in contravention of the Canadian Bill of Rights (Fenge, 1978). The issue was resolved through amendments to the National Parks

Act that designated the three parks as national park reserves pending the resolution of Aboriginal land claims. The Act also enshrined the rights of Aboriginal people to hunt, trap, and fish in northern national parks. In essence, the Native people were not giving up their claim to lands to which they asserted Aboriginal title; they simply agreed to allow the federal government to administer parks on their land until such time as their land claim agreement was ratified by both Parliament and the Inuit. The claim itself would establish final park boundaries and management conditions.

The amendments to the National Parks Act established the precedent for all other national parks that are subject to Aboriginal land claim agreements. To establish new national parks, the government must now negotiate agreements both with the provincial or territorial governments and with Aboriginal people. This signalled a new era in national park designation where, for the first time, Aboriginal people had some legal power to influence decisions (Dearden and Berg, 1993). Hence, parks such as South Moresby, Pacific Rim, and the Mingan Archipelago in southern Canada are designated national park reserves pending the settlement of land claims. In October 2000, Auyuittuq and Ellesmere Island were both legally changed from being national park reserves to national parks with the conclusion of a formal agreement between the federal government and Inuit and the passage of legislation only days before the 2000 federal election.

Northern Canada and New National Parks

Justice Thomas Berger's Mackenzie Valley Pipeline Inquiry of 1974–7 set the tone for the founding of new national parks for the next two decades. Among other things, Berger drew attention to the need to protect the northern wilderness and to 'do so now'. Berger argued that, in northern Canada, 'Withdrawal of land from any industrial use will be necessary in some instances to preserve wilderness, wildlife species and critical habitat' (Berger, 1977: 31). To that end, Berger recommended the north slope of the Yukon as a national park to protect the calving grounds of the Porcupine caribou herd from industrial development.

When J. Hugh Faulkner became the minister in charge of parks in 1978, he acted on Berger's recommendations. In 1978, he withdrew the north slope of the Yukon from industrial development and announced plans to establish a national park in the area. He also announced the '6 North of 60' program to initiate public consultation on a plan to establish five new national parks in the territories and a Canadian Landmark to protect the Pingos of Tuktoyaktuk. (Pingos, topographical features unique to the Far North, are mounds or hills with an ice core.) While it took 25 years, four of the proposed areas are now protected in five national parks (northern Yukon by Ivvavik and, immediately to the south, Vuntut, Ellesemere Island by Quttinirpaaq National Park, Banks Island by Aulavik National Park, and Wager Bay by Ukkusiksalik National Park); the other, Bathurst Inlet, was dropped because of extensive mineral staking and a lack of local support, with Tuktut Nogait National Park being established instead to represent this natural region. While it took some time to achieve these parks, Faulkner's announcement clearly gave the park establishment program some concrete areas to focus on, and remained a preoccupation of Parks Canada for several decades (the Pingos of Tuktoyaktuk remains as Canada's only designated Canadian Landmark).

As the 1985 centennial year of Banff National Park drew close, announcements on new national parks were reduced to a trickle because of budget cuts, the onset of a recession, and an overall decline in public and political interest both in the environment and in wilderness protection. The lack of local, Aboriginal, and provincial support for many of the planned parks was also an important factor. New monies for park expansion were in short supply, partly because of the need to spend money developing facilities in the new parks. Many of the 10 new national park agreements signed by Chrétien from 1968 to 1974, when he was Minister of Indian Affairs and Northern Development, called for a large investment of federal dollars in the development of recreational and tourism infrastructure. For example, $22 million was spent to build a 62-km stretch of road in La Mauricie National Park (Lothian, 1987). But there was some progress in the years following Chrétien's tenure. Two park agreements were signed between 1974 and 1984: Pukaskwa in 1974 and Grasslands in 1981, although the latter agreement failed to produce a park until a new agreement between Canada and Saskatchewan was approved in 1988. In the dying days of the Liberal government of Prime Minister John Turner in 1984, two additional parks were established: the Mingan Archipelago National Park Reserve in Quebec and the Northern Yukon National Park.

The establishment of the Northern Yukon National Park (now called Ivvavik) was the first park established as part of the comprehensive land claims settlement process, setting a precedent for future northern national parks (Sadler, 1989). Both the government of Canada and the Committee for Original People's Entitlement, representing the Inuvialuit of the Western Arctic, agreed to the park because it met their respective objectives: it represents several natural regions of the national parks system; it prohibits any industrial development within the calving grounds of the Porcupine caribou herd, which supports the traditional way of life of Aboriginal people; and Parks Canada and the Inuvialuit work together to co-operatively manage the national park.

Between 1968 and 1984, the Liberal governments of Prime Minister Trudeau and his short-term successor, John Turner, made a substantial contribution to the preservation of Canadian wilderness, particularly in the creation of 13 new national parks and the protection of more than 64,000 km^2 of wilderness lands: a land mass greater than the combined size of Nova Scotia and Prince Edward Island. But with the election of Brian Mulroney's Conservative government in September 1984, park advocates wondered if the Mulroney government would work to finish and expand on the many Liberal park initiatives that were left uncompleted.

PARKS ON THE POLITICAL AGENDA (1984–PRESENT)

The centennial year of the national parks should have been a chance to celebrate the achievements of the past century and to plan for their future. Instead, it proved to be a low point in the history of Canada's national parks. The new Progressive Conservative parks minister, Suzanne Blais-Grenier, quickly angered Canadians when she cancelled all guided walks in national parks and suggested she would not rule out logging and mining in them. The public outcry that followed was tangible proof that Canadians rejected the industrial exploitation of park resources, and was cited as a major reason for her loss of the portfolio in August 1985 (Bercuson, 1986).

In the centennial year no new national parks were established, nor was any plan to complete the system adopted, but it was not for lack of trying. In the dying months of the Liberal government, Parks Canada had been developing plans to expand the national parks system as its centrepiece for the 1985 centennial year. It was seeking cabinet approval to complete the national parks system by the year 2000 and for an allocation of $495 million to fund 20 new terrestrial and 10 new marine parks. Blais-Grenier did not approve the plan and it never saw the light of day (McNamee, 1992).

The public clearly demonstrated in 1985 the extent to which government action to protect wilderness was an urgency. Participants in the Canadian Assembly Project, a citizens' celebration of 100 years of heritage conservation, identified over 500 natural areas in need of protection. At the top of the list was a wilderness area on the far western shores of Canada that would increasingly dominate Ottawa's national parks agenda—South Moresby.

Also dominating the agenda that year was the future of Banff National Park itself. Environmentalists charged that a new management plan for the park proposed to reduce the amount of wilderness lands and to increase recreation and tourism development on park lands adjacent to the present wilderness zones (Alberta Wilderness Association, 1987: 225). The federal tourism minister, Tom McMillan, was also attacking parks policy: 'The national parks are a major tourism attraction and parks policy is tourism policy . . . too often park policy proceeds in the ends of conservation and the environment' (ibid.). Ironically, within several months, McMillan went from being the tourism minister to become the environment minister.

Post-1985 Action on New Parks

McMillan took over the parks portfolio in late 1985 and made it a high priority. Under McMillan, five new national parks were created, Parliament approved the first comprehensive amendments to the National Parks Act since 1930, the National Marine Parks policy was adopted, the first National Marine Park was established at Fathom Five in Ontario, and the federal Task Force on Park Establishment was appointed to examine new strategies to facilitate the creation of new national parks.

The Task Force concluded that Canada must take decisive action to protect its disappearing wilderness. It called on McMillan and Parks Canada to develop a strategic plan to ensure substantial progress by the year 2000 in completing the national parks system. Substantial progress was required because, in 1985, the national parks system was less than half complete (Dearden and Gardner, 1987). However, McMillan was not interested in developing a blueprint to complete the system (McNamee, 1986). His priorities were the preservation of the South Moresby wilderness archipelago in BC as a national park and the completion of several unfinished park initiatives.

Thus, McMillan turned his attention to negotiating an end to the logging of the temperate rain forest of the South Moresby archipelago on the Queen Charlotte Islands, an area that saw the arrest in 1985 of 72 Haida defending their native homeland from logging. Backed by a unanimous motion of the House of Commons, McMillan achieved an agreement signed by the Prime Minister and British Columbia's Premier in July 1987. McMillan also oversaw the completion of political negotiations to the establish Ellesmere Island, Pacific Rim, Grasslands, and Bruce Peninsula national parks.

Responding to requests from several conservation groups to make the protection of natural resources in parks the priority, McMillan approved and Parliament amended the National Parks Act to state: 'Maintenance of ecological integrity through the protection of natural resources shall be the first priority when considering park zoning and visitor use in a management plan' (Canada, 1988). This amendment was significant for two reasons. First, it clearly established the chief purpose of national parks to be the protection of natural resources. Second, in order to maintain the ecological integrity of the national parks, the government must now take action to define and eliminate the range of internal and external threats to park resources. Thus, the legislation compels the government to act against threats to park resources that emanate from areas outside the parks. This amendment prompted Parks Canada during the 1990s to define and implement programs that emphasized the protection of park resources over the traditional emphasis on developing and using parklands for tourism and recreation.

But the images of Haida elders being arrested while acting in the defence of their homeland, and the success of McMillan with the support of thousands of Canadians in preserving South Moresby, may have proved a turning point in the wilderness preservation movement (Dearden, 1988). John Broadhead (1989: 51), a leader of the South Moresby lobby, observed that:

> South Moresby had shaken the national tree. A profound moral dilemma had crystallized in the Canadian conscience, and it could no longer be ignored. It was this: which is more important—the integrity of the earth and the spiritual recreation of future generations, or short-term legal responsibilities to corporations and their shareholders? More to the point, what kind of system is this that renders the two mutually exclusive?

The federal government's lack of interest in developing a plan to complete the job of representing each of its 39 natural regions spurred the conservation community to lobby for the political commitment necessary to achieve this goal. They were backed by the release of the landmark report of the World Commission on Environment and Sustainable Development (WCED, 1987) that, among its many recommendations for actions that were fundamental to ensuring the survival of the planet, called on all nations to complete protected area networks that represented their diversity of ecosystems.

The Endangered Spaces Wilderness Campaign

In response to the Commission's call for action on protected areas, and spurred on by the success of the South Moresby campaign, World Wildlife Fund Canada and the Canadian Parks and Wilderness Society in 1989 launched the Endangered Spaces Campaign. The campaign goal was to persuade the federal, provincial, and territorial governments to complete their parks and protected areas systems by the year 2000 in order to represent each of the nation's approximately 350 natural regions. This stimulated action to expand the national parks system as the federal government became the first to endorse the campaign goal. In fact, Dearden and Dempsey (2004) suggest that the 1990s witnessed more change than any other decade for protected areas in Canada, and the Endangered Spaces Campaign was a catalyst for many of these changes.

In 1989, Lucien Bouchard, then minister in charge of national parks, announced that the federal government would complete the national parks system by the year 2000 because 'the very fragility of the planet compels the expansion of the national parks system' (McNamee, 1992). The federal cabinet confirmed this as a government-wide commitment when it released Canada's Green Plan, a federal environmental strategy, in December 1990. The Green Plan established targets to meet this commitment: at least five new terrestrial national parks were to be established by 1996, and agreements on the additional 13 parks required to complete the system would be achieved by 2000. It also allocated over $40 million to help plan and operate the new parks.

The campaign played a key role in prompting significant achievements by the federal government, as well as efforts that continue to this day. Between 1989 and 2000, Parks Canada established five new national parks, adding over 66,700 km^2 to the system. Unfortunately, at the end of the campaign, 14 of Parks Canada's 39 natural regions still lacked a national park. But several years later, in 2002, Prime Minister Jean Chrétien announced at the World Summit on Sustainable Development in South Africa, as well as in the House of Commons, that the federal government would work to create ten new national parks and five new national marine conservation areas, and to expand three existing national parks, within five years. Many of the candidate sites were identified in the Green Plan document itself (e.g., Torngat Mountains, Labrador) and during the Green Plan years, as Parks Canada identified new sites for action (e.g., Manitoba Lowlands). The 2003 budget allocated $144 million over five years and $29 million annually to establish and operate the new parks and marine conservation areas.

At the start of the twenty-first century, the following sites were established:

- Gulf Islands National Park Reserve in southern British Columbia to represent the Strait of Georgia Lowlands natural region by protecting approximately 26 km^2 of land spread out over 29 sites on 15 islands, including over 30 islets and reefs, as well as conserving the endangered Garry Oak ecosystem.
- Ukkusiksalik National Park, named after the soapstone found within its boundaries, is a 20,500 km^2 national park in Nunavut that is home to caribou, muskoxen, wolves, polar bears, barren-ground grizzlies, and Arctic hares, and is important to local Inuit communities who travel there to hunt and fish. This park represents the Central Tundra Natural Region.
- Torngat Mountains National Park Reserve protects 10,000 km^2 of Inuit homeland in the northern reaches of Labrador, conserving land that is home to polar bears, caribou, and a unique population of tundra-dwelling black bears, along with breathtaking fjords and rugged mountains. The Inuit of both Labrador and Quebec, who have an overlapping land claim to northern Labrador, have signed agreements with Parks Canada respecting the co-operative management of Canada's newest national park.

During this time, a landmark announcement was made in October 2006, when the environment minister travelled to the community of Lutsel K'e to sign a Memorandum of Understanding with the chief of the Lutsel K'e Dene First Nation that launched a three-year study for a national park on the East Arm of Great Slave Lake. It was historical for two reasons: this same community told Parks Canada officials in 1970 to take their maps and go home, putting a 37-year hiatus on any serious park planning efforts. However, upon reviewing Parks Canada's practices in working with Aboriginal people in a more collaborative fashion to ensure their role, the community not only endorsed a study, it and Parks Canada agreed to a study area of over 33,000 km^2, a nearly fivefold increase over the initial proposal of 7,700 km^2.

National Parks as Endangered Spaces

While the loss of unprotected wilderness lands has captured public attention, concern is growing over the degradation of existing parklands and their natural resources. In the 1960s and 1970s, the prevailing concerns were the impact of too many visitors and the inappropriate development of tourism facilities within and on the verge of parks. In the 1980s, increasing evidence suggested that developments in and around national parks were isolating them and causing a decrease in the environmental quality of the parks. The 1990s saw a greater effort to identify actions to protect the ecological integrity of national parks, and to develop the greater national park ecosystem concept as a way to co-ordinate the management of parklands and adjacent areas.

The first indication that many of Canada's national parks were at risk came in a 1987 report by Parks Canada that concluded 'that the magnitude and frequency of transboundary concerns will increasingly become a problem because of continuing development and pollution' (Irvine, 1987: vii). Four years later, in 1991, the first *State of the Parks Report* confirmed that none of the parks was immune to internal and external threats, citing water pollution, poaching, and logging on lands adjacent to park boundaries as some of the major threats to the integrity of parklands.

A decade later, Parks Canada confirmed that 13 of 36 national parks reported that the significant impact of human activities on park ecosystems was increasing relative to a similar analysis in a 1992 report. In essence, there was a measured decline in the health of natural ecosystems in one-third of our national parks over this brief period. Only three national parks reported that the impact is decreasing, and only one—Vuntut—could report its ecosystems in a pristine state. National parks that were part of efforts to manage the broader landscape for sustainability, such as biosphere reserves, fared no better. None of the three national parks that were part of a biosphere reserve—Waterton Lakes, Riding Mountain, and Bruce Peninsula—could report a decrease in the cumulative impact of human activities.

For most of their history, national parks were wild spaces located within larger expanses of wilderness. More recently, industrial and agricultural development on adjacent lands has moved right up to national park borders. The boundaries of Riding Mountain National Park in Manitoba are clearly visible from space because agricultural development has removed the boreal forest right up to the straight-line boundaries of the park. The *State of the Parks 1990 Report* described Fundy National Park as 'an ecological island in [an] area of intensively managed forest land' (Canadian Parks Service, 1991: 49). It also reported that Pacific Rim National Park Reserve is a narrow strip of wilderness and predicted that 'logging adjacent to [the] park could adversely affect [park] resources [and] watershed' (ibid., 223). In effect, human activities outside parks are reducing them to 'endangered spaces'. Furthermore, these 'endangered spaces' play a disproportionately important role in the protection of endangered species: while the national parks cover under 3 per cent of Canada's land area, they contain over 50 per cent of threatened vascular plant species and close to 50 per cent of threatened vertebrates (Dearden, 2000).

The growing consensus that Canada's national parks were under increasing assault led to two landmark studies, one on the future of Banff National Park and another on the future of the national park system. The motivation for both studies was similar—a need to secure independent substantiation that Banff and the rest of the national parks were under threat, and a need to develop an action plan to save these natural areas in order to ensure that, as Parliament called for in 1930, they would be passed on to future generations unimpaired.

In 1994, the federal minister for national parks appointed an independent Banff–Bow Valley Task Force to provide direction on the management of human use and development within the Bow Valley watershed of Banff National Park. After two years, the Task Force concluded that if the current trends continue 'it will lead to the destruction of the conditions in the Banff–Bow Valley that are required for a National Park' (Banff–Bow Valley Study, 1996: 18). Furthermore, the Task Force stated that Banff 'is clearly at a crossroads and changes must come quickly if the Park is to survive' (ibid.). The Task Force report produced a number of immediate actions. A new management plan for the park was released, legally protected wilderness areas were designated by the federal cabinet, and a number of park developments were removed. The report continues to influence the direction of the park as it works to restore its ecological integrity (see Chapter 10).

Several years later, the federal government appointed the Panel on the Ecological Integrity of Canada's National Parks to review the health of the entire national park system and to recommend directions on how best to protect it (see Chapter 9). Its conclusions were dramatic: 'Ecological integrity in Canada's national parks is under threat from many sources and for many reasons. These threats to Canada's sacred places present a crisis of national importance' (Parks Canada Agency, 2000: 1–9).

In releasing the Panel's report, Sheila Copps, the minister responsible for parks, informed Canadians that the report 'will not be gathering dust' and that she was 'asking Parks Canada to find ways of implementing all of the Panel's recommendations, if humanly and legally possible' (Lopoukhine, 2000). In response, Parks Canada released an Action Plan concentrating on four major targets: making ecological integrity central in legislation and policy; building partnerships for ecological integrity; planning for ecological integrity; and renewing Parks Canada to support the ecological integrity mandate. A year later, a progress report on Parks Canada's actions on all 127 recommendations was released to the first meeting of the Round Table that is required to meet every two years pursuant to the Parks Canada Agency Act (Parks Canada Agency, 2001).

Some political action was taken by Parliament to implement the Panel's report. Shortly before the 2000 federal election, a new National Parks Act was passed with a revised mandate that now states: 'Maintenance or restoration of ecological integrity, through the protection of natural resources and natural processes, shall be the first priority of the Minister when considering all aspects of the management of parks' (Parliament, 2000). The new section broadens the direction to Parks Canada both to maintain ecological integrity and to restore it in degraded parks. It also directs the minister to consider ecological integrity in all aspects of park management, as recommended by the Panel. Finally, the 2003 and 2005 federal budgets allocated a total of $135 million over five years and an additional $40 million every year thereafter to Parks Canada for it to implement an enhanced ecological integrity program with the goal being to improve Canada's national park system by March 2008 (Parks Canada Agency, 2006).

In the fight to protect Canada's national parks, environmental groups have broadened their tool kit by turning increasingly to the Federal Court of Canada. The first example of success came when the Canadian Parks and Wilderness Society took the government to court in 1992, alleging that logging in Wood Buffalo National Park was in contravention of the National Parks Act. Several months later, the Federal Court declared a 1983 contract that allowed logging in the park, and the Order-in-Council approving the contract, to be 'invalid and unauthorized by the provisions of the National Parks Act' (Federal Court of Canada, 1992). The Society was back in court in 2001 when it launched another legal action against the government, this time for Parks Canada's decision to allow the clearing and opening of an abandoned road through Wood Buffalo National Park. In October 2001, the Federal Court ruled against the Society, stating, in part, that the new ecological integrity clause afforded no higher level of protection to the national parks.

Environmental groups have also used the courts to protect national parks from external threats. In 1998, the Federal Court ordered the federal and Alberta governments to redo their environmental assessment of a proposed open-pit coal mine on lands immediately adjacent to Jasper National Park. The Court found that the environmental assessment failed to consider the cumulative impact of the coal mine or a series of other approved and proposed mineral and timber projects in the larger region. Shortly after the second series of hearings were concluded, the proponent announced that for financial reasons the project was on hold. Once again, public advocacy groups proved a potent force in shaping the government's national parks policy and preserving the Canadian wilderness.

But, for all the money, legislative and policy initiatives, and public advocacy efforts, success in implementing ecological integrity programs will only come with actual changes to the health of Canada's national parks. Perhaps the most dramatic example is the reintroduction of bison to the prairie ecosystem of Grasslands National Parks in the spring of 2006 after a 120-year absence. As part of the park's 'Prairie Persists' initiative, 71 plains bison were released into the west block of Grasslands National Park to re-establish a grazing regime through the return of large herbivores (see Chapter 5). Equally impressive was the transfer to Parks Canada of lands by five families, Ontario's St Lawrence Parks Commission, and the Nature Conservancy of Canada that resulted in the doubling of the size of St Lawrence Islands National Park and the protection of some lands of high biodiversity and rare species habitat. These and a growing number of examples in other national parks attest to the positive impact that comes from public support, political direction, financial investment, and staff and stakeholder commitments to passing Canada's national parks on to future generations unimpaired.

CONCLUSION

When Parliament passed the National Parks Act in 1930, it declared that the national parks are 'dedicated to the people of Canada for their benefit, education and enjoyment and such parks shall be maintained and made use of so as to leave them unimpaired for the benefit of future generations'. Administrators of the parks for many years interpreted this as being support for a recreational mandate. Today, however, national parks provide environmental protection for shrinking wilderness and wildlife habitat, offer opportunities for people to experience wild places, and serve as benchmarks against which to measure the impact of society's activities on the landscape. Much more emphasis is now being placed on the notion of maintaining and passing on the national parks to future generations in an unimpaired state, and on ensuring that visitors to national parks have an experience founded on an appreciation of the land itself.

History has shown that governments do not act in a benevolent fashion when it comes to wilderness protection. Politics and public pressure are what drive the park establishment process, and will continue to do so, because no law requires the establishment of parks or the preservation of wilderness areas. Therefore, we must understand more fully the political process, and seek to influence it with better information on the full range of national park and wilderness values so that politicians will act more decisively to preserve wilderness.

History has also demonstrated that wilderness preservation is a non-partisan political issue that requires the commitment of the minister in charge of the national parks portfolio, today, the Minister of the Environment. As ministers for national parks, Jean Chrétien, from the Liberal Party, and Tom McMillan, from the Progressive Conservative Party, shared a common determination to achieve political deals that, together, resulted in the creation of 15 national parks and the preservation of almost 100,000 km^2 of wilderness. It will require other politicians with the determination of Chrétien and McMillan to establish the remaining national parks. Public advocacy groups will have to ensure that each successive parks minister takes action to create new national parks, to stop the incremental loss of national park lands to commercial development, and to ensure that the experience remains true to the national parks idea.

Perhaps Chrétien said it best when he opened Kejimkujik National Park in Nova Scotia in 1969: 'Our national parks are part of the original face of Canada, inviolable spots which provide sanctuaries for man as well as nature. But it is man who must extend and preserve them. This is the task that lies ahead' (Lothian, 1976: II, 122).

ACKNOWLEDGEMENT

The author kindly acknowledges Jacinthe Seguin for her insight, comments, and help in revising this chapter. It is dedicated to the memory of Glen Davis, a Canadian philanthropist whose commitment to preserving Canada's wilderness has left an enduring legacy for future generations.

REFERENCES

Alberta Wilderness Association. 1987. 'Wilderness in Alberta: The need is now', in R.C. Scace and J.G. Nelson, eds, *Heritage for Tomorrow: Canadian Assembly on National Parks and Protected Areas*, vol. 3. Ottawa: Environment Canada Parks, 201–92.

Anon. 1962. 'A plan for Canada's centennial', *Canadian Audubon* 24, 3: 72–5.

Anon. 1972. 'Well done, Mr. Chrétien!', *Park News* 8, 3: 2.

Banff–Bow Valley Study. 1996. *Banff–Bow Valley: At the Crossroads*, Summary Report of the Banff–Bow Valley Task Force. Ottawa: Auditor General of Canada.

Bella, L. 1987. *Parks for Profit*. Montreal: Harvest House.

Bercuson, D., J.L. Granatstein, and W.R. Young. 1986. *Sacred Trust? Brian Mulroney and the Conservative Party in Power*. Toronto: Doubleday Canada.

Berger, T. 1977. *Northern Frontier, Northern Homeland: The Report of the Mackenzie Valley Pipeline Inquiry*. Ottawa: Minister of Supply and Services.

Bill, R. 1982. 'Attempts to establish national parks in Canada: A case history in Labrador from 1969 to 1979', MA thesis, Carleton University.

Broadhead, J. 1989. 'The All Alone Stone Manifesto', in M. Hummel, ed., *Endangered Spaces: The Future for Canada's Wilderness*. Toronto: Key Porter, 50–62.

Canada. 1969. *National Parks Policy*. Ottawa: Queen's Printer.

———. 1982. *Parks Canada Policy*. Ottawa: Minister of Supply and Services.

———. 1988. *An Act to Amend the National Parks Act Bill C-30*. Ottawa: Minister of Supply and Services Canada.

———, Canadian Environmental Advisory Council. 1991. *A Protected Areas Vision for Canada*. Ottawa: Minister of Supply and Services.

Canadian Parks and Wilderness Society. 1988. *Park News*, 25th Anniversary Issue and Park News Final Edition, 23.

Canadian Parks Service. 1991. *State of the Parks 1990 Report*. Ottawa: Minister of Supply and Services.

Chrétien, J. 1969. 'Our evolving national parks system', in J.G. Nelson and R.C. Scace, eds, *The Canadian National Parks: Today and Tomorrow*. Calgary: University of Calgary Press, 7–14.

———. 1985. *Straight from the Heart*. Toronto: Key Porter.

Craig-Brown, R. 1969. 'The doctrine of usefulness: Natural resource and national parks policy in Canada, 1887–1914', in J.G. Nelson, ed., *Canada Parks in Perspective*. Montreal: Harvest House, 46–62.

Dearden, P. 1988. 'Mobilising public support for environment: The case of South Moresby Island, British Columbia', in *Need-to-Know: Effective Communication for Environmental Groups*. Proceedings of the 1987 Annual Joint Meeting of the Public Advisory Committees to the Environment, Council of Alberta, 62–75

———. 2001. 'Endangered species and terrestrial national parks', in K. Beazley and R. Boardman, eds, *Politics of the Wild: Canada and Endangered Species*. Toronto: Oxford University Press, 75–93.

——— and L. Berg. 1993. 'Canada's national parks: A model of administrative penetration', *Canadian Geographer* 37: 194–211.

——— and J. Dempsey. 2004. 'Protected areas in Canada: Decade of change', *Canadian Geographer* 48: 225–39.

——— and J. Gardner. 1987. 'Systems planning for protected areas in Canada: A review of caucus candidate areas and concepts, issues and prospects for further investigation', in R.C. Scace and J.G. Nelson, eds, *Heritage for Tomorrow: Canadian Assembly on National Parks and Protected Areas*, vol. 2. Ottawa: Environment Canada Parks, 9–48.

Federal Court of Canada. 1992. *Canadian Parks and Wilderness Society v. Her Majesty the Queen in Right of Canada*. Trial Division, Vancouver.

Fenge, T. 1978. 'Decision-making for national parks in Canada north of 60', Working Paper #3, President's Committee on Northern Studies, University of Waterloo.

Foster, J. 1978. *Working for Wildlife: The Beginning of Preservation in Canada*. Toronto: University of Toronto Press.

Henderson, G. 1969. 'The role of the public in national park planning and decision-making', in J.G. Nelson, ed., *Canadian Parks in Perspective*. Montreal: Harvest House, 329–43.

Harkin, J.B. 1957. *The History and Meaning of the National Parks of Canada*. Saskatoon: H.R. Larson.

Irvine, M.H. 1987. 'Natural resource management problems, issues and/or concerns in Canadian national parks', Natural Resources Branch, Environment Canada Parks (unpublished manuscript).

Johnston, M.E. 1985. 'A club with vision: The Alpine Club of Canada and conservation 1906–1930', *Park News* 21: 6–10.

La Forest, G.V., and M.K. Roy. 1991. *The Kouchibouguac Affair: The Report of the Special Inquiry on Kouchibouguac National Park*. Fredericton, NB.

Lohnes, D.M. 1992 'A land manager's perspective on science and parks management', in J.H.M. Willison et al., eds, *Science and the Management of Protected Areas*. Proceedings of an international conference held at Acadia University, Nova Scotia, 14–19 May 1991. Amsterdam: Elsevier, 19–24.

Lopoukhine, N. 2000. Presentation to the IVth meeting of SAMPA, University of Waterloo, 19 May.

Lothian, W.F. 1976. *History of Canada's National Parks*, 2 vols. Ottawa: Parks Canada, Minister of Indian and Northern Affairs.

———. 1987. *A Brief History of Canada's National Parks*. Ottawa: Minister of Supply and Services Canada.

MacEachern, Alan. 2001. *Natural Selections: National Parks in Atlantic Canada, 1935–1970*. Montreal and Kingston: McGill-Queen's University Press.

McNamee, K. 1986. 'Tom McMillan: Our friend in court', *Park News* 21: 40–1.

———. 1992. 'Overcoming decades of indifference—the painful process of preserving wilderness', *Borealis* 3: 55.

Marty, S. 1984. *A Grand and Fabulous Notion: The First Century of Canada's Parks*. Ottawa: Minister of Supply and Services Canada.

Nash, R. 1969. 'Wilderness and Man in North America', in J.G. Nelson and R.C. Scace, eds, *The Canadian National Parks: Today and Tomorrow*. Calgary: University of Calgary Press, 35–52.

Parks Canada Agency. 2000. *Unimpaired for Future Generations? Protecting Ecological Integrity with Canada's National Parks, Vol. 2: Setting a New Direction for Canada's National Parks*. Report of the Panel on the Ecological Integrity of Canada's National Parks. Ottawa.

———. 2001. *Parks Canada First Priority: Progress Report on Implementation of the Recommendations of the Panel on the Ecological Integrity of Canada's National Parks*. Ottawa: Minister of Public Works and Government Services Canada.

———. 2006. *Performance Report for the Period Ending March 2006*. Ottawa.

Parliament of Canada. 2000. An Act respecting the National Parks of Canada. Statutes of Canada 2000. Chapter 32, Second Session, Thirty-sixth Parliament, 48–9 Elizabeth II, 1999–2000.

Rodney, W. 1969. *Kootenai Brown: His Life and Times.* Sidney, BC: Gray's Publishing.

Sadler, B. 1989. 'National parks, wilderness preservation and Native peoples in northern Canada', *Natural Resources Journal*: 185–204.

Waiser, W. 1989. *Saskatchewan's Playground: A History of Prince Albert National Park.* Saskatoon: Fifth House.

World Commission on Sustainable Development (WCSD). 1987. *Our Common Future.* Oxford: Oxford University Press.

KEY WORDS/CONCEPTS

amendment to the National Parks Act (1988)
Banff–Bow Valley Task Force
CPR
Dominion Forest Reserves and Parks Act (1911)
dual mandate
Endangered Spaces Campaign
James Harkin
John Muir
multiple use
National Policy
National Park Policy (1964)
National Park Policy (1979)
National Parks Act (1930)
National Parks System Plan
Revised National Parks Act (Bill C-27)
Rocky Mountain Park Act (1887)
Silent Spring
Spray Lake controversy
tourism

STUDY QUESTIONS

1. Discuss the connection between the construction of the Canadian Pacific Railway (CPR) and the early development of national parks in Canada.
2. Comment on the National Policy of Sir John A. Macdonald and how this influenced the management of Canada's first national parks.
3. Compare the Canadian view of national parks in the late 1800s (e.g., Banff) with the American view (e.g., Yellowstone).
4. Compare the Spray Lakes debate (Banff) with the Hetch Hetchy debate (Yosemite).
5. Discuss the role of James Harkin on the evolution of national parks in Canada.
6. What was the significance of the public response to the creation of a new national park at Kouchibouguac, New Brunswick? Discuss the role of local communities in the establishment and management of national parks.
7. Discuss the role of the National and Provincial Parks Association of Canada (now the Canadian Parks and Wilderness Society—CPAWS) on the development of national parks in Canada.
8. Comment on the role of provincial governments in the development of new national parks over the past 50 years.
9. What was the purpose of the Endangered Spaces Campaign?
10. Why study the history of parks? Of what relevance is this history to the management of parks today?

CHAPTER 3

Protected Areas: The Provincial Story

Chris Malcolm

INTRODUCTION

This chapter presents an overview of protected areas established and managed by provincial governments in Canada. Unlike federal protected areas, which are chiefly represented by national parks (IUCN Category II), provincial protected areas are represented by a variety of types, including provincial parks, ecological reserves, wilderness areas, wildlife management areas, and private areas, that range from IUCN Categories I to VI (Chapter 1). To complicate the issue, types of protected areas are not consistent from province to province. Even the term 'provincial park' does not denote the same purpose or level of ecological protection in all provinces. Some provinces possess 'ecological reserves' while others do not. It is therefore easier to refer to IUCN categories when discussing and comparing provincially administered protected areas in Canada. This chapter will focus primarily on IUCN Categories I and II, as 86 per cent of protected areas in Canada fall into these two categories. However, wildlife management areas, which are classified as IUCN Category IV, are also important in terms of protected area coverage in some provinces (e.g., Prince Edward Island) and will be considered as well. In particular, this chapter addresses the types, numbers, and areas covered by terrestrial protected areas for each province, the difference between national and provincial parks, the issue of various protected area categories, provincial marine protected areas, and future management.

HISTORY: THE EXAMPLE OF ONTARIO

The history of provincial protected area systems in Canada is as varied as the number of provinces in our country. The first provincial parks were established in Ontario and historical documentation of the province's protected area network is well developed. Its history includes public environmental movements, using parks for profit, and political battles, and will serve as an example of provincial protected area development in Canada.

In 1893, Canada's first provincial park, Algonquin, was established in central Ontario as 'a public park and forest reservation, fish and game preserve, health resort and pleasure ground' (Ontario, 2006a). From the beginning, regional public pressure was instrumental; Alexander Kirkwood published a pamphlet in 1886 calling for the creation of

'Algonkin Forest and Park', and Kent County residents pushed for protection of the Rondeau Peninsula on Lake Erie, which resulted in Rondeau Park in 1894 (Killan, 1992; Ontario, 2006a). Recognition by the government of Ontario that protected areas were in demand resulted in the Ontario Parks Act in 1913.

Protected areas in Ontario began their evolution primarily as preservation for recreation. In 1929 the Quetico–Superior Council, a US group, was formed to lobby for protection of the Rainy Lake watershed from large-scale forestry and hydroelectric development. These developments would have destroyed significant areas of both Minnesota and northern Ontario, including Quetico Park (established in 1913), a haven for canoe enthusiasts. In 1949, a Canadian-based group, the Quetico–Superior Committee, reorganized as the Quetico Foundation (QF) in 1954, was formed to join the battle.

Largely ignored by the Canadian public, the QF, composed of economic and political elite, fought the Ontario provincial government for the creation of a formal US–Canada treaty to protect the area. The movement ultimately failed and no treaty was established. However, in 1960, the Ontario and Minnesota governments agreed to co-operate in protecting the area through informal agreements (Wareki, 2000). Quetico Provincial Park and the Boundary Waters Canoe Area remain today to protect portions of the watershed on the Canadian side of the border.

The Quetico movement pushed for the protection of multiple-use protected areas. Recreation was the driving force for protection; still, there was an acceptance that timber harvesting was an economic necessity for the region. The QF was thus mainly concerned with protection of forests that bordered canoeing waterways. Other groups, however, formed to push for protection based on biocentric and scientific values of wilderness. The Federation of Ontario Naturalists (FON) was formed in 1931 to co-ordinate naturalist, angling, and sportsman associations, forestry clubs, and bird protection societies throughout Ontario to influence provincial nature legislation. The FON pushed for protection of large areas of habitat using an ecosystem-based approach (well before the term was invented), rather than single elements such as game fish or prime canoeing rivers. The main goal of the FON was to create what they termed 'sanctuaries', and later 'nature reserves', particularly within existing parks such as Quetico and Algonquin, that would mandate preservation from logging (Wareki, 2000). They drew on the writings of American naturalists such as Aldo Leopold and publicized the issue of nature protection in order to create a popular movement that could pressure politicians.

The work of the QF and FON had an effect on Frank A. MacDougal, an Ontario civil servant who became a follower of the philosophies of Gifford Pinchot, the first chief of the US Forest Service and an advocate of multiple and wise use of forest resources. MacDougal was the first real political ally for proponents of wilderness protection in Ontario. As park superintendent of the Department of Lands and Forests, MacDougal instituted a shoreline reserve policy for all canoe routes in Algonquin (1939) and protected two areas of pine forests in the park (1940). As deputy minister he instituted the same shoreline reserve policy in Quetico Park (1941) and established a 7,700 ha wilderness area in Algonquin Park (1944).

By 1954 there were eight provincial parks in the province, each established separately from regional initiatives, but conserved in a utilitarian sense, where watersheds and wildlife were often protected but logging was frequently permitted. There was still a fundamental disagreement between the Ontario government and groups such as the FON over the acceptance of commercial operations in provincial parks. Continual pressure from the FON eventually resulted in the Wilderness Areas Act (1959), which provided legislation for the creation of wilderness areas, and furthered the nature reserve concept.

Outdoor recreation was also seen as a profit-making enterprise for parks, and the rise of the automobile had led to several parks located along the proposed Trans-Canada Highway in the 1930s and 1940s. The Ontario government recognized the great demand for recreation in natural areas close to urban centres, and in 1954 the Parks Branch was created within the Department of Lands and Forests. The mandate of the new division was the rapid development of parks along the shores of the southern Great Lakes (Killan, 1992). By 1960, there were 72 provincial parks in Ontario, and by 1970, 108.

The environmental movement of the 1960s and 1970s finally brought wilderness protection to the attention of the general public, as well as a framework for protected areas in Ontario. In 1967, recognizing the conflicting goals of commercial profit from forest resources (from logging or commercial recreation) versus preservation, the Ontario government established a classification scheme for its protected areas: natural environment (Figure 3.1), nature reserve, primitive, recreation, and wild river. Natural environment and recreation parks would allow for commercial profits, while primitive and nature reserves would preserve habitat from commercial land use. All existing and future parks were placed within this scheme. The first primitive park, Polar Bear Provincial Park, a vast area (1,800,000 ha) of the Hudson Bay Lowland, was established in 1968; at the time, Polar Bear Provincial Park was the second largest protected area in Canada, after Wood Buffalo National Park. Much to the disappointment of groups such as the FON, Quetico, Algonquin, Killarney (on Georgian Bay), and Lake Superior provincial parks were classified as natural environment parks, where commercial logging was permitted (ibid.).

The allowance of commercial logging in these parks, coupled with the technological advances in logging practices that permitted year-round harvest of virtually all trees and the development of extensive road networks for large transport trucks, was in part responsible for bringing the concept of conservation to the general public. The National and Provincial Parks Association of Canada (now the Canadian Parks and Wilderness Society [CPAWS]), founded in 1963, and the Algonquin Wildlands League (now the Wildlands League, a chapter of CPAWS), created in 1968, became important voices to continue the battle to rid logging from Ontario provincial parks. These groups, unlike the QF and FON, were able to connect with the public, due to the increased public awareness and demand for nature experiences, and draw support for their movements on a much larger scale. Logging was banned in Killarney (1971) and Quetico (1974) and both parks were redesignated as primitive parks. Logging, however, continued in Algonquin and Lake Superior (ibid.).

The six years between 1978 and 1983 were extremely important in forming the provincial park system that exists today in Ontario. In 1978 the Ontario government produced the *Ontario Provincial Parks Planning and Management Policies Manual*, an

FIGURE 3.1 Kakabeka Falls Provincial Park is a natural environment park in northern Ontario, 30 km west of Thunder Bay along the Trans-Canada Highway. This park protects an area of ecological and historical significance, while allowing public recreation such as vehicle camping. *Photo: C. Malcolm.*

official framework for the development of a protected areas network based in scientific and recreation research, written by protected areas professionals. The manual also reorganized the protected areas classification system into the categories used today, adding historical parks, and renaming primitive to wilderness parks and wild river to waterway parks.

The planning efforts that developed the policy manual eventually led to the Ontario Strategic Land Use Plan (SLUP) in 1981. SLUP set out a co-ordinated planning framework, based on a comprehensive analysis of the province's natural resources and land-use allocations. Novel for the time, extensive public consultations were also included. A set of objectives was established for a comprehensive network of protected areas, with the aim to achieve them by the year 2000 (Ontario, 1982).

In 1983, following up on SLUP, and in an apparent flurry of conservationist-oriented zeal, the government of Ontario established 155 new provincial parks, including four wilderness parks (Killan, 1992). Seventeen years later, as the year 2000 approached, Premier Mike Harris's government outdid its predecessors by announcing *Ontario's Living Legacy*, establishing an astounding 378 new provincial parks, comprising 2.4 million ha (Ontario, 1999). Ninety per cent of these parks were immediately placed under regulation of the Ontario Parks Act and Public Lands Act (Ontario, 2005).

Given Ontario's recent rate of protected area establishment, as opposed to its initial struggles, it would seem that creating provincial parks has become an easy process. The majority of Crown lands in Ontario and the rest of Canada's provinces (but not the territories) are owned by the provinces, meaning that provincial governments already own substantial tracts of land that could potentially become protected areas. This differs for the creation of national parks, where the federal government must negotiate with the provinces to acquire land, which is often a lengthy process (see Chapter 9). However, even though the provinces own vast tracts of land that could be set aside as protected areas, competition from extractive resource uses on these lands, forestry and mining in particular, have always been important limiting factors. When Alan Pope, then Minister of Natural Resources, announced the 155 new provincial parks in 1983, it was soon discovered, after the initial excitement had waned, that the sizes of the new parks were much smaller that originally planned. In addition, 90 of the originally proposed parks were never established, 76 of which were proposed to have been nature reserves (Killan, 1992).

Competing land uses not only limit available land, but also fragment natural areas (Chapters 4 and 5). To be effective, protected areas should preserve representative samples of natural areas, including ecological processes and wildlife habitat, in a connected network. Numbers of protected areas alone are not sufficient. Ontario now appears to be taking the next step towards creating such a network. In July 2006, the provincial government passed a new Provincial Parks and Conservation Reserves Act, which 'make[s] ecological integrity a first priority when planning and managing within parks and conservation reserves' (Ontario, 2006b). This Act puts Ontario ahead of all other provinces in terms of legal protection of wilderness.

NATIONAL OVERVIEW

Canada currently has 9.9 per cent of its terrestrial area set aside for protection. The provinces administer 49.3 per cent of this area (Table 3.1). In terms of numbers of protected areas, the provinces govern 86.7 per cent of the 8,593 protected areas in Canada. There are 653 IUCN Category I provincial protected areas in Canada, represented by entities such as 'nature reserves', 'wilderness areas', and 'ecological reserves', totalling 18,602,310 ha (95.2 per cent fall within Category Ib) (Table 3.2). There are 1,144 Category II provincial protected areas, represented by areas such as 'provincial parks', totalling 12,918,639 ha. The average size of provincial Category I protected areas is 28,487.5 ha, while provincial Category II protected areas have an average size of 11,292.5 ha. Many provincial parks, which serve important recreation roles, are located in the southern parts of the provinces, close to the majority of Canada's population and competing land uses, and are therefore smaller in size. Wilderness areas and nature reserves are often located in the hinterland and larger in size.

There are exceptions to this tendency. One of the largest provincial parks in Canada, Algonquin Provincial Park, is 772,300 hectares, and located in south-central Ontario, less than 300 km from both Toronto and Ottawa. However, logging still occurs in the park. Ecological reserves (most often Category I) are often very small in size and are established to protect a particular ecological entity, process, or critical wildlife habitat, such as seabird nesting habitat (Figure 3.2).

Tables 3.2 and 3.3 illustrate some interesting aspects of provincial protected area networks in Canada. British Columbia possesses the most provincial protected area, with

TABLE 3.1 Management Agencies and Associated Amount of Protected Areas in Canada

Management Agency	% of Canada's Protected Areas	Number of Protected Areas	Area Protected (ha)
Provinces and territories	49.3	7,447	49,059,873
Federal (Parks Canada)	30.8	46	30,662,883
Federal (Environment Canada)	14.4	144	14,292,921
Federal (Indian Affairs and Northern Development)	3.2	1	3,174,140
Federal (Agriculture Canada)	0.8	65	755,864
Aboriginal	1.2	5	1,147,769
Private ownership	0.4	631	416,362
Other	0.01	254	11,880

Source: Adapted from Environment Canada (2006).

12.1 per cent of the province protected. BC is the only province to surpass the 12 per cent protection goal advocated by the World Commission on Environment and Development in *Our Common Future* (WCED, 1987) and (unofficially) adopted into the Endangered Spaces Campaign (Chapter 2), without including federally protected areas.

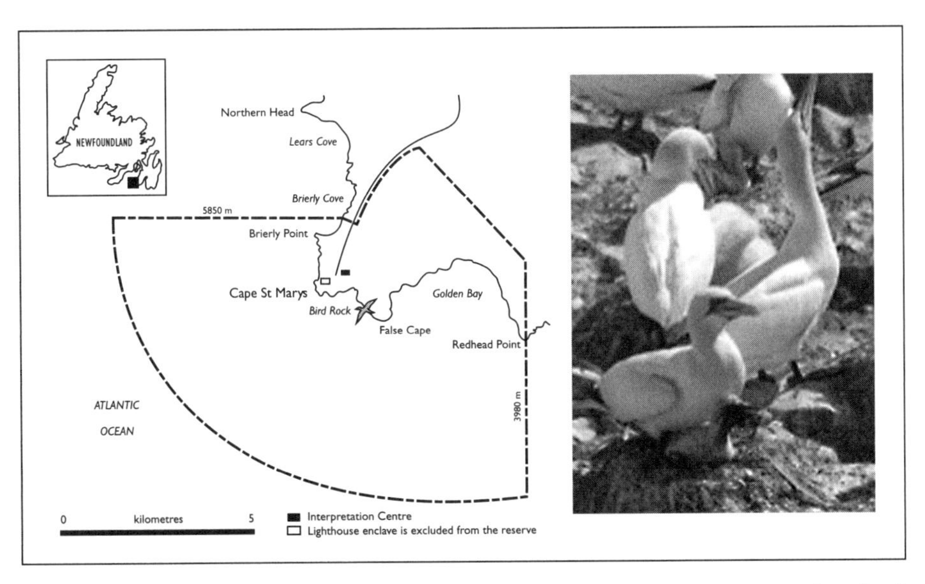

FIGURE 3.2 The Cape St Mary's Ecological Reserve, on the Avalon Peninsula in Newfoundland, comprises only 1,000 ha of terrestrial habitat (plus 5,400 ha marine), yet protects the southernmost breeding colonies of northern gannets (*Morus bassanus*) (pictured) and thick-billed murres (*Uria lomvia*) in the world. *Map: Government of Newfoundland and Labroador, 1983. Photo: C. Malcolm.*

TABLE 3.2 Protected Areas and Area Protected, Provinces and Territories, by IUCN Categories

	IUCN Category							
Province/	Ia		Ib		II		III	
Territory	No.	Area	No.	Area	No.	Area	No.	Area
BC	147	236,783	44	5,010,231	355	5,809,026	106	99,725
Alta	14	31,792	30	2,179,569	193	414,540	15	4,859
Sask.	62	659,810	4	455,856	11	678,763	385	5,908
Man.	17	6,286	6	1,571,910	24	1,116,051	11	48,050
Ont.	106	113,554	8	4,822,920	479	3,682,244	16	7,513
Que.	131	96,404	0	0	23	712,468	195	2,816,889
NB	20	3,022	0	0	18	174,911	0	0
NS	11	3,160	33	296,000	11	10,605	0	0
PEI	0	0	0	0	0	0	34	3,383
NL	16	18,054	2	396,500	27	94,291	6	287
YT	0	0	1	521,340	2	222,614	2	18,522
NWT	0	0	1	2,179,119	1	3,126	0	0
Nunavut	0	0	0	0	0	0	5	9,025
Totals	**524**	**1,168,865**	**129**	**17,433,445**	**1,144**	**12,918,639**	**775**	**3,014,161**

BC is also the only province to have substantially implemented its Protected Areas Strategy (Environment Canada, 2006, see below).

The extensive protected areas developed by the BC government have resulted in a protected areas network in which 94.7 per cent of the protected area in the province is administered by the provincial government (Table 3.3). The Ontario government administers an even larger percentage (97.6 per cent) of its protected areas. This is due in part to a substantial provincial protected area network, but also to limited federally protected area. Although there are five national parks in Ontario, four of them are extremely small: Point Pelee (2,000 ha), Georgian Bay Islands (1,300 ha), Bruce Peninsula (15,400 ha), and St Lawrence Islands (1,900 ha); only Pukaskwa on the north shore of Lake Superior, with 187,800 ha, is of significant size. Quebec administers the highest percentage of total provincial protected area, at 98.4 per cent. This is also partially due to a small number of federally protected areas. However, only 11 per cent of the provincially administered protected areas in Quebec are IUCN Categories I and II; 38 per cent are Category III and a further 42 per cent are designated as 'unclassified'. In

TABLE 3.2 (continued)

	IUCN Category							
Province/	IV		V		VI		IUCN Unclassified	
Territory	No.	Area	No.	Area	No.	Area	No.	Area
BC	0	0	0	0	1	24,368	182	265,853
Alta	0	0	1	9,701	14	102,225	0	0
Sask.	1,486	78,277	149	51,701	2,002*	1,733,892	0	0
Man.	36	82,595	8	1,129	0	0	0	0
Ont.	0	0	0	0	0	0	0	594,860**
Que.	78	594,999	0	0	387	52,582	238	3,130,956
NB	0	0	0	0	0	0	0	0
NS	2	401	0	0	0	0	0	0
PEI	87	8,754	7	816	0	0	0	0
NL	0	0	0	0	5	137,523	1	94
YT	2	105,550	0	0	2	341,400	0	0
NWT	0	0	1	10,400	1	627,506	0	0
Nunavut	0	0	2	142,058	0	0	0	0
Totals	**1,691**	**870,576**	**168**	**215,805**	**2,412**	**3,019,496**	**421**	**3,991,763**

*Saskatchewan Watershed Authority as well as Fish and Wildlife Development Fund lands (measured in number of quarter sections or portions thereof) are included in IUCN Category VI protected areas.
**This area represents the recreation/utilization zone of Algonquin Provincial Park and therefore has no number associated with it.
Source: Adapted from Environment Canada (2006).

comparison, 96 per cent of British Columbia's and 93 per cent of Ontario's provincial protected areas fall under IUCN Categories I and II.

At the other end of the scale, the Alberta and Newfoundland and Labrador governments control only 33.2 per cent and 38.8 per cent, respectively, of the protected area within their provinces, where, in both cases, there are large national parks. In Alberta, there are five national parks: Banff, Jasper, Wood Buffalo, Elk Island, and Waterton Lakes; and, in Newfoundland and Labrador there are three: Gros Morne, Terra Nova, and Torngat Mountains Reserve. However, the government of Newfoundland and Labrador is currently planning the addition of six wilderness reserves, each greater than 100,000 ha, to protect woodland caribou (*Rangifer tarandus caribou*) habitat, which will significantly add to the province's total (Newfoundland and Labrador, n.d).

There are no provincially administered IUCN Category I or II protected areas in Prince Edward Island. Only Prince Edward Island National Park (1,823 ha) represents these top categories. However, 8,754 ha of wildlife management areas are classified as IUCN Category IV, which represent almost 50 per cent of the protected area in the

TABLE 3.3 Total Numbers of Protected Areas and Area Protected, by Province and Territory

Province/ Territory	Provincial Protected Areas	Area Administered by Prov./Terr. (ha)	% of Total Area Protected by Prov./Terr.	% of Total Protected Area Administered by Prov./Terr.
BC	835	11,445,986	12.1	94.7
Alta	267	2,742,686	4.1	33.2
Sask.	4,099	3,664,207	5.6	68.7
Man.	102	2,826,021	4.3	66.1
Ont.	609	9,221,091	8.5	97.8
Que.	1,052	7,404,298	4.9	98.4
NB	38	177,933	2.5	76.7
NS	57	310,166	5.6	68.1
PEI	128	12,953	2.3	80.7
NL	57	646,749	1.6	38.0
YT	9	1,209,426	2.5	23.1
NWT	4	2,820,151	2.1	28.0
Nunavut	7	151,083	0.1	0.7
Totals	**7,264**	**42,632,750**		

province. There are also 53 privately administered protected areas in PEI, which currently fall outside the IUCN classification scheme, and these represent 6.4 per cent of the protected area in the province. The non-profit NGO, Island Nature Trust (see Chapter 16), is an important player in protecting habitat within the province. For most provinces, privately owned protected areas are negligible. In the three territories, Yukon, Northwest Territories, and Nunavut, there is a much larger percentage of federal Crown land, leaving less land for the territorial governments to set aside for protection. In addition, land claim and mineral extraction issues complicate territorial protected area development.

Even with the large numbers of provincial protected areas established in Canada, particularly in Ontario and British Columbia, no provincial government has completed the 1992 Statement of Commitment, signed by the federal, provincial, and territorial Ministers of Environment, Parks and Wildlife, to complete a representative network of protected areas (Environment Canada, 2006). Some provinces are moving towards these targets, however. Nova Scotia, for example, passed the Environmental Goals and Sustainable Prosperity Act in April 2007, which requires the province to protect 12 per cent of its terrestrial area by 2015 (Nova Scotia, 2007). In the next section, limiting factors to protected area development are presented.

PROVINCIAL VERSUS NATIONAL PARKS

What is the difference between a provincial and national park? At first glance, they both protect habitat and often provide recreational opportunities, including vehicle camping. Many people may not realize that they are in one type or the other. However, at the most fundamental level, national parks protect landscape of national significance. That being said, because of the vast size of Canada there are many provincial parks that are also of national significance. National parks are established by Parks Canada, a federal government agency within the Ministry of the Environment, through the National Parks Act. National parks are placed on federal Crown land and can be in any province or territory of Canada. Parks Canada has developed a National Parks System Plan (Parks Canada, 1997), in which national parks are organized into 39 terrestrial natural regions (Chapter 2). As mentioned above, the provincial governments own most of the land in Canada's 10 provinces. This means that in order to establish a national park, the federal government usually has to negotiate with the provincial government to obtain the land (see Chapter 9); this can often be a tedious and expensive process.

Provincial parks are established and managed by the provincial governments in which the parks are situated. Similar to the federal natural region classification, all provinces have divided their lands into regions based on natural features. However, the method by which the classifications are determined may differ from province to province. Some of the first provincial parks in Canada were established based on similar reasoning to the first national parks (see Chapters 2 and 10), mainly as tourism destinations for economic profit. It is easy to understand, then, why provinces, historically, have been and can be reluctant to cede land to the federal government, as they lose potential long-term economic revenue. Sometimes, a better option for the provincial government is to retain the land and create a provincial park. In doing so the province retains tourism revenue, as well as options for resource extraction, such as logging or mining, in the future. Income from these activities can be put towards operating the provincial park system.

There are advantages and disadvantages, beyond economics, to development of provincial park systems. Some of the advantages may include:

- The provinces assess their landscapes at a smaller scale than does the federal government and, consequently, identify more 'natural regions' within their boundaries. As a result the provinces can identify a greater number of representative areas in which to place protected areas. As such, provincial protected area systems may be more ecologically diverse and therefore more resilient.
- Local management (provincial government), as opposed to federal management, based in Ottawa, can mean that local concerns are more likely be taken into account, and local communities may be more able to take part in management issues.

There are potential disadvantages as well:

- There can be less emphasis on ecological conservation within provincial systems than is the case for national parks. Some provincial governments have weak conservation mandates and, therefore, provincial park systems are often less prone to protect against recreation and extraction activities (which can counteract the first advantage above).

- Multiple protected area agencies (e.g., Parks Canada, Environment Canada, Fisheries and Oceans Canada, as well as provincial government agencies) may cause inefficiencies in protected area establishment and management on a national scale, due to lack of communication, duplication of protected area types, and diverse conservation goals.
- The general public may not be aware whether they are visiting a national or provincial protected area. This can weaken the conservation message.
- National parks have very strong legislation, policy, management capability, and accountability mechanisms. Provincial parks often find themselves at the whim of provincial politicians and may lack adequate resources for effective protection.

The issue of multiple agencies being responsible for habitat protection in Canada is particularly important due to the number of difficult management issues that exist in our attempts to create efficient protected area networks. Some of these issues are discussed in the next section.

MANAGEMENT ISSUES

Across Canada, the provinces have developed a diverse array of protected area types. They are given labels such as conservation area, conservation reserve, ecological reserve, heritage rangeland, migratory bird sanctuary, natural area, nature reserve, provincial park, recreation area, wilderness park, wildland park, wildlife area, wildlife management area, wildlife park, and wildlife reserve. None of these terms are common across all provinces (even the term 'provincial park' is problematic, as PEI does not have any). Varying levels of consideration are given to ecological protection between provincial protected area types and networks.

This assortment of designations can make comparisons between provinces, or assessment of provincially protected areas on a national scale, difficult. The Canadian Council on Ecological Areas (CCEA), Natural Resources Canada, and Environment Canada have co-operated to classify into IUCN categories and map all protected areas in Canada. These are the classifications reported in this chapter and found in Environment Canada's *Canadian Protected Areas Status Report* (2006). It is noted in the report, however, that classifications may 'vary by jurisdiction' (ibid., 4). This means that a 'provincial park' or 'ecological reserve' in one province may not serve the same purpose or afford the same level of protection in another province.

For example, 'provincial parks', while consistently serving important recreational opportunities such as camping, hiking, and fishing, vary with respect to ecological protection. In New Brunswick, provincial parks are managed by the Department of Tourism and Parks, and the purpose of the park system is first and foremost to attract tourists and provide recreational opportunities for them. In 2004, the New Brunswick Department of Tourism and Parks undertook an extensive review of its provincial park system. Tourism and Parks Minister Joan MacAlpine, in announcing the review, stated: 'this is a brand new day for our parks system—a day which will begin to lead us on a path of revitalization and development. . . . Through this process we hope to rejuvenate each of the parks with additional programming and infrastructure development' (Communications New Brunswick, 2004). The *New Brunswick's Provincial Parks* guide states that New Brunswick provincial parks offer 'unique experiences from the highest peak in the Maritimes to warm salt water swimming and fascinating discovery beaches. What they do have in common is breathtaking scenery, cultural and recreational activities and camping second to none' (New Brunswick, 2006: 2). In the *2005–2008 Strategy Plan for Tourism and Parks* there is no mention of ecological protection (New Brunswick, 2005).

In contrast, the government of Saskatchewan states that its parks system 'helps to protect and preserve our natural and cultural heritage' (Saskatchewan, 2006) and the government of British Columbia states: 'the provincial system of parks is dedicated to the protection of natural environments for the inspiration, use and enjoyment of the public' (British Columbia, 2007). As mentioned earlier, the government of Ontario recently passed a new Provincial Parks and Conservation Reserves Act with an emphasis on ecological integrity in planning and management (Ontario, 2006b).

These examples illustrate a range of interpretations of the term 'provincial park'. And while 'provincial park' is the most recognized term associated with protected areas on a provincial scale, it may go unnoticed that New Brunswick has a 152,351 ha of 'protected natural areas' governed under the Protected Natural Areas Act (New Brunswick, 2004), 98 per cent of which is protected yet available for recreational activities, similar to 'provincial parks' in other provinces (the other 2 per cent falls into IUCN Category Ia, in which only scientific research is permitted).

To try and make sense of the variety of types and management regimes of protected areas within and between Canadian provinces, organizations such as the CCEA and the Canadian Parks and Wilderness Society (CPAWS) have been working to establish methods of assessing protected areas across jurisdictional boundaries. The CCEA has developed a Conservation Areas Reporting and Tracking System (CARTS), which is a publicly accessible web-based portal for standardized collection, analysis, and mapping of protected areas at all levels of government (www.ccea.org/carts.html).

CPAWS was established in 1963 to promote the establishment of protected areas through public awareness and by encouraging provincial and federal agencies to commit to protected area network development (Chapter 2). CPAWS maintains chapters in every province and a national co-ordination office in Toronto. The CPAWS organizational framework allows it to assess the effectiveness of protected areas networks on provincial scales (Box 3.1), while maintaining a national mandate.

Whatever the character of provincial protected area networks, there are some common difficulties among provinces in their ability to manage them. Environment Canada (2006) reports that currently the majority of Canadian provinces are not able to effectively manage or monitor their terrestrial protected area networks. The factors limiting provincial abilities can be grouped roughly into two categories:

1. *knowledge-based restrictions*, including inadequate information required to maintain natural ecological processes or preserve habitat for wide-ranging species;
2. *ecological impacts*, including incompatible adjacent land uses, habitat fragmentation, exotic invasive species, and increasing visitor use. Habitat fragmentation occurs when contiguous habitat becomes separated into smaller fragments by roads, logging, or other land development projects. Often, smaller fragments of habitat are not large

enough to support sustainable populations of wildlife. In addition, the human-made features that separate fragments can be dangerous for wildlife as they come into contact with human activities.

BOX 3.1 Differences in the Status of Alberta Protected Areas

Alberta has 504 provincial parks and protected areas covering 4.2 per cent or 2,762,800 ha of the province. However, the number of sites within each of eight designations and their represented area protected are variable. Seventy-five per cent of sites are recreation and natural areas, both of which contain the least amount of legal protection and compose less than 8 per cent of the protected land base. The largest parks—the wilderness areas and Willmore Wilderness Park—have a comparatively higher degree of protection, and, with only four sites, add up to 20 per cent of Alberta's protected area.

The Alberta parks and protected areas classifications are designed to meet primarily a conservation and/or recreation objective and determine how each area is managed, budgeted, and legislated. These different site types range from 'intensively developed recreation areas to pristine wilderness' (ATPRC, 2007). The eight classifications provide varying degrees of protection and a range of outdoor recreation opportunities and are managed under three pieces of legislation: the Provincial Parks Act, the Wilderness Areas, Ecological Reserves, Natural Areas and Heritage Rangelands Act, and the Willmore Wilderness Park Act. While the types vary in management objectives, they all must adhere to the vision, mission, and goals of the Alberta Parks Division. In other words, each must contribute to the primary goal of preservation and be balanced with the other goals of heritage appreciation, outdoor recreation, and heritage tourism.

Only one of Alberta's spectrum of sites—wilderness area—offers the legal protection required under 'the national parks and similar reserves (Categories I–III)' (IUCN, 1983; see www.cpawsnab.org). The three wilderness areas (0.15 per cent of the province) are the only protected areas that restrict all of commercial timber cutting, mining, hydroelectric and other dam types, industrial facilities, commercial fishing, sport and commercial hunting, farming and grazing of domestic animals (as per IUCN, 1983). They are also the only parks within Alberta's network that prohibit oil and gas activities. Sport hunting and grazing of domestic animals are restricted in the ecological reserves and provincial parks designations. The wilderness areas, together with the ecological reserves, are also the only classifications that have clear and consistent restriction of off-highway vehicles (OHVs), i.e., all-terrain vehicles and snowmobiles.

Only certain park legislation clearly prohibits OHV use (see Table 3.4). In addition, there is no specific reference within the goals of the Division, or within the goals of each designation, as to what the definition of 'outdoor recreation' within parks means. This ambiguity, together with lobbying from the off-road vehicle associations, has begun to open up a significant amount of the protected land base to legal (designated trails) and illegal (random use) motorized recreation.

TABLE 3.4 Site Spectrum of Alberta's Parks and Protected Areas

Site Type	Description of Site	Number of Sites	Area (km^2)	% of Alberta Protected Land Base	OHV Legislation
Ecological Reserves	Preserve and protect natural heritage in an undisturbed state for scientific research and education.	16	294.43	0.70	Prohibited
Wilderness Areas	Preserve and protect natural heritage, where visitors are provided with opportunities for non-consumptive, nature-based outdoor recreation.	3	1,009.89	0.15	Prohibited
Willmore Wilderness Park	Established under its own legislation in April 1959; it is similar in intent to wildland parks.	1	4,596.71	0.70	Prohibited
Provincial Parks	Preserve natural heritage; they support outdoor recreation, heritage tourism, and natural heritage appreciation activities that depend on and are compatible with environmental protection.	73	2,182.73	0.33	Prohibited but exceptions are being made
Heritage Rangelands	Preserve and protect natural features that are representative of Alberta's prairies; grazing is used to maintain the grassland ecology.	1	77.6	0.01	Prohibited but an exception has been made
Wildland Provincial Parks	Preserve and protect natural heritage and provide opportunities for backcountry recreation.	32	17,298.10	2.62	Permitted on designated trails
Natural Areas	Preserve and protect sites of local significance and provide opportunities for low-impact recreation and nature appreciation activities.	149	1,323.78	0.20	Permitted
Recreation Areas	Support outdoor recreation and tourism; they often provide access to lakes, rivers, reservoirs, and adjacent Crown land.	229	844.95	0.13	Permitted on designated trails
Total		**504**	**27,628.19**	**4.18**	

Box contributed by Rebecca Reeves, ParksWatch Program Co-ordinator, Canadian Parks and Wilderness Society, Edmonton Chapter.

The greatest limitations reported by provincial park agencies are limited abilities to undertake inventory and monitoring studies, identify appropriate indicators, and assess ecological stressors (ibid.). For example, lack of scientific information and competing land uses are likely the major contributors to limited planning for inclusion of biodiversity hotspots and critical habitat for species at risk within provincial protected areas networks. DeGuise and Kerr (2006), whose analysis included all protected areas in Canada, reported that there is no relationship between the location of protected areas and the densities of species at risk. In addition, the areas with the highest densities of species at risk have few or no protected areas, particularly in southern Canada (Warman et al., 2004; DeGuise and Kerr, 2006).

Habitat fragmentation is another important management concern with respect to wildlife species that require large ranges. Sixty-four per cent of the protected area in Canada is scattered among approximately 1,500 protected areas less than 1,000 ha in size (Environment Canada, 2005). Jurisdictional boundaries, whether provincial or international borders or divisions between federally and provincially controlled land, add to this problem. A number of provincial parks have a southern boundary shared with the Canada–US international border (e.g., Cathedral Provincial Park in British Columbia, Turtle Mountain Provincial Park in Manitoba, and Quetico Provincial Park in Ontario). However, some adjacent provincial governments have made co-operative initiatives to protect contiguous habitat on either side of their borders. For example, Duck Mountain Provincial Park in Manitoba and Duck Mountain Provincial Park in Saskatchewan are essentially one protected area (joined by Duck Mountain Provincial Forest in Manitoba). So, too, are Woodland Caribou Provincial Park in Ontario and Atikaki, Nopiming, and Whiteshell provincial parks in Manitoba.

The governments of Alberta and Saskatchewan have gone one step further, creating Cypress Hills, Canada's only interprovincial park, in 1989. This unique agreement allows the two provinces to jointly manage an environment that includes rough fescue grasslands, which are only found in the Cypress Hills plateau region in Canada (Alberta, 2004). Environment Canada (2006) notes, though, that these co-operative initiatives are few in number, and most adjacent provincial governments are not currently working together to protect ecological units, such as watersheds, that span their boundaries.

There are also opportunities for federal and provincial protected area agencies to co-operate. Currently, Mount Robson and Humber provincial parks in British Columbia are continuations of Jasper National Park in Alberta. Mount Assiniboine Provincial Park in BC joins Kootenay National Park, to the west, and Banff National Park, to the east. These are examples of protected areas that adjoin each other, affording opportunities for joint management. However, there are currently no examples of jointly managed protected area networks, comprised of national and provincial protected areas, that are spatially separated yet form important habitat units. In Manitoba, from north to south, Porcupine Provincial Forest, Duck Mountain Provincial Park and Forest, Riding Mountain National Park, Spruce Woods Provincial Park and Forest, and Turtle Mountain Provincial Park form a disjunct chain of protected areas, separated by agriculture. Recent research suggests that these protected areas form important habitat units for wolves and elk, and should therefore form a single wildlife management unit (Chapter 13). Following up on management issues such as the Manitoba example, and co-operating to create more interprovincial protected areas such as Cypress Hills, would provide opportunity for the development of ecosystem-based protected area networks.

Most provinces have now developed protected areas strategies for their jurisdictions. These strategies attempt to address some of the limiting factors discussed above, and outline processes for developing representative protected area networks for the provinces. Environment Canada (2006) reports that British Columbia is the only province to have substantially implemented its protective area strategy, although not all natural regions in the province are well represented (Table 3.5). All other jurisdictions (except Newfoundland and Labrador, the Northwest Territories, and the Yukon) are in the process of implementation; Alberta and Ontario have, for all intents and purposes, implemented their strategies; however, they continue to add components to them. Newfoundland and Labrador is still in the process of developing a protected areas strategy and Nunavut does not have a strategy at all (Chapter 11).

A crucial element of developing a protected areas strategy for each province and territory is to assess the current state of protection, classified by the ecological units present (the term used for ecological units varies between provinces as well, e.g., ecoregion, ecodistrict, natural region). Similar to the federal system of 39 representative regions, each province has identified its representative ecological units, based on relatively homogeneous areas of soil type and/or vegetative communities (i.e., biogeoclimatic zones). Each province has recently reported the percentage of its ecological units protected

TABLE 3.5 Reported Percentages of Protected Ecological Units, by Province and Territory

Province/Territory	Reported % Protected
BC	33% high, 19% moderate, 48% low
Alta	66% high, 19% moderate, 15% little or none
Sask.	no data
Man.	27% high, 9% moderate, 42% partial, 22% none
Ont.	no data[1]
Que.	46% high, 15% moderate, 39% low
NB	14% high, 71% moderate, 14% low
NS	34% fully represented
PEI	no data
NL	17% high, 69% have study areas, 14% none
Yukon	33% represented, 33% partial, 33% none
NWT	62% represented, 33% partial (<10%) or none
Nunavut	no data[2]

[1]Ontario is in the process of collecting the necessary data.

[2]Nunavut does not have a protected areas strategy and therefore no representative ecological unit protection targets.

Source: Adapted from Environment Canada (2006).

(Table 3.5). However, the classification criteria are, again, inconsistent, making it difficult to compare at a national scale. Several provinces have not completed the analysis process. This is a good example of the need for consistency across the country, towards which agencies such as the CCEA and CPAWS are working.

MARINE PROTECTED AREAS

Establishment of marine protected areas (MPAs) in Canada has been primarily a federal endeavour (for a discussion MPA establishment and management on a federal scale, see Chapter 15). Canada currently has 0.5 per cent of its territorial marine habitat set aside for protection. Contrary to their role in protecting terrestrial habitat, Canadian provinces are responsible for only 12 per cent of Canadian MPAs (Table 3.6). Provincial MPAs tend to be marine extensions of terrestrial provincial parks or ecological reserves, such as the Cape St Mary's Ecological Reserve in Newfoundland (Figure 3.2).

Most of the provincial MPAs have been established by British Columbia and Quebec. Together, these two provinces, along with PEI and Newfoundland and Labrador, have protected 505,870 ha of marine habitat. There are currently no provincial MPAs in Nova Scotia, New Brunswick, Manitoba, Ontario, or the territories, although Nunavut is currently developing a Territorial Parks Act, which is expected to include marine habitat (ibid.).

Some provincial governments have been working in co-operation with federal agencies to establish MPAs. For example, Quebec, in co-operation with Parks Canada, protects 113,800 ha in the Saguenay–St Lawrence Marine Park, the only MPA of its kind in Canada (Figure 3.3). British Columbia collaborated with Parks Canada to restrict seabed resource extraction in the Gwaii Haanas National Marine Conservation Area Reserve, an area of rich marine biodiversity and cultural significance for the Haida First Nation. BC's protected areas strategy mandates the establishment of an MPA network and the province is currently working with Environment Canada to develop a special agreement within the federal Oceans Action Plan to identify a regional framework that will establish provincial and federal responsibilities to do so. Some recent research examining grey whale use of existing MPAs in BC could help lay the foundation for a network of MPAs to conserve vital links in the marine food web (Chapter 15).

PEI co-operated with Fisheries and Oceans Canada (DFO) to establish the Basin Head Marine Protected Area in 2005, an inshore lagoon system that supports an ecologically and commercially significant community of Irish moss (*Chondrus crispus*). The PEI government retains a role in the management of the MPA through participation on the Basin Head Lagoon Ecosystem Conservation Committee. New Brunswick co-operated with DFO to establish the Musquash Estuary MPA in 2007 by ceding land to the federal government in order to include intertidal habitat (DFO, 2007).

Establishment of provincial MPA networks is slow, even though all the coastal provincial governments agreed, in the 1992 Statement of Commitment to Complete Canada's Network of Protected Areas, to accelerate MPA networks, and recommitted to this in 2000 (Environment Canada, 2006). However, the limiting factors of minimal resource inventories and inadequate scientific understanding are even greater obstacles for marine ecosystems than for terrestrial. In addition, jurisdictional complexities and limited finan-

TABLE 3.6 Management Agencies and Associated Amount of Marine Protected Areas in Canada

Administrative Jurisdiction	Type of MPA	No. of MPAs	Area Protected (ha)	% of Total Canadian MPAs	
BC	Marine portions of terrestrial protected areas (provincial parks and ecological reserves)	114	181,450	5.5	
Que.	Waterfowl gathering areas	352	195,333	6.0	12.0
PEI	Marine portion of a terrestrial protected area	1	87	0.003	
NL	Marine portions of ecological reserves	6	15,200	0.5	
Parks Canada and Que.	Saguenay–St Lawrence Marine Park	1	113,800	3.5	
Parks Canada	Marine portions of national parks	11	938,000	28.6	32.5
Parks Canada	National marine conservation area	1	11,500	0.4	
Environment Canada	Marine portions of national wildlife areas	13	152,317	4.6	47.9
Environment Canada	Marine portions of migratory bird sanctuaries	51	1,417,145	43.3	
Fisheries and Oceans Canada	Marine protected areas (no-take zones)	5	253,530	7.7	7.7

Source: Adapted from Environment Canada (2006).

FIGURE 3.3 The Saguenay–St Lawrence Marine Park is jointly managed by Quebec and Parks Canada. It protects the biologically productive confluence of the Saguenay and St Lawrence Rivers, which attracts numerous whale species such as blue, fin, beluga, and minke (pictured) whales. These are viewed in a regulated whale watching industry. *Photo: C. Malcolm.*

cial resources hamper provincial MPA development (see Chapter 15 for more detailed discussions of these issues). In the future, with increased federal attention placed on MPAs through the Marine Protected Areas Strategy, it is likely that continued provincial–federal co-operation will be the vehicle by which MPAs are established in Canada.

FUTURE MANAGEMENT

Two important issues, until recently, have not been included in the 'to-do' list for provincial park management. The first is appropriate and effective conservation of *freshwater ecosystems* within provincial park networks. The second is the potential impact of *climate change* on provincial park ecosystems.

No province in Canada has undertaken an inventory of freshwater systems within its protected area network. As a result, the amount of freshwater habitat within provincial protected areas and its requirement for protection on a national scale is largely unknown. Yet, the need for this knowledge is clear. During the twentieth century, damming, drainage, pollution, and sedimentation led to widespread declines in the health of North American freshwater fish populations, and in most areas the number of freshwater species at risk in a given habitat is higher than expected when examined (Groombridge and Jenkins, 1998). The World Wildlife Fund recently reported that freshwater vertebrate species suffered greater population declines than either terrestrial or marine vertebrates between 1970 and 2000 (WWF, 2004). The WWF calculated a Living

Species Index, which tracked population trends of 1,100 vertebrate species populations around the world. While terrestrial and marine population indexes each declined by 30 per cent during the 30-year time period, the freshwater index declined by 50 per cent.

Currently, five provinces (British Columbia, Alberta, Manitoba, Ontario, and Quebec) are planning to include freshwater conservation within their protected area networks. Environment Canada (2006) reports that British Columbia's protected area network includes 13.1 per cent of the province's freshwater habitat. This coverage is predominantly unplanned, however, as the BC protected areas strategy is focused on forested and alpine habitat. The coverage is also minimal in the southern part of the province (Rae and Bifford, 2004). Further, that which is currently within protected areas may not help to conserve species due to poor reserve design (McPhail and Carveth, 1993) (e.g., not based on watershed geography). Ontario has a provision for the protection of freshwater systems through its waterway parks, although this classification is mainly for recreation and historical purposes. However, 42 of Ontario's 71 ecodistricts currently contain a waterway park, which can serve as a basis for a protective network of freshwater habitat.

Climate change also poses significant potential management issues for provincial parks in Canada. Average yearly temperatures in Canada have increased 1.3°C since 1945, nearly twice the global average over the same period (Environment Canada, 2007). Due to its northern geographical location, temperature changes in the near future for Canada are also projected to be substantially higher than the global average, with mean annual temperature increases of 3.1 to 10.6°C projected by the end of the twenty-first century (PCIC, 2007). The Intergovernmental Panel on Climate Change (IPCC, 2007) predicted that global average temperature increases exceeding 1.5 to 2.5°C during the twenty-first century will likely cause major changes in ecosystem community structure and function, species' ecological interactions, and species' geographic ranges, resulting in an increased risk of extinction for 20–30 per cent of currently documented species. The Millennium Ecosystem Assessment (2005) stated that these alterations induced by climate change will become one of the primary causes of global biodiversity loss. Some researchers have already documented species responses to climate change, such as plants flowering, birds migrating and/or breeding earlier in the year, and, in mountain regions, butterflies and birds found at higher altitudes (for reviews of these changes, see, e.g., Hughes, 2000; McCarty, 2001).

These predicted changes in ecological communities may lead to a future in which the environments, habitats, and species that protected areas were created to conserve no longer exist within their boundaries. While the unknown by far outweighs the known consequences of climate change, British Columbia, Alberta, Saskatchewan, and Ontario have begun to assess the potential impacts to their park systems. Ontario is likely the most advanced in this process (Box 3.2).

Most provincial jurisdictions have realized the need to adopt an integrated management approach to address the litany of issues they face in establishing and managing protected areas. To this end, British Columbia, Alberta, Saskatchewan, Manitoba, and Ontario have initiated integrated landscape management (ILM) processes. ILM incorporates the holistic approaches of ecosystem management, ecological management, the ecosystem approach, watershed management, and integrated coastal zone management, stressing the importance of appropriate spatial and temporal scales to each situation (Canadian Integrated Landscape Management Coalition, 2005). The multi-stakeholder approach is also a core concept. Through this process, gaps in protected area requirements can hopefully be identified and addressed, while these needs are integrated with resource management planning in general.

Current examples of provincial ILM projects include the Great Bear Rainforest management process in British Columbia and the Athabasca Land Use Planning Process in Saskatchewan. The combined Central Coast and North Coast Land and Resource Management Plans focus on 6.4 million ha of land on the central and north coast of BC. The plans recommend more than 100 new provincial parks, which will protect 1.8 million ha, including 200,000 ha of Kermode 'Spirit Bear' habitat, 'biodiversity areas', where some resource use is to be allowed while sustaining ecological integrity, and ecosystem-based management in 'operating area zones', which focus on community stability and economic diversification. The plan was developed using a multi-stakeholder approach

BOX 3.2 Climate Change and Ontario's Protected Areas

The existing state of parks and protected areas throughout Canada has largely been rationalized on the notions of ecological representation and *stable* heritage assets. This has resulted in a fixed assemblage of lands and waters housing elements of biodiversity usually within a defined political, ecoregional, and/or ecodistrict context. Such approaches to conservation, designed to protect specific natural features, species, and ecological communities in situ, have not taken into account potential shifts in ecosystem composition, structure, and function that could be induced by global climatic change. As Scott and Suffling (2000) emphasize, climate change represents an '*unprecedented challenge*' to protected areas in Canada.

Ontario's system of protected areas consists of over 600 provincial parks, conservation reserves, and wilderness areas and protects over 9.4 million ha (approximately 9 per cent) of the province's terrestrial area. In partnership with the Ontario Ministry of Natural Resources (MNR) Climate Change Program, Ontario Parks is moving towards integrating climate change into several protected area programs. In 2003, in collaboration with Ontario Parks, the Parks Research Forum of Ontario (PRFO) hosted a Climate Change State-of-the-Art Workshop for park planners and managers (Beveridge et al., 2005). The workshop explored the evidence for climate change, the implications for protected areas management, the uncertainties involved, and the measures that might be taken to adapt to them. Since then, a collaborative scoping analysis specific to Ontario Parks has been completed by researchers from the University of Waterloo, the MNR, and Ontario Parks (Lemieux et al., 2007). The scoping analysis revealed the possibility of a broad range of climate change impacts (e.g., changes in ecosystem composition, structure, and function, increased forest fire severity, and the possible extirpation of polar bears from Polar Bear Provincial Park), with significant policy, planning, and management implications at the system and individual park level. These potential impacts are outlined in Table 3.7.

TABLE 3.7 Implications of Climate Change for Ontario Parks' Current Policy, Planning, and Management Frameworks

Policy, Planning, and Management Issue	Implications
System planning	Climate change will alter the boundaries of ecosystems upon which ecological representation is based. As these ecosystems change, protected area representations that have been accomplished may be compromised.
Park establishment	Some ecosystems and habitats on the intervening landscapes between established parks, conservation reserves, and other formally protected areas may emerge as areas important to meet protection commitments and conserve aspects of biodiversity.
Managed landscapes	The role and importance of some ecosystems and habitats in protecting Ontario's biodiversity on the intervening landscapes may increase with rapid climate change.
Protected areas habitat	Habitats located inside some protected areas may no longer be accessible to species that depend on them. For example, some species may lose the ability to physically access traditional habitat (e.g., polar bears in Polar Bear Provincial Park) or may lose the ability to physiologically or phenologically respond to new, emerging climate regimes in the protected area (e.g., rare arctic-alpine plant species that are widely disjunct from their principal range, such as alpine chickweed and Drummond's mountain avens currently situated within Lake Superior, Michipicoten Island, and Sleeping Giant provincial parks and Slate Islands Natural Environment Park).
Invasive species	Some species from other ecosystems likely will find and occupy ecological niches in existing protected areas in a changed climate.
Recreation and tourism resources	The availability of some recreational activities will decline while new opportunities will emerge.
Fire planning and management	Many ecosystems within Ontario's protected areas depend on fire for renewal. However, under changing climatic conditions, natural resource managers will find it increasingly difficult to achieve a balance among protecting socio-economic values (e.g., property and human health), protecting representative natural values (e.g., rare or endangered species and ecosystems), and promoting the use of fire in restoring and maintaining ecosystem health in Ontario's protected areas.
Monitoring and reporting	Long-term monitoring and reporting will be necessary if Ontario's protected areas institutions and organizations are to manage for climate change.

Source: Adapted from Lemieux et al. (2007).

Box contributed by C.J. Lemieux and D.J. Scott, Ontario Ministry of Natural Resources.

that included First Nations, resource industries, environmental non-governmental organizations (e.g., Greenpeace and the BC Chapter of the Sierra Club), and local governments and citizens (Environment Canada, 2006; British Columbia, 2006).

The Athabasca Land Use Plan covers 12 million ha in northern Saskatchewan. The plan calls for the development of guidelines to design both protected areas and sustainable development of natural resources. During the planning stages there are restrictions of cottage development, outfitting, and other commercial enterprises (although existing mineral extraction rights are not restricted). This project is also strongly rooted in the multi-stakeholder model, including three Dene First Nations of the Prince Albert Grand Council, CPAWS, the Saskatchewan Mining Association, local communities, and the provincial government (Athabasca Interim Advisory Panel, 2006a, 2006b).

SUMMARY

The development of provincial protected areas in Canada, similar to the history of federal protected areas, has had to struggle with the problem of use versus protection since the inception of protected areas under the jurisdiction of provinces. Naturalists and outdoor recreation enthusiasts began a fight against provincial governments and logging and mining companies for protection of Canada's wilderness in the late 1880s, and this struggle continues today in some provinces. Because there is more provincial than federal Crown land in southern Canada, provincial governments have always had a large amount of control over the establishment and management of protected areas. However, this has led to a wide variety of approaches to protected area establishment and management across the country. Provinces such as British Columbia and Ontario have developed extensive protected area networks, while provinces such as PEI and Alberta have not. Only four of eight provincial governments with an ocean coastline have established marine protected areas under their jurisdiction. However, the practice of collaboration between provincial and federal governments in the establishment and management of federal MPAs is more prevalent than for terrestrial protected areas. Considering the lack of scientific knowledge and financial resources for MPAs at the provincial level, this relationship is likely to continue.

While a lack of understanding with respect to ecological processes and the difficulty of establishing and maintaining inventory and monitoring programs have been recognized, very little has been accomplished with respect to freshwater ecosystem representation or the potential impact of climate change on provincial protected area networks. Several provinces are just beginning to investigate these extremely important issues.

Most provinces have also recognized the need to integrate the management of protected areas with larger land management processes that address problems surrounding resource use. In this regard, some provinces have developed ILM frameworks and have begun projects within them.

REFERENCES

Alberta. 2000. Wilderness Areas, Ecological Reserves, Natural Areas, and Heritage Rangelands Act. Edmonton: RSA, c. W-9.

———. 2000. Willmore Wilderness Park Act. Edmonton: RSA, c. W-11.

———. 2004. *Cypress Hills Interprovincial Park: Managing the Resources.* Edmonton: Department of Tourism, Parks, Recreation and Culture. At: <www.cd.gov.ab.ca/enjoying_alberta/parks/featured/cypresshills/parkmanage>.

Alberta Community Development. 2003. *Regulations: Provincial Parks and Recreation Areas.* Edmonton: Government of Alberta.

———. 2004. *Alberta Recreation Survey: Summary of Results.* Edmonton: Government of Alberta.

Alberta Tourism, Parks, Recreation and Culture (ATPRC). 2007. *Managing the Network.* Edmonton: Government of Alberta. At: <www.cd.gov.ab.ca/preserving/parks/managing/spectrumsites.asp#wildland>.

Athabasca Interim Advisory Panel. 2006a. *Draft Athabasca Land Use Plan: Stage One.* Regina: Saskatchewan Environment.

———. 2006b. *Appendices—Draft Athabasca Land Use Plan: Stage One.* Regina: Saskatchewan Environment.

British Columbia. 2006. *Central Coast Land and Resource Management Plan.* Victoria: Integrated Land Management Bureau. At: <ilmbwww.gov.bc.ca/lup/lrmp/coast/cencoast/index.html>.

———. 2007. *BC Parks.* Victoria: Ministry of the Environment. At: <www.env.gov.bc.ca/bcparks/index.html>.

Communications New Brunswick. 2004. 'Provincial parks to undergo extensive review', 24 Aug. At: <www.gnb.ca/cnb/news/tp/2004e0900tp.htm>.

DeGuise, I.E., and J.T. Kerr. 2006. 'Protected areas and prospects for endangered species conservation in Canada', *Conservation Biology* 20: 48–55.

Environment Canada. 2005. *Biodiversity and Protected Areas.* Ottawa. At: <www.ec.gc.ca/soer-ree/English/Indicator_series/new_issues.cfm>.

———. 2006. *Canadian Protected Areas Status Report, 2000–2005.* Gatineau, Que.: Environment Canada.

———. 2007. *Climate Trends and Variation Bulletin.* At: <www.msc-smc.ec.gc.ca/ccrm/bulletin/national_e.cfm>.

Fisheries and Oceans Canada. 2007. 'Musquash Estuary is Canada's newest Marine Protected Area (MPA)', 7 Mar. At: <www.dfo-mpo.gc.ca/media/infocus/2007/20070307_e.htm>.

Hughes, L. 2000. 'Biological consequences of global warming: Is the signal already apparent?', *Trends in Ecology and Evolution* 15: 56–61.

Intergovernmental Panel on Climate Change (IPCC). 2007. *Climate Change 2007: Climate Change Impacts, Adaptation and Vulnerability.* Working Group II Contribution to the Intergovernmental Panel on Climate Change Fourth Assessment Report. Summary for Policymakers. Cambridge: Cambridge University Press.

International Union for the Conservation of Nature and Natural Resources (IUCN). 1983. 'Bali Declaration', in *Proceedings of the Third World National Parks Conference.* Gland, Switzerland: IUCN.

Jones, B., and D.J. Scott. 2006. 'Implications of climate change for visitation to Ontario's provincial parks', *Leisure* 30, 1: 233–61.

Kennett, S. 1995. 'Special places 2000: Protecting the status quo', *Resources: The Newsletter of the Canadian Institute of Resources* (Calgary), No. 50.

Killan, G. 1992. 'Ontario's provincial parks, 1893–1993: "We make progress in jumps"', in L. Labatt and B. Littlejohn, eds, *Islands of Hope: Ontario's Parks and Wilderness*. Willowdale, Ont.: Firefly Books, 20–44.

Lemieux, C.J., and D.J. Scott. 2007. Personal communication.

———, ———, P.A. Gray, and R.G. Davis. 2007. *Climate Change and Ontario's Provincial Parks: Towards an Adaptation Strategy*. Ontario Ministry of Natural Resources, Climate Change Research Report 06. Sault Ste Marie, Ont.

McCarty, J. 2001. 'Ecological consequences of recent climate change', *Conservation Biology* 15: 320–31.

McPhail, J.D., and R. Carveth. 1993. *A Foundation for Conservation: The Nature and Origin of the Freshwater Fish Fauna of British Columbia*. Victoria: BC Ministry of Environment, Lands and Parks Fisheries Branch.

Millennium Ecosystem Assessment. 2005. *Ecosystems and Human Well-being: Biodiversity Synthesis*. Washington: World Resources Institute.

New Brunswick. 2004. Protected Natural Areas Act (OC 2004-200). Fredericton: New Brunswick Regulation 2004-57.

———. 2005. *Tourism and Parks Strategic Plan, 2005–2008*. Fredericton: Department of Tourism and Parks.

———. 2006. *New Brunswick's Provincial Parks*. Fredericton: Department of Tourism and Parks.

Newfoundland and Labrador. n.d. *A Framework for a Protected Areas Strategy*. St John's: Parks and Natural Areas Division. At: <www.env.gov.nl.ca/parks/apa/pas/framework.html>.

Nova Scotia. 2007. Bill No. 146: Environmental Goals and Sustainable Prosperity Act. 1st Session, 60th General Assembly, royal assent, 13 Apr. Halifax: Ministry of Environment and Labour.

Ontario. 1982. *Northwestern Ontario Strategic Land Use Plan*. Peterborough, Ont.: Ministry of Natural Resources.

———. 1999. *Ontario's Living Legacy Land Use Strategy*. Peterborough, Ont.: Ministry of Natural Resources.

———. 2005. *Ontario's Biodiversity Strategy*. Peterborough, Ont.: Ministry of Natural Resources.

———. 2006. *Government and the Tourist Industry: Growth of a Parks System*. Toronto: Queen's Printer for Ontario. At: <www.archives.gov.on.ca/english/exhibits/tourism/government_parks.htm>.

———. 2006. *Proposed Legislation for Provincial Parks and Conservation Reserves—Background Information*. Toronto: Queen's Printer for Ontario.

Ontario Ministry of Natural Resources. 1992. *Ontario's Provincial Parks: Planning and Management Policies*. Peterborough, Ont.: Ministry of Natural Resources.

Parks Canada. 1997. *National Parks System Plan*. Ottawa: Parks Canada, Canadian Heritage.

Pacific Climate Impacts Consortium (PCIC). 2007. PCIC Scenario Access Database (BETA). At: <www.pacificclimate.org/tools/>.

Rae, R., and D. Bifford. 2004. 'Are BC's protected areas representing freshwater ecosystems?', in N.W.P Munro, P. Dearden, T.B. Herman, K. Beazley, and S. Bondrup-Nielsen, eds. *Making Ecosystem-Based Management Work: Connecting Managers and Researchers*. Proceedings of the Fifth International Conference on Science and Management of Protected Areas, 11–16 May. Wolfville, NS: SAMPAA.

Reeves, R. 2007. Personal communication. Canadian Parks and Wilderness Society, Edmonton Chapter.

Saskatchewan. 2006. *2006 Saskatchewan Parks Guide.* Regina: Saskatchewan Environment.

Scott, D., and R. Suffling. 2000. *Climate Change and Canada's National Parks.* Toronto: Environment Canada.

Swinnerton, G. 1993. 'The Alberta park system: Policy and planning', in P. Dearden and R. Rollins, eds, *Parks and Protected Areas in Canada: Planning and Management.* Toronto: Oxford University Press.

Wareki, G.M. 2000. *Protecting Ontario's Wilderness: A History of Changing Ideas and Preservation Politics, 1927–1973.* New York: Peter Lang.

Warman, L.D., D.M. Forsyth, A.R.E. Sinclair, K. Freemark, H.D. Moore, T.W. Barrett, R. L. Pressey, and D. White. 2004. 'Species distributions, surrogacy, and important conservation regions in Canada', *Ecology Letters* 7: 374–9.

World Commission on Environment and Development. 1987. *Our Common Future.* Oxford: Oxford University Press.

World Wildlife Fund. 2004. *Living Planet Report 2004.* Gland, Switzerland: World Wildlife Fund for Nature.

KEY WORDS/CONCEPTS

climate change
ecological units
freshwater ecosystems
integrated landscape management
IUCN categories
jurisdictional boundary
marine protected areas
national park
protected area strategy
provincial park

STUDY QUESTIONS

1. Why are the IUCN protected area categories used to provide comparisons between provinces?
2. In Ontario, 90 of the new parks proposed in 1983 were never established. Why was this the case, and what lessons were learned?
3. Table 3.3 indicates that British Columbia has allocated a high percentage (12.1 per cent) of provincial land to protected area, but Table 3.5 indicates that only 33 per cent of ecological units had high representation. What does this say about the protected area strategy for BC?
4. Comment on the distinction between national parks and provincial parks. In your region of the country, select a national park and a provincial park and compare the management, policy, and legislation for each.
5. What are the factors that limit the ability of provinces to effectively manage or monitor their protected areas? What do you think could be done to improve this situation?
6. Establishment of marine protected areas in the provinces has been slow. Why is this so, and what can be done in this regard?
7. Examine in your province the status of freshwater ecosystem conservation, and discuss what needs to be done.
8. Why is climate change an issue for provincial protected areas? In your province, provide specific consequences of climate change that will effect the management of protected areas.
9. In what ways does integrated landscape management advance our approach to protected area management?

PART II

Conservation Theory and Practice

> The one process on-going . . . that will take millions of years to correct is the loss of genetic and species diversity by the destruction of natural habitats. This is the folly that our descendants are least likely to forgive us.
>
> *E.O. Wilson, Harvard University*

This section supplies the background for understanding the ecological basis for protected area planning and management. An understanding of these principles is necessary to answer questions such as, Where should parks be located? How many parks are needed? Where should boundaries be placed? What criteria should be used for determining the size and shape of parks, and the impacts of adjacent land management? If the protection of biodiversity is to be an important objective of parks, or the most important objective of parks, then we need to examine the basic principles of conservation biology and determine how they can be applied to the management of parks and protected areas (Chapter 4).

Managing for ecological integrity represents a significant and controversial 'paradigm shift' from the thinking that parks should be managed always to allow for natural processes (i.e., minimal interference). A number of factors contribute to an active approach to management: the small size of many parks; previous management actions (such as fire suppression); considerations regarding the safety of communities and resources in adjacent communities; and the fluid nature of ecosystem functioning, which rarely recognizes park boundaries. These factors create less than ideal conditions in most parks and require at least short-term measures. Since many park agencies are moving towards management policies that place a priority on ecological integrity, it is imperative to consider the issues around the implementation of such policies. Chapter 5, which includes many specific examples, outlines Parks Canada's approach to active management of national parks.

CHAPTER 4

Application of Ecological Concepts to the Management of Protected Areas

Jeannette C. Theberge and John B. Theberge

INTRODUCTION

Increasingly, a more environmentally conscious society views parks and other protected areas as places that perform vital ecological functions and services. Green spaces moderate water cycles, absorb pollutants, regulate atmospheric gases, buffer the spread of crop and livestock diseases, capture energy, stabilize populations, act as gene pools, and provide habitat for wildlife. They do these things best when they function naturally. To that end, the primary goal of Canada's national parks is the maintenance of the highest possible level of ecological integrity on the lands they manage. Similar goals are expressed in many provincial and other protected area systems in Canada and worldwide.

Because of this paramount ecosystem-management goal, politicians and the public look to ecologists for direction. The responsibility ecologists bear is daunting, because ecosystems are the most complex entities known in the universe. The stakes for Canada are high, because 22 per cent of the world's remaining wilderness lies within Canada (Parks Canada Agency, 2000). The stakes are high, too, because we have lost so much already.

The focus of this chapter is to present the various ideas and theories that are useful in protecting park ecosystems. As ecological theories have advanced through the past decades, so has the management of protected areas. Today, the generally accepted management objectives for protected areas, derived from current ecological theory, include representation of native ecosystem types, maintenance of viable populations, maintenance of ecological and evolutionary processes, and continuance of ecosystem resilience in the face of human pressures both in the short and long terms (Noss, 1996). Applying ecological theory to meet these objectives, however, is a substantial challenge and even somewhat of an art form. In this chapter we frame several 'ways of thinking' to aid in the application of ecological theory to the effective management of protected areas.

Several terms, which we use frequently, require definition. 'Protected areas' refers to all kinds of jurisdictions with varying mandates, ranging from national parks, which protect ecological integrity, to provincial parks, which vary in levels of multiple and extractive uses, to other areas such as wildlife refuges, conservation areas, and ecological reserves. In this chapter we will use the terms 'protected areas' and 'parks' interchangeably.

'Ecological integrity' is considered to be a condition characteristic of a natural region and, apart from external influences, is likely to persist, and includes abiotic components and the composition and abundance of native species and biological communities, rates of change, and supporting processes (Canada National Parks Act, 2000). 'Ecosystem-based management' is an approach that integrates scientific knowledge of ecological relationships within a complex socio-political and values framework towards the goal of protecting native ecosystem integrity over the long term (Grumbine, 1994). This type of management achieves its goal by recognizing ecological boundaries, using adaptive and interdisciplinary methods, and accommodating human use within these constraints (see Chapter 13). While other goals are adopted by agencies involved in managing protected areas, these two concepts are considered to be at the forefront of effective ecological protection (Quinn and Theberge, 2003).

THINKING BIG

Any application of ecological thinking to parks today must extend beyond park boundaries to include the region in which the park is situated. Air, soil, wildlife, and plant propagules are not bounded by parks. The boundaries of protected areas are constructs of human perception, not impermeable ecological barriers. Good planning during park establishment will follow a set of ecological objectives to try to ensure that the most important ecological functions are kept intact (see Box 4.1). While these objectives are generally understood and accepted among park managers, institutional and societal barriers present a daunting challenge to their application (Chapter 13). The establishment of protected areas and their subsequent effective management require broad vision through thinking big, particularly regarding the regional context of the park, the dynamics of metapopulations across boundaries, and the consideration of issues related to scale. Each is discussed below.

Regional Context

Because protected areas are never isolated ecologically from surrounding lands or waters, ecologists have suggested that biodiversity might be conserved best through an ecosystem-based management (EBM) approach (Wright, 1996) that includes consideration of the regional setting, sometimes called the greater park ecosystem (GPE). For many parks, the regional context is complex: many jurisdictions with opposing mandates; different land uses of varying intensity; scattered human settlements and cultural sites; and scattered biological hot spots such as key nesting or migration locations or high levels of biodiversity. It is difficult for the park manager to obtain a bird's-eye view of this regional context in order to understand how the park may be influenced by activities in the region, and vice versa.

To overcome this problem, some park managers create a synthesis of the greater park ecosystem through an exercise involving mapping and public discussion (e.g., Skibicki et al., 1995; Komex, 1995). One such method is called the ABC (abiotic, biotic, and cultural) Resource Survey (Grigoriew et al., 1988). This method combines information from research studies, inventories, and human activities into a multi-layered GIS (geographic information system). The process delineates a GPE where ecosystem components and stresses (such as developments) have the greatest influence on each other. The final product helps park managers focus their efforts on issues within the GPE.

BOX 4.1 Ecology-based Boundaries for Protected Areas

During the creation of the boundaries of a protected area, the following guidelines (Theberge, 1989) should be addressed to minimize transboundary ecological problems. Likewise, for a park that is already established, a park manager can use these guidelines to identify existing or potential ecological problems. Ecosystem-based management is enhanced when the boundaries of protected areas are established with ecological rather than political criteria.

Abiotic Guidelines

1. Boundaries should sever drainage areas as little as possible.
2. Boundaries should not leave out headwater areas.
3. Boundaries should consider subsurface trans-basin water flow.
4. Boundaries should not cross active permafrost terrain.
5. Boundaries should include and not threaten rare geomorphic and hydrological features and processes.

Biotic Guidelines

1. No rare or unique community in the candidate natural area should be severed with a boundary.
2. Boundaries should not sever highly diverse communities, especially wetlands, ecotones, and riparian zones, or lakes or marine coastal zones to compensation depth.
3. Boundaries should not sever communities with a high proportion of dependent faunal species.
4. Boundaries should not jeopardize the ecological requirements of either numerically rare or distributionally rare (or uncommon) species.
5. Boundaries should not jeopardize the ecological requirements of niche specialists.
6. Boundaries should not jeopardize populations of spatially vulnerable species: those that migrate locally, are space demanding, seasonally concentrating, or limited in powers of dispersal.
7. Boundaries should not jeopardize populations of K-selected (low fecundity) species.
8. Boundaries should not jeopardize populations of range-edge or disjunct species.
9. Boundaries should take into special account pollution-susceptible species.
10. Boundary delineation should take into special account the ecological requirements of ungulate species (including their predators).

This big regional thinking has proved useful, for example, in Pukaskwa National Park on the north shore of Lake Superior. The ABC analysis aided in the establishment of a GPE that helped park staff identify relevant regional land-users (i.e., local people; industries such as timber, mining, and tourism) (Skibicki, 1994). In the following years, discussions occurred between the park's staff and local timber companies about the extent of landscape change in the GPE. In 2003, Pukaskwa entered into a dispute resolution process with a logging company over the proposed development through an otherwise roadless wilderness of a primary forestry road and connectors within five kilometres of the park boundary. The impacts of the proposed roads were considered to be substantial. Specifically, this road system would fragment the entire 25-km length of the northern park boundary, cause potential impacts to the wolf population, impede caribou movements, increase illegal access to the park in isolated areas, increase poaching, encourage encroachment of invasive species, result in the eventual loss of connectivity to other protected areas, and impact wildlife populations with transboundary movements due to changes in harvesting pressure (Theberge, 2003). In 2004, the logging company agreed to several mitigation measures. A Zone of Co-operation was established between the two parties extending five kilometres from the park boundary into the provincial lands in which future forestry operations and prescribed burns will be jointly determined (ibid.) (Figure 4.1).

Metapopulation

One of the easiest ways to comprehend how protected areas are interconnected with the larger region is to think about the daily movements of individual animals. Large-bodied animals such as bears and moose frequently have home ranges that cross jurisdictional boundaries. Combined, these individuals within a species make up a population. But if

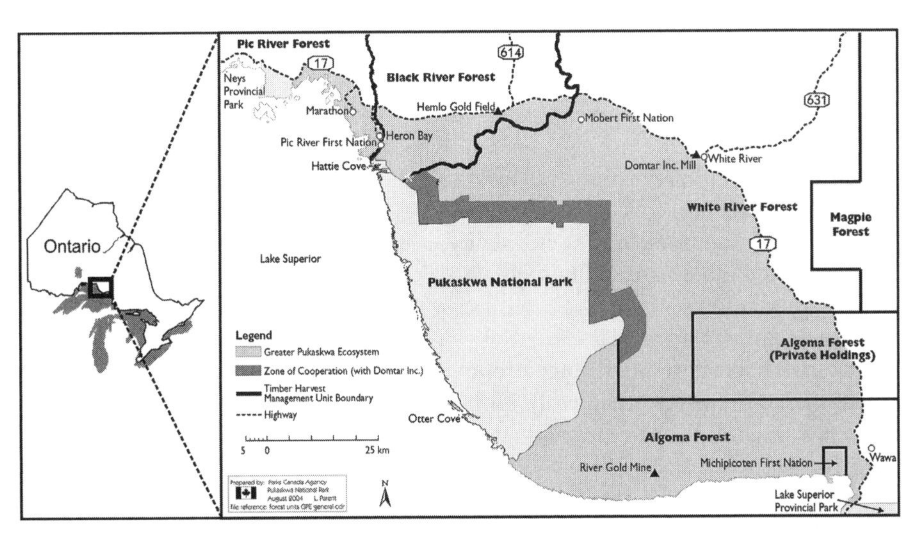

FIGURE 4.1 A proposed Greater Pukaskwa National Park Ecosystem and Zone of Co-operation.

you take an even broader perspective, groups of populations may interact together. This is known as a metapopulation—a population of populations linked by dispersal.

Thus, a single population of a species may, in reality, consist of many subpopulations—with flows between them—that live in sometimes vast, semi-discontinuous habitats. For example, one could view caribou as comprising one metapopulation living in northern British Columbia, the Yukon, and parts of Alaska. This herd ebbs and flows across the landscape, spawning temporary, seemingly discrete herds here and there that may either persist or decline.

The key concepts related to metapopulation revolve around having populations that are spatially discrete with various probabilities of extinction in different population patches (McCullough, 1996). Most isolated small populations eventually go extinct, whereas connected populations, even ones that experience only periodic exchange of members, are more likely to persist. Sophisticated metapopulation models have been developed to account for the complexity of variables such as variations in distances between populations, habitat quality of different patches, rates of demographic change and population turnover, and ecological processes affecting subpopulations (i.e., competition or predator–prey interactions) (Hastings and Harrison, 1994).

Although metapopulations occur naturally, populations are increasingly becoming disjunct due to human activities (McCullough, 1996). Therefore, it is imperative that the management of protected areas goes beyond park boundaries to consider population welfare in this broad context. This concept of metapopulation drastically alters the way management prescriptions should be applied and forces park managers into multi-jurisdictional dialogue to protect park species.

Scale

The topics described above are inherently linked to the concept of scale. Scale is understood as a combination of grain (i.e., the resolution that one picks to view the landscape) and extent (i.e., the dimensions of the outer boundaries of the area under study). Zooming into a small extent (or area) with a fine grain (or resolution), landscape patterns look different than at a larger extent or coarser grain (Figure 4.2).

The park management prescription for a particular ecological issue could vary depending on the scale of investigation. This is because the mechanisms and constraints that produce ecological phenomena at one scale might be different at another scale. For example, Keddy (1991) concluded that tree species richness was strongly correlated with (1) evapotranspiration and energy availability at the continental scale; (2) biomass, stress, disturbance, and dominance at the regional scale (i.e., among vegetation types); and (3) species regeneration potential at the local scale (i.e., within vegetation types).

Likewise, mammal presence is correlated with different attributes at different scales. Female grizzly bears respond to environmental conditions beyond their immediate vicinity (i.e., 300 metres), frequently selecting different landscape features at different scales (i.e., high vegetation diversity at immediate scales, and high levels of terrain ruggedness at 3-km scales) (Theberge, 2002). Partly because of this large perceptual scale, grizzly bear management in the regional ecosystem surrounding Banff National Park and Kananaskis Country is based on large 9 km^2 blocks (Herrero, 2005).

FIGURE 4.2 A heterogeneous landscape in Banff National Park, Alberta. In the foreground (at a fine degree of resolution) the environment is a complicated mix of grassland and open spruce/fir forest. At a coarser scale (the distant walls of the valley) the landscape is comprised of conifer forest, rock, and avalanche chutes.

THINKING CONNECTED

Landscapes are a tapestry of ecosystems and interconnections. As discussed in the previous section, metapopulations interact across large landscapes. So do ecosystem components, such as hydrology, air circulation, and biogeochemical cycles. This section investigates some of the most important concepts of landscapes as they relate to park management and addresses three questions. Why does it matter if a park becomes isolated from other natural wildlands? What types of landscape components might keep parks connected to each other or to wildlands? How does fragmentation of regions influence park health?

Isolation and Island Biogeography

As terrestrial lands become increasingly developed, converted, recreated, and polluted, protected areas appear as havens of nature and refuges for wildlife. Significant attention has been paid by ecologists to the attributes of protected areas that are at risk if they become surrounded by human developments.

The theory of island biogeography has provided an analogy for isolated terrestrial environments and suggests that small islands are unable to support as many species as large islands of similar habitat (Diamond, 1975; Simberloff, 1974). As well, the number of species surviving on an island represents equilibrium between species immigration and extinction, which depends on its distance from a colonizing source (MacArthur and Wilson, 1967). In theory, the number of species on an island will be greater if the island is large and sources of immigration are close.

Although the application of this theory has received substantial debate (summarized in Doak and Mills, 1994), the analogy between islands and isolated terrestrial patches remains appealing because of similarities between terrestrial habitat fragments and islands. Indeed, western North American parks have experienced extinction rates that are inversely related to park size (Newmark, 1995).

Research has resulted in several principles regarding the optimum pattern for a system of reserves. These principles also guide park managers in the placement of facilities inside parks and assist discussions with other agencies about sensitive areas in the regional landscape. These principles include: (1) blocks of habitat close together are better than blocks far apart; (2) habitat in contiguous blocks is better than fragmented habitat; and (3) interconnected blocks of habitat are better than isolated blocks (Noss et al., 1997).

Corridors of Connectivity

Landscapes are heterogeneous. Blocks of habitat (or vegetation patches) are interspersed at different distances or connected with fingers of vegetation. Landscapes are composed of patches, corridors (connections between vegetation patches), and a matrix (the dominant vegetation type) (Forman and Godron, 1986) in various proportions and spatial arrangements. For example, a vegetation patch might be surrounded by extensive agricultural land (i.e., the matrix) and be connected to other patches through a strip of natural vegetation (i.e., the corridor). Or, a protected area might be considered a patch surrounded by a matrix of other human land uses. Connectivity is maintained by (1) protecting multiple large patches with a large core, which allows dispersal or recolonization, and (2) maintaining corridors between these patches across the landscape (Forman, 1997).

A park with ecological integrity would contain not only all the vegetation species and communities that characterized the natural region but a spatial arrangement of patches, corridors, and matrix vegetation types representative of a system's configuration caused by natural disturbance. If that configuration is altered by human disturbance, through activities such as fire suppression, avalanche control, or insect control, then the integrity of the spatial arrangement of landscape components is impaired.

Parks or refuges connected by corridors maintain higher species diversity by allowing reciprocal immigration (Simberloff and Cox, 1987), lowering extinction rates, and minimizing the effects of catastrophes upon populations (Simberloff et al., 1992). However, corridors have disadvantages, too; they can contribute to the spread of disease, disrupt local genetic adaptation by facilitating outbreeding, increase susceptibility to fire, or make poaching easier (Noss, 1987). Consequently, the types and extents of corridors should be decided on the basis of the purpose that the corridor would serve for the species expected to use them.

Today, the objective of connecting parks with corridors is seen as vital to maintaining not only parks but also ecological functions across broader landscapes. Initiatives like the Yellowstone to Yukon project (Locke, 1998) are underway to establish and maintain a corridor of sufficient naturalness that populations such as grizzly bears and wolves—two space-demanding species—can persist.

Fragmentation: Losing Connection

Natural and human-caused disturbances lead to the creation of an increasingly complex mosaic of patches. This process is called fragmentation. Any portion of the landscape may become fragmented, including an individual patch, or a corridor, or the background matrix. Human-caused fragmentation is considered a damaging process (Wilcox and Murphy, 1985).

Species that are especially susceptible to the effects of fragmentation include those that need undisturbed habitats and long-distance migrants. Other vulnerable species are those with short dispersal distances, short life cycles, or low productivity as well as those that depend on unpredictable resources, large patches, or large territories or habitats that are in some way specialized (Noss and Csuti, 1997).

Increasingly, roads are recognized as levying one of the greatest negative impacts to the ecological integrity of natural ecosystems. The impacts of roads include habitat loss, introduction of non-native species, increased human access and use, and fragmentation of the regional landscape (Forman et al., 2003). Protected areas usually have low road densities. However, because parks are too small to sustain the populations or metapopulations of many species, park managers need to be concerned about the road density in the greater region. For example, in the case of wolf persistence in the Great Lakes region, a road density of less than 0.6 km/km^2 has been documented as necessary (Mladenoff et al., 1995). Managers of protected areas in that region that have wolves with transboundary movements, such as Pukaskwa National Park and Lake Superior Provincial Park, could discuss with regional land agencies the benefits of reduced road densities. In particular, it was deemed appropriate to recommend that road density be reduced to half of the minimum threshold, to 0.3 km/km^2, within 10 km of Pukaskwa National Park (Theberge, 2003). That distance of 10 km was chosen because it is approximately half the average territory diameter of wolves in that region (Forshner, 2000).

THINKING ABOUT SPECIES

Many people travel in protected areas hoping to catch a glimpse of wildlife. Park managers and the public alike express concern for the population status of individual species. Significant park conservation efforts revolve around individual species. However, ecosystem-based management requires any concern over individual species to be wrapped in the appropriate ecosystem blanket, specifically their habitat and trophic web of interactions.

This section explores some major concepts related to species persistence as relevant to park managers. To think about species within a park management context it is important to consider these ecological questions. How is this ecosystem organized and which species are its primary drivers? Which species are vulnerable to population declines? When should rare species receive attention over other species? How does one determine if a population is too small? These concepts are discussed below.

Top-Down–Bottom-Up Ecosystem Organization

The interconnections within ecosystems are vast. Most relevant are those interconnections that influence, or are caused by, functional ecologically dominant species, variously defined as those that exert the greatest influence by virtue of energy capture, space, biomass, habitat modification, or control. A park ecologist must recognize the major trophic interconnections among species. Recent interest has focused on a surprisingly basic question: are ecosystems controlled primarily from the bottom of trophic webs upward, or from the top down? If control is being exerted from the top down, is there a cascade of trophic consequences?

'Control' here refers to influence on the distribution, abundance, and in some cases composition of species. Energy and nutrients obviously flow from the bottom up in food webs, but the control systems among populations may exert themselves either way. A top-down example would be wolves limiting the density of a herbivore species, whose lack of browsing in turn alters the composition of tree species. Alternatively, from the bottom, something else in the environment may be acting to limit the size of a herbivore population, such as food supply, thereby constraining the carnivore population.

In terrestrial systems, relevant research has taken place in parks. With relatively less human impacts, basic relationships within ecosystems are more apparent. On Isle Royale, tree-ring data from balsam fir showed suppressed growth when moose numbers were high. Coupled with an interpretation that moose density was largely determined by predation, the researchers concluded that here was an example of top-down control, and a trophic cascade (McLaren and Peterson, 1994). Similarly, at Banff, where the wolf predation rate was high, elk numbers were lower than on the control site. As well, willow and aspen growth was greater, as was songbird diversity and abundance. The conclusion was drawn that a top-down influence prevailed in a trophic cascade (Hebblewhite et al., 2005).

In Algonquin Park, however, the data did not support a strong top-down influence. Wolves did not have a major impact on changing the numbers of their prey or altering their distribution and movements. Instead, deer and moose killing by humans in the region lessened the proportional impact of wolves, and logging was altering the natural herbivore–forest relationship (Theberge and Theberge, 2004).

In Yellowstone, an apparent increase in willow and cottonwood coincided with a decline in elk, which happened subsequent to the reintroduction of wolves in 1994 and 1995. This apparent trophic cascade was augmented by the reappearance of beaver (Ripple and Beschta, 2003). Then, a landmark publication in 2005 displayed convincing evidence that wolf predation on Yellowstone elk between 1995 and 2004 had exerted little influence on elk numbers. Low summer precipitation, severe winters, and hunter kill accounted for most of the elk decline (Vucetich et al., 2005). Therefore, there had been no top-down influence on herbivore numbers but the presence of wolves changed the distribution of elk, and vegetation recovery may have been related to the change (Fortin et al., 2005).

Only where wolves are shown to limit their prey can top-down predator effect be claimed, occurring in about half the studies in a 1985 compilation (Theberge and Gauthier, 1985). Wolf-caused limitation may only occur in landscapes with relatively little disturbance, natural or human-caused (McLaren and Peterson, 1994; Theberge and Theberge, 2004). Beyond this generalization, the question of whether ecosystems are controlled from the top down or the bottom up may be too simplified. Ecosystems consist of webs of relationships, and the best we can do is try to pick out the dominant ones—which almost always shift even over short time spans because ecosystems are dynamic.

Vulnerability

Some species appear to be more vulnerable to stress or change than others. A suite of biological traits influences the probability of local or widespread extinction. Species exhibiting these traits need special consideration from park managers, specifically to ensure that functions within the park are not detrimental. Also, park managers need to collaborate with other regional agencies to ensure that species with such traits persist in the landscape (e.g., multi-agency efforts in the 40,000 km^2 Banff region to increase the likelihood of grizzly bear persistence).

1. *K-strategists* are species that produce few offspring, invest in a great deal of parental care in their welfare, and are long-lived. Their low reproductive rate makes them vulnerable. Sometimes they are habitat specialists, as well, which limits their ability to switch habitat (Shaw, 1985). Large mammals and large birds tend to be K-strategists. Many amphibians and reptiles, while fecund, show habitat specificity, which can be a threat to them.
2. *Summit predators* feed at the top of food chains and hence depend on all the lower links (Wootton, 1994; Morin and Lawler, 1995). Under top-down conditions, the presence of populations of summit predators can have a profound impact on the structure and productivity of other trophic levels (see previous section). As well, summit predators may suffer from the concentration of toxins in food chains. Most vertebrate summit predators, for example, birds of prey, are K-strategists, making them doubly vulnerable.
3. *Spatially concentrated species* are vulnerable because a large number or even significant portions of regional populations can be wiped out by local environmental events such as an oil spill or industrial development. Such species include seabirds, geese, swans, and other congregating waterfowl, as well as muskoxen and caribou (Smith et al., 1986).

4. *Migratory birds* are vulnerable because of destruction of migratory or tropical wintering habitats and accumulation of toxins along migration routes.
5. *Long-distance migratory mammals* are vulnerable if they transect jurisdictional boundaries and co-operative management is not in place. Big-game species migrating across park boundaries commonly are open to exploitation outside the park.
6. *Large-bodied species* are often vulnerable because of generally low reproductive rates (Vemreij, 1986) and extensive ranges. Large-bodied carnivores have large home ranges, ranging from 150 km^2 for black bears to over 2,000 km^2 for wolf packs (reviewed in Noss et al., 1996). Some species with large home ranges have such extensive movements that they will frequently or seasonally enter several management jurisdictions each year, such as grizzly bears in the Banff National Park region (Herrero, 1994) or wolves in Algonquin Provincial Park (Forbes and Theberge, 1996).

Vulnerability to stress or change often leads to rarity. In Canada, rare species are protected on federal lands (including parks) by the Species at Risk Act (SARA). Many provinces have also adopted the spirit of SARA. Many rare species are not limited to parks, and individuals or populations may move frequently across park boundaries onto unprotected lands. For these species to recover, it is paramount that regional agencies work together to research the problem leading to rarity, such as habitat loss, and devise environmental mitigations. Of course, it would be preferable (and easier for park management) if species did not become rare in the first place. To that end, park managers and biologists need to anticipate threats and stresses to the regional ecosystem and provide advice to governments, developers, and the public.

Many ecologists have advocated that the best approach to safeguard against the decline of vulnerable and rare species is to identify and protect critical areas. For example, in applying a species filter to vertebrates in Algonquin Provincial Park, the second author identified two suites of species at risk that grouped into wetlands and old-growth habitats (Theberge, 1995), which meant that management could be targeted to these ecosystem types. Likewise, priority areas for protection have been identified across Canada. Details on these areas are provided by such programs as the Boreal Forest Campaign of the Canadian Parks and Wilderness Society, the conservation program of the World Wildlife Fund, the Inland Rainforest Campaign of the Valhalla Wilderness Society, the Yellowstone to Yukon Conservation Initiative, and the Okanagan Ecoregional Assessment of the Nature Conservancy of Canada. A marine example of identifying critical areas is provided by Morgan et al. (2003).

When Is 'Life Support' Appropriate for Rare Species?

Science has illustrated that most components in an ecosystem are interrelated. Consequently, a single-species approach to conservation may lead to an unintended effect for other species. For example, in the case of a declining mountain caribou herd in the Columbia Mountains of British Columbia, proposals are afoot in multiple-use lands to institute predator control. The mountain caribou are an ecotype of woodland caribou (*Rangifer tarandus caribou*). They are considered 'threatened' by SARA. In the herds that are in the GPE of Mount Revelstoke National Park, 51 per cent of mortality of radio-collared caribou was caused by predation (spread across grizzly and black bear, cougar, wolf, and wolverine in descending order), while 49 per cent was caused by other factors (unknown, avalanche, vehicle collision) (Parks Canada and BC Ministry of Forestry, unpublished data, 1992–2006). In this case, significant questions remain regarding factors affecting the population decline—it is still not known if the predation is additive or compensatory to other forms of mortality, or what proportion of fetuses make it through pregnancy, birthing, and their first year.

The management of predator populations to support caribou populations brings up many questions. Is there enough information to suggest that predator control would affect this caribou herd? Is it justifiable to reduce predator populations, particularly ones that are already considered at risk? (In this instance, for example, grizzlies and wolverines are considered 'threatened' by the Species at Risk Act, and both are 'Blue-listed and of special concern' in British Columbia.) Is it mismanagement to place significant funds towards the management of proximate factors (i.e., predation) if ultimate factors (i.e., habitat change) are not radically altered? These are difficult questions. Their answers are likely dictated by the spirit of the policy of the protected area agency.

Managing rare species needs to occur within a broad ecosystem perspective—such is the whole intent of ecosystem-based management. By placing rare species on last-minute life support (i.e., placing them at a high priority, allocating significant funds to their recovery, and manipulating ecosystem functions such as predation, insect disturbance, or fire disturbance), it is important to realize that other components of the ecosystem may be put at risk or other conservation issues may be neglected.

Viable Populations

One other question that park managers often think about regarding species is: how many individuals are necessary for population persistence? This topic has received attention since the early 1980s through various calculations of 'minimum viable population', or MVP. Often, MVP is calculated for the most space-demanding species in an environment, with the hope that the space requirements of other species will be met within that area. Hence, calculations often focus on large carnivores such as the gray wolf (Theberge, 1983) or the tiger (Tilson and Seal, 1987). As well, some have suggested that MVP analyses target critical or keystone species (Power et al., 1996), such as the major herbivore (Soule, 1987).

Two conceptual bases exist for calculating MVP: genetics and population demography. Regarding genetics, at least 50 free-breeding adults are necessary to prevent more than 1 per cent of inbreeding per generation, an arbitrary minimum threshold. Population size must be adjusted upward for non-breeding animals such as juveniles or those excluded from breeding by social behaviour (Soule, 1980). Minimum population size and space requirements per individual then determine the minimum size of a park or reserve.

For large carnivores these calculations show a need for very large reserves. Hummel (1990) provides rough estimates for a minimum population size for wolves of about 150. With an approximate average density of 100 km^2 per wolf in western parts of the species' range (based on a survey made by the second author of provincial and territorial wildlife management agencies in 1990), this results in a need for 15,000 km^2. Few parks in North America are that large. Given no immigration, Franklin and Soule (1981) calculated that large carnivores (10–100 kg) can be expected to survive the next century in only 0–22 per cent of the world's parks, and in none after 1,000 years.

Regarding demography, calculations of MVP produce probabilities of extinction over specified periods by considering such population parameters as birth rate, mortality, and reproductive age (Gilpin and Soule, 1986). Minimum areas calculated in these ways tend to be even larger than those calculated through genetics. Especially at risk of extinction through demographic processes are large-bodied, long-lived species with low rates of turnover, such as elephants and redwood trees, compared with small-bodied, short-lived species such as shrews and annual plants. For example, elephants require a minimum area of 10,000 km^2 for a 99 per cent probability of persistence for 1,000 years, whereas shrews require 1,000 km^2 (Belovsky, 1987).

Today, MVP calculations have evolved into a more sophisticated PHVA—population and habitat viability analysis. Using a computer program called Vortex, a team of specialists from the Conservation Breeding Specialist Group of the World Conservation Union travels worldwide to host workshops on populations at risk. This program requires data not only from genetics and demographics, but also environmental variables such as frequency of catastrophe (meaning sudden change) and habitat alteration (Boyce, 1992). Calculation of the size of a viable population is only one product of these workshops. More important are conservation plans developed from the analysis of vulnerable characteristics in the ecology of the population. PHVA workshops have been run in several Canadian parks—for example, for grizzly bears in Banff (Herrero et al., 2000) and wolves in Algonquin (Box 4.2).

THINKING ABOUT HEALTH AND INTEGRITY

A healthy ecosystem has a relatively high level of ecological integrity—that is, it has fully functional ecological communities with natural rates of energy capture and flow, nutrient uptake and cycling, with intact food webs and undisturbed mechanisms of population regulation. Unnatural levels of stress can erode ecosystem health and integrity. This section explores the concept of stress, and how monitoring components of an ecosystem over time can assist park managers to identify and deal with it.

Stress

Ecosystems have evolved with many natural stressors (i.e., climate variation, succession, fire, disease) differing in intensity, duration, and frequency of occurrence. Stress plays an integral and ongoing role in the organization, evolution, and functions of ecosystems that may influence ecosystems additively, synergistically, or in multiple ways. Without a periodic disruption, the process of ecological succession stagnates as resources are immobilized by their structure. Bursts of growth and high net productivity usually follow disturbances, and rejuvenated systems replace senescent systems.

Widespread agreement exists that ecosystems exhibit common patterns of response when stressed by either natural or human causes (Freedman, 1989). Natural stressors may preadapt ecosystems to human-caused stresses. Among characteristics exhibited by stressed ecosystems are:

1. changes in nutrient cycles, including increased leakiness;
2. changes (normally increases) in net primary productivity;

BOX 4.2 Ecological Effects on a Park Wolf Population of Killing by Humans Adjacent to the Park

FIGURE 4.3 Algonquin Park wolf.

Among Ontario's Algonquin Provincial Park wolf population, between 1988 and 1999 an average of 67 per cent of wolf deaths were caused by human killing. This mortality took place in townships adjacent to the park, affecting packs that held transboundary territories and that migrated annually to a white-tailed deer wintering area outside the park. Snaring and shooting were the most common causes of death.

The ecological impact of this killing was to contribute to an average annual mortality of 35 per cent, which was beyond the productivity of the population, causing it to drop by one-third during the study. A 'Population and Habitat Viability Analysis' workshop was held by the World Conservation Union to address the conservation problem faced by this population, and using a predictive model

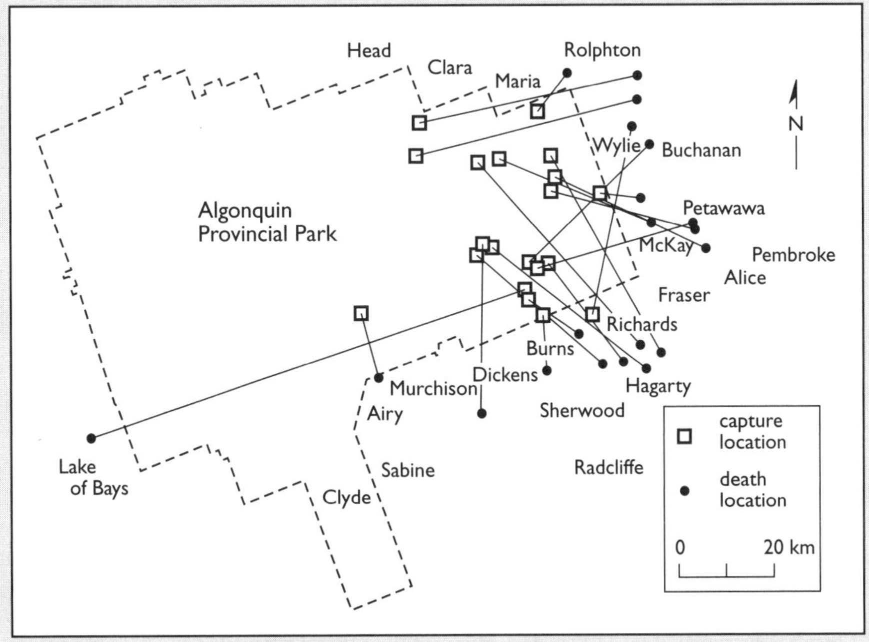

FIGURE 4.4 Map of wolf deaths caused by humans in the townships surrounding Algonquin Provincial Park. The individual territories were largely inside the park.

called 'Vortex', the population was shown to be unable to sustain itself more than a few decades (CSBG, 2000).

As serious as the killing by humans was on the population demography of Algonquin Park wolves, an equivalent concern was the threat of accelerated hybridization with coyotes that live adjacent to the park and at times invaded the park and interbred with park wolves. Just as hybridization with an expanding coyote population doomed the exploited and fragmented red wolf populations in the southern United States by the late 1960s, hybridization had eliminated the wolf south of Algonquin Park and begun to threaten the park population. Field data indicated that coyote-like animals tended to invade or show up in the park where wolf packs had been eliminated or fragmented.

Genetic analysis (Wilson et al., 2000) showed that Algonquin wolves are red wolves, not gray wolves as formerly thought. The North American evolved red wolf has the capacity to interbreed with coyotes, which are believed to have diverged from each other in relatively recent times—150,000 to 300,000 years ago. The gray wolf, with a different evolutionary history played out over a long time in Eurasia, does not appear to interbreed with coyotes.

Thus, killing by people contributed to both demographic and genetic deterioration of a supposedly protected park wolf population. In 2002, a permanent protection zone for both wolves and coyotes was placed around the park. This is the first example of such a zone in North America for any park's large carnivore population.

3. changes in species composition, including loss of late successional stage species and a greater proportion of small-bodied, rapidly reproducing, hardy species (from a list of 18 characteristics by Odum, 1985, and by Rapport et al., 1985).

Feedback allows ecosystems to cope with the effects of stress. For example, some ecosystems react to stress by replacing their more sensitive species with functionally similar but more resistant species. But stress can go too far and also cause ecosystem collapse. Monitoring is essential to detect the effects of stress on ecosystems.

Monitoring

Protected areas are subjected to human-caused stress from both outside and within their boundaries. A basis for ecosystem monitoring has emerged from the exploration of stress ecology and ecosystem health (Woodley and Theberge, 1992) that focuses on tracking components of biodiversity, ecosystem function, and stressors. The intent of monitoring is to provide early warning of undesirable trends in the ecosystem so that appropriate mitigation may prevent deterioration. Determining which ecosystem components, or indicators, to monitor is a challenge (Box 4.3). Ecosystem functions can be difficult to describe. Stressors can be abundant. Biodiversity is exceedingly complex with its four levels of organization (genetic, population, ecosystem, landscape) (Noss, 1990).

Once indicators of environmental change have been selected, objectives need to be established. For example, to monitor grizzly bear populations, the objective may be to detect a 3 per cent change in the population size over 10 years. To assist management response to monitoring trends, ecologists identify the point at which change is unacceptable, usually based on research results, and a more precautionary point at which management action is triggered. For example, management action might be triggered when water acidity reaches a pH of 7.0 because freshwater aquatic life may not be viable at a pH less than 6.5.

THINKING ABOUT UNCERTAINTY

Only the arrogant promote ecology as a fully predictive science. Like climate, ecosystems exhibit non-equilibrium and non-linearity. A high degree of interconnections creates new relationships and makes them dynamic. Succession never rebuilds exactly what was there before. Stochastic (unpredictable) events drive change. Uncertainty arises from many different sources, such as unanticipated influences of an activity on an unintended target, synergistic effects of two or more stresses (e.g., interactions between toxins), activities in one jurisdiction affecting the health of an adjacent ecosystem, and environmental stochasticity (i.e., random, unpredictable change such as a catastrophe).

Catastrophe theory has significantly altered earlier concepts of stability in ecosystems. It has placed a greater premium on resilience in ecosystems, that is, on their capacity to bounce back. Catastrophe—sudden change such as a population crash or eruption or a wholesale change in the ecosystem through fire or flood—is natural in most ecosystems. It only carries a pejorative connotation if resource management demands a steady state. A steady system is not the objective in most protected areas, where natural ecological forces are supposed to prevail.

The most frequently cited examples of catastrophe are fire and insect outbreak (Holling, 1973; Gawalko, 2003). In all the examples, predisposing environmental factors increase in intensity until a catastrophe becomes inevitable. Similar to this idea, Kaufman et al. (1998) postulated that there may be a critical level of biodiversity at which ecosystems are highly susceptible to mass extinction.

BOX 4.3 Establishment of an Ecological Monitoring Program in Glacier National Park

The Canada National Parks Act (CNPA) specifies that the 'state of the parks' will be reported every two years (see Chapter 9). Contributing to this assessment, the Ecological Integrity Monitoring Program is focused on two questions: (1) What is the state of ecological integrity and how is it changing? (2) How are park management activities changing ecological integrity? Each park must measure a number of carefully chosen ecological factors across all major ecosystem types to provide a defensible assessment of the state of ecological integrity. Table 4.1 shows an example of such a monitoring program being developed for Glacier National Park in the Columbia Mountains of British Columbia. Environmental categories were refined into subcategories and reduced to topics/indicators. Proposed measurements are listed in the last column. These measurements are under consideration because they relate to perceived threats to biodiversity, ecosystem functions, or specific stressors.

TABLE 4.1 Monitoring Program for Ecological Integrity Being Established at Glacier National Park

Environmental Category	Sub-category	Topic or Indicator	Measurements
Native biodiversity	Abundance of native species	Caribou population	Population size, body condition, cause of mortality
	"	Terrestrial birds diversity	Diversity and abundance of species
	"	Fish index of biotic integrity	Change in native & non-native species
	"	Mountain goat population	Index of population change
	"	Amphibian abundance	Species diversity, population index
		Benthic invertebrate species	Diversity and abundance
	Abundance of large carnivores	Grizzly bear population	Population index, adult female mortality, amount of area in secure habitat
	Density of exotic vegetation	Non-native plant populations	Extent of invasion, invasion into sensitive ecological sites
Terrestrial ecosystem condition	Habitat structure	Terrestrial vegetation structure	Species abundance and structure
	Productivity	Primary productivity	NDVI
	Human impacts	Human footprint in terrestrial ecosystems	Spatial extent of disturbance
	Soil nutrients and physical properties	To be determined	To be determined
Aquatic ecosystem condition	Water chemistry	Water quality	Metals, nutrients
	Stream habitat	To be determined	To be determined
	Productivity	Primary productivity	To be determined
	Human impacts	Aquatic connectivity	Culvert barriers to species movements
		Salt application to roads	Volume of salt in sensitive ecosystems
Climate/atmosphere	Precipitation & temperature	Precipitation	Rainfall, snowpack, air temperature
	Air quality	Airborne pollutants	To be determined
	Glacier cover	Glacier retreat	Mass balance, visual change
Landscapes	Natural disturbance rate	Area of natural disturbances	Insects/disease extent, landscape composition
		Fragmentation	Fragmentation, road density
	Area of old forest	Area of forest types	Forest age class distribution

Few examples exist of comprehensive ecosystem monitoring. Several agencies are establishing programs, namely the US National Park Service Vital Signs Program, Alberta Biodiversity Monitoring Program, Environment Canada's Ecological and Assessment Network, and Parks Canada's Ecological Integrity Monitoring Program (Chapter 5). Some examples of monitoring programs in Canadian protected areas are described by Cameron (2003) and Ure and Beazley (2003). Monitoring ecosystem change only aids in identifying threats, the first step in ecosystem management. Bureaucratic and political will, budgets, and public concern all need to be in place or monitoring will not be translated into remedial actions and, thus, will be of little consequence (Figure 4.5).

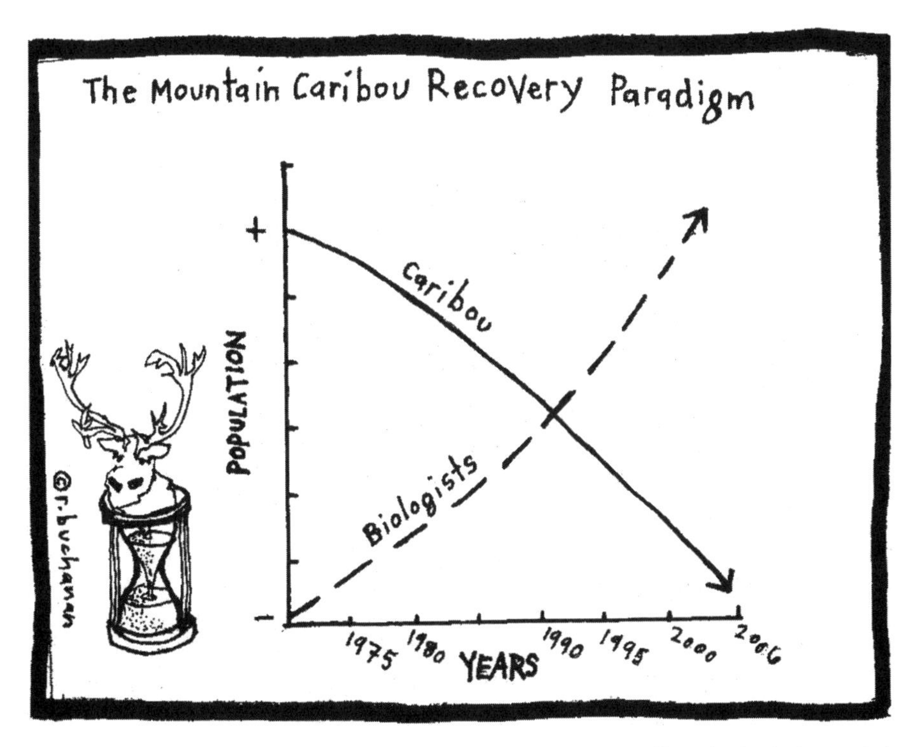

FIGURE 4.5 An undesirable relationship between monitoring efforts by biologists and the state of the species being monitored. While monitoring is a valuable tool to gather information objectively, it cannot prevent ecosystem deterioration unless it is accompanied by a commitment to implement mitigations by land management agencies and society as a whole. *Cartoon: R. Buchanan.*

Sometimes it is possible to incorporate catastrophe and uncertainty into population models, thereby informing managers of the potential outcomes of various management actions (Marshall et al., 1998). In models of minimum viable population sizes, the results are more realistic if we incorporate catastrophes with some predictable periodicity (based on probabilities). Normally, this process magnifies the necessary size of reserves and indicates the need for the protection of more than one representative sample of an ecosystem, and for multiple reserves to protect rare and endangered species.

Uncertainty in ecosystems forms a rationale for the application of 'adaptive management' as a general approach. Rather than taking a normative planning approach that, on the basis of best existing information, invokes management prescriptions for a fixed and usually lengthy period, adaptive management involves viewing management as a continuous experiment, as discussed in more detail in Chapter 5. Thus, adaptive management embodies the capacity to change prescriptions as ongoing research dictates. The concept has been around for a while (Holling, 1978). It has often been abused by representing a 'muddling through' rather than being based, as it should be, on testing falsifiable scientific hypotheses (Theberge et al., 2006). While often espoused by governments, it has been practised less often than might be expected because of the increased costs of the research component and added complexity to management programs. Nonetheless, its incorporation into park management, despite its inherent admission of uncertainty, is an antidote to misjudgement and error.

Uncertainties of Climate Change

Potential changes in climate introduce significant uncertainty to the future makeup of ecosystems at a magnitude never before seen. UNESCO, an organization that appoints the protection of natural sites of international significance, has acknowledged that 'climate change is one of the most significant global challenges facing the environment today' and that World Heritage Sites and Biosphere Reserves will be affected (UNESCO, 2006).

Predictive models of climate and vegetation change have been created for many bioregions that contain parks. However, significant uncertainty accompanies these models. Limitations and precision vary among them (e.g., Hamman and Wang, 2005). Also, it is not known how accurately the models will predict actual ecosystem response. Impacts to protected areas have been postulated (Box 4.4), including for individual national parks in Canada (Suffling and Scott, 2002). For now, however, managers can only speculate about future community structure, the persistence of vulnerable species, or even how much effort they should place in trying to save species at risk whose habitat will be lost.

Generally, protected areas across Canada (at various levels of government) are not viewed by the public as demonstrating leadership in the preparation for, or mitigation against, climate change. In contrast, many parks have conducted research or monitoring that investigates the ecological outcomes of such change (e.g., change in snow cover in Glacier and Sirmilik national parks in BC and Nunavut, respectively). Park staff have developed minor policies to reduce emissions (e.g., reduction of vehicle fleet and no idling of cars by the staff of Mount Revelstoke National Park in BC).

To adjust to climate change, land management agencies should implement trans-boundary mitigations and regional adaptation strategies that reduce the vulnerability of protected areas (UNESCO, 2006). For example, protected areas may be more resilient to climate change by reducing non-climatic sources of stress or by redesigning boundaries and buffer zones to facilitate migration of species. A comprehensive set of technical guidelines to assess climate change impacts and response strategies in general is available from the Intergovernmental Panel on Climate Change (Carter et al., 1994; Parry and Carter, 1998) and other agencies (Hansen et al., 2003; Barber et al., 2004).

THINKING BEYOND THE OBVIOUS

Ecosystems are more than the sum of their parts. Properties emerge by virtue of the complex interconnections and interdependencies among species. Some properties that emerge by virtue of the self-organized complexity in ecosystems include stability (ability to resist perturbations) and resilience (ability to bounce back from perturbations). A host of ecological processes contribute both positively and negatively to these properties so that various ecosystems may exhibit a natural level and a range around them that may be expressed at different stages of succession (Tables 4.2 and 4.3). For example, stability is conferred by trophic interconnections and reduced by the frequency of natural disturbance. Resilience is enhanced by a preponderance of r-selected (high reproductive rate) species in the ecosystem, and is reduced by erosional processes.

For all park ecosystems, it would be a constructive exercise to think through the processes, and human influences on them, that, in turn, influence ecosystem stability and resilience. The result would form a foundation for ecosystem monitoring based on emergent ecological processes. More thinking on this topic could form the next wave of conceptual advances in the science of park management.

BOX 4.4 Types of Terrestrial Ecosystems at Risk from Climate Change

- Small and/or isolated protected areas.
- Protected areas with high-altitude environments.
- Protected areas with low-altitude environments.
- Protected areas with rare or threatened species with restricted habitats or home ranges.
- Protected areas with species at the limits of their latitudinal or altitudinal range.
- Protected areas with abrupt land-use transitions outside their boundaries.
- Protected areas without usable connecting migration corridors.
- Protected areas with rare or threatened species near the coast.
- Protected areas with interior wetlands.

Source: UNESCO (2006), from a larger list of impacts of climate change to natural world heritage. 'Issues related to the state of conservation of World Heritage properties: The impacts of climate change on World Heritage properties', in Convention concerning the Protection of World Cultural and Natural Heritage, thirtieth session, Vilnius, Lithuania, 8–16 July 2006, WHC-06/30.COM/7.1, Annex 4, 27, at http://whc.unesco.org/archive/2006/whc06-30com-07.le.pdf.

CONCLUSIONS

Park management programs based on broad ecological thinking will be most successful if they can identify problems at early stages; as in human medicine, early detection results in a much enhanced probability of successful cures. The results from such anticipatory research can often help resolve regional land-use issues, and in fact may sometimes be the only information deemed credible or objective in inter-agency negotiations. Unfortunately, park budgets for ecosystem management often are so small that only crisis topics receive attention. This needs to change.

Ecological theory has pointed out that protected areas in Canada, and worldwide, are too small and too few to withstand external and internal assaults indefinitely or to protect viable populations—especially the functionally dominant large predators and migratory species—over the long term. As a consequence, parks and reserves must be seen for what they are—only parts of regional landscapes—and so should be parts of regional conservation strategies. Because parks and reserves are not ecologically self-sufficient, regional environmental management is absolutely necessary. This means that today's park biologists and managers need not only be fluent in the appli-

TABLE 4.2 Ecosystem Stability

Stability (resistance to change) means relatively constant:

- set of species (low extinction rates, few colonizers).
- abundance of those species (absence of fluctuations).
- energy capture and partitioning of flow through trophic levels.

Stability is an ever-adjusting state that reflects a number of conditions:

Condition	Effect
1. trophic web complexity	high = stable
2. species diversity	high = stable
3. stage of succession	climax = stable
4. structural complexity	complex = stable
5. cybernetic population mechanisms in dominant, keystone, and summit species	presence = stable
6. size of area	large = stable
7. distance to re-colonizers (new species)	far = stable
8. extent of non-lethal interspecific relationships (symbiosis, mutualism, commensalism, parasitism)	more = stable
9. number and frequency of catastrophes	few and infrequent = stable
10. proportion of r & K strategists as influencing fecundity	K strategists = stable in terms of infrequent eruptions
11. proportion of generalists/specialists	more generalists = stable
12. geographic location	community central in its range = stable

TABLE 4.3 Ecosystem Resilience

Resilience is the ability to bounce back after disturbance. In any ecosystem, resilience is an ever-adjusting state that reflects a number of conditions:

Condition	Effect
1. succession	recurs frequently = resilient early successional stage = resilient
2. distance to re-colonizers (new species)	close = resilient
3. proportion of r & K strategists nearby	more r-strategists = resilient
4. residual soil fertility	high = resilient
5. proportion of asexually reproducing plant propagules	high = resilient
6. survival of bacteria, seeds, and spores in the soil	high = resilient
7. species diversity	low = resilient (replacement time)
8. population sizes of keystone species	above persistence threshold = resilient

cation of ecological theory to park management, but also must be comfortable in the art of collaboration, negotiation, and communication with other agencies and stakeholders in the region. Park biologists and managers must take the lead in the formation, and long-term application, of regional conservation strategies. More than the parks are at stake.

REFERENCES

Barber, C., K. Miller, and M. Boness. 2004. *Securing Protected Areas in the Face of Global Change: Issues and Strategies. A Report by the Ecosystems, Protected Areas, and People Project.* Gland and Cambridge: IUCN. At: <www.iucn.org/themes/wcpa/pubs/theme.htm#climate>.

Belovsky, G.E. 1987. 'Extinction models and mammalian persistence', in M.E. Soule, ed., *Viable Populations for Conservation.* Cambridge: Cambridge University Press, 3–57.

Boyce, M.S. 1992. 'Population viability analysis', *Annual Review of Ecology and Systematics* 23: 481–506.

Cameron, R. 2003. 'Lichen indicators of ecosystem health in Nova Scotia's protected areas', in Munro et al. (2003).

Carter, T.R., M.L. Parry, H. Harasawa, and S. Nishoika. 1994. 'IPCC technical guidelines for assessing climate change impacts and adaptations', Department of Geography, University College London.

Conservation Breeding Specialist Group (CBSG). 2000. *The Wolves of Algonquin Park, PHVA: Final Report.* Apple Valley, Minn.: CBSG.

Diamond, J.M. 1975. 'The island dilemma: Lessons of modern biogeographic studies for the design of nature reserves', *Biological Conservation* 7: 129–46.

Doak, D.F., and L.S. Mills. 1994. 'A useful role for theory in conservation', *Ecology* 75, 3: 615–26.

Forbes, G.J., and J.B. Theberge.1996. 'Cross-boundary management of Algonquin Park wolves', *Conservation Biology* 10, 4: 1091–7.

Forman, R.T.T. 1997. 'Designing landscapes and regions to conserve nature', G.K. Meffe and C.R. Carroll, eds, *Principles of Conservation Biology*, 2nd edn. Sunderland, Mass.: Sinauer Associates, 331–2.

——— et al. 2003. *Road Ecology: Science and Solutions*. Washington: Island Press.

——— and M. Godron. 1986. *Landscape Ecology*. New York: John Wiley and Sons.

Forshner, A. 2000. 'Population dynamics and limitation of wolves (*Canis lupus*) in the Greater Pukaskwa ecosystem, Ontario'. Master's thesis, University of Alberta.

Fortin, D., H.L. Beyer, M.S. Boyce, D.W. Smith, T. Duchesne, and J.S. Mao. 2005. 'Wolves influence elk movements: Behaviour shapes a trophic cascade in Yellowstone National Park', *Ecology* 86: 1320–30.

Franklin, O.H., and M.E. Soule. 1981. *Conservation and Evolution*. Cambridge: Cambridge University Press.

Freedman, W. 1989. *Environmental Ecology: The Impacts of Pollution and Other Stresses on Ecosystem Structures and Functions*. San Francisco: Academic Press.

Gawalko, L. 2003. 'Management of the mountain pine beetle epidemic in British Columbia's parks and protected areas', in Munro et al. (2003).

Gilpin, M.E., and M.E. Soule. 1986. 'Minimum viable populations: Processes of species extinction', in Soule (1986: 19–34).

Grigoriew, P., J.B. Theberge, and J.G. Nelson. 1988. 'Park boundary delineation manual, the ABC resource survey approach', *Occasional Paper # 4, Heritage Resources Program*, University of Waterloo.

Grumbine, R.E. 1994. 'What is ecosystem management?', *Conservation Biology* 8, 1: 27–38.

Hamman, A., and T.L. Wang. 2005. 'Models of climatic normals for genecology and climate change studies in British Columbia', *Agriculture and Forest Meteorology* 128: 211–21.

Hansen, L.J., J.L. Biringer, and J.R. Hoffmann. 2003. 'Buying time: A user's manual for building resistance and resilience to climate change in natural systems', WWF Climate Change Programme, Berlin. At: <www.worldwildlife.org/climate/pubs.cfm>.

Hastings, A., and S. Harrison. 1994. 'Metapopulation dynamics and genetics', *Annual Review of Ecology and Systematics* 25: 167–88.

Hebblewhite, M., et al. 2005. 'Human activity mediates a trophic cascade caused by wolves', *Ecology* 86: 2135–44.

Herrero, S. 1994. 'The Canadian national parks and grizzly bear ecosystems: The need for interagency management', *International Conference on Bear Research and Management* 9, 1: 7–22.

———, ed. 2005. *Biology, Demography, Ecology, and Management of Grizzly Bears in and around Banff National Park and Kananaskis Country: The Final Report of the Eastern Slopes Grizzly Bear Project*. University of Calgary, Faculty of Environmental Design.

———, P.S. Miller, and U.S. Seal, eds. 2000. 'Population and habitat viability assessment for the grizzly bear of the central Rockies ecosystem (*Ursus arctos*)', Eastern Slopes Grizzly Bear Project, University of Calgary, and Conservation Breeding Specialist Group, Apple Valley, Minn.

Holling, C.S. 1973. 'Resilience and stability of ecological systems', *Annual Review of Ecological Systems* 4: 1–23.

———. 1978. *Adaptive Management Assessment and Management*. London: John Wiley and Sons.

Hummel, M. 1990. *Conservation Strategy for Large Carnivores in Canada*. Toronto: World Wildlife Fund Canada.

Kaufman, J.K., D. Brodbeck, and O.R. Melroy. 1998. 'Critical biodiversity', *Conservation Biology* 12, 3: 521–32.

Keddy, P.A. 1991. 'Working with heterogeneity: An operator's guide to environmental gradients', in J. Kolasa and S.T.A. Pickett, eds, *Ecological Heterogeneity*. New York: Springer-Verlag, 181–201.

Komex International. 1995. *Atlas of the Central Rockies Ecosystem: Towards an Ecologically Sustainable Landscape, A Status Report to the Central Rockies Ecosystem Interagency Liaison Group*. Calgary: Komex International.

Locke, H. 1998. 'Yellowstone to Yukon Conservation Initiative', in N.W.P. Munro and J.H.M. Willison, eds, *Linking Protected Areas with Working Landscapes Conserving Biodiversity*. Proceedings of the Third International Conference on Science and Management of Protected Areas. Wolfville, NS: SAMPAA, 255–9.

MacArthur, R.H., and E.O. Wilson. 1967. *The Theory of Island Biogeography*. Princeton, NJ: Princeton University Press.

McCullough, D.R. 1996. 'Introduction', in McCullough, ed., *Metapopulations and Wildlife Conservation*. Washington: Island Press.

McLaren, B.E., and R.O. Peterson. 1994. 'Wolves, moose and tree rings on Isle Royale', *Science* 266, 2: 1555–8.

Marshall, E., R. Haight, and F.R. Homans. 1998. 'Incorporating environmental uncertainty into species management decisions: Kirtland's warbler habitat management as a case study', *Conservation Biology* 12, 5: 975–85.

Mladenoff, D.J., T.A. Sickley, R.G. Haight, and A.P. Wydeven. 1995. 'A regional landscape analysis and prediction of favorable gray wolf habitat in the northern Great Lakes region', *Conservation Biology* 9: 279–94.

Morgan, L., P. Etnoyer, T. Wilkinson, H. Hermann, F. Tsao, and S. Maxwell. 2003. 'Identifying priority conservation areas from Baja California to the Bering Sea', in Munro et al. (2003).

Munro, N.W.P., P. Dearden, T.B. Herman, K. Beazley, and S. Bondrup-Nielsen, eds. 2003. *Making Ecosystem-Based Management Work*. Proceedings of the Fifth International Conference on Science and Management of Protected Areas, Victoria, BC, May. Wolfville, NS: SAMPAA. At: <www.sampaa.org/publications.htm>.

Newmark, W.D. 1995. 'Extinction of mammal populations in western North American national parks', *Conservation Biology* 9: 512–26.

Noss, R.F. 1987. 'Corridors in real landscapes: A reply to Simberloff and Cox', *Conservation Biology* 1: 159–64.

———. 1990. 'Indicators for monitoring biodiversity: A hierarchical approach', *Conservation Biology* 4: 355–64.

———. 1996. 'Protected areas: How much is enough?', in R.G. Wright, ed., *National Parks and Protected Areas: Their Role in Environmental Protection*. Cambridge, Mass.: Blackwell Science, 121–32.

——— and B. Csuti. 1997. 'Habitat fragmentation', in G.K. Meffe and C.R. Carroll, eds, *Principles of Conservation Biology*, 2nd edn. Sunderland, Mass.: Sinauer Associates, 269–304.

———, H.B. Quigley, M.G. Hornocker, T. Merrill, and P.C. Paquet. 1996. 'Conservation biology and carnivore conservation in the Rocky Mountains', *Conservation Biology* 10, 4: 949–63.

———, M.A. O'Connell, and D.D. Murphy. 1997. *The Science of Conservation Planning: Habitat Conservation under the Endangered Species Act*. Washington: Island Press.

Odum, E.P. 1985. 'Trends expected in stressed ecosystems', *Bioscience* 35: 419–22.

Parks Canada Agency. 2000. *Unimpaired for Future Generations? Protecting Ecological Integrity with Canada's National Parks, vol. 2: Setting a New Direction for Canada's National Parks*. Report of the Panel on the Ecological Integrity of Canada's National Parks. Ottawa.

Parry, M., and T. Carter, 1998. *Climate Impact and Adaptation Assessment: A Guide to the IPCC Approach*. London: Earthscan.

Power, M.E., et al. 1996. 'Challenges in the quest for keystones', *Bioscience* 46, 8: 609–20.

Quinn, M.S., and J.C. Theberge. 2003. 'Ecosystem-based management in Canada: Trends from a national survey and relevance to protected areas', in Munro et al. (2003).

Rapport, D.J., H.A. Reiger, and T.C. Hutchinson. 1985. 'Ecosystem behavior under stress', *American Naturalist* 125: 617–40.

Ripple, W.J., and R.L. Beschta. 2003. 'Wolf reintroduction, predation risk, and cottonwood recovery in Yellowstone National Park', *Forest Ecology and Management* 184: 299–313.

Shaw, J.H. 1985. *Introduction to Wildlife Management.* New York: McGraw-Hill.

Simberloff, D.S. 1974. 'Equilibrium theory of island biogeography and ecology', *Annual Review of Ecology and Systematics* 5: 161–82.

——— and J. Cox. 1987. 'Consequences and costs of conservation corridors', *Conservation Biology* 1: 63–9.

———, J.A. Farr, J. Cox, and D.W. Mehlman. 1992. 'Movement corridors: Conservation bargains or poor investments?', *Conservation Biology* 6, 4: 493–504.

Skibicki, A.J. 1994. *Preliminary Boundary Analysis of the Greater Pukaskwa National Park Ecosystem Using the ABC Resource Survey Approach.* Waterloo, Ont.: University of Waterloo, Heritage Resources Centre.

———, J.G. Nelson, and W.R. Stevenson. 1995. 'A summary of a preliminary boundary analysis of the greater Pukaskwa National Park ecosystem using the ABC resource survey approach', in T.B. Herman, S. Bondrup-Nielsen, J.H.M. Willison, and N.W.P. Munro, eds, *Ecosystem Monitoring and Protected Areas.* Proceedings of the Second International Conference on Science and Management of Protected Areas, 16–20 May 1994. Wolfville, NS: SAMPAA, 384–90.

Smith, P.G.R., J.G. Nelson, and J.B. Theberge. 1986. 'Environmentally significant areas, conservation and land use management in the Northwest Territories', Technical Paper Number 1, Heritage Resources Centre, University of Waterloo, Waterloo, Ont.

Soule, M.E. 1980. 'Thresholds for survival: Maintaining fitness and evolutionary potential', in M.E. Soule and B.A. Wilcox, eds, *Conservation Biology: An Evolutionary-Ecological Perspective.* Sunderland, Mass.: Sinauer Associates, 151–69.

———, ed. 1986. *Conservation Biology: The Science of Scarcity and Diversity.* Sunderland, Mass.: Sinauer Associates.

Suffling, R., and D. Scott. 2002. 'Assessment of climate change effects on Canada's national park system', *Environmental Monitoring and Assessment* 74, 2: 117–39.

Theberge, J.B. 1983. 'Consideration in wolf management related to genetic variability and adaptive change', in L.N. Carbyn, ed., *Wolves in Canada and Alaska.* Canadian Wildlife Service Report Series, Number 45. Ottawa, 86–9.

———. 1989. 'Guidelines to drawing ecologically sound boundaries for national parks and nature reserves', *Environmental Management* 13: 695–702.

———. 1995. 'Vertebrate species approach to trans-park boundary problems and landscape linkages', in T.B. Herman, S. Bondrup-Nielsen, J.H.M. Willison, and N.W.P. Munro, eds, *Ecosystem Monitoring and Protected Areas.* Proceedings of the Second International Conference on Science and Management of Protected Areas, 16–20 May 1994. Wolfville, NS: SAMPAA, 526–36.

——— and D.A. Gauthier. 1985. 'Models of wolf-ungulate relationships: When is wolf control justified', *Wildlife Society Bulletin* 13: 449–58.

——— and M.T. Theberge. 2004. 'The wolves of Algonquin Park, a twelve year study', *Department of Geography Publication Series Number 56.* Waterloo, Ont.: University of Waterloo.

———, ———, J.A. Vucetich, and P.C. Paquet. 2006. 'Pitfalls of applying adaptive management to a wolf population in Algonquin Provincial Park, Ontario', *Environmental Management* 37: 451–60.

Theberge, J.C. 2002. 'Scale-dependent selection of resource characteristics and landscape pattern by female grizzly bears in the eastern slopes of the Canadian Rockies', Ph.D. dissertation, University of Calgary.

———. 2003. 'Lessons learned during interagency negotiation regarding wolf conservation and forestry', in Munro et al. (2003).

Tilson, R.L., and U.S. Seal, eds. 1987. *Tigers of the World: The Biology, Biopolitics Management, and Conservation of an Endangered Species*. Park Ridge, NJ: Noyes Publications.

UNESCO (United Nations Educational, Scientific and Cultural Organization). 2006. 'Issues related to the state of conservation of World Heritage properties: The impacts of climate change on World Heritage properties', in Convention Concerning the Protection of World Cultural and Natural Heritage, World Heritage Committee. Thirtieth Session, Vilnius, Lithuania, 8–16 July 2006: World Heritage WHC-06/30.COM.7.1. At: <whc.unesco.org/archive/2006/whc06-30com-07.1e.pdf>.

Ure, D., and K. Beazley. 2003. 'Selecting indicators for monitoring aquatic integrity at Kejimkujik National Park and National Historic Site', in Munro et al. (2003).

Vucetich, J.A., D.W. Smith, and D.R. Stahler. 2005. 'Influence of harvest, climate, and wolf predation on Yellowstone elk, 1961–2004', *Oikos* 111: 259–70.

Wilcox, B.A., and D.D. Murphy. 1985. 'Conservation strategy: The effects of fragmentation on extinction', *American Naturalist* 125: 879–87.

Wilson, P.J., et al. 2000. 'DNA profiles of the eastern Canadian wolf and the red wolf provide evidence for a common evolutionary history independent of the gray wolf', *Canadian Journal of Zoology* 78, 12: 2156–66.

Wright, R.G. 1996. *National Parks and Protected Areas: Their Role in Environmental Protection*. Cambridge, Mass.: Blackwell Science.

Woodley, S., and J.B. Theberge. 1992. 'Monitoring for ecosystem integrity in Canadian national parks', in J.H.M. Willison et al., eds, *Science and the Management of Protected Areas*. New York: Elsevier.

Wootton, J.T. 1994. 'The nature and consequences of indirect effects in ecological communities', *Annual Review of Ecology and Systematics* 25: 443–66.

KEY WORDS/CONCEPTS

adaptive management
catastrophe theory
climate change
connectivity
corridors
ecological integrity
ecology
ecosystem-based management
ecosystem monitoring
ecosystem organization
fragmentation
greater park ecosystem
island biogeography
metapopulation
resilience
scale
species at risk
stability
stress ecology
top-down–bottom-up
uncertainty
viable population
vulnerability

STUDY QUESTIONS

1. What are ecological integrity and ecosystem-based management?
2. What sort of ecologically based attributes should be considered in park establishment or management?
3. Discuss how island biogeography theory relates to arguments about the size and location of protected areas.
4. How does fragmentation of regions influence park health?
5. Why are roads of particular concern for conservation biologists?
6. What are the advantages to investigating the organization of ecosystems?
7. What are the traits of species that are vulnerable to stress and change?
8. What is a PHVA and how does it differ from MVP?
9. What are the characteristics of ecosystems that are experiencing stress?
10. What is the intent of an ecosystem monitoring program, and what are examples of measurements?
11. What are the pitfalls and advantages of adaptive management?
12. How should protected areas prepare for climate change?
13. What is the difference between ecosystem stability and resilience?
14. Discuss the rationale for adopting a greater ecosystem approach to park issues.

CHAPTER 5

Planning and Managing for Ecological Integrity in Canada's National Parks

Stephen Woodley

INTRODUCTION

Throughout most of the history of national parks and equivalent protected areas, it was assumed that placing legal boundaries around areas was sufficient to protect them. The result would be healthy, self-regulating ecosystems unaffected by the impacts of the planet's 6.7 billion people. This 'preservation' stage of park management was accompanied by the perception that national parks were 'natural' and/or 'wilderness' and thus required no management, or at least very little management. With over 100 years of experience in protected areas management, it is now understood that a hands-off approach to park management will not work in many, if not most, protected areas. The reasons for this are that most protected areas are too small, there are conflicting external land-use practices outside parks, and these areas have been already historically modified. Many protected areas lack key ecological processes, such as fire or predation. Many parks are too small and unconnected to maintain viable populations of species, especially area-demanding species. Many parts of southern Canada have seen extirpations of keystone predators, such as the wolf, with resultant hyper-abundant populations of herbivores such as deer and moose. Fire regimes have been altered through much of Canada because of fire suppression and the loss of Aboriginal burning. Finally, most ecosystems protected as wilderness have thousands of years of human management history, conducted by Aboriginal peoples. In many cases, simply leaving parks alone will often not result in the conservation of those values that the parks were established for in the first place.

This chapter focuses on active management in Canadian national parks and the case study examples are from Parks Canada. However, the underlying principles regarding the need for active management apply to protected areas of all kinds, in Canada and elsewhere.

The revisions to the Canada National Parks Act (www.gov.pe.ca/law/statutes/pdf/n-01.pdf) in 1988 introduced the term 'ecological integrity' (see Chapter 9). This concept provides a new model for managing protected areas. As a concept, ecological integrity supersedes the notion of 'natural' as a management end point, and this has eral important ramifications. One of the key implications is that active management might be required in parks to maintain or restore ecological integrity.

During the early years of park establishment, Canadian national parks often were islands of civilization in a sea of wilderness. Park boundaries were only lines on a map and indistinguishable from adjacent lands on the landscape. Today, the impacts of forestry, agriculture, tourism, and urbanization have effectively isolated most parks as islands in a human-dominated landscape. Large-scale ecological insults, such as acidic precipitation and climate change, have no consideration for park boundaries. Ecosystem boundaries are difficult, if not impossible, to define. No matter where a park boundary is drawn, there will always be flows across the boundary. The flows may be water, nutrients, animals, or pollutants. The essential fact is that parks are connected to a larger landscape and management must take that into account. It has become increasingly obvious that instead of being natural, self-regulating ecosystems, many parks and protected areas are remnant islands assaulted by a variety of human-caused stresses, originating both within and outside (for a Canadian review, see Environment Canada, 2006).

Historically, parks were considered to be 'natural' areas, which were set aside for purposes of protecting and conserving representative, or unique, species and ecosystems. The stated goal of park management was simply to let nature take its course. It was implicit that management would be minimal. However, there were always many exceptions to the idea of minimal management. As late as the 1960s, predators, including bears, cougars, and eagles, were shot in national parks. The stated goal of the day was to maintain healthy populations of elk and mule deer, which were then considered threatened by the predators. Even today, wildfires continue to be suppressed in most protected areas in Canada and elsewhere.

Many landscapes in which parks are situated, especially in southern Canada, have been highly altered from their historical condition. Active management is often needed to allow species or ecosystems to persist in parks where otherwise they might be lost. To the extent that a park may be the last stronghold for a particular species, if lost from the park that species could be lost from the larger region, too. Thus, if parks are to include species and ecosystems characteristic of the surrounding natural region, park landscapes and species populations may have to be actively managed in order that species may persist.

To compensate for past or current actions, active management is frequently required in such areas as fire restoration, species and community restoration, harvest management, management of hyper-abundant native species, and elimination of non-native species. Active management should occur where there are reasonable grounds to believe that maintenance or restoration of ecological integrity will be compromised without it. Because of the difficulty in predicting ecosystem response, active management should be undertaken in national parks using adaptive management techniques (Box 5.1).

ECOSYSTEM INTEGRITY AND DYNAMIC ECOSYSTEMS

The use of ecological integrity as a goal in protected area management recognizes that ecosystems are inherently dynamic and that park managers, therefore, will be forced to make choices about managing for a particular ecosystem state or states. Most often,

BOX 5.1 Adaptive Management

Adaptive management is a formal process of 'learning by doing' for continually improving management policies and practices by learning from their outcomes. As Walters (1999) explains:

> It involves more than simply better ecological monitoring and response to unexpected management impacts. In particular, it has been repeatedly argued (Holling 1978, Walters 1986) that adaptive management should begin with an effort to integrate existing interdisciplinary experience and scientific information into dynamic models that attempt to make predictions about the impacts of alternative policies. This modeling step is intended to serve three functions: (1) problem clarification and enhanced communication among scientists, managers, and other stakeholders; (2) policy screening to eliminate options that are most likely incapable of doing much good, because of inadequate scale or type of impact; and (3) identification of key knowledge gaps that make model predictions suspect.

For protected areas, adaptive management offers a way to deal with the large amount of uncertainty surrounding ecosystem management. At its best, adaptive management integrates learning into its planning processes, to continually improve management for the protection of ecological integrity. Essentially, then, it is a cyclical and ongoing process, as shown in Figure 5.1.

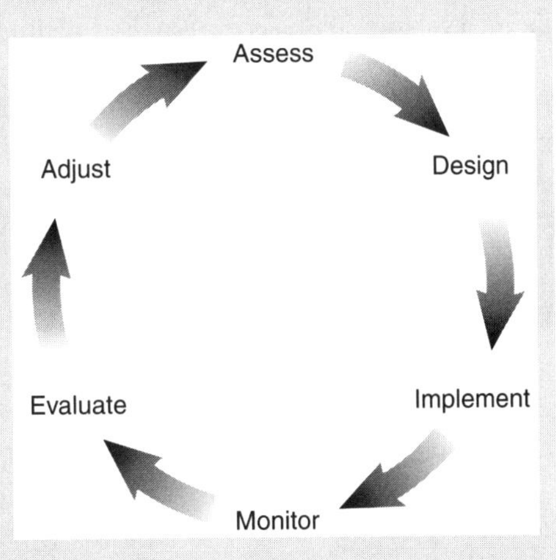

FIGURE 5.1 Recurring steps in the adaptive management process.

decisions are taken by default. Suppressing wildfire and watching a park ecosystem become all old-growth is a default decision that has enormous implications for biodiversity. Similarly, allowing high ungulate populations in a park where predators are absent is a 'default' decision that will result in very different ecological outcomes than in an ecosystem where predators are present. Like it or not, most park managers are faced with making difficult management choices.

No one particular state or era in time is necessarily correct or 'natural'. This was recognized in the Leopold Report (Leopold et al., 1963) done for US national parks, where the goal of scientifically based park management was recommended as a way to 'protect vignettes of primitive America'. However, the Leopold Report missed the fact that

America was not really 'primitive'. We now understand that the pre-Columbian Americas were populated by millions of Aboriginal peoples with cities, roads, and engineering structures (for a complete review, see Mann, 2005). Even outside the highly populated areas, Aboriginal peoples were keystone predators, regulating levels of ungulate populations and modifying ecosystems through complex fire use (see Pyne, 1983). Thus, ecological integrity must be understood in the context of the ongoing role of humans in ecosystem dynamics.

Protected areas management cannot attempt to recreate one particular era because most ecosystems are too dynamic and too complex. Ecosystem goals need be established while considering disturbance history, levels of herbivores and predators, the changing regional land-use patterns, etc. In some cases, it may be possible to restore and maintain a similar kind of ecosystem, but it will never be exactly the same. For example, fire may be used to maintain a successional stage, such as a grassland valley bottom, which, if left alone, would eventually fill in with trees. However, rather that recreating a particular era or point in time, the open grassland seral stage is best maintained for defined ecological objectives, such as maintaining a specific plant community or providing winter forage.

As a management end point, ecological integrity is a significant advance from the notion of 'natural' in that it forces the use of ecosystem science, in combination with societal wishes, to define and decide on ecosystem goals. Moreover, the concept of ecological integrity inherently recognizes that ecosystems are dynamic, self-organizing entities.

WHEN TO ACTIVELY MANAGE

It is a difficult transition for park managers to go from being protectors of nature, where nature always knew best, to actively managing ecosystems for a defined ecological goal. Although the national parks' *Guiding Principles* (Parks Canada, 1994) are clear on the need for active management, it has been a challenging concept to put into operation across the national park system. Current policy on active management in national parks is as follows:

> National park ecosystems will be managed with minimal interference to natural processes. However, active management may be allowed when the structure or function of an ecosystem has been seriously altered and manipulation is the only possible alternative available to restore ecological integrity. (Ibid., 34)

Active management in protected areas is complex. If the goal of a protected area is to have ecological integrity, by definition there is a requirement to have populations of native species and the processes upon which they depend. Active management is aimed at maintaining or restoring a process, species, or community because (1) it is likely to disappear or has disappeared, or (2) it is not functioning within the expected range of variation for that ecosystem. Problems regarding disappearance are easier to understand and include species reintroductions, as has occurred with the peregrine falcon in Fundy National Park and with bison in Grasslands National Park, as well as reintro-

BOX 5.2 Defining Active Management

Active management is any prescribed course of action directed towards maintaining or changing the condition of cultural, physical, or biological resources to achieve the Parks Canada Agency objectives. Active management includes ecosystem restoration, and the two terms are often used interchangeably. Active management requires taking actions in such potentially controversial areas as fire restoration or controlling hyper-abundant species. In complex ecological systems, there will typically be debate about why changes are occurring and whether or not such changes are detrimental to ecological integrity. To avoid gridlock, a successful approach is to use adaptive management. Under an adaptive management framework, actions can be taken simultaneously with testing the alleged effects on ecological integrity. Through feedback loops, results of the actions can be used to adapt or change future actions for improved results. Active adaptive management to restore fire, manage hyper-abundant native or non-native species, or upgrade infrastructure—where there are reasonable grounds that maintenance or restoration of ecological integrity will be compromised without active management—needs to be an ongoing part of the management of most parks.

duction of processes such as fire. Functional problems often pose more of a challenge. In this category, park managers must decide when a population or process is outside some acceptable limit for the park ecosystem in question. Active management seeks to restore the ecosystem component or process to a predefined value or range of values. Examples include managing white-tailed deer populations in Point Pelee at levels that allow the survival of the rare vegetation in the park.

RESTORING POPULATION AND COMMUNITIES

Restoring ecological integrity is one type of ecological restoration, which can be described as an intentional activity that initiates or accelerates ecosystem recovery with respect to its function (processes), integrity (species composition and community structure), and sustainability (resistance to disturbance and resilience). The Society for Ecological Restoration International (2004) defines ecological restoration as '*the process of assisting the recovery of an ecosystem that has been degraded, damaged, or destroyed*'. Ecological restoration in Canada's national parks attempts to recognize the dynamic nature of ecosystems and focuses on the development of resilient, self-sustaining ecosystems that are characteristic of the protected areas natural region.

In restoration activities in national parks, the concept of ecological integrity anchors the policies and practices of Parks Canada and is central to the development of an ecological restoration program. Parks Canada defines ecological integrity as 'a condition that is determined to be characteristic of its natural region and is likely to persist, including abiotic components and the composition and abundance of native species and biological communities, rates of change and supporting processes' (Canada National Parks Act, 2000). Although ecological restoration may be considered to be accomplished once project goals are met, ongoing ecosystem management may be required to maintain ecological conditions. That is, restoration is linked to other management practices aimed at preventing recurrent degradation of restored ecosystems that may occur due to ongoing alterations in the environment (e.g., repeated invasion of alien species) or anthropogenic changes (e.g., in land-use patterns). Such activities are related to the maintenance rather than the restoration of ecological integrity. In this sense 'ecosystem restoration' is synonymous with 'active management'.

The Society for Ecological Restoration International (2004) recognizes that some ecosystems evolved with human management and such management is required to maintain those ecosystems. As an example, in much of Canada, ecosystems have a historical fire regime that is a mix of both lightning and human ignitions (Van Wagner et al., 2007). The restoration of fire in such ecosystems must understand and incorporate both ignition sources. But beyond ignition sources, fire management must consider the season and intensity of burning. In many areas, Aboriginal fires were traditionally conducted in spring and fall, outside the midsummer lightning period. Spring and fall fires tended to be of lower intensity. Because biodiversity in a region is adapted to the historic fire regime, fire managers are faced with considerable complexity in designing a prescribed burn program for ecological restoration.

Ecosystem restoration needs to consider other indigenous ecological management practices in addition to fire. Aboriginal hunting practices, population movements, and silviculture are all considerations in protected areas management. Ecological restoration encourages, and may indeed depend on, long-term participation of local and Aboriginal people. Parks Canada recognizes that ecological integrity should be assessed with an understanding of the regional evolutionary and historic context that has shaped the system, including past occupation of the land by Aboriginal people.

Parks Canada has tried many types of ecological restoration in its parks. In some cases it is matter of reintroducing a native species back into the ecosystem, as with the return of bison to Grasslands National Park (Box 5.3). In other cases, park managers must completely mimic an ecological process, as has been done in Kejimkujik National Park in Nova Scotia with the piping plover (Box 5.4).

Restoring populations and communities in protected areas requires considerable scientific understanding of the ecosystem and its dynamics. In addition, restoration almost always involves diverse partnership with land managers outside protected areas, industry, First Nations and other Aboriginal groups, and non-government organizations. Restoration requires public support, which only occurs after consultation, engagement, and education. In many cases restoration programs are highly controversial, at least at the beginning of a project. Restoration projects can also challenge traditional ways of managing protected areas, resulting in questions from park staff and the public (Box 5.5). Finally, restoration requires monitoring to determine if the ecosystem results were as expected. All the above considerations mean that restoration requires considerable expertise, staff, money, and time. Very few protected area organizations currently have the resources to meet all these requirements.

BOX 5.3 Case Study: Reintroducing Bison to Grasslands National Park

After 120 years of absence, plains bison (*Bison bison*) were reintroduced in December 2005 to Grasslands National Park in southern Saskatchewan. Bison are large (up to 730 kg) keystone herbivores that modify prairie ecosystems through grazing, wallowing, trampling, and acting as a food source for a range of predators and scavengers.

Prior to European settlement, the prairies were home to millions of free-roaming bison. However, by the 1880s many changes had occurred on the landscape and the large herds that once roamed the prairie grasslands were nearly gone, the result of high demand in the buffalo robe trade and of the extremely organized and massive buffalo hunts. It is to the credit of conservation-minded people like Samuel Walking Coyote, Charles Allard, Michel Pablo, and 'Buffalo' Jones, who helped maintain small bison herds, that this species still exists today.

The bison were returned to Grasslands National Park as part of a larger prairie restoration program. Large herbivore grazing is an ecological process that has been missing from the prairie park for a number of years. Bison grazing patterns are somewhat different from those of domestic livestock; the bison graze heavily in some areas and lightly in others. This pattern creates a vegetation community that is diverse and therefore attractive to a variety of native species not found in the surrounding rangeland. Using prescribed fire and watering holes to facilitate bison movement, the park is aiming to create a specific grazing prescription, aimed at maintaining a range of prairie biodiversity.

Grasslands National Park initially relocated 71 bison, including 30 male calves, 30 female calves, and 11 female yearlings, from Elk Island National Park in east-central Alberta, which has been a 'bank' for Canadian bison for many years, where the 'capital' can be maintained and the 'interest' used to repopulate other parks (Chapter 1). The source bison herd has a long record with no reportable diseases, extensive health data, and no cattle genes.

In addition to the ecological benefits, the return of the bison to Grasslands will provide a wonderful experience for visitors. Bison are symbolic of the prairies and, as a keystone species, modify the ecosystem along with fire, drought, and flooding.

FIGURE 5.2 Bison in Grasslands National Park. *Photo: © Parks Canada/Wayne Lynch/08.80.10.01 (0.1).*

BOX 5.4 Case Study: Creating Habitat to Enhance Piping Plover Populations in Kejimkujik National Park

Eastern Canadian populations of the piping plover (*Charadrius melodus melodus*) are endangered and are listed under Canada's Species at Risk Act. These shorebirds nest in small beach depressions just above the high-tide line on exposed sandy or gravelly ocean beaches, sand spits, or barrier beaches. Key to recovering plover populations are protection of nests from human disturbance, ensuring that predator populations (crows, gulls, foxes, and raccoons) are not attracted and enhanced by human garbage, and maintaining sufficient nesting habitat. In Kejimkujik National Park, staff are working on all three aspects in an effort to enhance populations. However, most unique in this program is the creation of nesting habitat by plowing up marram grass beaches.

FIGURE 5.3 Piping plover on nest. *Photo: Nick Kontonicolas/1000birds.com.*

The restoration effort is underway at the Kejimkujik Seaside Adjunct, the coastal portion of Kejimkujik National Park. While resource conservation staff have successfully controlled predators and people within plover nesting areas, the number of nesting pairs still declined from a high of 27 pairs in 1977 to just four pairs in 2003. Researchers noted that dense mats of vegetation (mostly marram grass) had begun to threaten suitable plover nesting habitat. To help the Atlantic piping plover recover, plows were used to remove dense mats of marram grass. The habitat restoration has allowed plovers to successfully raise chicks.

BOX 5.5 Case Study: Restoring Kootenay's Original Dry Grasslands and Open Forests

The southwestern corner of Kootenay National Park in British Columbia is a dry, low-elevation valley that supports rich biodiversity and critical wildlife habitat. This area contains the only example of dry Douglas fir/ponderosa pine/wheatgrass vegetation in Canada's national parks and provides important winter range for wildlife, including the Rocky Mountain bighorn sheep (*Ovis canadensis canadensis*). It is also the site where most of the human activity in the region occurs.

For thousands of years, fires of both lightning and Aboriginal origin maintained a variety of habitats in the Columbia Valley, creating a healthy mixture of young, middle-aged, and old forests, shrub lands, open meadows, and dry grassy slopes. However, over a century of fire suppression and loss of Aboriginal burning have dramatically changed the region's ecology. Without the regenerative benefits of periodic, low-intensity surface fires, the Columbia Valley has been transformed into an even-aged blanket of mature forest that is encroaching on and dominating the original mosaic of species and habitats. Moreover, the dense forest now overtaking the area sets the stage for catastrophic wildfires, such as those seen in the summer of 2003.

To return ecological integrity to the valley and reduce wildfire risk, Parks Canada is restoring the grasslands and open forest biodiversity of the South Kootenays through the Redstreak Restoration Project. The dramatic first step in restoration is the mechanical harvesting of trees, followed by carefully planned and managed burns. The first phase of the project (2002–3) focused on tree harvesting and removal to decrease the amount of fuel available for the prescribed fires.

Because tree removal requires heavy equipment that can damage soils, small plants, and shrubs, Phase 1 of the project took place during winter, when the ground was frozen and covered with snow. This minimized the damage to the land and to small flora. The harvested trees were then sold, and the revenues returned to support the project. Hundreds of hectares have already been restored, using mechanical harvesting as a preparatory step for controlled fires. While mechanical tree harvesting is an odd sight to witness in a national park, it is an essential component of ecosystem restoration efforts. Periodic low-intensity fires were returned in the spring of 2005, using prescribed fire. This was only possible after the fuel load was dramatically reduced by mechanical thinning.

The intention of project managers is to restore between 350 and 400 hectares of fire-maintained open forests and grasslands in the Columbia Valley. Measures of success include winter use of the area by bighorn sheep, restoration of grasslands forbs, and minimal levels of invasive exotic species.

FIGURE 5.4 Aerial view of Radium Hot Springs, showing restoration in the benchland. *Photo:© Parks Canada/Wayne Lynch/10.100.03.31 (49).*

MANAGING HYPER-ABUNDANT POPULATIONS

A global problem faced by many protected areas is that of hyper-abundant populations, from elephants in some African parks to the ubiquitous white-tailed deer in North America. Populations can be defined as hyper-abundant when their numbers clearly exceed the upper range of variability that is characteristic of the ecosystem. Hyper-abundance or overabundance should be judged in context and could occur when a species causes ecosystem dysfunction or depresses or eliminates native species (see McShea et al., 1997). This can happen when predators are removed from the ecosystem, or when there is a food subsidy, such as high raccoon populations caused by available garbage.

There has been considerable debate (ibid.) about the management of hyper-abundant species, both inside and outside protected areas. The debate can be traced to the early days of modern wildlife biology and the famous example of deer in the Arizona's Kiabob Plateau (Leopold, 1943). The determination of what constitutes a hyper-abundant species is difficult and an ongoing challenge to park managers. In several parks, Parks Canada is routinely managing hyper-abundant populations, as discussed in Boxes 5.6 and 5.7.

Some parks have lost key species that may promote the maintenance of others. In some parks, a reduced abundance of large carnivores such as wolves has led to hyper-abundant populations of such prey species as elk and moose, and to significant changes in the abundance of other species (Kay et al., 1999). For example, Banff and Jasper have large problems with town-adapted elk and a dysfunctional predator–prey system—with resulting impacts on vegetation. Other species may be hyper-abundant because parks, as last enclaves, afford protection and attract regional populations. In many cases large populations are subsidized by extensive alternate food sources outside of the park. This is the case with deer in southern Ontario. Other large populations, such as deer in Gwaii Haanas National Park (in the Queen Charlotte Islands of BC), were introduced to islands with both abundant food and few predators.

BOX 5.6 Case Study: Managing Ungulates in Elk Island National Park

The story of managing ungulate populations in Elk Island National Park in Alberta begins with the construction of a 2.2-metre-high fence surrounding the park in 1905. The original purpose of the fence was to keep animals, especially bison, within the park and prevent wild ungulates from feeding on neighbouring croplands. The park is a now rare mix of aspen parkland and prairie, supporting high densities of ungulates. Elk Island is only 194 km^2, too small to maintain a population of large carnivores such as wolves. An occasional wolf is seen in the park, but the surrounding agricultural landscape is inhospitable to wolves and grizzly bears. Thus, the challenge for the park staff is to manage the ecosystem in the absence of large carnivores. In the absence of predators, ungulate population numbers vary widely, with extremely high numbers causing soil erosion and eliminating some species of plants (Bork, 1993; Vujnovic, 1998).

The philosophy of managing ungulate populations has been changing since the park's inception. At the beginning, managers would allow numbers to increase and then, based on the manager's field assessment, a surplus number was picked for slaughter. Historically, there was a moose slaughter about every four years. Also in the past, concern for diseases, such as brucellosis, resulted in large-scale test and slaughter operations. More recently, the concept of ecological carrying capacity (allowing ungulates to self-regulate by minimizing annual surplus removals) was explored to reduce management effort (Blyth and Hudson, 1987). Under this management regime, the total ungulate population went from an estimated 1,982 in 1985 to 3,463 in 1999, an increase of 178 per cent (Cool, 1999). This left the park with some of the highest ungulate densities known to exist in North America, resulting in excessive and chronic trampling, topsoil removal, the replacement of native plant species with non-native plants, and a complete removal of mid-story and mature shrubs. Some age classes of certain species, such as red osier dogwood, were completely lacking, and some species, such as the once common rough fescue, were eliminated (ibid.).

In 1997, the park hosted an ecosystem-based management workshop, with a multidisciplinary team of experts and scientists from across North America. The key recommendation from the workshop was that the park should adopt more of an adaptive landscape management approach to its program and thereby allow landscape processes as a primary focus. Reduction of ungulates was recommended to allow recovery of rangeland that appeared to be overutilized; vegetation targets also were proposed to manage for a gradient of vegetation across the park; and it was recommended that a science-based program should continue to be developed. In co-operation with provincial wildlife managers, surplus populations of elk and bison are typically rounded up into holding facilities and released in non-park areas where wild populations are low. Surplus animals are also used in reintroductions (e.g., the Mackenzie bison sanctuary in the Northwest Territories).

The target numbers for the park will be adjusted through an adaptive management process. Populations will be surveyed by air. The condition of vegetation, ungulate-caused erosion, and incidence of disease will also be monitored. A 3–5-year time period will allow adequate time to monitor the response, and it will allow some flexibility to adjust targets correspondingly if required.

It is never easy to manage ungulate populations within a relatively small area. High ungulate densities have a higher predisposition to the spread of both epidemic and exotic diseases than lower densities. There is also concern about chronic diseases such as liver fluke and avian tuberculosis. These diseases have the potential to cause rapid population declines. The presence of diseases makes the managed ungulates less desirable for transplanting to other wild areas.

Ungulates attract a high level of public interest and concern around their management. It is extremely difficult to round up and transport wild ungulates. With any handling of large wild ungulates, there is a high potential for injury to both animals and park staff. Capture myopathy, i.e., death caused by stress due to capture, is an always-present side effect of handling wildlife. There is also public concern regarding an ungulate population crash due to diseases.

All the above factors illustrate the practical difficulties of actively managing large ungulates. However, in the absence of having a protected area large enough to maintain viable predator populations, there is little other option but to manage populations. Populations that are too high will have undesirable impacts on vegetation, cause erosion, and increase the incidence of disease transmission. If populations are too low there is a chance that they might disappear simply due to random stochastic processes. Small

BOX 5.7 Case Study: White-tailed Deer Management in Point Pelee National Park

Point Pelee is a very small national park in southern Ontario, only 15 km^2 in area. Point Pelee is typical of many protected areas in southern Ontario and eastern North America in that problems have been caused by very abundant white-tailed deer populations. In the case of Point Pelee, high densities of deer caused undesirable changes to the vegetation, including elimination of several species of rare Carolinian flora and the invasion of many exotic species of plants. Exotic species constitute 60 per cent of the entire park flora and dominate many disturbed sites. In 1992 Point Pelee National Park began a program of reducing deer populations in the park, following a detailed assessment of the situation and considerable public consultation. The reduction is now an ongoing part of park management, with regular culling conducted by park staff. Overall, the park has been extremely successful at reducing locally abundant deer in the park with a series of culls over several years.

There are several reasons for the park's successful management of deer. First, there was a clearly articulated vision of ecological integrity in the park's management plan, with an obvious indication that high deer populations are inconsistent with protecting ecological integrity. The park is to be, so far as possible, representative of a functioning Carolinian ecosystem. As part of the vision, there were numerical targets for the post-cull population. Culls are conducted solely by staff and the park is closed during a cull to ensure public safety. This avoided the policy implications of allowing sport hunting in a national park. Park managers also invested in research into alternative methods of control, indicating clearly that they were aware of public sensitivities to shooting deer.

populations are also subject to undesirable genetic changes, such as inbreeding. So park managers, especially in smaller parks, are forced into a difficult juggling act, with a need to actively manage to keep the ecosystem within a desired state of ecological integrity.

The management of hyper-abundant species is fraught with many practical and policy pitfalls, as the case studies illustrate. Before any management is conducted the reasons for the hyper-abundance must be well understood and clear objectives and numerical targets for the control program, as well as a prediction of the impacts of the control measures, established. There should be a monitoring system in place to examine the causes of hyper-abundance, the dynamics of the population being controlled, and the predicted impacts of the control measures. Finally, the management program should be conducted under an adaptive management framework where the original assumptions are subject to review.

RESTORING FIRE AS AN ECOLOGICAL PROCESS

Excluding the Arctic and the wet coastal parks, a long period of fire suppression has been identified as causing significant impacts to ecological integrity in most Canadian national parks. Fire is a key ecological process in the boreal forest, grasslands, montane Cordillera, and even the Carolinian and Acadian forests. Yet for most of the history of those parks, fire has been actively excluded and fought as a destroyer of ecosystems. Paradoxically, it is now abundantly clear that biodiversity exists because of fire, rather than in spite of fire.

Historically, protected area agencies have treated fire by active suppression. In national parks, the Warden Service was developed largely to fight wildfires. However, the large size of the parks and slow transportation networks initially limited the amount of actual fire control. Beginning in the 1920s, new equipment was developed, hundreds of kilometres of fire roads were constructed, and networks of fire towers were erected. During this period it is likely that fire control began to alter the historical fire regime.

By the 1970s, managers came to realize that parks were not always self-regulating, natural ecosystems. Instead of being 'natural', park ecosystems were increasingly seen as 'impaired' and active management was considered necessary to correct the impaired condition. Attitudes towards fire began to change as fire was viewed as an important dynamic element in the ecosystem. The Canadian Parks Service (now Parks Canada) responded to the changing attitudes with a 1979 policy permitting active management or manipulation of the ecosystem, under certain well-defined conditions. With a new directive produced in 1986 and a comprehensive fire policy review called 'Keepers of the Flame', the Canadian Parks Service embarked on a new relationship with fire. For the first time, fire was officially recognized as an important element in the ecosystem and it was to be restored to its 'natural role' by active management. Unregulated wildfire was considered impossible in most parks because of the values at risk, including public safety, protection of property (including neighbouring lands), and rare species or habitat. Thus, the Canadian Parks Service began to use prescribed fire with an aim to restore the 'natural' fire regime. In most parks, unregulated natural wildfire continued to be eliminated.

While there are many examples of early use of prescribed fire in grasslands, the Canadian Parks Service began to formally use prescribed fire in the mid-1980s (White and Pengelly, 1992). Presently, prescribed burns have been conducted in 17 national parks in Canada. As a basis for the prescribed burn program, parks are required to prepare fire management plans and vegetation management plans. These plans are based on a historical assessment of the role of fire in the park and a detailed set of vegetation management objectives

As the fire-use program develops in national parks, lessons are learned but many

BOX 5.8 Case Study: Prescribed Fire and the Mountain Pine Beetle Epidemic in British Columbia and Alberta Parks

The mountain pine beetle (*Dendroctonus ponderosae*) is a native insect of the western lodgepole forests. Epidemic population outbreaks are characteristic ecological processes that contribute to forest diversity. However, the current 12-year beetle infestation, which is the largest insect epidemic in the province's history, is having a huge economic impact on the forest industry and communities of British Columbia. National parks were asked to partner to help contain the massive outbreak, specifically by halting the spread of populations over the Rocky Mountains into Alberta.

Beetles become epidemic where there are large tracts of older age-class lodgepole pine. Banff and Jasper have historically not seen large epidemics of mountain pine beetles because those ecosystems had frequent fires that diversified the forest age classes. However, with fire suppression for the last 80 years, the parks have become better beetle habitat.

Large-scale ecological issues require large-scale partnerships to manage. The management of mountain pine beetle issues required a partnership that included Parks Canada, federal and provincial government ministries, municipal governments, industry partners, First Nations, and non-profit organizations. Bringing diverse interests and stakeholders together was the first step in efforts to protect both the economic value of the provincial forests and the ecological integrity of the affected national and provincial parks. A program that began with annual monitoring for mountain pine beetles has now evolved into a co-ordinated, regional ecosystem approach, in partnership with all land managers.

Mountain pine beetles and fire are ecologically connected. As lodgepole forests age, there is an increasing probability that the biomass will be recycled by either fire or mountain pine beetles. The end result is always the same—a young lodgepole pine forest. The management goal for all mountain national parks is to restore 50 per cent of the historic fire cycle. Over the last 20 years the parks have been carefully building a fire management program. The successful implementation of prescribed fire and the management of wildfire have been building public support for using fire as a management tool. There are numerous benefits related to controlled fire, including directly reducing mountain pine beetle populations and beetle habitat, renewing forest health, improving wildlife habitat, and reducing susceptibility to wildfire.

In Jasper National Park, approximately 27,000 hectares of prime-age lodgepole pine beetle habitat were burned in a managed wildlfire, providing an effective fireguard on the south side of the Athabaska River Valley. In Banff National Park, the Fairholm prescribed fire in 2004 was over 5,000 ha in size and located between the towns of Banff and Canmore. Not only did this halt the spread of pine beetles in a key outbreak area, it also met the park's ecological goals for fire restoration. Both the Jasper and Banff burn areas have become an important staging ground for ongoing scientific research to better understand mountain pine beetle ecology, ecosystem management processes, and their effects on both the natural environment and public perception.

Thus far, the pine beetle program has strengthened inter-agency and industry working relationships and the effective management of public lands. The expansion of the beetle populations in the mountain national parks has been slowed, bringing al least short-term protection to Alberta's commercial forests. The hope of those involved with the program is that by dealing with the growth of beetle populations early, further epidemics can be effectively prevented.

A key outcome of the mountain pine beetle program is that public awareness and understanding of the beetle and of the prescribed fire program have grown. The Fairholme prescribed burn in Banff required two years to prepare a one-km-wide firebreak. The firebreak was constructed through forest thinning using commercial logging equipment, cutting to a very careful national park standard. Because the burn was conducted beside the towns of Banff and Canmore, there were increased issues surrounding public safety and smoke management. With such complex resource management issues, public engagement and understanding are essential to proceed with such radical management actions on public lands. This kind of resource management requires a team of people, including consultation specialists, educators, fire managers, and biologists.

debates remain unresolved. One challenge relates to scale. First-generation fire management plans, fuel maps, and other planning aids all stopped at the park boundary, despite the obvious ecological and even operational irrelevance of the line. However, with the need to manage on a broader, more relevant scale, the question becomes 'how big?' There is a dramatic change in the economic value of a tree, for example, when crossing a park boundary. On most lands surrounding national parks, fire is actively suppressed in favour of maintaining timber values. Under such conditions, the idea of ecosystem-based management becomes seriously challenged.

A central debate in the role of fires in parks is over the need to duplicate natural or historical fire regimes. The exact role of fire becomes mired in the question of the historical role of fire. Some researchers argue that fire frequencies have not been altered by human fire suppression (see Johnson and Larsen, 1991). This line of argument assumes that fire is a process driven by climate and that present-day fire frequencies are unaltered, despite multi-million dollar suppression efforts. The argument further states that a few very large, intense fires make up the vast majority of acreage burned. Therefore, prescribed burns are irrelevant and not in keeping with historical fire regimes. Other researchers (e.g., Martel, 1994), however, argue that human fire suppression and fire protection have definitely altered the fire regime and that prescribed fire is essential to create historical vegetation patterns. This dichotomy of viewpoints can paralyze an agency trying to develop a fire management program.

Increasingly, it is recognized that no correct formula exists for fire management. It is not simply a matter of conducting a fire history study and then preparing a fire management plan to duplicate some historical fire frequency. Parks must seek individual solutions to fire management, depending on their own unique situations. In some cases it will be possible to have entire parks, or large zones in parks, where fire frequencies are unaltered by humans (discounting global warming). In these areas, lightning-caused fires can burn without any intervention. They will simply be monitored. This is the case in Nahanni National Park, situated in the northern boreal forest of the Northwest Territories with little development on its boundaries. Some parks will have a mix of observation zones, full suppression zones, and evaluation zones. This kind of scenario is being developed in Wood Buffalo National Park in Alberta and the NWT. Other parks, especially in the more developed regions of the country, will have to ensure that fire is an ecosystem process by using only prescribed fire, as in Elk Island National Park.

As an ecological process, wildfire is still below its historical range in most parks. In virtually all provincial and some national parks, full fire suppression is still the rule.

Despite many successes, the combination of prescribed fire and wildfire is still at only 10 per cent of the historical long-term average. The internal goal within Parks Canada for fire restoration has been set at 50 per cent of the long-term historical average. Under an adaptive management framework, this appears to be a reasonable place to start. However, there will have to be huge changes in funding, staff, and public education in order to reach that target.

MANAGING ALIEN AND EXOTIC SPECIES

The majority of national parks in southern Canada report that 'exotic' organisms (invertebrates, fish, birds, mammals, and vegetation) cause major ecological effects (Parks Canada, 1997). Reported effects include elimination of species (e.g., native bull trout by exotic speckled trout in Rocky Mountain streams) and changes in native species abundance. In many national parks, exotic species make up 50 per cent of the total flora. Several parks have successfully removed invasive alien organisms that threaten ecological integrity. For example, Gwaii Haanas National Park Reserve successfully restored native vegetation on a few offshore islands by eliminating introduced mammals (Box 5.9).It is often difficult to decide what constitutes an 'exotic' species and when an alien species should be of concern. Most park managers have not developed a priority list of exotic species, nor have they established a list of appropriate control actions. Exotic species must be clearly categorized as to their history, potential for invasion into native ecosystems, and potential to effect negative change to ecological integrity. Many species, currently expanding their range as a result of human intervention or climate change, may have eventually reached national parks in the natural course of events. However, many other species that were purposely introduced are now at levels that make them impossible to eradicate with current technology (e.g., moose were introduced to Newfoundland). If such an organism is

BOX 5.9 Case Study: Eliminating Rats in Gwaii Haanas National Park

Norway rats were unintentionally introduced onto 17 islands in the Haida Gwaii archipelago, which includes Gwaii Haanas National Park. As predators, rats devastated the breeding seabird populations, eating eggs, young, and even adults of some species. Recently, three islands at the north end of the archipelago (Langara, Cox, and Lucy) were cleared of rats and Gwaii Haanas staff have conducted a similar eradication program on St James Island at the extreme south end of the islands. In the past it was considered impossible to eliminate rats from islands and this program is a breakthrough. The program involved persistent trapping and the use of poison baits (Taylor et al., 2000). Monitoring for the return of rats at Cape St James is ongoing and there was no evidence of rats at baited traps or chew-sticks. No rat feces, tracks, or other sign of rats were found.

The aim of active removal of the Norway rat will allow recovery of native seabird populations. The recovery of these populations will be monitored to evaluate the success of the program.

shown to negatively affect ecological integrity, the 'exotic' designation should be retained in case an appropriate removal technology is developed in the future.

Understanding the effect of exotic species on the ecological integrity of protected areas, especially under conditions of projected climate change, is of global importance. The spread of exotic or alien species is predicted to increase dramatically (see review by Bright, 1998). A working definition of an 'alien organism', developed by the Alien Species Focus Group at Environment Canada in 1994, is: 'An alien species is one that enters an ecosystem beyond its historic range, including any organisms transferred from one country or province to another' (in Mosquin, 1997).

This definition, modified from the US National Park Service, implies no positive or negative impact by the alien organism. The definition includes organisms entering through natural range extension and dispersal, through deliberate or inadvertent introduction by humans, and as a result of habitat changes caused by human activity. Exotic species do not necessarily impair ecological integrity, so a further distinction is warranted, to the effect that alleged negative effects of invasive species are evaluated and demonstrated in order to aid prioritization of exotic species designated for active management.

Determining the effect of exotic species on ecosystem structure and function is imperative. These organisms can have neutral, negative, or positive ecosystem effects. Many exotic species, especially plants, are relatively benign—they do not invade and alter native ecosystems. From a management perspective it would be most efficient to

BOX 5.10 Case Study: Eradicating Invasive Species in Garry Oak Ecosystems

Garry oak ecosystems are a mosaic of woodlands, meadows, and grasslands that are rare globally and have high biological diversity. Thousands of plant, animal, and insect species inhabit these ecosystems, among them the Garry oak, which is the only oak native to British Columbia. The Garry oak ecosystem is home to over 100 species at risk. Of these species, 23 are threatened or endangered through their global range, and 21 are listed by the Committee on the Status of Endangered Wildlife in Canada (COSEWIC) as being at risk nationally.

In Canada, Garry oak ecosystems are found only on southern Vancouver Island, the nearby Gulf Islands, and in two small stands on the mainland. Within these areas, most of the original Garry oak ecosystems have been cleared and converted to agricultural, residential, and industrial uses. Over 95 per cent of the original plant cover has already been lost. Encroaching suburbs and invasive species continue to threaten what remains of this diverse habitat. Parks Canada manages Fort Rodd Hill National Historic Site, a 54-hectare site with a significant Garry oak ecosystem. At this site, like most Garry oak ecosystems, non-native plants now comprise more than 40 per cent of the vegetation, presenting a serious challenge to maintaining the native species. Daphne, Scotch broom, and other invasive species out-compete native plants for space, light, water, and nutrients.

Parks Canada staff and volunteers are working together to remove the exotic invasive plants. Since 2002, local community members, university students, and Scouts have worked alongside staff to hack, pull, and remove invasive species from the site. In 2003 alone, over 80 volunteers contributed 543 hours of labour. By the summer of 2004, about 12.5 tonnes of invasive species had been removed from the site. After exotic plant removal the site is restored by planting, with the aim of reconstructing tracts of Garry oak habitat. A newly cleared, 1.3-ha site was fenced to protect new seedlings from rabbits and deer. Seeds collected from native plants at Fort Rodd Hill are being grown in greenhouses and transplanted to the fenced site to restore native plant cover (www.pc.gc.ca/lhn-nhs/bc/fortroddhill/index_e.asp).

FIGURE 5.5 Volunteers removing invasive species at Fort Rodd Hill National Historic Site. *Photo: © Parks Canada/C.Webb/2004.*

Education and science are key components of this restoration effort. Education comes from using volunteers, establishing on-site interpretive signage, and publications. On the scientific front, site staff and Garry Oaks Ecosystem Recovery Team botanists conducted a plant inventory at Fort Rodd Hill in 2002, finding 336 plant species. Seven of these species were rare, two of them at risk nationally. Last seen in the 1960s, the rare deltoid balsamroot was rediscovered at the site in 2002. Seeds have been collected and are being grown in a greenhouse. They will be transplanted later to help this endangered species survive (ibid.).

Gulf Islands National Park Reserve was established in 2003, further aiding in the protection of Garry oak ecosystems.

be able to predict the probability that a newly detected exotic would invade and damage native ecosystems. Unfortunately, there is currently no way of predicting how invasive an exotic species may be. Only early detection via monitoring, with an evaluation of ecosystem effects, can determine whether a species should be removed. Whether it can actually be removed is not a foregone conclusion.

MANAGEMENT OF HARVEST IN PROTECTED AREAS

Most Canadians assume national parks and protected areas are protected from harvest or resource extraction. In reality, most parks have some kind of active harvest or extraction (Table 5.1). The most common type of harvest is that of sport fishing. However, there are many other kinds of exceptions based in individual park establishment agreements (e.g., snaring of snowshoe hare and cutting of firewood is permitted in Gros Morne National Park on Newfoundland's west coast) or the recognized rights of First Nations.

TABLE 5.1 Harvesting Activities in National Parks

Type of Harvest or Extraction	Number of Parks Reporting Harvest
Aboriginal wildlife hunting/trapping	8
Non-Aboriginal wildlife harvest	6
Sport fishing	22
Commercial fishing	4
Problem or surplus wildlife	10
Domestic grazing	5
Domestic wood harvest	1

Source: Parks Canada (1997).

Predicting the impacts of harvest of any population requires an ongoing assessment of the population levels, age-specific birth and death rates, an understanding of environmental variability, and a model projecting populations over time. This information is rarely, if ever, available for harvest or extraction in protected areas. Even for sport fishing, there is rarely any comprehensive assessment of fishing pressure or fish populations.

CONCLUSION

In an ideal world, protected areas would be very large in size and managed with no human interference. They would be true benchmarks against which we could assess impacts on ecosystems outside protected areas. This is not the case, however. Most protected areas are too small to contain populations of large area-demanding species or area-demanding processes. Even the largest parks have transboundary issues. The only solution to these problems is to use active management.

Active management is a major conceptual shift from the way protected areas have been historically managed and should not be taken lightly. There are many implications. Active management requires more precise definition of management end points. Many protected area agencies now use ecological integrity as a primary management goal. Every five years Parks Canada requires the preparation of State of Park Reports for each

BOX 5.11 Case Study: Managing Sport Fishing in La Mauricie National Park

La Mauricie National Park is an excellent example of a well-managed sport fishery. Only 30 of 150 park lakes are open to fish for lake trout, bass, and pike. The remainder are closed so there will be unexploited fish populations that can act as benchmarks. Because the demand for fishing is very high, fishing opportunities are allocated by a draw that takes place every morning during the May–Labour Day fishing season. In national parks, lead sinkers and jigs are strictly forbidden, as they can poison loons and waterfowl. Fishing limits are set based on age-specific assessments of fish populations as well as assessments of the harvest.

national park, complete with detailed indicators, measures, thresholds, and targets for management. These feed into the Park Management Plan for each park, which sets an ecological vision and the required management actions.

Active management is not a licence to do anything in a protected area. It is an acceptance of the fact that key species and processes are missing in some protected areas and must be considered in management. Active management will work best when there is clear prediction of cause and effect, a monitoring system, and an adaptive management framework that allows refinement and reconsideration. In an increasingly complex and human-dominated world, active management will be increasingly necessary.

REFERENCES

Bright, C. 1998. *Life Out of Bounds: Bioinvasion in a Borderless World.* New York: Worldwatch Institute and Norton.

Bork, E.W. 1993. 'Interaction of Burning and Herbivory in Aspen Communities in Elk Island National Park', M.Sc. thesis, University of Alberta.

Cool, N. 1999. 'Elk Island National Park Ungulate Issue Analysis', Elk Island National Park internal document.

Environment Canada, 2006. *Canadian Protected Areas Status Report 2000–2005.* Catalogue no. En81-9/2005E. At: <www.cws-scf.ec.gc.ca/publications/habitat/cpa-apc/index_e.cfm>.

Holling, C.S., ed. 1978. *Adaptive Environmental Assessment and Management.* New York: John Wiley.

Johnson, E.A., and C.P.S. Larsen. 1991. 'Climatically induced change in fire frequency in the southern Canadian Rockies', *Ecology* 72, 1: 194–201.

Kay, C.E., C.A. White, I.R. Pengelly, and B. Patton. 1999. *Long-Term Ecosystem States and Processes in Banff National Park and the Central Canadian Rockies.* Occasional Paper No. 9. Ottawa: Parks Canada.

Leopold, A. 1943. 'Deer irruptions', *Wisconson Conservation Bulletin* 8: 3–11.

———, S.A. Cain, C.M. Cottam, I.N. Gabrielson, and T.L. Kimball. 1963. *Wildlife Management in the National Parks: The Leopold Report.* Advisory Board on Wildlife Management appointed by Secretary of the Interior Udall. At: <www.cr.nps.gov/history/leopold.htm>.

McShea, W.J., H.B. Underwood, and J.H. Rappole. 1997. *The Science of Overabundance: Deer Ecology and Population Management.* Washington: Smithsonian Institution Press.

Mann, Charles C. 2005. *1491: New Revelations of the Americas before Columbus.* New York: Knopf.

Martell, D.L. 1994. 'The impact of fire on timber supply in Ontario', *Forestry Chronicle* 70, 2: 164–73.

Mosquin, T. 1997. *Management Guidelines for Invasive Alien Species in Canada's National Parks.* Hull, Que.: Parks Canada Documentation Centre.

Parks Canada Agency. 2000. *Unimpaired for Future Generations? Conserving Ecological Integrity with Canada's National Parks.* Vol. 1: *A Call to Action*; vol. 2: *Setting a New Direction for Canada's National Parks.* Report of the Panel on the Ecological Integrity of Canada's National Parks. Ottawa.

Parks Canada. 1994. *Guiding Principles and Operational Policies*, Catalogue no. R62-275/1994E. Ottawa: Minister of Supply and Services Canada.

———. 1997. *State of Parks Report.* Hull, Que.: Parks Canada Agency Documentation Centre.

Pyne, S.J. 1983 'Indian fires: The fire practices of North American Indians transformed large areas from forest to grassland', *Natural History* 92, 3: 6, 8, 10–11.

Society for Ecological Restoration International, Science and Policy Working Group. 2004. *The SER International Primer on Ecological Restoration* (Version 2, Oct.). At: <http://www.ser.org/content/adoption.asp>.

Taylor, R.H., G.W. Kaiser, and M.C. Drever. 2000. 'Eradication of Norway rats for recovery of seabird habitat on Langara Island, British Columbia', *Restoration Ecology* 8, 2: 151–60.

Van Wagner, C.E., M.A. Finney, and M. Heathcott. 2006. 'Historical fire cycles in the Canadian Rocky Mountain parks', *Forest Science* 52, 6: 704–17.

Vujnovic, K. 1998. 'Grasslands of the Aspen parkland of Alberta', M.Sc. thesis, University of Alberta.

Walters, C.J. 1986. *Adaptive Management of Renewable Resources.* New York: Macmillan.

White, C., and I.R. Pengelly. 1992. 'Fire as a natural process and a management tool: The Banff National Park experience', in D. Dickinson et al., eds, *Proceedings of the Cypress Hills Forest Management Workshop.* Medicine Hat, Alta, 3–4 Oct.

KEY WORDS/CONCEPTS

active management
adaptive management
alien organism
ecological integrity
ecosystem restoration
fire regime
hyper-abundant species
laissez-faire management
minimum management
park harvesting
wilderness

STUDY QUESTIONS

1. Why do some people feel park management should be guided by the principle of 'let nature take its course'? List advantages and disadvantages of this approach.
2. List advantages and disadvantages of 'active' management.
3. Under what circumstances is active management appropriate?
4. Why is adaptive management a key principle in managing for ecological integrity?
5. How was adaptive management applied to the problems with elk, deer, and moose in Elk Island National Park?
6. Why is the culling of wildlife populations controversial?
7. Why was fire suppression used in past years? List advantages and disadvantages of this approach.
8. Why are prescribed burns controversial? List advantages and disadvantages of this approach.
9. Why do some people favour the approach of letting natural fires burn? List advantages and disadvantages of this approach.
10. What is a fire management plan?
11. Present arguments for and against hunting in national parks.
12. Present arguments for and against sport fishing in national parks.
13. Present arguments for and against hunting and trapping in national parks by First Nations peoples.

PART III

Social Science Theory and Application

I'd rather wake up in the middle of nowhere than in any city on earth.

Steve McQueen

Parks provide opportunities for people to experience the spiritual, aesthetic, and challenging attributes of a wild, natural setting. Paradoxically, the single greatest threat to ecological integrity is human use—or more accurately, human misuse and overuse. Human settlements developed within parks, such as the Banff and Jasper townsites, have profoundly altered the functioning of those parks' ecosystems. Similarly, the environmental impacts of people hiking, skiing, canoeing, camping, or in other ways enjoying the backcountry areas of our parks can have a profound effect on ecological integrity. Loss of vegetation around campsites, deeply eroded trails, and garbage are easily recognized changes caused by visitor behaviour in protected areas. Not so easily recognized are some of the less observable consequences, such as changes in the abundance or behaviour of wildlife or in the composition of vegetative communities that occur in some places as a result of visitor behaviour.

Some people have gone so far as to suggest that park environments should be strictly protected, by forbidding any park visitation (this occurs in some park zones where features are particularly rare or endangered). On the other extreme are those who argue that humans have been a natural part of park ecosystems for thousands of years. However, the consensus of opinion is that some types of human visitation, at a low intensity and properly managed, are unlikely to harm park environments much. Further, it can be argued that park visitors become park advocates, the strongest supporters for the protection of parks—parks unused are parks unappreciated. At a time when we are experiencing declines in visitation to many parks, there is a growing concern that lower use levels may translate into a decline in public support for parks.

A second concern has to do with conflicts that sometimes occur between visitors. These may have to do with crowding. For example, the presence of too many other people in the same backcountry campground can take away from the opportunity of experiencing solitude or closeness to nature. Sometimes conflict is created by one type of activity interfering with another type of activity. This may happen when hikers find themselves dodging manure left by horses carrying people along the same trail. Conflict also can occur without direct contact being made with offending groups—just knowing they may be present in the same park can create animosity and antagonism. For example, some hikers may experience conflict just knowing that hunting is allowed in the same park, even if they do not hear or see hunters during their park visit.

Social science contributes theory and methods for better understanding visitor behaviour—why some people choose to visit parks, and other do not; why some people are satisfied with their experience and others are not; and why some people prefer more services and facilities than others (Chapter 6). This growing body of social science research has contributed also in the development and refinement of practical tools for managing visitors, as reviewed in Chapter 7. These approaches and concepts include: direct management, indirect management, carrying capacity, recreation opportunity spectrum (ROS), limits of acceptable change (LAC), visitor impact management (VIM), visitor experience and resource protection (VERP), and visitor activity management process (VAMP).

Another concern is that many people visiting parks leave with only minimal understanding and appreciation of the setting—its ecological significance; its historical or cultural significance—or of the efforts park agencies are making to protect these values. Further, there is a concern that people who do not actually visit parks may not fully understand or support the need for having parks. In order to better communicate with park visitors and to the general public, the practice of park interpretation has become more sophisticated in recent years, building on social science theory related to communication and approaches to learning (Chapter 8). Effective interpretation creates a better visitor experience and a more informed constituency of people supporting park values. Hence, this section of the book focuses on the application of social science theory and methods to address park management issues.

CHAPTER 6

Social Science, Conservation, and Protected Areas Theory

Mark D. Needham and Rick Rollins

INTRODUCTION

Over 80 per cent of Canadians participate in some form of nature-related activity such as camping and boating, and much of this activity takes place in national or provincial parks (Environment Canada, 1999). Nature-based tourism is a significant industry in Canada, employing many people and attracting considerable investments. From 2001 to 2006, about 12 million people visited a national park in Canada (Parks Canada Agency, 2006a), and in 2006 this translated into a contribution of $1.2 billion to Canada's gross domestic product (Parks Canada Agency, 2006b). Although there is some debate about appropriate recreation and tourism use in parks and protected areas, there is consensus that some forms of visitor use may be acceptable or desirable. The major issue, however, is how to manage this use effectively in ways that protect park resources, provide for satisfactory visitor experiences, and create a constituency of supporters for parks.

Why do people seek out places such as Gros Morne, Algonquin Park, the Nahanni River, Banff, or Pacific Rim? What kinds of activities do they pursue? What types of experiences and benefits are generated from participating in these activities? What impacts do park visitors create? In what ways do visitors contribute to or detract from environmental sustainability of parks? What types of visitor services and facilities are desirable or appropriate? What types of experiences should or should not be provided in park settings? What types of conflict occur between and within different user groups and why? To what extent are people willing to pay for parks through taxes or user fees? How much public support exists for protected landscapes compared to use for other purposes such as logging, ranching, or urban development? Questions such as these have been explored by social scientists conducting research in Canada and elsewhere, so the intent of this chapter is to provide an overview of the contribution of social science to management of protected areas. In this chapter, social science refers to theory and research that has been applied to park management from disciplines such as sociology, psychology, geography, economics, anthropology, tourism, and leisure studies. This literature has contributed to the ongoing development of techniques for visitor management that are described in subsequent chapters of this book.

Many parks are besieged with requests to provide more facilities such as trails, campgrounds, parking areas, marinas, and downhill ski areas. Pressure is also placed on park managers to increase overnight accommodation in parks and to include roofed structures such as alpine huts, hostels, hotels, and luxury resorts. There is increasing demand on parks to accommodate more visitors and different types of activities such as camping, backpacking, rock climbing, horseback riding, hunting, fishing, all-terrain vehicle use, canoeing, kayaking, sailing, waterskiing, downhill skiing, nordic skiing, and snowmobiling. In any given park, some of these activities can be considered, but it is not possible to provide all types of visitor activities, opportunities, services, and facilities—to do so would result in loss of natural character and conversion of parks to more developed landscapes. Park managers must decide what activities should be permitted, how much use should be allowed, where this use will be allowed, and how use will be managed. In the face of budget constraints and an expanding set of visitor demands on parks, managers are challenged to articulate what purpose or role a park is to fulfill and what balance between visitor use and resource protection is appropriate.

In addition to environmental impacts created by visitors in parks (see Chapter 12), managers must deal with a variety of social issues including crowding, vandalism, and conflict among user groups. These issues can extend beyond park boundaries and impact adjacent land and nearby communities. Communities such as Tofino near Pacific Rim National Park and Canmore near Banff experience many tourism benefits due to their close proximity to popular national parks. These communities, however, sometimes experience visitor-related problems such as traffic congestion and lineups at grocery stores, gas stations, banks, and hospitals.

Visitor management is complex. In this book, impacts of visitor use are examined within the topic of ecological integrity, where it is noted that most threats to ecological integrity stem from visitor activity within parks and/or human activity outside of parks impacting park ecosystems (Chapters 5 and 13). It is apparent, therefore, that maintaining ecological integrity requires an understanding of human behaviour. Social science data can assist in this effort. When dealing with an issue such as feeding of bears by park visitors, for example, managers need an understanding of the social sciences to influence or regulate behaviour of visitors, tourism operators, and other groups and agencies that bring visitors to parks and gateway communities.

THE BEHAVIOURAL APPROACH

Social science research into visitor behaviour in parks is described under a variety of headings such as outdoor recreation, adventure tourism/recreation, nature-based tourism, and ecotourism. What these terms have in common is the study of leisure behaviour: how people act and feel when not at work, and when activities are freely chosen and intrinsically satisfying (Manning, 1999). Park agencies seek to provide satisfying leisure experiences that minimize damage and unacceptable change to natural and social attributes of the area. Nevertheless, visitors sometimes describe their personal experiences as unsatisfactory. Dissatisfaction can take several forms, including concerns about crowding, litter, and damage to park environments. Sometimes visitors express concerns about noisy or rowdy behaviour of other visitors, or conflict with other types of users (e.g., hikers with horseback riders, skiers with snowmobilers). People also express concerns about facilities and services provided by park agencies, including complaints regarding upkeep of campgrounds or trails, quality of interpretive programs, or availability of park wardens.

Social science researchers have examined these issues to understand outdoor recreation behaviour and assist managers in their task of providing quality visitor experiences while protecting park environments. To summarize this research and show how it can be applied to park management issues, this chapter begins with a description of the 'behavioural approach' illustrated in Figure 6.1. The behavioural approach, which is also analogous to 'experience-based management' (Manfredo et al., 1983), proposes that people engage in specific activities in certain settings to fulfill motivations and realize a group of benefits that are known, expected, and valued (Manning, 1999). These benefits (e.g., satisfaction) occur when actual experiences or outcomes meet or exceed expectations or forces that push or pull people to seek out specific leisure activities and experiences. Researchers using motivational explanations (discussed below) are concerned with what arouses or activates leisure behaviour (i.e., forces that *push* people to engage in certain activities). Researchers have also examined characteristics of leisure activities and settings that *pull* people to select certain activities or settings over others (Mannell and Kleiber, 1997; Mannell, 1999). People, for example, may seek backpacking experiences in Jasper National Park because they are being 'pushed' by motivational factors such as the need to 'escape urban life' and 'be close to nature'. They may be 'pulled' by beliefs that the backcountry in Jasper is a relatively easily accessible natural setting devoid of urban characteristics and little crowding would be experienced. If these push and pull motivations were substantial, a person might select the activity of backpacking in a setting such as Jasper National Park. If outcomes of this experience turned out as expected in terms of these push and pull motivations, the person would be satisfied with the experience and the feedback loop might result in the person seeking similar experiences in the future. If the experience was not as expected, it is less

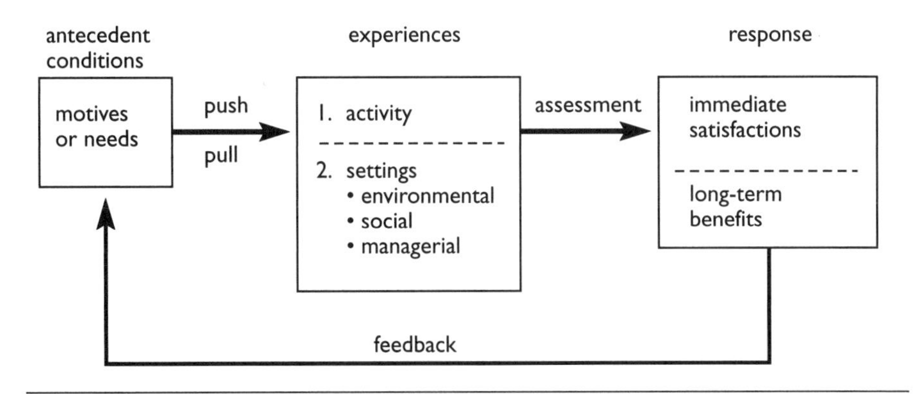

FIGURE 6.1 Behavioural model of outdoor recreation. After Mannell (1999).

likely that all benefits would be realized and the feedback loop may result in a lower probability that a similar experience would be sought.

This model assumes that motivations are shaped by the expectation that effort to participate will lead to performance (e.g., engage in certain activities in specific settings), which will lead to desired experience outcomes, benefits, and satisfaction (Manfredo et al., 1983). The model also assumes that individuals typically have multiple motives for leisure experiences in general and for outdoor recreation in particular (e.g., to develop skills, be close to nature, escape daily routines, get physical exercise). Experiences in this model are defined as the interaction between an activity and a setting. People vary in their preferences for type of activity. Some people, for example, may prefer backpacking rather than canoeing. People also vary in their preferences for different types of settings; a backpacking experience in Jasper National Park, for example, may differ substantially from a backpacking experience on the West Coast Trail segment of Pacific Rim National Park. A canoeing experience in Algonquin Provincial Park may differ from canoeing in Winisk River Provincial Park in Ontario. Recreation settings differ somewhat in appearance and character, and can be distinguished based on variability in three important parameters: (a) environmental conditions (e.g., modern to primitive); (b) social conditions (e.g., isolated to crowded); and (c) managerial conditions (e.g., few regulations to many regulations). Table 6.1 outlines these differences for two park activities, wilderness hiking and picnicking.

The behavioural model is comprised of two types of benefits. The first type of benefit is satisfaction with expected psychological or individual motivations (e.g., developing skills, affiliating with others, escaping daily routine, seeking adventure). The second type of benefit refers to ultimate or longer-term personal, societal, economic, and/or environmental benefits that result from engaging in recreation experiences

TABLE 6.1 Behavioural Model Illustrated with Wilderness Hiking and Family Picnicking

Level	Example 1	Example 2
1. Activity	wilderness hiking	family picnicking
2. Setting		
a. environmental setting	backcountry/wilderness	frontcountry
b. social setting	few people/groups	many people/groups
c. managerial setting	no restrictions no facilities	some restrictions many facilities
3. Motives	risk-taking challenge physical exercise	in-group affiliation change of pace
4. Benefits		
a. personal	enhanced self-esteem	family solidarity
b. societal	increased commitment to conservation	increased work efficiency

Source: After Mannell (1999).

(e.g., enhanced self-esteem and self-identity, personal growth, family cohesion, enhanced workplace efficiency). This second type of benefit is also known as 'benefits-based management' and extends the behavioural approach that initially focused primarily on benefits that accrue to the individual participant. Participation in an activity such as hiking, for example, can have: (a) personal benefits, such as enhanced self-esteem and physical exercise; (b) societal benefits, such as improved community health and solidarity; (c) economic benefits, such as lower health-care costs because people are engaging in physical exercise; and (d) environmental benefits, such as increasing interest in and commitment to the natural environment (Manfredo and Driver, 2002).

As shown in Table 6.1, wilderness hiking may take place in a backcountry setting with few other people, no facilities, and few restrictions. On the other hand, a family picnic could take place in a frontcountry setting used by several other groups and be provided with many facilities and a number of rules and restrictions regarding behaviour and use of the area. Both recreation experiences are influenced by various motives and lead to different types of benefits.

Outdoor recreation research in the 1960s focused primarily on participation levels in various recreation activities, but more recent studies have explored other aspects of

FIGURE 6.2 Raeside's cartoon reveals differences in visitors' expectations and the values of park managers. *Cartoon: Adrian Raeside.*

the behavioural model, including psychological benefits (e.g., Manfredo et al., 1983; Twynam and Robinson, 1997) and broader personal and societal outcomes (e.g., Haggard and Williams, 1991; Manfredo and Driver, 2002). A recent study of summer recreationists (e.g., sightseers, hikers, mountain bikers) at several sites within the Whistler Mountain ski area in British Columbia illustrates relationships among activities, settings, motivations, and benefits that constitute the behavioural approach (Needham et al., 2004b). Most visitors at the developed restaurant area near the top of the mountain were sightseers who were motivated to engage in on-mountain tours, view the alpine scenery, and visit the restaurant and gift shop. Hikers at more remote backcountry sites were motivated primarily to get exercise and view scenery. Preferences for facilities and services, and satisfaction with social (e.g., crowding) and environmental (e.g., erosion) conditions differed among sites.

The behavioural approach has advanced ways that visitor management is approached in many jurisdictions. In a landmark study, Clark and Stankey (1979) noted a consistent finding that people vary in preferences for different types of outdoor recreation settings, presumably as a consequence of differing motivations and/or activity preferences. On the basis of these findings, they concurred with Shafer (1969) that there was no such thing as the 'average camper' and reasoned that park agencies need to provide different kinds of recreation opportunities rather than uniform standardized settings. This led to development of the 'Recreation Opportunity Spectrum' (ROS), which, as discussed in Chapter 7, is a system of land allocation or zoning such that outdoor recreation settings could be arrayed along a continuum from primitive to modern based on level of setting modification and access, and visitors' activities, motivations, and experiences (Manning, 1999). In the ROS, different types of recreation opportunities are created by varying environmental, social, and managerial conditions. Consistent with the behavioural approach, ROS can be used to plan and manage parks and recreation settings for different types of users based on the mix of outcomes, activities, and settings sought by visitors.

VISITOR MOTIVATIONS

The behavioural model provides a useful outline for much of the kind of outdoor recreation theory, concepts, and research that have assisted the human dimensions of park management. Motivations, however, are more complex than those portrayed in Figure 6.2!

A leisure or recreation motivation is a reason for visiting an area or participating in an activity at a given time (Manfredo et al., 1996). Leisure motives are identified by people when asked what needs that they seek to satisfy through leisure involvement. Researchers typically provide study participants with a list of push and/or pull reasons (i.e., leisure motivations) and ask them to rate the importance of each motive for their participation in leisure activities. These reasons are generally referred to as 'expressed leisure motives' (Mannell and Kleiber, 1997), and are often only part of a larger and more complex picture of what motivates people to engage in leisure activities. Many of the motives reported for leisure engagement are based on physiological, learned, and cognitive motives, and these, in turn, are influenced by the interaction of inherited characteristics and socialization experiences. People are often unaware of these motives when reporting reasons why they engage in activities, so capturing these deeper motives presents a challenge to researchers and managers.

Despite the large number and types of motives that have been reported in many studies of leisure behaviour, there is broad agreement that a relatively small number of basic types are operative. The Paragraphs About Leisure (PAL) motivation scales (Driver et al., 1991), for example, involve 44 psychological needs that may be gratified by participation in recreation (e.g., achievement, relaxation). The PAL scales can be reduced to eight broad reasons for participating: self-expression, power, security, intellectual aestheticism, companionship, compensation, service, and solitude. The multiple (over 300) motivations in the widely used Recreation Experience Preference (REP) scales

BOX 6.1 Importance-Performance (I-P) Analysis: Linking Satisfaction and Motivational Factors

Measurement of visitor satisfaction in parks and protected areas has incorporated a number of methodologies reflecting the push and pull aspects of the behavioural approach. This figure illustrates how satisfaction was assessed in Yoho National Park in British Columbia by using the push element of motivation and the anticipated psychological benefits. Park visitors were asked how important they felt about each motivational factor and then how satisfied they were with each factor (Rollins and Rouse, 1993). The resulting importance-performance (I-P) matrix allows managers to identify important factors that are satisfied ('keep up the good work'), important factors that are not satisfied ('concentrate here'), unimportant factors that are satisfied ('too much effort here'), and unimportant factors that are not satisfied ('low priority'). The motivation of seeking solitude, for example, was important to most visitors (81 per cent), but many visitors (36 per cent) did not feel this had been achieved. The 'solitude' aspect of the Yoho experience is an area of concern and may require management attention. On the other hand, the motivation 'to be close to nature' was important to most visitors (93 per cent) and was achieved by most (80 per cent). Finally, the motive to 'meet new people' was viewed as not important by most visitors (84 per cent) so it may be viewed as irrelevant for visitor management.

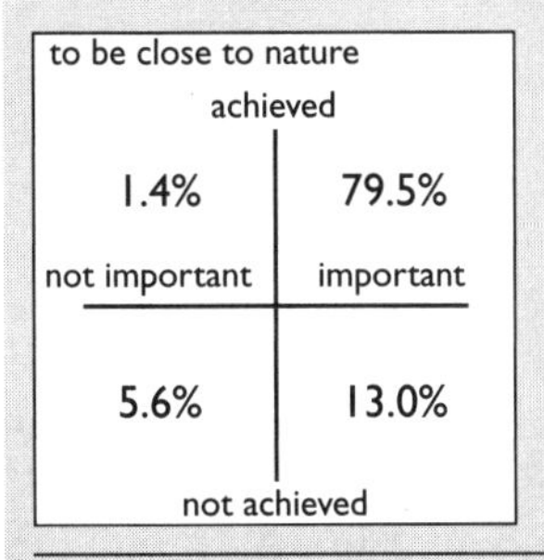

FIGURE 6.3 'Push' factors and satisfaction of visitors to Yoho National Park.

(e.g., Driver et al., 1991; Manfredo et al., 1996) have been reduced to 19 domains, eight of which have been shown to be important to most recreationists in parks and wilderness settings (Rosenthal et al., 1982): exploration, general nature experience, exercise, seeking exhilaration, escape from role overload, introspection, being with similar people, and escape from physical stressors. Both the PAL and REP motivation scales emphasize gratification of needs and pursuit of desired outcomes and benefits.

The work of Iso-Ahola (1982, 1989) has particular relevance to understanding outdoor recreation behaviour. Focusing on the social-psychological aspects of personal and interpersonal rewards, Iso-Ahola proposed that leisure participation is based on two motivational dimensions: (a) seeking (i.e., approach), and (b) escaping (i.e., avoidance). These two motivational forces simultaneously influence an individual's leisure behaviour. Activities may be engaged in because they provide opportunities for novelty or change from daily routines and stress. The 'escape' dimension is seen as a powerful leisure motive due to the constraining nature of a person's life, particularly from his or her work. This aspect of motivation is based on the need for optimal arousal in that individuals are considered to be constantly trying to escape from under-arousing or over-arousing experiences. The 'seeking' dimension is the tendency to search for psychological satisfactions from participation in leisure activities. These satisfactions can be divided into personal (e.g., self-determination, sense of competence, challenge, learning, exploration, relaxation) and interpersonal (e.g., social contact, connectedness) types. Iso-Ahola (1989) suggested that both seeking and escaping are forms of intrinsic motivations that are undertaken without concern for some form of external reward. In considering both seeking and escaping, Iso-Ahola proposed that individuals are motivated to participate in recreation if they perceive that the activity provides certain rewards (e.g., feelings of mastery, competence) and helps them leave everyday routine environments behind.

Recreation participation is a dynamic, multi-phase experience consisting of phases such as anticipation, travel-to, on-site, travel-back, and recollection. On-site phases are also dynamic and include experiences at various stages of an outing. Despite these phases, it is generally accepted that motivations tend to initiate recreation participation and satisfaction occurs as a result of this participation (see Manning, 1999).

VISITOR SATISFACTION

The behavioural approach suggests that people partake in recreation to fulfill their specific motivations and achieve certain benefits or outcomes. Satisfaction is one outcome that is a consistent goal of recreation management. Satisfaction is 'the positive perceptions or feelings that an individual forms, elicits, or gains as a result of engaging in leisure activities and choices; it is the degree to which one is content or pleased with his or her general leisure experiences and situations' (Beard and Ragheb, 1980: 22). Satisfaction is the difference between desired and achieved goals, or the congruence between expectations (i.e., motivations) and outcomes. According to Mannell (1989), satisfaction can be divided into 'global appraisal' (i.e., satisfaction with the entire experience) and 'facet appraisal' (i.e., satisfaction with various subcomponents

BOX 6.2 Measuring Visitor Satisfaction with 'Pull' Factors

An alternative approach to the measurement of visitor satisfaction focuses more on the 'pull' aspect of motivation, as illustrated in Figure 6.4. Here visitors were asked to rate the quality of conditions experienced on the West Coast Trail during the summer of 2000 (Rollins and Randall, 2000). Most visitors were satisfied with those aspects of the setting described at the top of the figure (e.g., condition of campsites, availability of fresh water), but setting characteristics listed at the bottom of the figure were not rated as positively (e.g., condition of boardwalks, availability of park staff, or trail signs).

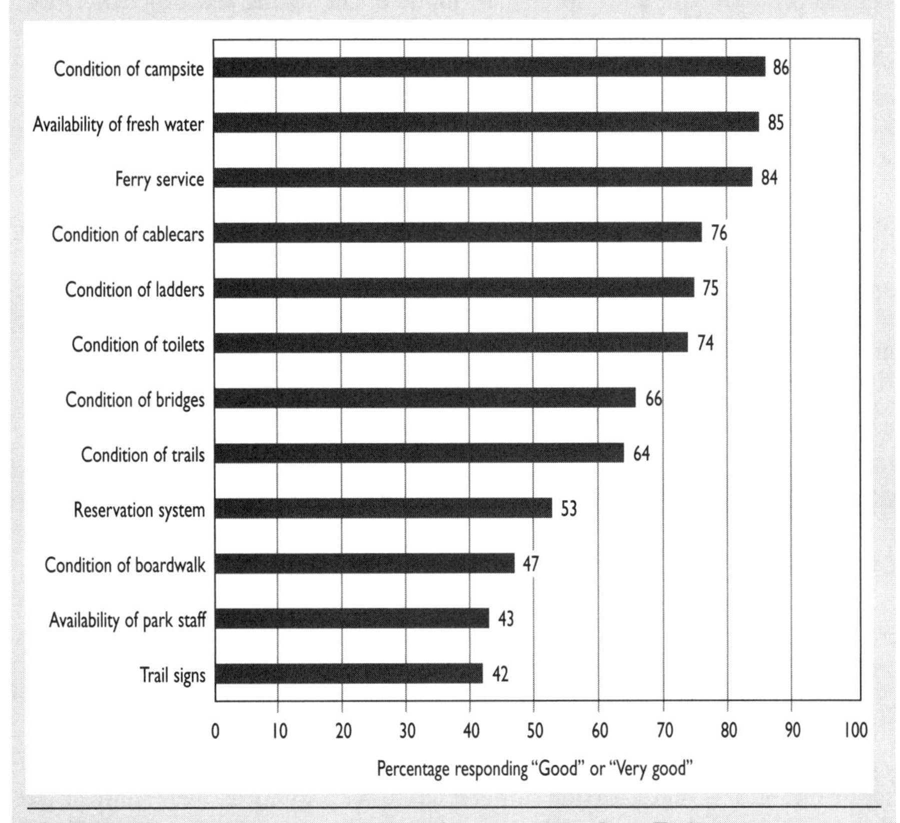

FIGURE 6.4 'Pull' factors in visitor satisfaction on West Coast Trail.

of the experience). Similarly, Jackson (1989) stated that satisfactions are divided between internal and external factors. Internal factors are shaped by motivations, past experience, and expectations; external factors involve specific setting attributes.

Hendee's (1974) 'multiple satisfactions' approach suggests that recreation resources offer people the opportunity for a range of experiences that, in turn, give rise to various human satisfactions. In other words, an individual's satisfaction with an activity or experience is complex; he or she may evaluate several aspects of the activity and experience (e.g., resource, social, managerial). Satisfaction is based on different experiences that often provide different types of satisfactions, and satisfaction is based on multiple factors that differ from person to person.

Despite recognition of the multiple satisfactions approach, researchers have typically measured global evaluations of the overall experience or outing (see Manning, 1999). This approach may be useful when comparing satisfaction across settings, times, or groups (e.g., consumptive versus non-consumptive recreationists), but there is often little variance in global measures because overall satisfaction tends to be uniformly high. Satisfaction with more specific attributes of the setting and experience (e.g., weather, crowding, fees, trails, litter), however, can vary and some satisfactions may outweigh others. Box 6.2, for example, shows that over 80 per cent of visitors were satisfied with the condition of campsites, availability of fresh water, and ferry services along the West Coast Trail, whereas less than 50 per cent of visitors were satisfied with the boardwalks, park staff, and signage. Compared to a single measure of overall satisfaction, therefore, examining users' satisfaction with multiple aspects of the setting and experience can be more meaningful for informing management. According to Pierce et al. (2001), it is important not only to measure overall satisfaction and satisfaction with components of the setting and experience, but also to determine the relative importance of these factors and components that motivate behaviour. 'Importance-performance analysis' is a useful tool for measuring relationships between motivations and satisfaction, and for revealing conditions that do or do not need management attention (Box 6.1) (Vaske et al., 1996a).

CROWDING, CARRYING CAPACITY, AND NORMS

Crowding

Overall use levels have declined in many parks and protected areas in Canada, but crowding still persists even in some areas where use levels have declined. Crowding is one factor that can influence outcomes of recreation participation and satisfaction. Crowding is a subjective negative evaluation that the number of people observed or number of encounters with other people, groups, or activities (i.e., reported encounters) is too many (Vaske and Donnelly, 2002). This concept has been measured in many visitor surveys on a 9-point scale from 1, 'not at all crowded', to 9, 'extremely crowded' (see Shelby et al., 1989). Increasing participation has resulted in perceptions of crowding in many parks in Canada. Visits to the West Coast Trail in Pacific Rim National Park, for example, increased from a few hundred people in 1969 to about 8,000 people by 1984, by which time 34 per cent of visitors reported that they felt crowded (Rollins, 1998). At the Whistler Mountain ski area in British Columbia, summer visitation (July to September) to the high alpine area increased from approximately 180,000 in 2000 to over 250,000 in 2004, with more than 50 per cent of visitors reporting that they felt crowded (Needham et al., 2004a, 2004b). Popularity of recreation in many natural settings in North America has led to concerns about crowding. It was hoped that social science research would provide managers with systematic scientific data from which it would be possible to reduce crowding problems, but actual research into crowding has produced mixed results. The following discussion describes recent developments in the understanding of crowding in park environments.

Early conceptualizations of crowding postulated that visitor perceptions of crowding would be directly proportional to the number of people in a setting at a given time. More people in a setting should create more reports of crowding; fewer people should result in less crowding. Results of several research studies, however, showed weak relationships between use levels and crowding. Researchers speculated that this unexpected result was due to faulty approaches in measurement and what should have been examined was level of contacts (i.e., number of reported encounters) rather than visitor numbers or densities (Shelby and Heberlein, 1986). It was reasoned that in the same park at the same time, the number of encounters might vary from place to place, with some people experiencing higher numbers of encounters than people in other parts of the same park. Recent studies, therefore, have examined relationships between crowding and the number of encounters that people experienced. Surprisingly, these results often did not turn out as expected either, with several studies reporting a weak relationship between contacts and crowding. Manning (1999) reviewed over 30 crowding studies that exhibited this weak relationship between use levels, reported encounters, perceived crowding, and satisfaction.

Reviews of these studies have provided a number of possible explanations for such unexpected results. First, some of these crowding studies may suffer from a type of sampling error whereby visitors who anticipate crowds decide to visit less crowded parks, at less crowded times, or are displaced by people who are more tolerant of higher crowding. As a result of this temporal or spatial 'displacement', less tolerant visitors may not be included in samples of users. Second, if a visitor encounters more people than expected, he or she might redefine the experience. This is known as 'product shift'. A wilderness area, for example, may be re-evaluated as a semi-wilderness area as a consequence of more encounters, and visitors may perceive a product shift and consequently may not feel crowded. Related to this product shift is a third concern described as a 'cognitive dissonance' effect or 'rationalization', which speculates that because recreation experiences are largely voluntary and self-selected, visitors will have invested time, money, and energy into their park experience. The last thing that visitors will want to admit to themselves or a researcher is that they felt crowded or dissatisfied with their experience. Displacement, product shift, and rationalization are behavioural responses to crowded conditions (see Shelby et al., 1988; Manning, 1999). A fourth explanation is that use levels in some studies are not high enough to have a major impact on visitor experiences. Finally, many visitors to settings such as parks are first-time visitors with little or no prior expectation for appropriate use levels. For these 'uninitiated newcomers', there may be a tendency to view existing conditions as appropriate, regardless of the level of contacts experienced (Manning, 1999).

FIGURE 6.5 A commercial raft launch site just below Bow Falls on the Bow River and adjacent to the Banff Springs Hotel. *Photo: Guy Swinnerton.*

Carrying Capacity

The number of encounters with other people that visitors experience and the extent to which they feel crowded are often used to inform social carrying capacity, which is discussed in more detail in the next chapter. Social carrying capacity, however, is generally defined as the level of use beyond which unacceptable impacts such as crowding occur to visitor experiences (Shelby and Heberlein, 1986). A parallel concept is the environmental or ecological carrying capacity approach, which is aimed at determining levels of use beyond which unacceptable impacts occur to park environments (e.g., water quality, vegetation loss, soil compaction, wildlife disruption). Yet another type of carrying capacity is managerial capacity, or the extent to which there are adequate facilities to accommodate users' needs (ibid.).

Defining acceptable conditions is central to carrying capacity and related frameworks described in Chapter 7 (e.g., Limits of Acceptable Change, Visitor Impact Management, Visitor Experience and Resource Protection, Visitor Activity Management Process). These frameworks necessitate measuring social, ecological, and managerial 'indicators' (e.g., crowding, litter) to reveal 'standards of quality' (e.g., encounter no more than 25 people) or thresholds at which indicator conditions reach unacceptable levels and are inconsistent with management objectives (Manning, 2004, 2007). Determining acceptable conditions, however, has been problematic, as illustrated in the crowding discussion above. The structural norm approach outlined in the next section has helped address some of these problems.

Norms

The 'structural norm approach' has provided a basis for measuring indicators and informing standards of quality. One line of research commonly defines 'norms' as standards that individuals use for evaluating activities, environments, or management strategies as good or bad, better or worse (Shelby et al., 1996; Needham et al., 2005). Norms clarify what people believe conditions or behaviour 'should' be (Heywood, 2002). Much of the normative work in parks and recreation is based on the 'return potential model' (see Vaske and Whittaker, 2004). This approach describes norms as evaluative standards using a graphic device called a 'social norm curve' or an 'impact acceptability curve'. Figure 6.6 represents the amount of indicator change increasing from left to right along the horizontal axis. The vertical axis represents the evaluative responses with the most positive evaluation at the top of the axis, the most negative on the bottom, and a neutral category in between. The majority of studies have used 'acceptability' as the evaluative response (see Manning, 1999, 2007).

An example of the structural norm approach is a study conducted in Gwaii Haanas National Park off the coast of British Columbia. Kayakers in the park were surveyed and shown photographs depicting the same marine setting, but the number of other kayakers was varied in each image (Vaske et al., 1996b; Freimund et al., 2002). After viewing each photograph, respondents indicated whether they felt the number of kayakers in the setting was acceptable or unacceptable. Using this method, a personal norm was computed for each kayaker. These individual results were then aggregated across the sample of kayakers to determine how much consensus or agreement existed among kayakers for different use levels. If a large degree of consensus exists, then it is possible to express this finding as a social norm.

The norm curve (i.e., curved line) in Figure 6.6 crosses the neutral position at the point when approximately nine kayakers would be encountered. This is known as the 'minimum acceptable condition' (Manning, 1999). If the number of encounters ever exceeds nine contacts, the experience would be viewed as unacceptable by a majority of kayakers (assuming reasonable consensus in opinion). Fewer than nine contacts would be more acceptable. The most desirable situation in these results (i.e., 'optimal condition', depicted by highest point on curve) occurs when the number of contacts with other kayakers is zero, but establishing a management standard of zero visitors is unrealistic in most park and recreation settings.

Validity of the normative approach depends on a number of factors. The first factor is the amount of agreement or consensus within the group, which is known as 'norm crystallization'. If a large amount of variability exists in acceptance of impacts (e.g., contacts with kayakers), it may be difficult to describe this curve as representing a social norm. When consensus does not exist, however, it may be possible to identify subgroups who share a higher level of consensus than the whole group taken together. The Gwaii Haanas data, for example, can be segmented by examining responses of subgroups such as motorboaters and kayakers (Figure 6.7). Given differences in norms between these two groups, results indicated little consensus when motorboaters and kayakers were grouped together. Box 6.3 shows how various subgroups hold different views about acceptable use levels in the Whistler Mountain/Garibaldi Provincial Park recreation area.

The structural norm approach for addressing issues related to carrying capacity has been used by several park agencies to address social impacts, including encounters and

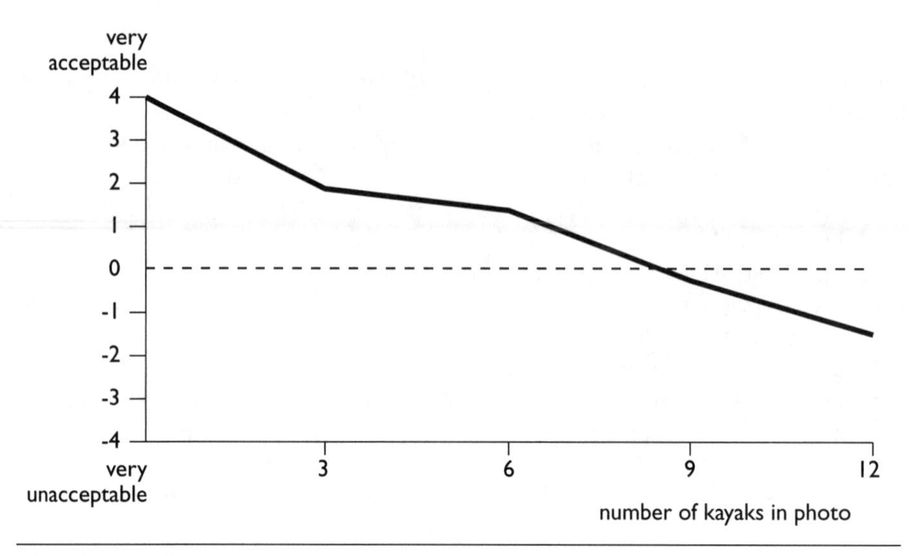

FIGURE 6.6 Kayaker norms for encountering other kayakers in a wilderness setting in Gwaii Haanas National Park Reserve. After Vaske et al. (1996b).

crowding, and resource impacts, such as litter and erosion (see Shelby et al., 1996; Vaske and Donnelly, 2002; Manning, 2007). Needham et al. (2004a), for example, found that many summer visitors at several sites at the Whistler Mountain ski area and adjacent Garibaldi Provincial Park reported crowding and encountered more people than they believed each site could adequately handle (i.e., their norm). The social carrying capacity of the sites was likely being exceeded. Directional trails, education, higher user fees, and zoning were management strategies supported for alleviating social impacts. Other studies have measured norms for indicator impacts and conditions in other protected areas in Canada, including the Columbia Icefield in Jasper National Park (Vaske et al., 1996c) and Broken Group Islands in Pacific Rim National Park (Randall, 2003).

Advantages of the structural norm approach are that it provides a proven applied and theoretical tool for managers to understand the extent to which indicator impacts are acceptable or unacceptable, identifies the importance of indicators, and describes the amount of consensus regarding acceptable indicator conditions (Vaske and Whittaker, 2004). A concern with conventional approaches for measuring crowding and related norms, however, is the failure to come to terms with a deeper understanding of crowding. People may feel crowded when they encounter people behaving in ways that interfere with their anticipated experiences, irrespective of the density of users or number of encounters experienced. Encountering a group of 10 backpackers at a campsite, for example, may be undesirable simply because of the anticipated noise level. It may not be the number of people per se that generates a crowding impact. If a group of 10 backpackers was behaving quietly, others may not feel crowded. This suggests that part of managing use and crowding involves managing behaviour to reduce user conflict.

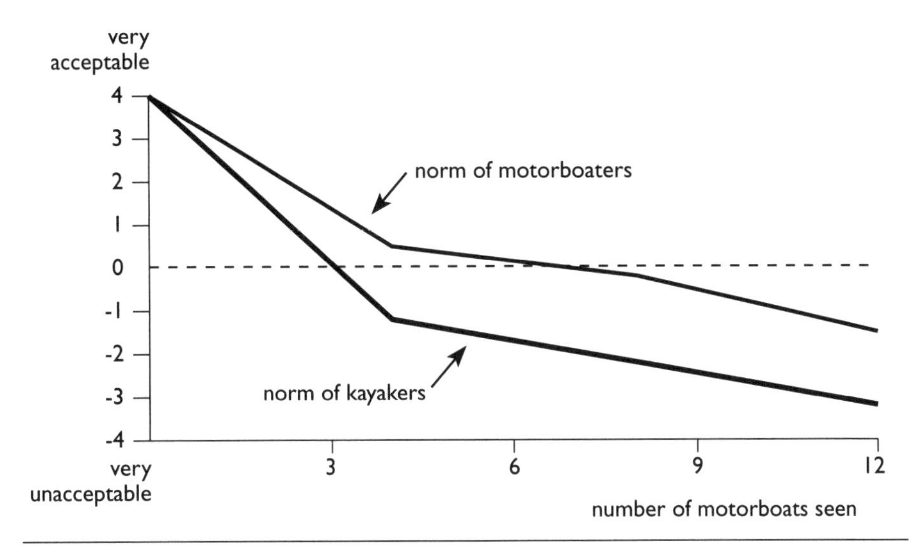

FIGURE 6.7 Kayaker norms versus motorboater norms for encountering motorboaters in a wilderness setting in Gwaii Haanas National Park Reserve. After Vaske et al. (1996b).

There are many indicators and standards of quality within the social, resource, and managerial dimensions that characterize parks and protected areas. As a result, another potential limitation of using the structural norm approach to inform decisions related to carrying capacity is that not all of these indicators can be measured and managed simultaneously, and trade-offs must be made, especially when visitor demand for parks and protected areas is high (Manning, 2007). Park managers are often faced with making trade-offs in planning and management decisions. Providing more access, for example, may allow more park visitors, but this might be detrimental to biophysical resources and cause more crowding. On the other hand, limiting access may reduce resource impacts and crowding, but would allow fewer people to experience parks and possibly erode public support for protected areas. Recent research, therefore, has used sophisticated methodological and analytical techniques such as 'stated choice modelling' and 'conjoint analysis' to examine normative trade-offs in parks and other outdoor recreation areas (see Manning, 2007). McCormick et al. (2003), for example, examined visitor trade-offs regarding backcountry experiences at Kluane National Park Reserve, Yukon. Trade-offs favoured solitude and quietude at campsites over trail encounters and managerial aspects such as fees and regulations. Data on trade-offs allow a deeper understanding of complex relationships among social, environmental, and managerial attributes that shape park experiences, and these data can assist managers in establishing priorities when faced with challenging decisions.

Summary of Crowding-Related Research

Research into crowding suggests that level of interaction with other people during outdoor recreation experiences is an important component of satisfactory outcomes.

People vary, however, in their preferences and acceptance for different contact levels. This variability may be due to different motivations. Some people may be more highly motivated than others to 'get away from other people', as predicted by the behavioural approach described earlier. Previous experience is another probable source of variability in contact preferences, with more experienced visitors likely to be more sensitive to higher use levels. Size, behaviour, and 'alikeness' of groups encountered are additional factors that may influence contact preferences and crowding (Manning, 2007).

Management responses to crowding vary. Quetico Provincial Park in Ontario is a large canoeing park characterized by a maze of lakes and rivers with endless possibilities for route choices. When crowding concerns emerged at Quetico, analysis of travel patterns revealed that crowding occurred along more heavily used routes and was related to higher visitor traffic through some access points. Managers used computer simulation models to predict likely contact levels that would result if some visitors were required to use other points of access into the park. After examining a series of computer simulations, a quota system was established at each access point. If a canoeing party arrived at a certain access point to begin their trip and the daily quota was filled, the group was directed to another access point where the quota was not filled. Evaluation of this approach demonstrated that reports of crowding diminished for Quetico while use levels actually increased through this more efficient spatial redistribution of visitors (Peterson et al., 1977).

In the West Coast Trail region of Pacific Rim National Park, reports of crowding compelled park managers to develop a quota system (Rollins and Bradley, 1986). Unlike the Quetico example, the West Coast Trail is a single trail with limited route options so a spatial redistribution strategy was not possible. Instead, West Coast Trail managers developed a temporal redistribution system. This involved a daily quota of 52 people, split between the two ends of the trail so that 26 people per day per trailhead were admitted into the park between 1 May and 30 September. This daily quota was computed by redistributing use from what had been a July–August concentration to that of a May–September season. Previous use levels were estimated to be approximately 8,000 people, so this total visitor level was divided by the number of days between 1 May and 30 September, and the result was a quota of 52 people per day. Annual use levels were kept constant, but daily use levels were reduced in the peak season by shifting more visitors to the shoulder seasons. Evaluation revealed high satisfaction with the quota system and encounter levels experienced while hiking, but some lingering concerns with encounter levels at campsites (Rollins, 1998). Quotas established on the West Coast Trail and in Quetico are examples of management efforts to reduce crowding and sustain quality experiences. Other possible approaches for managing encounters and crowding include: zoning, restricting or prohibiting some activity groups, advertising alternative recreation opportunities, advertising similar experiences found in other locations, fixed itineraries or directional trails, physical site alterations, education, user fees, and permits or reservation systems (see Chapter 7).

Although social science provides data that can be used to develop standards for various indicators and to inform crowding and carrying capacity-related decision-making, some element of management judgement must be exercised. What point(s) along a

BOX 6.3 Social Norms for Different Stakeholder Groups Regarding Acceptable Densities of Visitors in the Whistler Mountain Backcountry Area

North of Vancouver, the Whistler Mountain/Garibaldi Provincial Park area has received increasing use in the summer months as a consequence of ski lifts now operating from July to October, making the backcountry more accessible. A number of stakeholder groups were consulted to determine appropriate impacts and management actions for the area (Needham and Rollins, 2005).

To determine acceptable use levels, the structural norm approach was used. Respondents completed surveys containing a series of photographs depicting differing use levels in the area. For each photograph, respondents were asked to indicate how acceptable or unacceptable they felt about each scenario (level of density), using a 5-point scale of –2 = 'very unacceptable' to +2 = 'very acceptable'. Results for each stakeholder group are portrayed in Figure 6.8.

Five social norm curves are displayed, one for each stakeholder group. Where each norm curve crosses the neutral position in the graph (acceptability = 0) is the 'minimum acceptable condition' for that group. Clearly, private companies operating in the area are willing to accept higher densities of use compared to other stakeholder groups, particularly when compared to provincial and local government agencies.

These findings provide important insights into crowding and how crowding is perceived differently by various interest groups. Acceptable conditions are somewhat different for each stakeholder group in this example, an important consideration when applied to management frameworks such as Limits of Acceptable Change (LAC), which are described in Chapter 7.

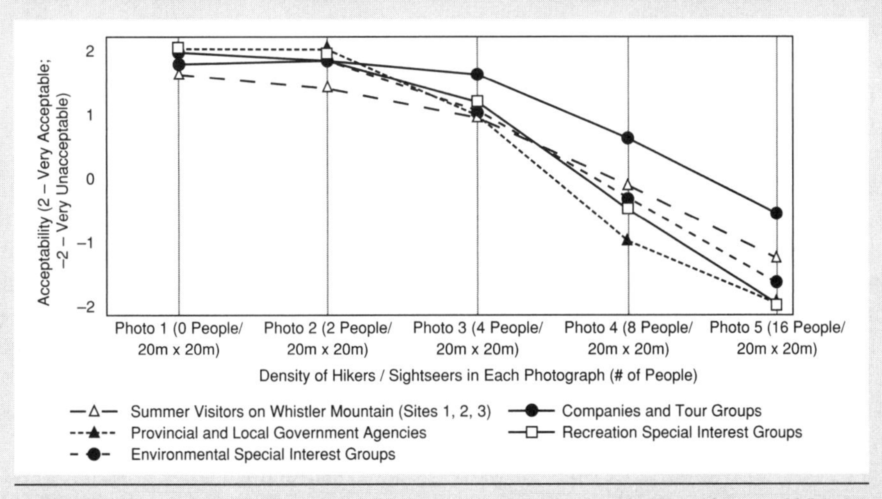

FIGURE 6.8 Social norm curves of stakeholder groups for density of use at Whistler Mountain.

range of standards across multiple indicators should be selected for management? This comes down to a management decision that takes into account additional factors such as the purpose, objectives, and significance of an area as defined by law and policy, significance of cultural and physical resources, historic precedent, extent to which financial resources and personnel are available for management, and influence of multiple stakeholders and interest groups (Manning, 1999). Management decisions about indicators and standards of quality are not, however, either/or decisions; providing an array of visitor recreation opportunities within and among parks may be a more plausible solution to minimizing impacts than simply regulating or prohibiting use (Manning, 2007). The Recreation Opportunity Spectrum (ROS) discussed above and in detail in the next chapter provides one approach for allocating opportunities within parks and other natural resource settings.

VISITOR CONFLICT

Like crowding, conflict is another factor that can influence the satisfaction of visits to park and recreation areas. Empirical research has revealed several different types of conflict that occur between people participating in similar or different types or styles of outdoor recreation (see Graefe and Thapa, 2004). 'One-way' or 'asymmetrical' conflict occurs when one activity group experiences conflict with or dislikes another group, but not vice versa. A study of snowmobilers and cross-country skiers in Alberta, for example, showed that skiers disliked encounters with snowmobilers, but snowmobilers did not mind skiers (Jackson and Wong, 1982). This finding is consistent with a more recent study of cross-country skiers and snowmobilers in two other alpine areas (Vaske et al., 2007). 'Two-way' conflict occurs when there is resentment or dislike in both directions, which has been demonstrated in recent studies of downhill skiers and snowboarders (Vaske et al., 2000; Thapa and Graefe, 2003). Conflict between users engaged in different activities (e.g., hikers versus mountain bikers) is known as 'out-group' conflict, whereas conflict between participants in the same activity (e.g., hikers versus other hikers) is known as 'in-group' conflict. Studies have examined different types of conflict among participants in activities such as: canoeing, hiking, hunting, motorboating, motorcycling, horseback riding, fishing, wildlife viewing, rafting, waterskiing, biking, ATV/OHV riding, skiing, and snowmobiling (see Manning, 1999; Graefe and Thapa, 2004; Vaske et al., 2007).

Most studies in parks and recreation areas have examined 'interpersonal' (i.e., goal interference) conflict where the actual physical presence or behaviour of an individual or group interferes with goals or expectations of another individual or group (Vaske et al., 2007). A skier, for example, may experience interpersonal conflict if he or she is cut off by or collides with a snowboarder. Jacob and Schreyer (1980) identified four factors that may influence this type of conflict. First, 'activity style' suggests that participants who are intensely involved in the activity are more likely to experience conflict because they place more importance on the activity and have well-defined goals, objectives, and expectations. These goals can range from quite general (e.g., to have a good time) to more specific (e.g., to spend quiet time with family in a remote setting). For example, people with the specific goal of spending quiet time with family in a remote setting are predicted to have greater potential to experience conflict when encountering a noisy group than someone for whom this goal is not important. Second, 'resource specificity' implies that visitors who are strongly attached to a resource such as a park (i.e., 'place attachment'; see Williams and Vaske, 2003) are more likely to experience conflict because they are more possessive of the site and consider its attributes to be exceptional and unique. Third, 'mode of experience' suggests that individuals who are 'focused' on the activity and resource have more sensitive perceptions of the environment around them, and consequently are more likely to experience conflict. Fourth, 'tolerance for lifestyle diversity' refers to acceptance or rejection of different lifestyles. Thus, visitors who are intolerant of lifestyles unlike their own, and who are less willing to share resources, are more prone to report conflict. Backpackers and skiers, for example, may report conflict with helihikers and heliskiers because they may be perceived as wealthy and flaunting their affluence. Studies have offered empirical support for some of Jacob and Schreyer's (1980) propositions (see Graefe and Thapa, 2004).

Interpersonal conflict is generally viewed as stress created when recreation behaviour of one group of people directly interferes with another group in the achievement of recreation goals or motivations. Defined in this way, crowding can be seen as a special case of recreation conflict, and both can be understood within the general behavioural model described above (Figure 6.1). When two groups of people decide to visit the same recreation setting to pursue different activities, the activities may interfere with each other because the two groups have different goals as determined by differing motivations. For example, a family may choose to go camping at a particular campground to achieve a family experience. Another group may choose the same campground as a venue for letting off steam and having a late-night party in a setting where they anticipate being free of restrictions common in more urban venues. Obviously, the potential for conflict between these two groups is high. Sometimes conflict is not equally perceived among groups. For this example, the family might be annoyed by the arrival of the partying group, whereas the partiers may be unaffected by the presence of the family and perhaps oblivious to the conflict created (i.e., one-way, asymmetrical conflict).

Jacob and Schreyer's (1980) model provides a framework for explaining interpersonal conflict, but it is likely that additional factors are involved, such as locus of control and anticipated consequences (Ewert et al., 1999). 'Locus of control' refers to the extent to which a person feels that he or she has control over events. People with high control are more likely to experience conflict as a precursor to taking actions to reduce conflict, whereas visitors with a lower locus of control may devise other ways of coping with conditions in a recreation setting. 'Anticipated consequences' are also thought to influence conflict. In Neck Point Park in British Columbia, for example, conflict arose between one group of park users (e.g., scuba divers, windsurfers) demanding road access to the waterfront area of the park, and a second group (e.g., birdwatchers, dog walkers) who wanted to keep the area roadless. Examination of perceived consequences of road access by the two groups revealed different expected consequences (Rollins et al., 2002). People supporting road access felt that more people would enjoy the park and that water-based activities would be enhanced and safer. People opposed to road access believed that a road would take away from the natural atmosphere, make the park less safe for pedestrians, and lead to crowding and rowdy behaviour. Based on this understanding of visitor perceptions, a satisfactory resolution was possible by providing a road on the southern periphery of the park, minimizing interaction of cars and pedestrians. Limited short-term parking was provided for only three vehicles to reduce concerns of crowding and depreciative behaviour.

Most conflict studies have examined interpersonal (i.e., goal interference) conflict. Recent research, however, has introduced and explored 'social values' conflict. Social values conflict occurs between groups who do not share similar opinions, norms, or values about an activity (Vaske et al., 1995). Unlike interpersonal conflict, social values conflict is defined as conflict that can occur even when there is no direct physical contact or interaction among groups (Vaske et al., 2007). For example, although encounters with horseback riders may be rare in park environments, visitors may philosophically disagree about the appropriateness of such animals in these settings. A study of wildlife viewers and hunters showed that viewers did not witness many hunters or hunting behaviours (e.g., see animals be shot, hear shots fired) in a particular backcountry setting because management regulations and rugged terrain and topography separated the two groups (Vaske et al., 1995). Regardless, wildlife viewers reported conflict with hunters simply because of a conflict in values regarding the appropriateness of hunting in the area.

Understanding the extent and type of conflict is important for managing parks and related recreation settings because some management strategies may be effective for addressing one type of conflict, but not another. When conflict stems from interpersonal conflict, for example, spatial zoning or temporal segregation of incompatible groups may be effective. When the source of conflict is a difference in social values, visitor information and education may be needed (Graefe and Thapa, 2004; Vaske et al., 2007). Managers need to understand the basis of visitor concerns to develop strategies for managing conflict.

VISITOR VALUES, BELIEFS, ATTITUDES, AND BEHAVIOUR

The extent to which conflict, satisfaction, and crowding occur in parks and related recreation settings is largely influenced by visitor evaluations of conditions and experiences. These evaluations are shaped by visitors' values, beliefs, and attitudes. It is important to measure and understand these cognitions and the relationships among them because they can influence behaviour, such as support of and receptivity towards specific park management actions.

Theory proposes that human thought is arranged in a hierarchy (Figure 6.9) consisting of general values, beliefs and value orientations, and more specific higher-order cognitions such as attitudes, intentions, and behaviour (Ajzen and Fishbein, 1980; Manfredo et al., 2004a). At the base of this hierarchy are 'values', which are abstract and enduring, and are concerned with desirable end-states (e.g., freedom, success) and modes of conduct (e.g., honesty, politeness). Values are basic modes of thinking shaped early in life by family or other peers, few in number, relatively stable over time, change slowly, guide life decisions, and transcend situations and objects (Fulton et al., 1996).

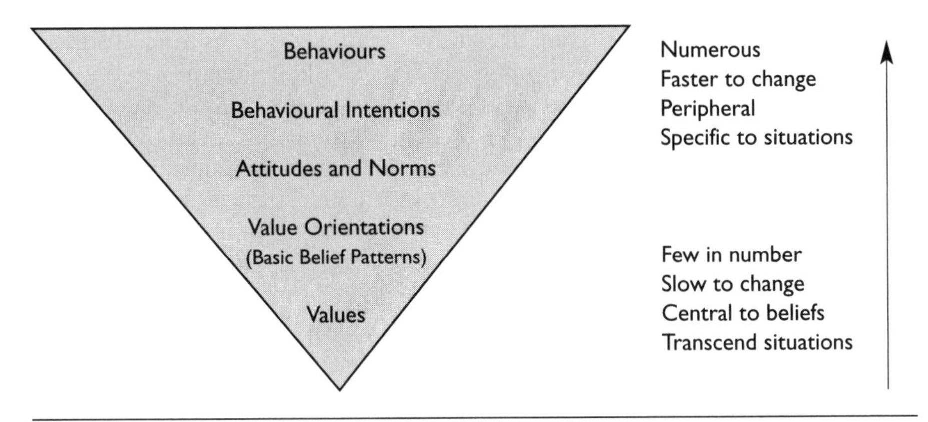

FIGURE 6.9 The cognitive hierarchy model of human behaviour. After Fulton et al. (1996) and Vaske and Donnelly (1999).

'Value orientations' reflect an expression of these basic values and are revealed through the pattern and direction of multiple basic beliefs that an individual holds regarding a more specific situation or issue (Manfredo et al., 2004a). Fulton et al. (1996), for example, asked individuals how strongly they agreed with several basic belief statements such as 'humans should manage wild animal populations so that humans benefit' and 'we should use wildlife to add to the quality of human life.' Taken together, these items indicated beliefs related to 'wildlife use'. Patterns of basic beliefs about wildlife use, hunting, and wildlife rights were combined into a value orientation scale called the wildlife protection–use continuum. Similar value orientations such as the anthropocentric–biocentric continuum have also been examined (Vaske and Donnelly, 1999). Values and value orientations can be used to identify groups with divergent preferences for management, and can help predict attitudes towards management and anticipate receptivity to and polarization over prevention and mitigation strategies (Manfredo et al., 2004a).

An 'attitude' is a tendency to evaluate a specific object, situation, or issue with some degree of favour or disfavour (Ajzen and Fishbein, 1980). Unlike values, we have many attitudes, which are more specific to particular objects. Sometimes 'attitude' is confused with 'satisfaction'. Leisure satisfaction refers to the 'after the fact assessment of an earlier [leisure] involvement or set of involvements' (Mannell, 1999: 238). Leisure attitudes usually refer to positive or negative opinions that people have regarding a leisure setting or activity. In the Neck Point Park example discussed above, attitudes were divided between people supporting road access into the park and people opposed to road access. Satisfaction, on the other hand, could be measured by examining actual experiences that people describe after the road is constructed.

Attitudes are thought to consist of cognitive, affective (i.e., emotional), and behavioural components (Ajzen and Fishbein, 1980). This can be illustrated by considering attitudes towards a camping fee system for forest recreation sites in British Columbia (Rollins and Trotter, 2000). The affective component refers to feelings of like or dislike for an 'attitude object', which in this case was user fees at recreation sites. Often, a single-item affective measure of like–dislike is used for measuring attitudes. However, information about why people hold certain attitudes can be identified by including measures of the cognitive component, referred to as attitudinal beliefs or perceptions (Ajzen and Fishbein, 1980). The cognitive component of attitudes in this example consisted of relevant beliefs that people held about consequences of establishing user fees at forest recreation sites. Positive attitudinal beliefs included 'would create more respect for sites', 'would lead to reduced vandalism', and 'would make people more willing to comply with rules and regulations.' Negative beliefs included 'would lead to confrontations between visitors and fee collectors', 'would detract from freedom', and 'cost to collect fees would be too expensive.' An example of the behavioural component was: 'I would camp less frequently if a user fee was introduced.' Attitudes towards fees at forest sites were determined by measuring the extent to which people agreed or disagreed with each of these types of belief statements. Analysis of responses indicated general support for user fees, although some people expressed concerns (agreed, but with negative attitudinal beliefs). In addition, users who would camp less if fees were introduced were more likely to have negative beliefs and attitudes towards fees. These results made it possible for the BC Forest Service to develop an approach to user fees that addressed many of these concerns.

This example illustrates established models of behaviour and decision-making such as the theory of reasoned action (Figure 6.10), which suggests that: (a) 'behaviour' is influenced by 'intention' to engage in that behaviour; (b) intention is a function of 'attitudes' and 'subjective norms' about the behaviour or issue (i.e., what you think other people think you should do, as determined by normative beliefs or judgements about what others feel is appropriate and motivation to comply with others); and (c) attitudes are a function of 'beliefs' that the issue or behaviour will lead to certain outcomes (i.e., cognitive) and favourable or unfavourable (i.e., affective, evaluative) 'evaluations' of these outcomes (Ajzen and Fishbein, 1980). Models such as this have helped predict behaviour for recreation and natural resource issues, including camping and hunting participation, support for wildfire management, preferences for user fees, and support for wildlife management (see Manfredo et al., 2004a; Vaske and Whittaker, 2004).

Another application of attitude theory is illustrated in a household survey conducted by BC Parks to determine attitudes towards setting aside more wilderness areas in British Columbia (BC Parks, 1994). Positive beliefs included protection of wildlife, preservation of biodiversity, places to conduct scientific studies, and stimulation of the economy by tourism. Negative beliefs included possible loss of jobs, reduction in government revenues through fees and taxes from industry, and restriction of recreation activities since no roads would be allowed into the areas. Results indicated that 61 per cent of respondents felt there was too little designated wilderness in BC, 3 per cent said there was too much wilderness, and 37 per cent said the amount of wilderness was about right. Repeated polling provided convincing evidence of public support for creating more wilderness parks in BC, and contributed to government actions in the last decade to increase the amount of protected area from about 5 per cent to 12 per cent (see Chapter 2) of the provincial land base.

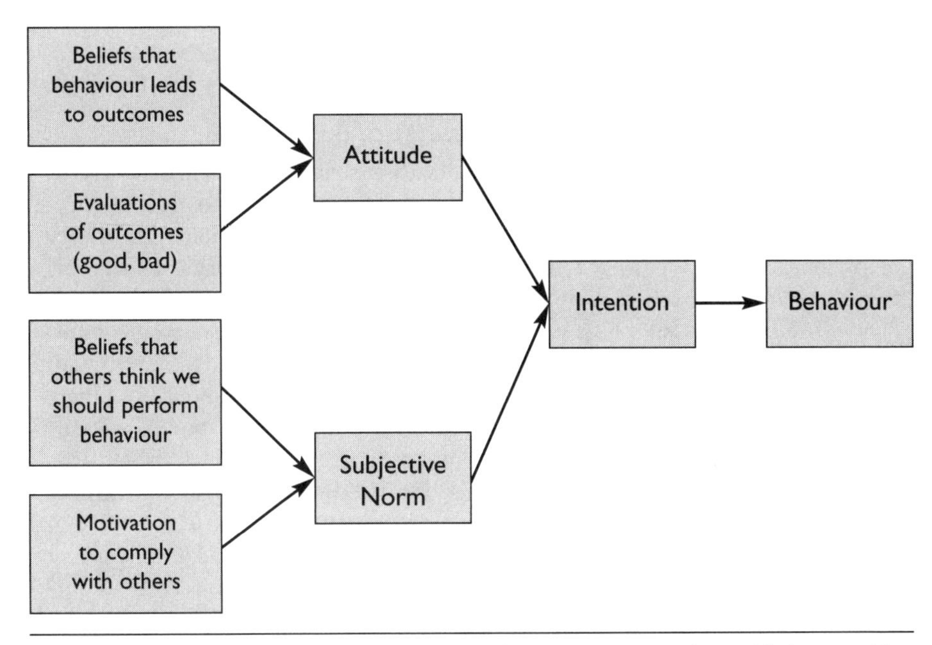

FIGURE 6.10 The theory of reasoned action for predicting attitudes and behaviour. After Ajzen and Fishbein (1980).

Attitude surveys, interviews, and public opinion polls can be useful for documenting support or opposition towards park management activities. This is important because managers often hold different perceptions of park environments than do park visitors, and managers are often unaware that visitors have different opinions and perceptions (Hendee et al., 1990; Needham and Rollins, 2005). Opinions of park visitors or the general public may be based on misperceptions or misunderstandings, and this kind of finding can be identified through social science research.

It is important to understand relationships among values, orientations, beliefs, and attitudes. A simplified example may help to illustrate. An individual may possess a value that it is important to respect life. As a result, he or she may agree with a belief statement such as 'animals should have similar rights to humans' (related to a protectionist or biocentric value orientation). This individual may hold normative beliefs that it is unacceptable to eat meat and humans should not eat meat. Consequently, he or she may have unfavourable or negative attitudes about hunting and would not intend to go hunting or actually engage in this behaviour. By understanding visitors' values, beliefs, and attitudes, park managers may be able to predict future behaviour and anticipate support or opposition towards management strategies and decisions.

Understanding these relationships among values, beliefs, and attitudes is important because it also allows a better understanding of visitor behaviour. Given that these cognitions can help predict intentions and behaviour, they can also be targeted for change by persuading and eliciting desirable behaviour (Box 6.4). This is important, especially

BOX 6.4 Applying the Theory of Reasoned Action To Understand Attitudes and Behaviour Regarding a Voluntary No-Fishing Policy in Pacific Rim National Park

In the Broken Group Islands segment of Pacific Rim National Park, concerns regarding overharvesting of rock cod by park visitors led the park to implement a voluntary no-fishing policy. In a study exploring compliance with this policy, the theory of reasoned action was used to explore attitudes, subjective norms, and beliefs thought to influence compliance behaviour of visitors to the park (Randall, 2003). Survey results indicated a high correlation between intentions regarding compliance, subjective norms, and attitudes. People who said they intended to comply with the voluntary no-fishing policy had positive attitudes towards the policy and these views were shared by important others (subjective norm). Examination of attitudinal beliefs (Table 6.2) revealed that those opposed to the policy held beliefs that reflected this position, as they more strongly endorsed possible negative outcomes (beliefs) and were less supportive of possible positive outcomes (beliefs).

Park managers can use results like these to target messages aimed at influencing attitudes and behaviour to be more supportive of management objectives (as outlined in Chapter 8). In this example, park managers could improve compliance with the voluntary no-fishing policy by providing information to enhance positive outcomes of the policy (i.e., will retain food sources for other creatures; protect marine life for future generations; reduce amount of litter caused by fishing). A second communication strategy could focus on discrediting negative outcomes or beliefs listed on the bottom half of the table.

TABLE 6.2 Attitudinal Beliefs Regarding Voluntary No-Fishing Policy, Pacific Rim National Park

	% Agreeing with Statement	
	Visitors Opposed to the Policy	Visitors Supportive of the Policy
Positive beliefs		
Will retain food sources for other creatures	47.4	91.2
Will protect marine life for the future	36.5	88.5
Will reduce litter caused by fishing	40.0	79.4
Negative beliefs		
Will have a low compliance rate	71.6	60.8
Will detract from visitor satisfaction	69.0	42.6
Will take away from my experience	57.2	9.6
Will have negative economic impact	48.8	30.0
Will decrease my food source	47.5	15.8

when designing and evaluating educational information and interpretation efforts in parks. Dual-process persuasion models such as the 'elaboration likelihood model' (ELM) and 'heuristic systematic model' have received attention in park and recreation settings for improving information and education campaigns, as discussed in Chapter 8.

VISITOR SEGMENTATION AND SPECIALIZATION

Many studies described in this chapter have illustrated that people vary tremendously in their attitudes, motivations and goals, activities selected, and preferred setting attributes. This makes it difficult for managers to plan for 'the average camper who doesn't exist' (Shafer, 1969) and has led to a number of studies aimed at improving understanding of park visitor diversity (see Manning, 1999). This type of investigation determines if it is possible to identify meaningful homogeneous subgroups or market segments of similar people within the more heterogeneous population of park visitors. Managers, therefore, may be equipped to understand and cater to needs and requirements of different market segments. By understanding and managing for subgroups, more visitors may have satisfactory experiences and fewer conflicts may occur.

One approach to segmentation is by visitor activity type (see Figure 6.7). This activity approach is the basis of the 'Visitor Activity Management Process' (VAMP) developed

FIGURE 6.11 An aircraft equipped with tundra tires lands beside the Firth River in Vuntut National Park Reserve with a group of rafters. Attitude studies have shown that aircraft accessibility is controversial in many parks. *Photo: P. Dearden.*

BOX 6.5 Segmenting Visitors to Canada's Mountain National Parks

To understand the types of visitors to Canada's mountain national parks, Parks Canada conducted surveys with visitors to Banff, Jasper, Kootenay, and Yoho national parks (McVetty, 2003). Using an analysis of trip motives, reported activities, and reported spending, it was possible to segment visitors into three groups:

- *Getaway visitors* (44 per cent): people staying for 2–3 days, focusing their visit on a specific activity or area, and spending less money than the other two segments;
- *Comfort visitors* (35 per cent): people tending more to use hotels and restaurants in the park, and spending more money than other segments;
- *Camping visitors* (21 per cent): people focusing on camping and recreational vehicle touring. This group attaches higher importance to learning about natural and historical heritage.

by Parks Canada and described in the next chapter. Another approach is to segment people by setting preferences. This approach was used in a study of backcountry visitors to Yoho and Kootenay national parks (Rollins and Rouse, 1993). In this study, three distinct user groups were identified: a 'purist group' wanting no backcountry facilities; a 'semi-rustic group' expressing preferences for shelters, huts, firepits, and picnic tables; and a 'rustic' group who expressed more ambivalent preferences for camping facilities, but were more supportive of horse facilities (e.g., corrals, grazing areas). A similar approach was used to examine preferences for activities, settings, and psychological outcomes of visitors to forest areas in northern Ontario (Twynam and Robinson, 1997). Distinct market segments were revealed and labelled as enthusiasts, adventurers, naturalists, and escapists. Escapists, for example, indicated a higher preference for remoteness, unaltered nature, and physically demanding and challenging activities such as climbing, canoeing, and kayaking. This group placed high importance on solitude, knowledge, and learning. Other approaches to segmentation include differentiating individuals based on demographic and socio-economic characteristics, site characteristics (e.g., frontcountry versus backcountry), beliefs and value orientations, and competing views of different interest groups and citizen advocacy organizations (e.g., Bright et al., 2000; Needham and Rollins, 2005; Box 6.3).

A common segmentation approach involves the concept of 'recreation specialization'. Specialization is defined as 'a continuum of behaviour from the general to the particular, reflected by equipment and skills used in the sport and activity setting preferences' (Bryan, 1977: 175). At one end of the continuum are novices or infrequent participants who do not consider the activity to be a central life interest or show strong preferences for equipment and technique. The other end includes more avid participants who are committed to the activity and use more sophisticated methods. Recreationists are thought to progress to higher stages along the continuum, reflected by increasing skill and commitment (Scott and Shafer, 2001).

The specialization concept has been examined relative to individuals engaged in a variety of activities in different settings. Highly specialized recreationists can differ from their less specialized counterparts on attributes such as motivations, management and setting preferences, perceived environmental impacts, crowding evaluations, and other related attitudes and opinions (see Manning, 1999; Scott and Shafer, 2001). An experienced canoeist who has paddled on several trips over a number of years in Algonquin Provincial Park, for example, is likely to be concerned if use levels become much higher; a novice canoeist travelling in the same area at the same time may be less concerned about use levels.

There is little consensus among researchers about how best to measure specialization (Scott and Shafer, 2001). Both single-item (e.g., frequency of participation; Ditton et al., 1992) and multi-dimensional approaches (e.g., Needham et al., 2007) have been employed to segment recreationists. Researchers generally agree, however, that specialization is a multi-dimensional concept consisting of behavioural, cognitive, and affective components. Behavioural variables include equipment investment and 'experience use history' (Schreyer et al., 1984), such as how often a canoeist has paddled in Algonquin Park and number of other canoe trips taken. Cognitive variables include skill level and knowledge. Indicators of affective attachment or commitment include enduring involvement and centrality to lifestyle (see Manning, 1999; Scott and Shafer, 2001). These dimensions, however, do not always increase linearly together in lock-step fashion, suggesting that specialization may be best suited for revealing styles of involvement and career stages in an activity rather than a single aggregate of dimensions and linear continuum of progression (Lee and Scott, 2004; Needham et al., 2007).

Specialization can be linked to the behavioural approach in a number of ways. Compared to novices or visitors who may be classified as less specialized, for example, people who are highly specialized have been shown to have more complex and developed goals and motivations that are important to them and not easily substituted (e.g., Kerstetter et al., 2001; Needham et al., 2007). Specialists are also more likely to develop a personal connection between anticipated goals and specific setting characteristics and site choices (e.g., Cole and Scott, 1999). A specialized climber, for example, might be disappointed to find an alpine hut in a favourite climbing area, and feel that the preferred experience of self-sufficiency has been diminished by this new facility. Conversely, a novice climber who is just developing connections between climbing, personal motivations, and setting characteristics may not feel the same impact when encountering this alpine hut. In a study of wilderness use in Clayoquot Sound, British Columbia, concerns regarding visible logging increased as a function of specialization, with levels of concern increasing from 52 per cent expressed by the low specialization group to 92 per cent for the high specialization group (Rollins and Connolly, 2001). In a study of vehicle campers in Alberta, McFarlane (2004) reported that more specialized campers chose campgrounds that were more remote and demanded higher self-reliance and decreased dependency on facilities and services.

As with other approaches to visitor segmentation, managers can apply the specialization concept to identify subgroups within a population of park visitors. Each subgroup will be more similar in their views and expectations, and may warrant somewhat different opportunities and management responses. Management approaches such as ROS or LAC (described in the next chapter), for example, might be employed such that zones are created in a park to provide a wide selection of beginner to more challenging routes that allow visitors of differing degrees of specialization to find appropriate route choices.

SOCIAL IMPACTS OF PARKS AND PROTECTED AREAS

Most research on protected area management in Canada has been directed to issues related to ecological integrity or visitor management in parks. Comparatively less attention has been given to social impacts of parks on nearby communities. Since parks attract many visitors who are not residents of the area, these tourists are likely to spend at least some time in nearby host or gateway communities. Examples of host communities include Tofino near Pacific Rim National Park, Marathon near Pukaskwa National Park in Ontario, and Pangnirtung near Auyittuq National Park on Baffin Island. Examining relationships between parks and adjacent communities also reflects thinking about ecosystem management discussed in Chapter 13 and ecotourism discussed in Chapter 12.

While visiting host communities, tourists may purchase goods and services such as groceries, fuel, camping supplies, restaurant meals, and accommodation in hotels or motels. In addition, tourists may meet and interact with local residents in ways that enrich the lives of local residents. However, not all interactions are positive. Visitor numbers may stress local services not designed to handle the surge of visitors and create congestion and related problems, and sometimes the behaviour of visitors is offensive to residents of host communities. These issues are described in greater detail in Chapter 12, but can also be related to the behavioural model (Figure 6.1) discussed earlier in this chapter. Conflicts between park visitors and residents of host communities can be seen as a variation of the behavioural model, with the experience dimension (i.e., activities, settings) occurring within the host community rather than within the park. Presumably, conflict occurs when behaviour of visitors blocks goals and expectations that residents have developed regarding community values and ideals.

CONCLUSIONS

This chapter presented some of the major areas of social science theory and research that address visitor management issues in parks and protected areas. The behavioural model (Figure 6.1) suggests that visitor behaviour can be understood in terms of visitor motivations, psychological goals that visitors develop as a consequence of these motivations, and how various activities and settings are perceived as facilitating the achievement of important goals, outcomes, and benefits. Visitor satisfaction is seen as the achievement of recreational goals, that is, the degree of congruence between expectations and actual experiences. Issues such as crowding and conflict between groups can be explained, in part, through this model.

Social science research has demonstrated that people vary considerably in their motivations, preferences for different activities and settings, experiences, and attitudes towards conditions and management. This diversity suggests that quality of visitor experiences can be enhanced through management strategies such as zoning that attempt to provide various opportunities aimed at serving different market niches. These market niches can be described and refined through various approaches to visitor segmentation, including the specialization approach. These insights have contributed to development of a number of approaches to visitor management described in the next chapter. Park managers, however, cannot act upon all visitor demands or preferences, no matter how well documented, if park resources are threatened by such actions.

Finally, protection of park ecosystems should take precedence over provision of visitor experiences. Protection of park ecosystems, however, requires support and co-operation of park visitors, some of whom may be asked to do without certain facilities or services, or the opportunity to participate in certain types of activities, in order to reduce environmental stresses. Protection of park environments also requires support of other constituents, including local communities located nearby or sometimes within parks. Involvement of visitors and the general public in the resolution of park issues can be facilitated by selection of appropriate social science techniques such as public meetings, expert panels, focus groups, surveys/questionnaires, and referendums. Increasingly, park managers are required to use these social science methods as part of the process of resolving park issues and gaining constituent support.

REFERENCES

Ajzen, I., and M. Fishbein. 1980. *Understanding Attitudes and Predicting Social Behaviour.* Englewood Cliffs, NJ: Prentice-Hall.

BC Parks. 1994. *Wilderness Issues in British Columbia.* Victoria: Ministry of Environment, Lands, and Parks.

Beard, J.G., and M.G. Ragheb. 1980. 'Measuring leisure satisfaction', *Journal of Leisure Research* 12: 20–33.

Bright, A.D., M.J. Manfredo, and D.C. Fulton. 2000. 'Segmenting the public: An application of value orientations to wildlife planning in Colorado', *Wildlife Society Bulletin* 28: 218–26.

Bryan, H. 1977. 'Leisure value systems and recreation specialization: The case of trout fishermen', *Journal of Leisure Research* 9: 174–87.

Clark, R., and G. Stankey. 1979. 'The recreation opportunity spectrum: A framework for planning, management, and research', USDA Forest Service Research Paper PNW-98.

Cole, J.S., and D. Scott. 1999. 'Segmenting participation in wildlife watching: A comparison of casual wildlife watchers and serious birders', *Human Dimensions of Wildlife* 4: 44–61.

Ditton, R.B., D.K. Loomis, and S. Choi. 1992. 'Recreation specialization: Re-conceptualization from a social worlds perspective', *Journal of Leisure Research* 24: 33–51.

Driver, B.L., H.E. Tinsley, and M.J. Manfredo. 1991. 'The paragraphs about leisure and recreation experience preference scales: Results from two inventories designed to assess the breadth of perceived psychological benefits of leisure', in B.L. Driver and G.L. Peterson, eds, *Benefits of Leisure.* State College, Penn.: Venture Publishing, 263–86.

Environment Canada. 1999. *The Importance of Nature to Canadians: Survey Highlights.* Ottawa: Minister of Public Works and Government Services Canada.

Ewert, A.W., R.B. Deiser, and A. Voight. 1999. 'Conflict and the recreation experience', in E.L. Jackson and T.L. Burton, eds, *Understanding Leisure and Recreation: Mapping the Past, Charting the Future.* State College, Penn.: Venture Publishing, 335–45.

Freimund, W.A., J.J. Vaske, M.P. Donnelly, and T.A. Miller. 2002. 'Using video surveys to access dispersed backcountry visitors' norms', *Leisure Sciences* 24: 349–62.

Fulton, D.C., M.J. Manfredo, and J. Lipscomb. 1996. 'Wildlife value orientations: A conceptual and measurement approach', *Human Dimensions of Wildlife* 1: 24–47.

Graefe, A.R., and B. Thapa. 2004. 'Conflict in natural resource recreation', in Manfredo et al. (2004b: 209–24).

Haggard, L.M., and D.R. Williams. 1991. 'Self-identity benefits of leisure activities', in B.L. Driver and G.L. Peterson, eds, *Benefits of Leisure.* State College, Penn.: Venture Publishing, 103–20.

Hendee, J.C. 1974. 'A multiple-satisfaction approach to game management', *Wildlife Society Bulletin* 2: 104–13.

———, G.H. Stankey, and R.C. Lucas. 1990. *Wilderness Management.* Golden, Colo.: North American Press.

Heywood, J.L. 2002. 'The cognitive and emotional components of behavior norms in outdoor recreation', *Leisure Sciences* 24: 271–81.

Iso-Ahola, S.E. 1982. 'Toward a psychological theory of tourism motivation: A rejoinder', *Annals of Tourism Research* 12: 256–62.

———. 1989. 'Motivation for leisure', in E.L. Jackson and T.L. Burton, eds, *Understanding Leisure and Recreation: Mapping the Past, Charting the Future.* State College, Penn.: Venture Publishing, 247–79.

Jackson, E.L. 1989. 'Perceptions and decisions', in G. Wall, ed., *Outdoor Recreation in Canada.* Toronto: Wiley, 76–132.

——— and R. Wong. 1982. 'Perceived conflict between urban cross-country skiers and snowmobilers in Alberta', *Journal of Leisure Research* 14: 47–62.

Jacob, G.R., and R. Schreyer. 1980. 'Conflict in outdoor recreation: A theoretical perspective', *Journal of Leisure Research* 12: 368–80.

Kerstetter, D.L., J.J. Confer, and A.R. Graefe. 2001. 'An exploration of the specialization concept within the context of heritage tourism', *Journal of Travel Research* 39: 267–74.

Lee, J., and D. Scott. 2004. 'Measuring birding specialization: A confirmatory factor analysis', *Leisure Sciences* 26: 245–60.

McCormick, S., W. Haider, D. Anderson, and T. Elliot. 2003. 'Estimating wildlife and visitor encounter norms in the backcountry with a multivariate approach: A discrete choice experiment in Kluane National Park and Reserve, Yukon, Canada', in N. Munro and P. Dearden, eds, *Science and Management of Protected Areas: Making Ecosystem-Based Management Work.* Victoria, BC: University of Victoria.

McFarlane, B.L. 2004. 'Recreation specialization and site choice among vehicle-based campers', *Leisure Sciences* 26: 309–22.

McVetty, D. 2003. 'Understanding visitor flows in Canada's mountain national parks: The patterns of visitor use studies in Banff, Jasper, Kootenay, and Yoho national parks', in N. Munro and P. Dearden, eds, *Science and Management of Protected Areas: Making Ecosystem-Based Management Work.* Victoria, BC: University of Victoria.

Manfredo, M.J., and B.L. Driver. 2002. 'Benefits: The basis for action,' in M.J. Manfredo, ed., *Wildlife Viewing: A Management Handbook.* Corvallis: Oregon State University Press, 43–69.

———, ———, and P.J. Brown. 1983. 'A test of concepts inherent in experience-based setting management for outdoor recreation areas', *Journal of Leisure Research* 15: 263–83.

———, ———, and M.A. Tarrant. 1996. 'Measuring leisure motivation: A meta-analysis of the recreation experience preference scales', *Journal of Leisure Research* 28: 188–213.

———, T.L. Teel, and A.D. Bright. 2004a. 'Application of the concepts of values and attitudes in human dimensions of natural resources research', in Manfredo et al. (2004b: 271–82).

———, J.J. Vaske, B.L. Bruyere, D.R. Field, and P.J. Brown, eds. 2004b. *Society and Natural Resources: A Summary of Knowledge.* Jefferson, Mo.: Modern Litho.

Mannell, R.C. 1989. 'Leisure satisfaction', in E.L. Jackson and T.L. Burton, eds, *Understanding Leisure and Recreation: Mapping the Past, Charting the Future.* State College, Penn.: Venture Publishing, 281–302.

———. 1999. 'Leisure experience and satisfaction', in E.L. Jackson and T.L. Burton, eds., *Leisure Studies: Prospects for the Twenty-First Century.* State College, Penn.: Venture Publishing, 235–52.

——— and D.A. Kleiber. 1997. *A Social Psychology of Leisure.* State College, Penn.: Venture Publishing.

Manning, R.E. 1999. *Studies in Outdoor Recreation: Search and Research for Satisfaction.* Corvallis: Oregon State University Press.

———. 2004. 'Recreation planning frameworks', in Manfredo et al. (2004b: 83–96).

———. 2007. *Parks and Carrying Capacity: Commons without Tragedy.* Washington: Island Press.

Needham, M.D., and R.B. Rollins. 2005. 'Interest group standards for recreation and tourism impacts at ski areas in the summer', *Tourism Management* 26: 1–13.

———, ———, and J.J. Vaske. 2005. 'Skill level and normative evaluations among summer recreationists at alpine ski areas', *Leisure/Loisir: Journal of the Canadian Association for Leisure Studies* 29: 71–94.

———, ———, and C.J.B. Wood. 2004a. 'Site-specific encounters, norms and crowding of summer visitors at alpine ski areas', *International Journal of Tourism Research* 6: 421–37.

———, J.J. Vaske, M.P. Donnelly, and M.J. Manfredo. 2007. 'Hunting specialization and its relationship to participation in response to chronic wasting disease', *Journal of Leisure Research* 39: 413–37.

———, C.J.B. Wood, and R.B. Rollins. 2004b. 'Understanding summer visitors and their experiences at the Whistler Mountain ski area, Canada', *Mountain Research and Development* 24: 234–42.

Parks Canada Agency. 2006a. 'Parks Canada attendance 2001–02 to 2005–06'. At: <www2.parkscanada.gc.ca/docs/pc/attend_E.pdf>.

———. 2006b. *Corporate Plan 2006/07–2010/11.* At: <www.pc.gc.ca-docs-pc-plans-plan2006-2007-cp_0607-E.pdf>.

Peterson, G.L., R.F. de Battencourt, and D.K. Wong. 1977. 'A Markov-based linear programming model of travel in the Boundary Water Canoe Area', in *Proceedings: River Recreation Management and Research Symposium.* St Paul, Minn.: USDA Forest Service North Central Forest Experiment Station, 342–56.

Pierce, C.L., M.J. Manfredo, and J.J. Vaske. 2001. 'Social science theories in wildlife management', in D.J. Decker, T.L. Brown, and W.F. Siemer, eds, *Human Dimensions of Wildlife Management in North America.* Bethesda, Md: Wildlife Society, 39–56.

Randall, B.C. 2003. 'An examination of visitor management issues within the Broken Group Islands, Pacific Rim National Park Reserve', Master's thesis, University of Victoria.

Rollins, R. 1998. 'Managing for wilderness conditions on the West Coast Trail area of Pacific Rim National Park', in N.W.P. Munro and J.H.M. Willison, eds, *Linking Protected Areas with Working Landscapes: Proceedings of the Third International Conference on Science and Management of Protected Areas.* Wolfville, NS: SAMPAA, 643–51.

——— and G. Bradley. 1986. 'Measuring recreation satisfaction with leisure settings', *Recreation Research Review* 13: 22–7.

——— and S. Connolly. 2001. 'Visitor perceptions of Clayoquot Sound: Implications from a recreation specialization model', in S. Bondrup-Nielsen, N.W.P. Munro, G. Nelson, J.H.M. Willison, T.B. Herman, and P. Eagles, eds, *Managing Protected Areas in a Changing World: Proceedings of the Fourth International Conference on Science and Management of Protected Areas*. Wolfville, NS: SAMPAA, 1401–12.

———, R. Harding, and M. Mann. 2002. 'Resolving conflict in an urban park setting: An application of attitude theory', *Leisure/Loisir: Journal of the Canadian Association for Leisure Studies* 26: 135–46.

——— and C. Randall. 2000. *West Coast Trail Visitor Survey*. Ucluelet, BC: Pacific Rim National Park.

——— and J. Rouse. 1993. 'Segmenting backcountry visitors by setting preferences', in J.H.M. Willison, S. Bondrup-Nielsen, H.T.B. Drysdale, and N.W.P. Munro, eds, *Science and Management of Protected Areas*. Wolfville, NS: SAMPAA, 485–98.

——— and W. Trotter. 2000. 'Public attitudes toward user fees in provincial forest lands', *Leisure/Loisir: Journal of the Canadian Association for Leisure Studies* 24: 139–59.

Rosenthal, D.H., D.A. Waldman, and B.L. Driver. 1982. 'Construct validity of instruments measuring recreationists' preferences', *Leisure Studies* 5: 89–108.

Schreyer, R., D. Lime, and D. Williams. 1984. 'Characterizing the influence of past experience on recreation behavior', *Journal of Leisure Research* 16: 34–50.

Scott, D., and C.S. Shafer. 2001. 'Recreation specialization: A critical look at the construct', *Journal of Leisure Research* 33: 319–43.

Shafer, E., Jr. 1969. 'The average camper who doesn't exist', USDA Forest Service Research Paper NE-142.

Shelby, B., N.S. Bregenzer, and R. Johnson. 1988. 'Displacement and product shift: Empirical evidence from Oregon rivers', *Journal of Leisure Research* 20: 274–88.

——— and T.A. Heberlein. 1986. *Carrying Capacity in Recreation Settings*. Corvallis: Oregon State University Press.

———, J.J. Vaske, and M.P. Donnelly. 1996. 'Norms, standards, and natural resources', *Leisure Sciences* 18: 103–23.

———, ———, and T.A. Heberlein. 1989. 'Comparative analysis of crowding in multiple locations: Results from fifteen years of research', *Leisure Sciences,* 11: 269–91.

Thapa, B., and A.R. Graefe. 2003. 'Level of skill and its relationship to recreation conflict and tolerance among adult skiers and snowboarders', *World Leisure* 45: 15–27.

Twynam, G.D., and D.W. Robinson. 1997. *A Market Segmentation Analysis of Desired Ecotourism Opportunities*. Sault Ste Marie, Ont.: Natural Resources Canada, Canadian Forest Service, Great Lakes Forestry Centre, NODA/NFP Technical Report TR-34.

Vaske, J.J., J. Beaman, R. Stanley, and M. Grenier. 1996a. 'Importance performance and segmentation: Where do we go from here?', *Journal of Travel and Tourism Marketing* 5: 225–40.

———, P. Carothers, M.P. Donnelly, and B. Baird. 2000. 'Recreation conflict among skiers and snowboarders', *Leisure Sciences* 22: 297–313.

——— and M.P. Donnelly. 1999. 'A value-attitude-behavior model predicting wildland voting intentions', *Society and Natural Resources* 12: 523–37.

——— and ———. 2002. 'Generalizing the encounter-norm-crowding relationship', *Leisure Sciences* 24: 255–70.

———, ———, W.A. Freimund, and T. Miller. 1996b. 'The 1995 Gwaii Haanas visitor survey', HDNRU Report No. 26. Fort Collins, Colo.: Colorado State University.

———, ———, and J.P. Petruzzi. 1996c. 'Country of origin, encounter norms, and crowding in a frontcountry setting', *Leisure Sciences* 18: 161–76.

———, ———, K. Wittmann, and S. Laidlaw. 1995. 'Interpersonal versus social values conflict', *Leisure Sciences* 17: 205–22.

———, M.D. Needham, and R.C. Cline Jr. 2007. 'Clarifying interpersonal and social values conflict among recreationists', *Journal of Leisure Research* 39: 182–95.

——— and D. Whittaker. 2004. 'Normative approaches to natural resources', in Manfredo et al. (2004b: 283–94).

Williams, D.R., and J.J. Vaske. 2003. 'The measurement of place attachment: Validity and generalizability of a psychometric approach', *Forest Science* 49: 830–40.

KEY WORDS/CONCEPTS

attitudes
behaviour
behavioural approach
beliefs
benefits
benefits-based management
carrying capacity
cognitive dissonance
community impacts
constraints
crowding
crystallization
displacement
encounters
environmental impacts
experience-based management
importance-performance analysis
interpersonal conflict
intrinsic motivation
leisure behaviour
Limits of Acceptable Change (LAC)
locus of control
minimum acceptable condition
motivations
multiple satisfactions
nature-based tourism
product shift
push/pull motivations
Recreation Opportunity Spectrum (ROS)
satisfaction
segmentation
social impacts
social interference
social science
social values conflict
specialization
structural norm approach
theory of reasoned action
value orientations
values
zoning

STUDY QUESTIONS

1. Describe the behavioural approach, using an activity familiar to you (e.g., skiing, mountain biking, scuba diving, fishing).
2. Discuss how the behavioural approach provides the conceptual underpinning of ROS.
3. Discuss why an understanding of visitor motivations is important for a park manager.
4. Why is a single overall or global measure of satisfaction problematic for informing park and recreation management?
5. What is importance-performance (I-P) analysis and how does this help inform park management?
6. Crowding is a frequently reported concern, yet it is difficult to determine how to manage to reduce crowding in parks. Discuss.
7. Explain the normative approach to the measurement of crowding-related indicators.
8. Compare and contrast one-way, two-way, in-group, out-group, social values, and interpersonal conflict.
9. Define, with examples, the different components of the cognitive hierarchy. How can this information be useful for park management?
10. Using an outdoor activity familiar to you, describe how specialization could be involved within this activity and how it could be measured. How might this specialization influence the selection of preferred setting characteristics?

CHAPTER 7

Visitor Planning and Management

Wolfgang Haider and Robert J. Payne

INTRODUCTION

The use of parks and protected areas for tourism and recreation is as old as the concept of protected areas itself. For example, when Banff National Park was created, the attraction of scenery and hot springs for rail-based tourism was one crucial criterion for its establishment. This chapter will look at the approaches that have emerged for visitor planning and management. It will become obvious that managing for use adds an additional layer of complexity to parks and protected areas management, as it introduces the necessity for social research (as outlined in Chapter 6) that needs to be integrated with the concepts of ecosystem management and adaptive management. Parks, as well as other protected areas, are in the unique position of protecting representative and significant natural areas of importance, and at the same time of offering visitors with opportunities to understand, appreciate, and enjoy natural and cultural heritage. Although the current Parks Canada Act (Parliament of Canada, 2000) states that ecological integrity is the 'first' priority of Parks Canada, the 'dual mandate' between protection and use remains. It is in this context that visitor planning and management is significant, and the concept of carrying capacity emerges as its most basic notion.

'Carrying capacity', which implies limiting use levels in parks in order to reduce visitor impacts and crowding, is an attractive idea. It possesses an intuitive appeal that has led many people to place it at the centre of theories and prescriptions focusing on the human use of natural environments. For environmental managers especially, it promises scientific justification for difficult decisions that inevitably involve competing human interests as well as incomplete knowledge of the natural environment. US researchers polled National Park Service managers who were responsible for backcountry management in national parks (Manning et al., 1996). Among the problems that managers reported was the deterioration of campsites and trails. The researchers (ibid., 144–5) identified three trends:

- Backcountry impacts were primarily related to recreational use.
- Negative visitor experiences and crowding were becoming issues.
- Carrying capacity was 'a pervasive but unresolved issue'.

Based on the concept of carrying capacity, several visitor management frameworks have been developed over the past 30 years and will be discussed in this chapter: the Recreation Opportunity Spectrum (ROS), Limits of Acceptable Change (LAC), Visitor Impact Management (VIM), Visitor Activity Management Process (VAMP), and Visitor Experience and Resource Protection (VERP). The approach used is to compare each of these management frameworks to the traditional carrying capacity model.

CARRYING CAPACITY AND PROTECTED AREAS

Traditionally, the literature on social carrying capacity is characterized by a concern with wilderness and backcountry recreation situations, a fact that reflects the management responsibilities of American protected areas agencies, especially the US Forest Service (Hendee et al., 1990). The changes in thinking about carrying capacity in wilderness areas mirror changes in thinking about wilderness itself. Implicit in this thinking is that wilderness can be thought of in terms of the visitor experience, or in terms of the naturalness of the setting. With the passage of the American Wilderness Act in 1964, managers in the US were required to define 'wilderness' in meaningful terms for management. Social scientists pointed out that people's perceptions and understandings of wilderness differed and that defining wilderness ought to take such differences into account (Stankey, 1973; Stankey and McCool, 1984). A good deal of research was conducted into crowding, revealing that a number of demographic, social, and economic variables influenced whether people felt crowded or in conflict with other users in recreational settings (see Chapter 6). It became clear that it was ineffective to base management actions on definitions that were so changeable. Something more firmly based in science was required. The following definition appears in the US Wilderness Act (s. 2[c]):

> A wilderness, in contrast with those areas where man and his own works dominate the landscape, is hereby recognized as an area where the earth and its community of life are untrammeled by man, where man himself is a visitor who does not remain. An area of wilderness is further defined to mean in this Act an area of undeveloped Federal land retaining its primeval character and influence, without permanent improvements or human habitation, which is protected and managed so as to preserve its natural conditions and which (1) generally appears to have been affected primarily by the forces of nature, with the imprint of man's work substantially unnoticeable; (2) has outstanding opportunities for solitude or a primitive and unconfined type of recreation; (3) has at least five thousand acres of land or is of sufficient size as to make practicable its preservation and use in an unimpaired condition; and (4) may also contain ecological, geological, or other features of scientific, educational, scenic, or historical value.

Note that the addition of the size criterion gives an operational dimension to the definition that allows measurement and determination of what does not constitute wilderness as well as what might. The notion of carrying capacity emerged out of this early thinking about wilderness and how to manage wilderness and wilderness experiences.

Carrying capacity was introduced into the parks, recreation, and tourism literature by Wagar (1964). Three variations of carrying capacity might be applied in national park settings. One, *design (or physical) carrying capacity*, is an architectural/engineering adaptation that specifies particular levels of use in or for facilities. In national park settings, it might be used to manage roads in frontcountry areas. For example, if roads were designed for light automobile traffic but were being used by many more cars than expected or by many large recreational vehicles, it would be reasonable to say that the carrying capacity of those roads had been exceeded, with the consequent problems of maintenance and public safety. In another example, a visitor centre, the numbers of visitors at one time may be such that fire safety becomes an issue because the building was not designed with the number of exits necessary for the larger-than-expected number of visitors at one time.

In 'mid-country' or even backcountry areas, design carrying capacity may assist managers in meeting ecological integrity goals and satisfying visitor expectations. In Point Pelee National Park, for instance, a boardwalk gives many visitors access to areas of the marsh they would otherwise not see and does so in a manner that minimizes their effects on marsh ecology. In Pacific Rim National Park, boardwalks and ladders help visitors to experience the backcountry qualities of the West Coast Trail while serving to minimize the environmental impacts of their visits. Especially in frontcountry settings, design carrying capacity offers managers of parks and protected areas alternatives to actions that limit human use. Design carrying capacity also illustrates clearly the importance of the purpose of an area in determining a threshold beyond which degradation occurs or problems arise. Although design carrying capacity offers useful solutions, it also might predispose managers towards modifying the natural environment in favour of human use, regardless of the objectives declared for the setting. When design carrying capacity solutions are implemented in wilderness settings, they can facilitate the decline of the values they were implemented to protect—wilderness values.

A second variant, *ecological carrying capacity*, is more familiar. In national parks, ecological carrying capacity refers to the capability of the natural environment to withstand human use. Investigation of ecological carrying capacity can take several forms. One direction investigates the effects of visitors on the ecology of the visited area. The other direction evaluates proposals for new developments in a park or protected area. In this latter case, for example, Parks Canada is required to investigate whether the development will have environmental effects and, if so, must take steps to mitigate them. While environmental impact assessment may appear straightforward, there are thorny scientific and management issues that require attention.

A third form, *social carrying capacity*, has received a good deal of attention in the parks' literature and in management practice. Social carrying capacity acknowledges that visitors respond to the social, as well as the natural, environment in parks, as discussed in Chapter 6 (the 'behavioural model'). Social carrying capacity focuses on the experiences that visitors have in the park. For some visitors, too many other people, people who are different in their interests, or people who are too different in their behaviours constitute negative influences. Such negative influences disturb their own enjoyment. Managing social carrying capacity requires that managers not only appreciate the interactions of visitors in national parks, but also that they have ideas about what those interactions should be.

The following discussion will focus on ecological and social carrying capacity, and link these concepts to visitor management frameworks discussed later in the chapter.

The notion of facility carrying capacity is subsumed in the discussion of management frameworks.

Ecological Carrying Capacity

Determining ecological carrying capacity involves finding answers to three questions:

- What is the ecological impact of human use in a particular ecosystem?
- Does the impact change the character of the ecosystem?
- Is the change, if any, an acceptable change?

The first of these questions may be answered by scientific research, assuming that it is possible to develop a causal relationship between human use and ecological impact. The research field of recreation ecology has contributed plenty of insights to this question (Hammitt and Cole, 1998; Leung and Marion, 2000). For example, a Canadian study on trail impacts has been completed in British Columbia's Mt Robson Provincial Park (Nepal and Way, 2007).

The second question requires more data and information. What was the condition of the ecosystem before it was disturbed by human use? Was that previous condition a 'natural' state or was it also caused by some human interference? Research on this question goes beyond the recreation context, and Parks Canada is now undertaking several large projects on these issues of benchmarking. Is the cause of the disturbance something that national park managers may influence through their actions? Nelson (1968) raised several of these questions in relation to Banff National Park four decades ago. The recent Banff–Bow Valley Study (Page et al., 1996) has served only to sharpen them.

The third question requires something completely different if it is to be answered. It is not a scientific question; rather, it is a 'values' question (Becker et al., 1984). An 'acceptable change' implies that national park managers have established a range of desired conditions and can then judge the change, determining if it is 'good' or 'bad'. While Parks Canada has produced management plans for national parks for a number of years, it is evident from the Banff–Bow Valley experience (Page et al., 1996) that Parks Canada has not been prepared until recently to judge whether the changes to park ecosystems have been desirable or undesirable. Consideration of values inevitably introduces the need for objectives in the planning and management process.

Kachi and Walker (1999) reviewed work being conducted in Canadian national parks on human use management. Their results indicate that a focus on ecological carrying capacity is common. For example, in Gwaii Haanas National Park Reserve and Haida Heritage Site, staff examined visitor impacts on backcountry campsites in a subjective manner, determining that 52 of 75 sites required management intervention. A part of that intervention includes setting standards, which, when not met, demand remedial actions by managers. These findings and subsequent standards form the backbone of the *Gwaii Haanas Backcountry Management Plan* (Parks Canada, 1999a).

Similarly, managers at Kejimkujik National Park in Nova Scotia were concerned about the ecological impacts of backcountry visitation and camping. They responded in a unique manner by constructing tent pads, fire boxes, and pit privies on each backcountry campsite. In addition, they ensured that a supply of firewood was present at each site. Managers hoped that, by taking these actions, impacts would be limited to campsites and trails. Kingston, Carr, and Payne (2002) investigated whether these management efforts had been effective in terms of vegetation and found that effects were limited to areas immediately adjacent to trails and campsites, regardless of the levels of use. However, levels of use and disturbed campsite areas were found to be closely related. The authors concluded that the park managers' actions to concentrate impacts seem to have been effective.

Social Carrying Capacity

Compared to outdoor recreation research on ecological carrying capacity, social carrying capacity research has received much more attention. Like the ecological variant, it concerns itself with three questions:

- What is the social impact of human uses of a particular ecosystem?
- Does that impact change the nature of recreation experiences available in that ecosystem?
- Is the change, if any, an acceptable one?

Social carrying capacity focuses on the manifold relationships among users of a park or protected area. That these users may be having an ecological impact is not an issue for social carrying capacity. The possibility, however, that the numbers or the behaviours of some visitors may affect the recreational experience of other visitors is at the heart of social carrying capacity.

One important contribution of social carrying capacity in the operationalization of this definition is the elicitation of 'encounter norms' (Vaske et al., 1986) through visitor surveys (see Chapter 6). These norms serve as standards by which managers might evaluate and monitor crowding in backcountry or wilderness settings. More recently, a need has been recognized to address social carrying capacity in frontcountry settings, especially in those occupied by day users. Some such studies (e.g., Heberlein et al., 1986) have encountered a significant barrier, namely, the considerable diversity in motivations, expectations, and knowledge among people in frontcountry settings. Consequently, attempts to set standards for conditions of visitor experience have been daunting. Recent research in frontcountry and urban recreation settings has shown that carrying capacity issues in these environments are much more complex. Minimum and maximum standards of crowding may exist, and the evaluation by survey respondents is also influenced by other situational variables, such as activity type, size of groups, and dog management (Arnberger and Haider, 2007).

Social carrying capacity has been widely connected to recreational conflict—that is, when two groups of recreationists are so different in themselves, in their activities, and/or in their attachments to the natural environment, one group is dissatisfied or even displaced (Chapter 6). Duffus and Dearden (1990, 1993), for example, suggest that as visitor numbers grow for whale-watching activities at Robson Bight Ecological Reserve on Vancouver Island, the nature of the visitors changes due to social carrying capacity issues. Initially, when visitor numbers are low a high proportion of specialized whale-watchers are less concerned with facilities and services. As numbers increase, the proportion of generalists, those demanding higher levels of services and facilities, increases, changing the nature of the experience. Consequently, the more specialized visitors are displaced as their social carrying capacity is reached. Jackson et al. (2003) describe the conflict between motorized (snowmobilers) and non-motorized (cross-country skiers) winter users on the Chilkoot Trail National Historic Site. An understanding of their different motivations and expectations has led to a 'Winter Use Recreation Strategy', which limits use to non-motorized activities only on certain weekends.

Reflecting on Carrying Capacity

Carrying capacity is both a venerable and a popular idea with the public and protected area managers alike, but does carrying capacity have a role in national park management? Some writers, including Wagar (1974), have suggested that perhaps the focus should be on determining acceptable conditions in the environment (i.e., a qualitative approach) rather than the level of stress (i.e., a quantitative approach) being applied to the environment (Washburne, 1982; Vaske, 1994). Such a shift of focus would change carrying capacity, but might also render it more useful to park managers. Others are not in agreement, continuing to advocate a 'traditional' carrying capacity approach to managing human use of natural environments for recreation and tourism (e.g., Butler et al., 1993; Heberlein et al., 1986).

Shelby and Heberlein (1984) long ago suggested an approach to carrying capacity that still is worthy of attention. They defined carrying capacity as 'the level of use beyond which impacts exceed acceptable levels specified by evaluative standards' (ibid., 441). Their approach required managers to take three actions:

- Specify *management parameters* (i.e., something that managers could manipulate) for an area. An example would be the number of people allowed to use a park campground.
- Specify *impact parameters* relevant to an area. Impact parameters are the consequences related to the management parameter. For example, an environmental impact parameter could be the loss of vegetation at a park campground (presumably related to the number of people who visit the campsite). A social impact parameter could be the number of contacts that occur between parties in the campground.
- Determine *evaluative standards.* For example, in the campground situation outlined above, how much change in vegetation is acceptable, or how many contacts with other groups are acceptable. These assessments of acceptable change are grounded in values, and may vary between different visitors or interest groups.

Shelby and Heberlein maintained that only within this context is it reasonable to employ the idea of carrying capacity. Note that they did not differentiate between ecological and social carrying capacity. Rather, they felt the approach was as useful for one as the other. Moreover, they offered two other insights to this issue: that carrying capacity is tied to particular sites or landscapes; and that social and ecological expressions of carrying capacity may be closely related.

Elsewhere, the recognition of the multi-purpose nature of park management has led the US National Park Service to define carrying capacity as 'the type and level of visitor use that can be accommodated while sustaining the desired resource and social conditions that complement the purpose of a park unit and its management objectives' (US National Park Service, 1997: 30).

Carrying capacity is often regarded as consisting of three dimensions—the resource component, the managerial component, and the experiential component (Manning, 2007). The latter aspect, especially understanding desired experiences, has dominated much of the social science recreation research (Chapter 6). Management objectives are broad narrative statements. At one extreme, they may be defined in a top-down manner, mostly within the administrative setting of an agency; at the other, a bottom-up approach involves wide public and stakeholder participation.

Indicators are more specific, measurable variables reflecting the meaning of the objectives. Examples of indicators are: number of fire rings at a campsite; number of horse groups encountered; number of orca whales observed. *Standards* define what conditions are thought to be acceptable. An example of a standard is a '90 per cent chance of seeing no more than three horse parties in a day on a given trail'. Table 7.1 lists the characteristics of good indicators and standards. The application of these carrying capacity components is essential in most visitor management frameworks to be discussed below.

TABLE 7.1 Characteristics of Good Indicators and Standards

Good indicators are:

- *Specific*: they need to be more precise than saying, for example, that the trails in a particular park are 'challenging'.
- *Objective*: they should be measurable.
- *Reliable and repeatable*: measurement yields similar results under similar conditions.
- *Related to visitor use*: they should relate to some aspect of visitor use, such as location or number of users.
- *Sensitive*: they should be sensitive to visitor use over a short period of time (if an indicator changes only after impacts are substantial, it will not serve as an early warning mechanism, allowing managers to react in a timely manner).
- *Manageable*: they should be responsive to management action.
- *Efficient and effective to measure*: they should be relatively easy and cost-effective to measure.
- *Integrative or synthetic*: they should be measures for more than one component of parks and protected areas.
- *Significant*: they help to define the quality of park resources and the visitor experience.

Good standards are:

- *Quantitative and specific*: they should make reference, for example, to 'three encounters' rather than 'a small number of encounters'.
- *Time- or space-bounded*: they should provide a unit of measurement for time (e.g., hour, day) and refer to a specific location (e.g., campground, trailhead, along trail).
- *Expressed as a probability*: thus, a standard might state that 'no more than three encounters will be experienced 90 per cent of the time'.
- *Impact-oriented*: they should focus directly on the impacts that affect the quality of park resources and the visitor experience rather than on the management action used to keep impacts from violating the standards.
- *Realistic*: they should reflect conditions that are realistically attainable.

Source: Adapted from Manning (2007: 28–33).

The link between carrying capacity and management purpose has seldom been more overtly stated than in the US National Park Service definition cited above. Such a definition pays attention to management purposes, facilitating results-based management and accountability. Moreover, the definition hints at responses to carrying capacity issues as being more than limiting human use (Cole et al., 1987). Management responses may be 'direct' or 'indirect' (Hendee et al., 1990): direct responses refer to techniques aimed at regulating behaviour; indirect responses involve techniques aimed at influencing behaviour (Table 7.2). Direct approaches include zoning of incompatible uses, providing increased enforcement of regulations, limiting use levels, and restricting certain types of activities. Indirect approaches include hardening of campsites, providing information to visitors (i.e., educating visitors about appropriate behaviour), and charging fees.

Examples of the application of these management strategies in national parks are illustrated in Table 7.3 below.

TABLE 7.2 Indirect and Direct Techniques for Managing Visitor Use in Parks and Protected Areas

Indirect Management Techniques

Emphasize influencing or modifying behaviour. Individual retains freedom to choose. Control less complete, more variation in use possible.

1. Physical Alterations
   - Improve, maintain, or neglect access roads
   - Improve, maintain, or neglect campsites
   - Make trails more or less difficult
   - Build trails or leave areas without trails
   - Improve fish or wildlife populations or take no action (stock, or allow depletion or elimination)
2. Information Dispersal
   - Advertise attributes of wilderness
   - Advertise recreation opportunities in nearby areas, outside of the park
   - Educate users to basic concepts of ecology and care of ecosystems
   - Advertise underused areas and general patterns of use
3. Eligibility Requirements
   - Charge constant entrance fee
   - Charge differential fees by trail zones, season, etc.
   - Require proof of camping and ecological knowledge and/or skills

Direct Management Techniques

Emphasis on regulation of behaviour. Individual choice restricted. High degree of control.

1. Increased Enforcement
   - Impose fines
   - Increase surveillance of area
2. Zoning
   - Separate incompatible uses (e.g., separate zones for horse use and hiker use)
   - Prohibit use at times of high damage potential (e.g., no horse use in high meadows until soil moisture declines)
   - Limit length of stay at some campsites

TABLE 7.2 continued

3. Rationing Use Intensity
   - Rotate use (open or close access points, trails, campsites)
   - Require reservations
   - Assign campsites and/or travel routes to each camper group
   - Limit use via access point
   - Limit size of groups, number of horses, number of canoes, etc.
   - Limit camping to designated campsites only
   - Limit length of stay in area (maximum or minimum)
4. Restrictions on Activities
   - Restrict building campfires
   - Restrict horse use, hunting, or fishing
   - Restrict collecting flowers, shells, etc.

Source: Adapted from Hendee et al. (1990).

TABLE 7.3 Examples of Indirect and Direct Management Strategies

Indirect Techniques

Physical Alterations

1. *Improve or neglect access.* At Point Pelee National Park, a 1.25-km stretch of East Road was removed so that it can revert back to a natural state, resulting in an increase in natural habitats, reconnection of habitats, and a decrease in road kills. Elsewhere in the park, the East Road was improved by widening and resurfacing to accommodate two-way traffic.
2. *Improve or neglect campsites.* At Kejimkujik National Park each of the backcountry campsites is equipped with a fireplace, picnic table, outhouse toilet, and gravel tent pad. Camping is only permitted at these designated campsites.

Information Dispersal

1. *Advertise recreation opportunities in nearby areas.* At La Mauricie National Park, visitors are encouraged to visit Les Forges du Saint Mauricie National Historic Site. At Point Pelee National Park, visitors are informed of other birding opportunities, at sites such as Hillman Marsh Conservation Area, Wheatley Provincial Park, and Kopegaron Woods Conservation Area.
2. *Provide minimal-impact education.* Mingan Archipelago National Park Reserve has initiated an awareness program on the impact of interfering with seabirds and navigation ethics in the Mingan Archipelago.

Eligibility Requirements

1. *Charge constant entrance fees.* Mount Revelstoke National Park charges a $4 entry fee. Pacific Rim National Park Reserve charges a fee of $70 for hiking the West Coast Trail and an additional $25 for ferry services across two rivers.
2. *Charge differential fees.* Kouchibouguac National Park charges $13 for unserviced sites from 16 May to 27 June and 2 September to 13 October. The fee at these same sites from 28 June to 1 September is $16.25.

(continued)

TABLE 7.3 continued

Direct Techniques

Increased Enforcement

1. *Impose fines.* In any national park, a visitor who is convicted of hunting is subject to fines of up to $150,000, six months' imprisonment, or both.
2. *Increase surveillance.* The Gwaii Haanas Watchman program was developed out of concerns for protecting old Haida village sites from vandalism and other damage. Watchmen are posted at several sites to make visitors aware of the significance of the sites and how to visit the sites without leaving a trace.

Zoning

1. *Separate incompatible uses.* In Jasper National Park, horseback riders are not permitted at picnic sites, campgrounds, or car-accessible campgrounds.
2. *Temporal zoning.* At Bruce Peninsula National Park, motorboats are prohibited on Cyprus Lake from 15 June to 15 September, but canoes and sailboards may be used anytime.

Rationing Use Intensity

1. *Rotate use.* Quetico Provincial Park varies the number of backcountry permits provided at each access point.
2. *Require reservations.* Of the 50 backcountry permits issued daily for the Chilkoot Trail, 42 can only be obtained with a reservation. Similarly, the West Coast Trail has a daily quota of 52 backcountry permits, of which 40 are available by reservation.
3. *Assign campsites.* In Jasper National Park, backcountry camping is only allowed in designated campsites. The same approach is used in Kejimkujik National Park.
4. *Limit size of groups.* At Pacific Rim National Park Reserve, no more than 52 people per day are allowed to enter the West Coast Trail. In addition, no single group can have more than nine members.
5. *Limit length of stay.* Bruce Peninsula National Park has imposed a maximum length of stay of 14 days at campgrounds.

Restrictions on Activities

1. *Restrict type of use.* A fishing ban is in effect for all streams and rivers in Glacier National Park.
2. *Restrict camping practices.* On the Chilkoot Trail, visitors must pack out everything they pack in, make use of the grey water pits that are provided, and use camping stoves rather than campfires.
3. *Restrict collecting.* In the Broken Groups Islands unit of Pacific Rim National Park, visitors are not allowed to collect shells from the beaches.

CARRYING CAPACITY AND VISITOR MANAGEMENT FRAMEWORKS

Carrying capacity is clearly a well-established concept in the field of recreation, parks, and tourism management. Documented difficulties in its implementation, however, have led researchers and managers to look beyond carrying capacity (Stankey and McCool, 1989) in search of appropriate management methodologies. In response, several recreation and visitor management frameworks have been developed and applied consistently on public lands in North America, including its parks and protected areas systems.

Table 7.4 outlines the connections between the five visitor management frameworks (VMFs) and carrying capacity. Drawing on the published work of Payne and Graham

TABLE 7.4 Comparison of Visitor Management Frameworks

Visitor Management Framework	Related to Carrying Capacity?	Scope	Scale	Applications
Recreation Opportunity Spectrum (ROS)	Yes; social carrying capacity	Social	Landscape	Four mountain parks (Yoho, Kootenay, Banff, Jasper), Pukaskwa National Park
Visitor Activity Management Process (VAMP)	No	Design and social	Landscape and site	Various national parks; heritage canals; visitor risk management, appropriate activity assessment
Visitor Impact Management (VIM)	Yes	Social and ecological	Landscape and site	Columbia Icefields, Jasper National Park
Limits of Acceptable Change (LAC)	Yes; through its connection to ROS	Social and ecological	Landscape	Yoho National Park, Chilkoot Trail National Historic Park
Visitor Experience and Resource Protection (VERP)	Yes	Social and ecological	Landscape and site	None

Sources: Payne and Graham (1993); Nilsen and Tayler (1998).

(1993) and Nilsen and Tayler (1998), the intentions in this section are: (1) to outline five common visitor management frameworks; (2) to sketch their connections with carrying capacity; and (3) to consider their utility in park and protected area management.

All these frameworks, except the ROS, consist of a series of eight or nine steps, which are to be followed in their implementation. These steps operationalize the aforementioned components of carrying capacity, such as the setting of objectives, as well as the selection of standards and indicators. They differ in their emphasis on public participation, and in other details of implementation, which reflect the circumstances of their respective original implementation and the targeted management agency. The Recreation Opportunity Spectrum (ROS) was developed for recreation management on the vast holdings of the US Forest Service; the Limits of Acceptable Change (LAC) framework was conceived by the US Forest Service for the management of wilderness areas; the Visitor Impact Management (VIM) framework was developed for the US Parks Service, as was the Visitor Experience and Resource Protection (VERP) framework.

The Recreation Opportunity Spectrum Framework

The Recreation Opportunity Spectrum (ROS) (Clark and Stankey, 1979, 1990) was developed out of a concern for social carrying capacity (Driver et al., 1987). It does not, however, address ecological carrying capacity. Using standards that describe recreation 'settings' or areas, the ROS framework systematically divides a landscape along a continuum or spectrum of recreation opportunities ranging from primitive (wilderness) through to urban (developed). The supply of opportunities identified in this way compares visitors' demands for opportunities, allowing managers to match supply and demand where possible and permissible under legislation and policy. Often, however, ROS is only employed to describe the supply of opportunities and to help set associated management objectives. For example, the BC Ministry of Forests uses the ROS as a recreation inventory tool on all public forests of the province, but does little beyond this to actively manage visitor experiences (BC Ministry of Forests, 1998).

The ROS is based on the idea that people participate in recreational activities in specific settings to achieve desired experiences and benefits (see the discussion of the 'behavioural model' in Chapter 6). People vary in their preferences for setting characteristics, so the emphasis in ROS is to provide a variety of setting types, as defined by the types of factors listed in Figure 7.1. Hence, human modification, access, user interaction, and management regime lie at the heart of ROS. Variation among these factors determines the nature of the setting in which the activities occur. By manipulating these factors—for example, by providing trail access to an area previously without a trail—settings can be managed to produce the desired results.

The ROS is most effective at a landscape (rather than site) level, where its technique of applying social carrying capacity is especially effective in wilderness and backcountry environments. The ROS is also a systematic land planning and management framework, requiring that planners and managers who use it accept its underlying powerful rationale derived from studies of visitor behaviour and preferences. This management planning orientation is in direct contrast to the more problem-oriented approach adopted by the other frameworks reviewed here.

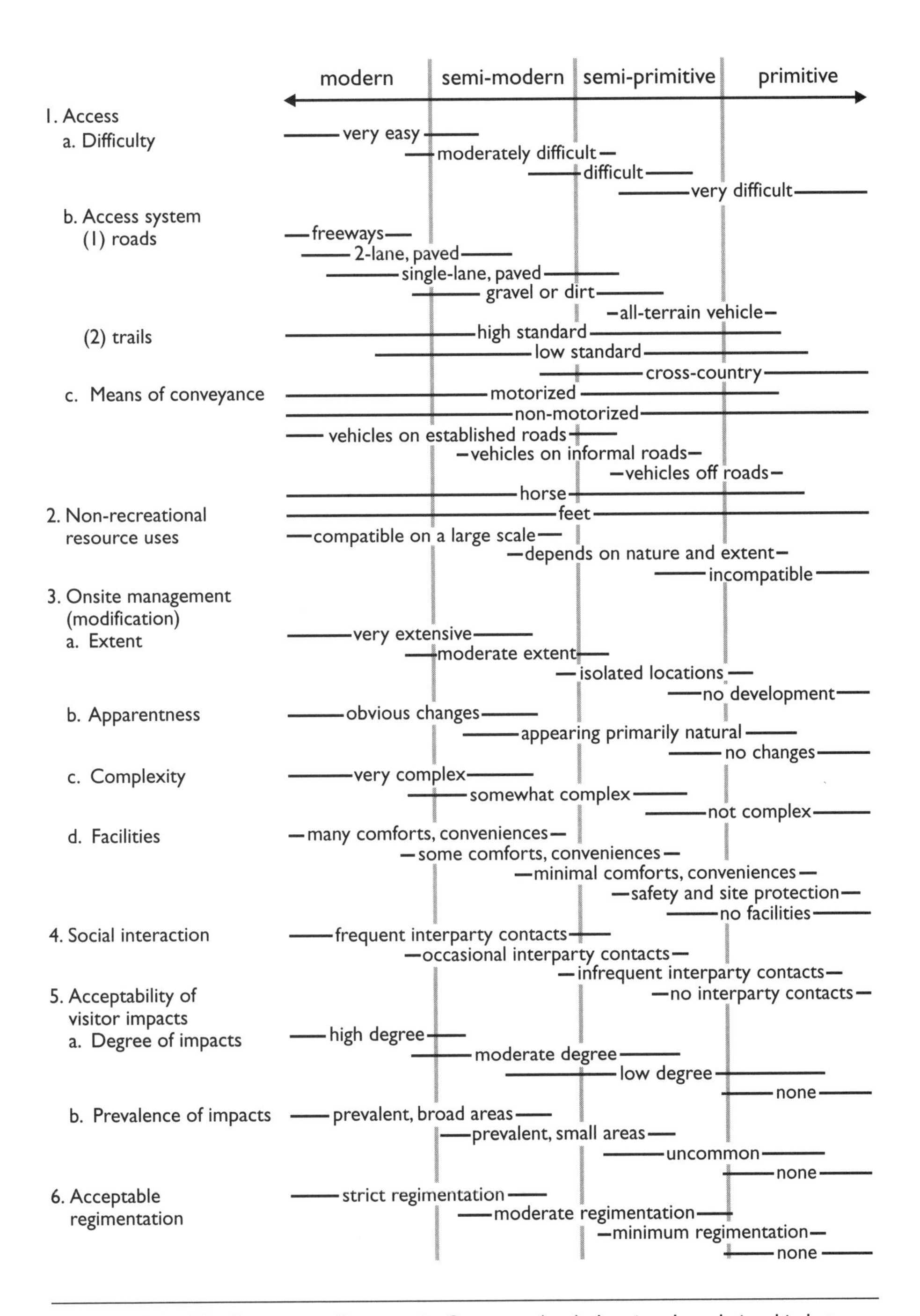

FIGURE 7.1 The Recreation Opportunity Spectrum (ROS) showing the relationship between the range of opportunity setting classes and management factors. *Source*: Clark and Stankey (1979).

BOX 7.1 Applying the Recreation Opportunity Spectrum in Yoho National Park

Parks Canada uses a zoning system that reflects resource sensitivity as well as recreation capability (see Chapter 9). However, in some parks the national park zoning system has been supplemented with a zoning system derived from the Recreation Opportunity Spectrum (ROS). Figure 7.2 illustrates how ROS has been applied in Yoho National Park.

This version of ROS consists of four recreation zones. The 'frontcountry' zone includes the Trans-Canada Highway, paved access routes into the park, and the town of Field, containing stores, restaurants, hotels, and a park visitor centre. Social interaction here is very high.

The 'semi-primitive' zone contains significant structures, including several warden's cabins and commercial roofed accommodations (e.g., Lake O'Hara Resort), a tea house at Twin Falls, and a number of trails, backcountry campgrounds, and trail shelters. Social interaction with other groups is relatively high, but not as high as in the frontcountry zone.

The 'primitive' zones have fewer facilities, consisting mainly of a few trails and backcountry campgrounds and one warden cabin. This zone has fewer services, facilities, and visitor numbers compared to the semi-primitive or frontcountry zones.

The 'wildland' zone provides the most pristine recreation experience in the park, with few trails, no established backcountry campgrounds, and three remote warden's cabins. Opportunities for solitude are highest in this zone.

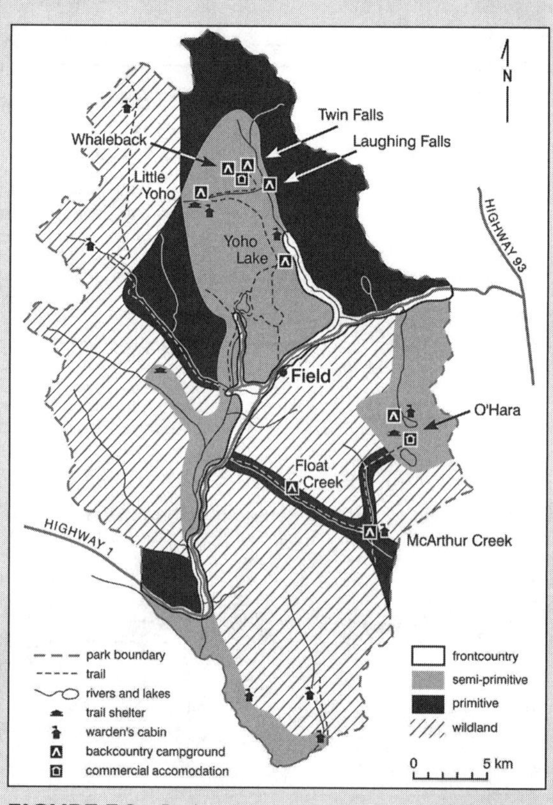

FIGURE 7.2 Backcountry opportunities at Yoho.

There are few explicit applications of the ROS in Canadian national parks. There is evidence of its use in backcountry areas of the four mountain parks (Canadian Parks Service, 1986); it was also employed more recently in assessing visitor opportunities in

Yoho and Pukaskwa national parks (Payne et al., 1997). Pacific Rim National Park used the ROS concept to develop a zoning system including semi-primitive, primitive, and wildland zones for the West Coast Trail (Parks Canada, 1994a).

It could be argued that the zoning system developed for national parks resembles ROS in spirit (see Chapter 9), albeit these zones do not follow the criterion of distance from road as strictly. The five zones prescribed by the Canada National Parks Act for the entire system of parks (Parks Canada, 1994b) are: Zone 1 (special protection); Zone 2 (wilderness); Zone 3 (natural environment); Zone 4 (outdoor recreation); and Zone 5 (park services). Clearly, these zones serve an important function in the planning and management for ecological integrity, especially in the five-year parks management plans, but they provide a range of visitor opportunities at the same time. The greatest visitor pressures occur in Zone 1 and Zone 2. ROS focuses primarily on visitor experiences, whereas the Parks Canada zoning system attempts to reflect both ecological integrity and visitor experiences. In our opinion, the Parks Canada zoning does not adequately address visitor experiences. For example, in Zone 2 (wilderness), both hiking and horse-back use might be acceptable activities, but visitor conflict may occur between these two activities. Applying ROS principles could lead to the designation of two ROS areas within a wilderness zone: one supporting horseback use, the other supporting hiking.

The Limits of Acceptable Change Framework

The Limits of Acceptable Change (LAC) framework was developed by the US Forest Service to complement the ROS (Stankey et al., 1985, McCool, 1990). Essential steps in the LAC process (Figure 7.3) are:

Step 1. Identify area concerns and issues.
Step 2. Define and describe management objectives.
Step 3. Select indicators of resource and social conditions.
Step 4. Inventory resource and social conditions.
Step 5. Specify standards for resources and social conditions.
Step 6. Identify alternatives.
Step 7. Identify management actions for each alternative.
Step 8. Evaluate and select an alternative.
Step 9. Implement actions and monitor conditions.

The LAC was developed as a complement and elaboration to the ROS. Like ROS, it is concerned with identifying opportunities for a variety of recreation experiences through the provision of a variety of settings. LAC goes beyond ROS by specifying indicators and standards for resource conditions and social conditions separately for each type of setting or zone. The requirement for monitoring commits the managing agency to specify a management response when monitoring suggests an 'unacceptable change'. Significant in the LAC process is the inclusion of 'stakeholders'—those individuals and organizations with an interest in the management of the area. These stakeholders work with the management team to determine the range of settings, indicators, standards, and management responses. As such, the LAC has emerged as the most generic of all

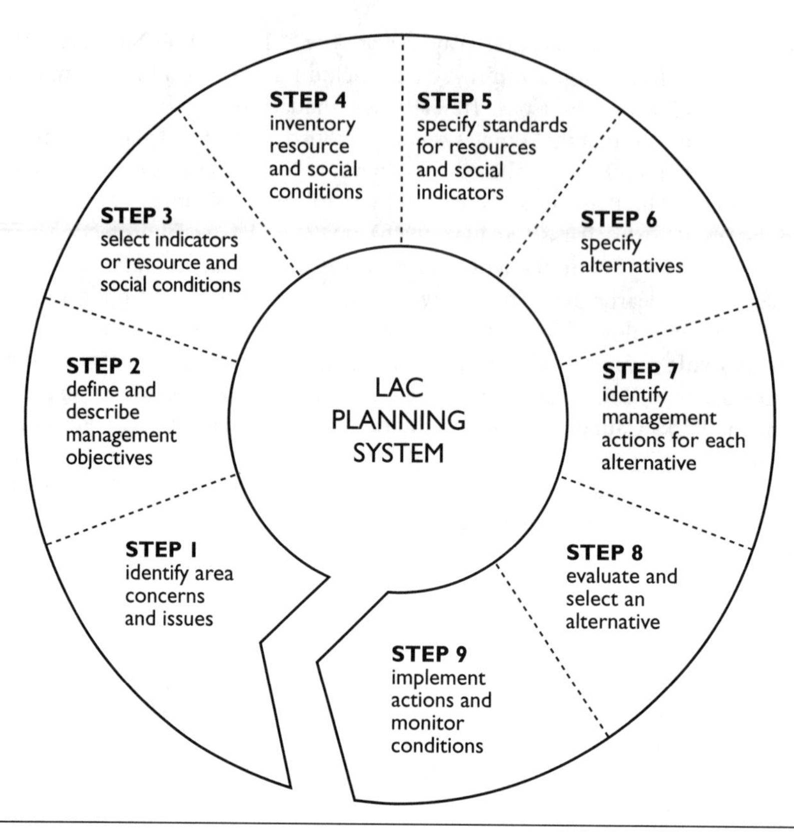

FIGURE 7.3 The Limits of Acceptable Change planning system. *Source*: Hendee et al. (1990). With permission from *Wilderness Management*, 2nd edn, by John C. Hendee, © 1990, Fulcrum Publishing, Golden, CO USA. All rights reserved. www.fulcrumbooks.com.

visitor management frameworks. It has been applied in hundreds of studies worldwide, and the concept has been adapted successfully to the modern management environment of GIS and the Internet. A thorough and modern application of LAC is ongoing in the Boone National Forest, which started in 2004 and still awaits completion (USDA Forest Service, 2007).

Like the ROS, LAC is most appropriately used at the landscape scale. LAC, however, differs from ROS in that it is problem-oriented. This orientation is well illustrated by an LAC application in Swan Lake Wilderness Area (Box 7.2) and in an application at the Chilkoot Trail National Historic Site (Elliot, 1994).

The Visitor Impact Management Framework

The Visitor Impact Management (VIM) framework was developed by researchers in concert with the National Parks and Conservation Association, an American non-government organization specializing in park issues (Graefe, 1990).

BOX 7.2 Swan Lake Wilderness Area

Located in northern British Columbia, the Swan Lake Wilderness Area has a management plan that incorporates many features of the LAC planning framework. The area is managed to provide four types of visitor experience, expressed in different zones:

Zone 1: **Primitive.** This is an undisturbed and unmodified natural environment, without trails, providing opportunities for solitude and isolation.

Zone 2: **Semi-primitive non-motorized I.** This zone is characterized as an essentially unmodified natural environment, moderately affected by the actions of users, with moderate opportunities for isolation and solitude.

Zone 3: **Semi-primitive non-motorized II.** This area is predominately natural appearing, but some locations may be substantially affected by human use (portage trail). Prolonged solitude and isolation may not be possible.

Zone 4: **Roaded resource land—natural environment.** This zone is a transportation corridor through the park, including the existing access road.

Examples of indicators, standards, and management actions are listed in Table 7.5. The social indicator is the likelihood of encountering others, per day. The environmental indicator is measured with a 'campsite impact index', derived from a rating of several conditions, including vegetation loss, tree damage, and root exposure.

TABLE 7.5 Swan Lake Wilderness Area Indicators, Standards, and Management Actions

Standards by Management Zone			
Zone 1	Zone 2	Zone 3	Zone 4
Social Indicator: Encounters Per Day			
75% chance of zero encounters per day	75% chance of 2 or fewer encounters per day	75% chance of 3 or fewer encounters per day	Not applicable
Environmental Indicator: Campsite Index			
Index no more than 30	Index no more than 49	No more than 50% of sites with index between 50–60	Not applicable

Note: The Campsite Condition Index is derived from a rating of nine campsite indicators (vegetation loss, bare soil increase, tree damage, root exposure, development, cleanliness, camp area, barren core camp area, social trails), which get weighted and summed. This particular index ranges from 20 (least impact) to 60 (most impact), and has been divided into three condition classes: minimally, moderately, and highly impacted.

If the social indicator for any zone is exceeded, the management actions include:

- Inform and educate users on behaviour to minimize obtrusiveness.
- Discourage use of crowded areas.
- Limit group size.

If the environmental indicator for any zone is exceeded, the management actions include:

- Inform and educate users on minimum-impact camping techniques.
- Discourage damaging actions, such as wood fires.
- Prohibit camping in certain areas.

Source: BC Ministry of Forests (1996). 'Swan Lake Wilderness Area'. Copyright © Province of British Columbia. All rights reserved. Reprinted with permission of the Province of British Columbia. www.ipp.gov.bc.ca.

As its name suggests, VIM exemplifies a concern for managing visitor impacts on the natural environment in parks. Consequently, the VIM framework displays a strong connection to carrying capacity and, especially, to ecological carrying capacity. However, VIM also attempts to deal with social carrying capacity (e.g., Vaske, 1994), a feature that

FIGURE 7.4 The Columbia Icefields, Jasper National Park. Management should try to accommodate different types of recreation and recreationists in national parks, but are there limits? *Photo: P. Dearden.*

BOX 7.3 Applying VIM in the Columbia Icefields

The application of VIM to the Columbia Icefields (Vaske, 1994) was intended to assist in the development of area plans for the Columbia Icefields, the most heavily visited day use in Jasper National Park. The Icefields are located adjacent to the heavily travelled Icefields Parkway, which provides access between Banff and Jasper. At the time of the study there was a Parks Canada information centre and commercial snowcoach tour operation to the Athabasca Glacier operated by the Brewster Corporation. The study provided input to a major redevelopment of the area, including the construction of a new Icefields Centre that contains snowcoach staging facilities and a Parks Canada visitor centre.

All seven steps of the VIM framework were applied. Ecological and facility impacts were not the limiting capacity indicators, especially in light of the new facility. Of greater concern was the social carrying capacity, particularly of snowcoach users. Specific management objectives concerning the number of snowcoach users and their learning experiences were identified. Measurable indicators for management objectives, with a key indicator being perceived crowding, were specified. Standards for the crowding indicator were set. Comparing existing conditions to the crowding standard using survey data revealed that crowding was not a problem for the intensively managed glacier experience. However, crowding was a concern in the visitor facilities.

The application of the VIM framework at the Icefields demonstrated that, while the level of crowding on the glacier had not exceeded the determined management standard, crowding at the visitor facilities was a problem. This situation has since been addressed through the new facility development.

reinforces the tentative links between social and ecological carrying capacity. The revision of carrying capacity presented by Shelby and Heberlein (1984) is also prominent. VIM features requirements for managers to specify ecological standards for park areas, to determine effective ways to monitor conditions in those areas, to identify problems when standards are not achieved, and to act to restore or maintain desired conditions. In this, VIM echoes features of LAC and anticipates those of the Visitor Experience and Resource Protection (VERP) framework.

Eight steps comprise the Visitor Impact Management framework (Figure 7.5):

Step 1. Review existing databases.
Step 2. Review management objectives.
Step 3. Select key impact indicators.
Step 4. Select standards for key impact indicators.
Step 5. Compare standards and existing conditions.
Step 6. Identify probable causes of impacts.
Step 7. Identify appropriate management strategies.
Step 8. Implement the best strategy.

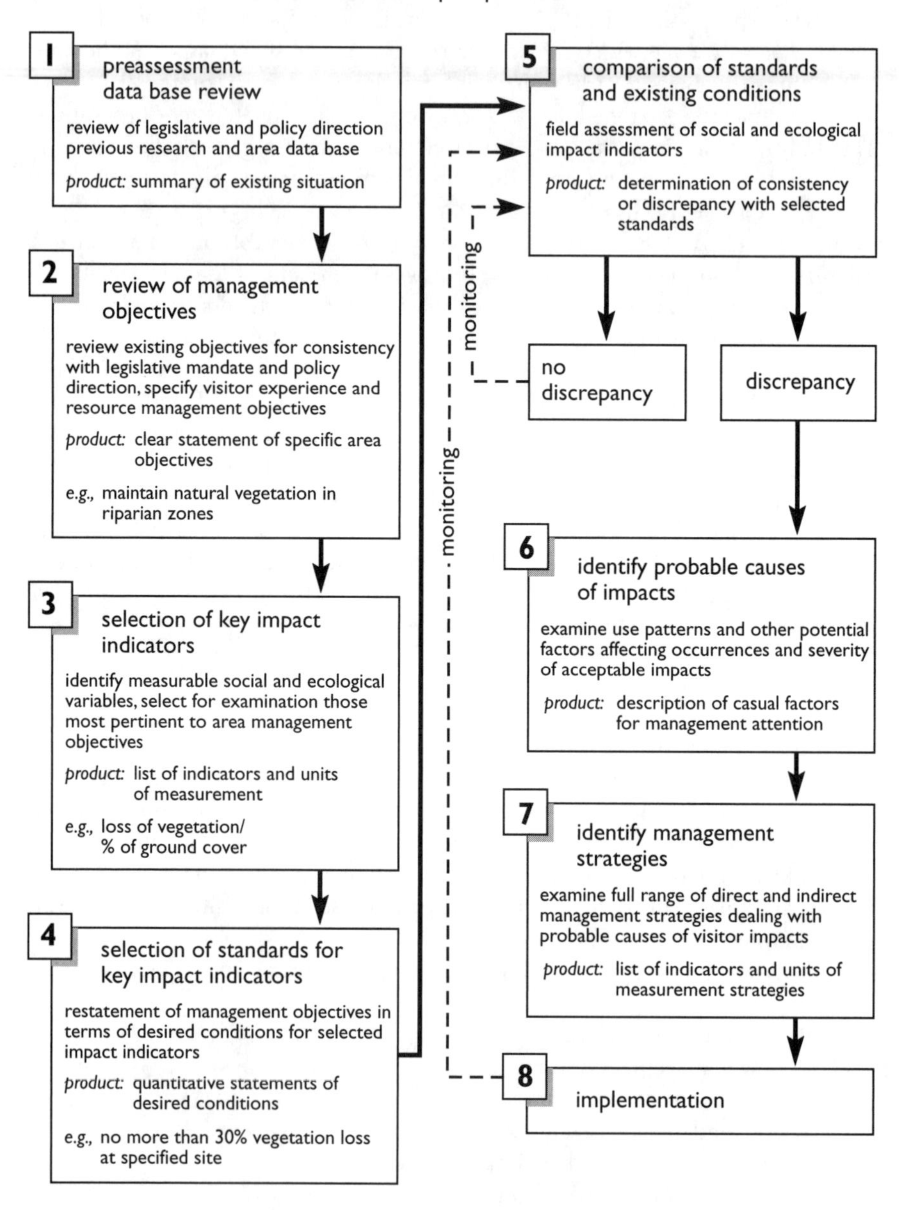

FIGURE 7.5 The Visitor Impact Management process. *Source*: Graefe (1990).

VIM is definitely a spatial framework and works best in site-specific situations rather than in landscapes, although it can be utilized with ROS or LAC at the landscape level. Applications in Canadian national parks are limited to Vaske's (1994) use of VIM at the Columbia Icefields in Jasper National Park (Box 7.3).

VIM appears to be an adaptation of the LAC towards the needs of the US Parks Service in the late 1980s, but was never applied extensively in the US. One main difference to the LAC seems to be the absence of an explicit participatory component. Its main strength is the acknowledgement of both ecological and social impacts associated with visitors to parks and protected areas.

The Visitor Experience and Resource Protection Framework

The Visitor Experience and Resource Protection (VERP) framework answers a need within the US National Park Service for an integrative, more participatory management process (Manning, 2007; Manning, 2001). The VERP framework builds on and combines several aspects of other management frameworks in attempting to integrate social and ecological carrying capacity issues with appropriate indicators and standards of quality. VERP contains nine components (Manning, 2001):

Step 1. Assemble a project team, consisting of planners, managers, and researchers.
Step 2. Develop a public involvement strategy.
Step 3. Develop clear statements of park purposes, significance, and primary interpretive themes. This step sets the stage for the rest of the process.
Step 4. Map and analyze the park's important resources and potential visitor experiences.
Step 5. Identify potential management zones that cover the range of desired resource and social conditions consistent with the park's purpose.
Step 6. Apply the potential management zones on the ground to identify a proposed plan and alternatives. The park's purpose, significant resources, and existing infrastructure are included at this stage of analysis.
Step 7. Select indicators of quality and associated standards for each zone. A monitoring plan is developed at this stage.
Step 8. Park staff compares desired conditions with existing conditions to address discrepancies (monitoring).
Step 9. Identify management strategies to address discrepancies (see Table 7.1). Strategies should favour indirect techniques where possible. Monitoring of conditions is ongoing.

VERP shares with LAC and VIM the requirement for park managers to specify social and ecological standards, to monitor conditions, to identify problems, and to find remedial solutions. VERP is an integrative tool and is clearly applicable at both landscape and site levels, features that make it an attractive instrument for park managers. Applying VERP, however, has proven to be difficult, primarily because managers find it taxing to identify both social and ecological conditions in park zones. While VERP has been applied at Isle Royale, Acadia, and Arches national parks in the US, no applications have yet occurred in Canadian national parks.

BOX 7.4 Implementing VERP in Acadia National Park

Jacobi and Manning (1999) applied the Visitor Experience Resource Protection (VERP) framework in Acadia National Park on Mount Desert Island near Bar Harbor, Maine, where roads originally established for horse and carriage had become the focus for a variety of visitor activities. The advent of the mountain bike exacerbated an already crowded situation and threatened to create recreational conflict. Complaints to park managers cited the crowding on the carriage roads and specified behaviours (e.g., high speed by mountain bikers when among walkers) that were rude and perhaps dangerous. The persistence of the complaints and the lack of action by park managers threatened to lower the quality of the recreational experience for many visitors.

Using the VERP framework, Jacobi and Manning confirmed that the problem was one of social rather than ecological carrying capacity. The solution that VERP helped to create centred on two questions: what level of visitor experience should the park seek to attain; and what percentage of visitors should have that level of experience. The first part of the solution—the assessment of acceptable use levels—was a social science question (see Chapter 6); the second part—percentage of visitors to have that experience—was a management decision.

The solution established a limit of 3,000 users per day on the carriage road system, implying higher likelihoods of encounters at nodes and lower likelihoods along sections of the network. In addition to answering the question of crowding, this limit would reduce the probability of conflict. However, additional indicators were designed for anti-social behaviours, affording another means to evaluate the success of the solution. Monitoring, the authors emphasize, is a critical activity to ensure the quality of the visitor experience.

The Visitor Activity Management Process

The Visitor Activity Management Process (VAMP) was developed by Parks Canada in the late 1980s to complement its existing Natural Resource Management Process (Figure 7.6). VAMP has largely been superseded by changes in Parks Canada's decision-making structures, particularly the Recreation Activity Assessment Process, described in Box 7.5.

While VAMP had no obvious roots in carrying capacity, several aspects combined to provide this framework with exceptional capability. The VAMP framework revolved around visitor activity profiles. A visitor activity profile connects a particular activity (for example, cross-country skiing) with the social and demographic characteristics of participants, with the activity's setting requirements, and with trends affecting the activity. For cross-country skiing, the visitor activity profile used by Parks Canada is composed of four sub-activities: recreation/day-use skiing; fitness skiing; competitive skiing; and backcountry skiing.

Differences among participants' socio-demographic characteristics, equipment, and motivations, as well as their setting needs, make each of these sub-activities—all involving cross-country skiing—quite unique (see Chapter 6 for discussion on

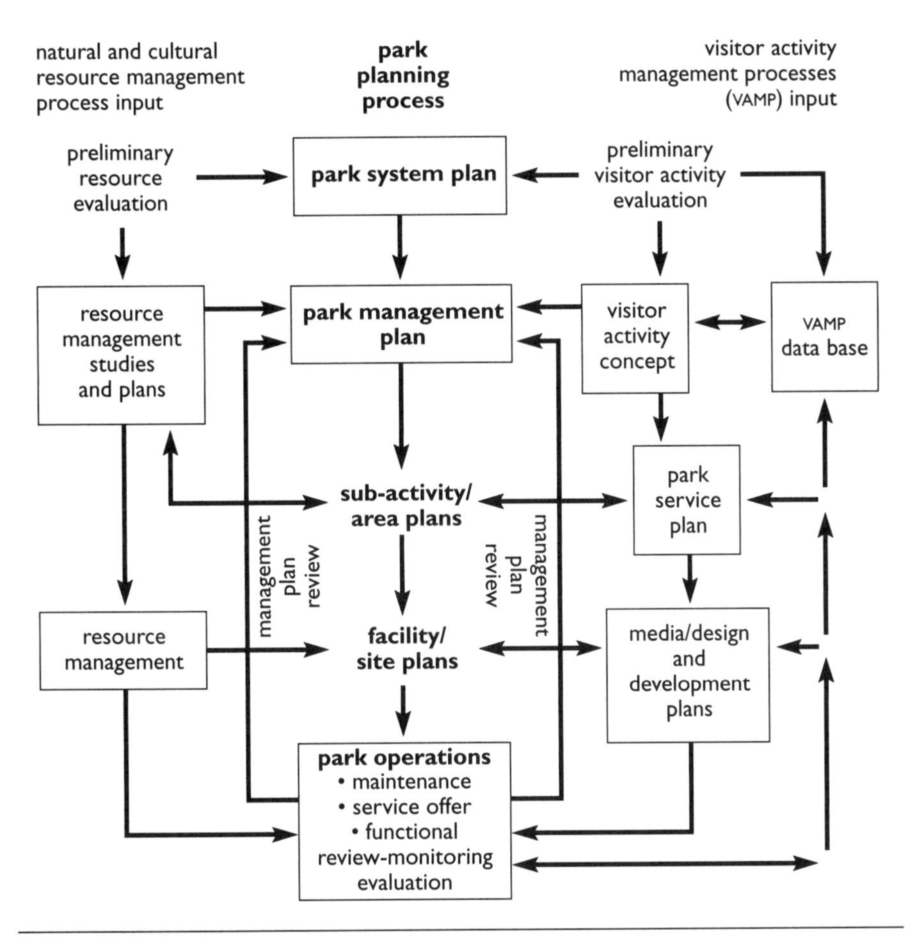

FIGURE 7.6 The national park planning process, showing the role of the Visitor Activity Management Process (VAMP). *Source*: Parks Canada (1986).

market segmentation). More importantly for planning and management, these distinct variations require different levels of service and have differing environmental effects. Competitive skiing, for instance, requires creating courses that meet standards set by cross-country skiing associations. On the other hand, backcountry skiing requires few changes to the landscape, since the natural setting is such a large part of what these skiers want.

VAMP's power was first in its focus on visitor activities as a means of understanding and managing human use of parks and protected areas. Using the VAMP framework, Parks Canada was able to tailor opportunities, programs, services, and facilities to specific visitor activity groups and, in turn, to specific areas. Parks Canada has used the VAMP framework in two other important initiatives, one focusing on risk management (Parks Canada, 1998a), the other on appropriate activity assessment (Parks Canada, 1994c).

BOX 7.5 Using VAMP in Appropriate Activity Assessment

The principle that not all types of activities are appropriate in parks and protected areas is well established for Canada's national parks. Appropriate activity assessment goes back to the early 1980s when both VAMP and an approach to appropriate activity assessment were under development (Nilsen, 1994). The most current iteration of appropriate activity assessment contains the following principles (Parks Canada, 2006).

Recreational activities in national parks, national historic sites, and national marine conservation areas will:

1. Sustain or enhance the character of the place.
   - Be consistent with aspirations of Canadians for national heritage places.
   - Be compatible with the unique and/or valued features of the place.
   - Sustain or enhance the value and respect that visitors accord the place.
   - Provide authentic and/or high-quality opportunities.
2. Respect natural and cultural resources.
   - Are compatible with and may benefit natural resource protection goals.
   - Are compatible with and may benefit cultural resource protection goals.
   - Positive and negative impacts of the activity are well understood.
   - Management practices to promote positive and address negative impacts are available.
3. Facilitate opportunities for outstanding visitor experiences.
   - Provide opportunities for visitors to have fun and enjoy their visit.
   - Respond to the needs and expectations of identified audiences.
   - The risk and level of public safety services can be communicated and the risk can be mitigated where appropriate.
   - The activity helps achieve Parks Canada's visitation objectives.
   - With appropriate mitigations, this activity can occur without adversely affecting the experience of other visitors.
4. Promote public understanding and appreciation.
   - Present opportunities to communicate key messages relevant to this unique place.
   - Inspire stewardship of our natural and cultural heritage.
5. Value and involve local communities.
   - There are appropriate opportunities to engage the local community in management of the activity.
   - The activity is considered acceptable within the cultural context.
   - The activity offers social and/or cultural benefits to local communities.
   - The activity contributes to the local economy.

While not explicitly spatial in orientation like the ROS, VAMP did provide an opportunity to define service objectives and levels of service (standards) in order to address social carrying capacity issues. It relied on the natural resource management process and the environmental impact assessment process to address elements more related to ecological carrying capacity. Like other visitor management frameworks, the ongoing challenge was to ensure integration of functional processes for overall park management goals and objectives.

By the 1990s, the focus inside Parks Canada had shifted to the question of the appropriateness of visitor activities—a much more complicated question requiring value judgements on the part of managers. The Parks Canada response, the appropriate activity assessment (Parks Canada, 1994c), depended on the understanding of activities gained from the VAMP work but also used the National Parks Act and national parks policy as filters to assess appropriateness (Box 7.5). The framework functioned in this way: hang-gliding was proposed for several mountain national parks. Examining the activity, staff recognized that the infrastructure needed to support the activity presented barriers to appropriateness. In order for the activity to occur, roads or hardened trails would be required if participants in hang-gliding were to get their equipment to mountaintops. Such construction would entail substantial disruption of areas that were without roads and/or hardened trails. On ecological grounds, and in respect to the protection mandate in legislation and policy, hang-gliding was deemed inappropriate for national parks.

Since the 1980s, VAMP has had a profound impact on managers in Canadian national parks, although the most important application in the future will likely occur through visitor activity assessment. While VAMP itself is now a historical curiosity, its roots are evident in Parks Canada's continuing struggles to manage visitors in a systematic way.

Summary of Visitor Management Frameworks

Several points concerning the visitor management frameworks bear emphasizing:

- Carrying capacity is at the root of all VMFs.
- LAC, VIM, and VERP incorporate the revised version of carrying capacity.
- Recent VMFs incorporate social and/or ecological standards.
- Recent VMFs require monitoring to track the achievement of standards.
- Recent VMFs function at both the landscape and site scales.
- Depending on the management situation, VMFs may be combined to produce a more powerful tool.

For the sake of completeness, one should also mention the only significant adaptation of these frameworks outside of North America, the Tourism Opportunity Management Model (TOMM), which was developed for Kangaroo Island in Australia (Manidis Roberts Consultants, 1997; Newsome et al., 2002). In spirit, this framework follows the LAC very closely, but it attempts to avoid the somewhat negative connotations of 'limits' and 'acceptable change', which might not necessarily be appreciated by the tourism industry. Essentially TOMM extends the LAC to parks and gateway communities by explicitly considering both commercial and community interests in all stages of implementation and monitoring. As such, it has potential for application in Canada, especially in locations such as Banff and Canmore, or Tofino and Ucluelet.

LINKING VISITOR MANAGEMENT AND ECOSYSTEM MANAGEMENT

The initiative to manage ecosystems in parks and protected areas can be traced to Agee and Johnson (1988), who introduced the term 'ecosystem management' and suggested what might be involved in implementing such a management philosophy (see Chapter 13). Ecosystem management has become the paradigm through which both the natural environment and people's use of the natural environment may be managed successfully. Nepstad and Nilsen (1993) make the point that any serious consideration of ecosystem management in parks and protected areas must include an understanding of human use patterns in the past and in the present. Others, such as Freemuth (1996), suggest that for ecosystem management to be effective, it will be necessary for it to reflect social values.

In Parks Canada, for example, there is a growing realization that many ecological problems in national parks are related to human uses and to facilities to support those uses both inside and outside park boundaries (Parks Canada, 1995; Page et al., 1996). The human use dimension is acknowledged in Parks Canada's definition of ecosystem management (Parks Canada, 1999b):

> Ecosystem management provides a conceptual and strategic basis for the protection of park ecosystems. It involves taking a more holistic view of the natural environment and ensuring that land use decisions take into consideration the complex interactions and dynamic nature of park ecosystems and their finite capacity to withstand and recover from stress induced by human activities. The shared nature of ecosystems also implies that park management will have effects on surrounding lands and their management.

Integrating Social and Ecological Dimensions

Merely stating that 'people are part of ecosystems and human conditions are shaped by, and in turn, shape ecosystems' (Driver et al., 1995) is insufficient if that belief is not reflected in management frameworks and practices. Parks Canada has begun to respond to the demands of legislation and the recommendations of the Ecological Integrity Panel in a way that may promote the integration of visitor management with ecosystem management. Having defined ecological integrity in legislation, the agency set about identifying park-specific indicators of ecological integrity and ways of monitoring those indications. The 'human use principles' suggested by the current *Banff National Park Management Plan* (Parks Canada, 2004) indicate movement in the right directions (Box 7.6).

Case studies follow the human use principles outlined for Banff National Park, where, for example, area planning employs grizzly bear habitat areas as critical spatial units. In Yoho National Park, grizzly bear–human contacts in the Lake O'Hara area forced Parks Canada to go beyond monitoring to examine the social and ecological dimensions of this problem and to suggest solutions (Petersen, 1998). Wright and Clarkson (1995) discuss a study of recreation use on major rivers in Jasper National Park

FIGURE 7.7 Recreational conflicts occur when different activities, such as fly-fishing and kayaking, come into contact. *Photo: Jim Butler.*

designed to determine ecological impacts of human use and to discover the extent to which such riverine activities contribute to visitors' understanding, appreciation, and enjoyment of their national park visits. Both studies address the reality facing park managers: understanding ecological impacts is not sufficient; human use also must be understood before managers respond. In addition, managers need to consider an array of direct and indirect strategies in developing responses, with carrying capacity being only one part of a solution.

More recently, Parks Canada has created an External Relations and Visitor Experience unit. Increased funding has allowed the hiring of a chief social scientist, as well as social scientists who are based in regional service centres. In a direct parallel with indicators and monitoring for ecological integrity, the new unit will identify park-specific experiences that are appropriate to the agency's legislation and monitoring to ensure that visitors are actually reporting such experiences. This two-pronged approach (ecological integrity and visitor experience) goes some way to stating the 'desired conditions' that are a part of the visitor management frameworks discussed here. At the same time, this approach meets the requirements of both legislation and policy for national parks.

It is important to point out that these indicator and monitoring methods may help to force a spatial focus in which ecological and social factors must be considered together rather than separately. The zones in park management plans come to mind as appropriate spatial units upon which such integration might be focused.

BOX 7.6 Principles for Human Use Management in the Banff National Park Management Plan

- A variety of techniques for managing human use will be applied—indirect methods such as education and facility design and capacity when possible, direct methods such as reservations and quotas when necessary.
- Human use management techniques will apply to commercial and non-commercial users and to park operations.
- A range of appropriate opportunities will be provided for users to enjoy the park, from remote wildland areas to intensively developed day use facilities.
- Collaboration on a regional basis is part of human use management.
- Public advice and input are integral to human use management.
- Human use decisions will be based on science and, when possible, supplemented with other information such as legislation, policy, financial implications, cumulative effects, and traditional and local knowledge.
- Scientific data will not always be available or complete and rarely are sufficient by themselves; informed decision-making will be the norm, using as much information and knowledgeable advice as possible.
- Integrity and common sense will underlie all decision–making.
- The precautionary principle will apply when the potential consequences are uncertain, coupled with monitoring and adaptive management; management responses will evolve with scientific research.
- Tangible targets will be established, when feasible, for measuring success in achieving human use goals and objectives; targets will rely on a suite of ecological and social indicators for measuring success in providing a quality opportunity (e.g., to reduce congestion) and for limiting ecological impacts.
- The implementation of the human use management strategy will be phased in.

Source: Adapted from Parks Canada (2004).

CONCLUSION

This chapter began as an attempt to come to terms with carrying capacity, a very attractive idea in the parks and protected areas field. Carrying capacity was shown to be complex, including variations such as ecological, social, and design carrying capacity. More than 30 years of academic and practical work have determined that if carrying capacity is to be a useful tool in a park manager's arsenal, it is necessary to recognize that it has both scientific and value bases. It is the latter that seems to occasionally intimidate managers. In its modern form, carrying capacity may offer an opportunity to confront the protection–use dilemma at the heart of managing parks and protected areas.

An investigation into the relationship between carrying capacity and visitor management frameworks (VMFs) revealed that several of the VMFs cover both ecological and social carrying capacity. Moreover, these frameworks operate from the same basis as the modern understanding of carrying capacity. The relationship between carrying capacity and the various visitor management frameworks discussed was demonstrated to be close.

Finally, visitor management was considered in relation to ecosystem management, a new paradigm in the management of parks and protected areas. Ecosystem management if implemented carefully may offer parks and protected agencies such as Parks Canada the opportunity to achieve an integration of management processes that has long been desired.

We opened this chapter by referring to the traditional protection–use dichotomy in parks and protected areas management. If agencies such as Parks Canada can successfully integrate ecological and social dimensions of ecosystem management in practice, the dual mandate that has been ascribed to parks and protected areas management may become a thing of the past.

REFERENCES

Agee, J.K., and D.R. Johnson. 1988. *Ecosystem Management for Parks and Wilderness*. Seattle: University of Washington Press.

Arnberger, A., and W. Haider. 2007. 'Would you displace? It depends! A multivariate visual approach to intended displacement from an urban forest trail', *Journal of Leisure Research* 39, 2: 345–65.

BC Ministry of Forests. 1996. *Swan Lake Wilderness Area: Swan Lake Management Plan*. Victoria.

———. 1998. *Recreation Opportunity Spectrum Inventory—Procedures and Standards Manual*. Victoria.

Becker, R., A. Jubenville, G.W. Burnett, and A.R. Graefe. 1984. 'Fact and judgement in the search for a social carrying capacity', *Leisure Sciences* 6, 4: 475–86.

Butler, R., D.A. Fennell, and S.W. Boyd. 1993. *Canadian Heritage Rivers System Recreational Carrying Capacity Study*. Ottawa: Canadian Parks Service.

Canadian Parks Service. 1986. *In Trust for Tomorrow: A Management Framework for Four Mountain Parks*. Ottawa: Supply and Services Canada.

Clark, R.N., and G.H. Stankey. 1979. *The Recreation Opportunity Spectrum: A Framework for Planning, Management and Research*. Seattle: Pacific Northwest Research Station, USDA Forest Service, General Technical Report PNW-98.

——— and ———. 1990. 'The recreation opportunity spectrum: A framework for planning, management and research', in Graham and Lawrence (1990: 127–58).

Cole, D.N., M.E. Petersen, and R.C. Lucas. 1987. *Managing Wilderness Recreation Use: Common Problems and Potential Solutions*. Ogden, Utah: Intermountain Research Station, USDA Forest Service, General Technical Report INT-230.

Driver, B.L., P.J. Brown, G.H. Stankey, and T.G. Gregoire. 1987. 'The ROS planning system: Evolution, basic concepts and research needs', *Leisure Sciences* 9: 201–12.

———, C.H. Manning, and G.R. Super. 1995. 'Integrating the social and biophysical components of sustainable ecosystems management', in J.E. Thompson, ed., *Analysis in Support of Ecosystem Management*. Washington: Ecosystem Management Analysis Center, USDA Forest Service.

Duffus, D.A., and P. Dearden. 1990. 'Non-consumptive wildlife-oriented recreation: A conceptual framework', *Biological Conservation* 53: 213–31.

Elliot, T.E. 1994. 'Attitudes toward limiting overnight use of the Chilkoot Trail National Historic Site', M.Sc. thesis, University of Montana.

Freemuth, J. 1996. 'The emergence of ecosystem management: Reinterpreting the gospel?', *Society and Natural Resources* 9, 4: 411–17.

Graefe, A.R. 1990. 'Visitor impact management', in Graham and Lawrence (1990: 213–34).

Graham, R., and R. Lawrence, eds. 1990. *Towards Serving Visitors and Managing Our Resources.* Waterloo, Ont.: Tourism and Recreation Education Centre and Environment Canada, Canadian Parks Service.

Hammitt, W., and D. Cole. 1998. *Wildland Recreation: Ecology and Management.* New York: John Wiley and Sons.

Heberlein, T.A., G.E. Alfano, and J.J. Vaske. 1986. 'Using a social carrying capacity model to estimate the effects of marina development at the Apostle Islands National Lakeshore', *Leisure Sciences* 8, 3: 257–74.

Hendee, J.C., G.H. Stankey, and R.C. Lucas. 1990. *Wilderness Management*, 2nd edn. Golden, Colo.: Fulcrum.

Jackson, S., W. Haider, and T. Elliot. 2003. 'Resolving inter-group conflict in winter recreation: Chilkoot Trail National Historic Site, BC', *Journal for Nature Conservation* 11: 317–23.

Jacobi, C., and R. Manning. 1999. 'Crowding and conflict on the carriage roads of Acadia National Park: An application of the Visitor Experience and Resource Protection Framework', *Park Science* 19, 2: 22–6.

Kachi, N., and K. Walker. 1999. *Status of Human Use Management Initiatives in Parks Canada.* Hull, Que.: Ecosystems Management Branch, Parks Canada.

Kingston, S.R., A.P. Carr, and R.J. Payne. 2002. *The Ecological Impact of Backcountry Recreation in Kejimkujik National Park: Final Report.* At: <bolt.lakeheadu.ca/~bpaynewww/keji2001finalrep2.pdf>.

Knopf, Richard C. 1990. 'The limits of acceptable change (LAC) planning process: Potentials and limitations', in Graham and Lawrence (1990: 201–11).

Leung, Y., and J. Marion. 2000. 'Recreation impacts and management in wilderness: A state-of-the-knowledge review', *Proceedings: Wilderness Science in a Time of Change.* USDA Forest Services Proceedings RMRS-P-15-Vol-5.

McCool, S.F. 1990. 'Limits of acceptable change: Evolution and future', in Graham and Lawrence (1990: 185–93).

Manidis Roberts Consultants. 1997. *Developing a Tourism Optimisation Management Model (TOMM) (Final Report).* Sydney, NSW.

Manning, R.E. 2001. Visitor experience and resource protection: A framework for managing the carrying capacity for national parks', *Journal of Park and Recreation Administration* 19, 1: 93–108.

———. 2007. *Parks and Carrying Capacity—Commons without Tragedy.* Washington: Island Press.

Nelson, J.G. 1968. 'Man and landscape change in Banff National Park: A national park problem in perspective', in J.G. Nelson, ed., *Canadian Parks in Perspective.* Montreal: Harvest House, 63–96.

Nepal, S.K., and P. Way. 2007. 'Comparison of vegetation conditions along two backcountry trails in Mount Robson Provincial Park, BC (Canada)', *Journal of Environmental Management* 82: 240–9.

Nepstad, E., and P. Nilsen. 1993. *Towards a Better Understanding of Human/Environment Relationships in Canadian National Parks.* Ottawa: Supply and Services Canada, National Parks Occasional Paper No. 5.

Newsome, D., S. Moore, and R. Dowling. 2002. *Natural Area Tourism: Ecology, Impacts and Management.* Channel View, UK: Clevedon.

Nilsen, P. 1994. *A Proposed Framework for Assessing the Appropriatness for Recreation Activities in Protected Heritage Areas.* Ottawa: Parks Canada.

——— and G. Tayler. 1998. 'A comparative analysis of human use planning and management frameworks', in N.W.P. Munro and J.H.M. Willison, eds, *Linking Protected Areas with Working Landscapes Conserving Biodiversity.* Wolfville, NS: SAMPAA, 861–74.

Page, R.S., D. Bayley, J. Cook, and B. Ritchie. 1996. *Banff–Bow Valley: At the Crossroads.* Hull, Que.: Supply and Services Canada.

Parks Canada. 1994a. *Park Management Guidelines: Pacific Rim National Park Reserve.*

———. 1994b. *Guiding Principles and Operational Policies.* Ottawa: Ministry of Supply and Services Canada.

———. 1994c. *A Proposed Framework for Assessing the Appropriateness of Recreation Activities in Protected Heritage Areas.* Ottawa: Department of Canadian Heritage.

———. 1995. *State of the Parks: 1994 Report.* Ottawa: Department of Canadian Heritage.

———. 1997. Lake O'Hara Trail Monitoring Program, Yoho National Park. At: <www.worldweb.com/ParksCanada-Yoho/TrailMonitor/English/index.htm>.

———. 1998a. *Parks Canada Visitor Risk Management Handbook.* At: <parkscanada.pch.gc.ca/library/risk/english/tdmvrme.html>.

———. 1998b. *State of the Parks: 1997 Report.* Ottawa: Department of Canadian Heritage.

———. 1999a. *Gwaii Haanas Backcountry Management Plan.* At: <www.pc.gc.ca/pn-np/bc/gwaiihaanas/plan/plan3_e.asp>.

———. 1999b. Ecosystem management. At: <parkscanada.pch.gc.ca/natress/ENV_CON/ECO_MAN/ECO_MANE.HTM>.

———. 2004. *Banff National Park Management Plan.* At: <www.pc.gc.ca/pn-np/ab/banff/docs/plan1/chap5/plan1e_E.asp#accom>.

———. 2006. *Recreation Activity Assessment Bulletin* (Dec.).

Parks Canada Agency. 2000. *Unimpaired for Future Generations? Protecting Ecological Integrity with Canada's National Parks, Vol. 2: Setting a New Direction for Canada's National Parks.* Report of the Panel on the Ecological Integrity of Canada's National Parks. Ottawa.

Parliament of Canada. 2000. An Act respecting the National Parks of Canada. Statutes of Canada 2000. Chapter 32, Second Session, Thirty-sixth Parliament, 48–9 Elizabeth II, 1999–2000.

Payne, R.J. 1997. *Visitor Information Management in Canada's National Parks.* Ottawa: Natural Resources Branch, Parks Canada.

———, A.P. Carr, and E. Cline. 1997. *Applying the Recreation Opportunity Spectrum (ROS) for Visitor Opportunity Assessment in Two National Parks: A Demonstration Project.* Ottawa: Natural Resources Branch, Parks Canada, Occasional Paper No. 8.

——— and R. Graham. 1993. 'Visitor planning and management in parks and protected areas', in P. Dearden and R. Rollins, eds, *Parks and Protected Areas in Canada: Planning and Management.* Toronto: Oxford University Press, 185–210.

Petersen, D. 1998. 'Allocation of land resources between competing species—humans and grizzly bears in the Lake O'Hara area of Yoho National Park', in N.W.P. Munro, J.H.M. Willison, and J.H. Martin, eds, *Linking Protected Areas with Working Landscapes Conserving Biodiversity.* Wolfville, NS: SAMPAA, 478–91.

Shelby, B., and T.A. Heberlein. 1984. 'A conceptual framework for carrying capacity determination', *Leisure Sciences* 6, 4: 433–51.

——— and ———. 1986. *Carrying Capacity in Recreational Settings.* Corvallis: Oregon State University Press.

Stankey, G.H. 1973. *Visitor Perception of Wilderness Recreation Carrying Capacity.* Ogden, Utah: USDA Forest Service, Intermountain Forest and Range Experimental Station, Research paper INT-142.

———, D.N. Cole, R.C. Lucas, G.L. Peterson, and S. Frissel. 1985. *The Limits of Acceptable Change System for Wilderness Planning.* Ogden, Utah: USDA Forest Service, Intermountain Research Station, General Technical Report INT-176.

——— and S.F. McCool. 1984. 'Carrying capacity in recreational settings: Evolution, appraisal and application', *Leisure Sciences* 6, 4: 453–74.

——— and ———. 1989. 'Beyond social carrying capacity', in E.L. Jackson and T.L Burton, eds, *Understanding Leisure and Recreation: Mapping the Past, Charting the Future.* State College, Penn.: Venture.

USDA Forest Service, Daniel Boone National Forest. 2007. Limits of acceptable change. At: <www.fs.fed.us/r8/boone/lac/>.

United States National Park Service. 1997. *The Visitor Experience and Resource Protection (VERP) Framework: A Handbook for Planners and Managers.* Denver: Denver Service Center, US Department of the Interior.

Vaske, J.J. 1994. *Social Carrying Capacity at the Columbia Icefield: Applying the Visitor Impact Management Framework.* Ottawa: Parks Canada, Department of Canadian Heritage.

———, B. Shelby, A.R. Graefe, and T.A. Heberlein. 1986. 'Backcountry encounter norms: Theory, method and empirical evidence', *Journal of Leisure Research* 18, 3: 137–53.

Wagar, J.S. 1964. *The Carrying Capacity of Wild Lands for Recreation.* Washington: Society of America Foresters, Forest Science Monograph 7.

———. 1974. 'Recreational carrying capacity reconsidered', *Journal of Forestry* 72: 274–8.

Washburne, R.F. 1982. 'Wilderness recreational carrying capacity: Are numbers necessary?', *Journal of Forestry* 80, 11: 726–8.

Wright, P.A., and P. Clarkson. 1995. 'Recreation impacts on river ecosystems: Assessing the impacts of river use on the biophysical and social environment', in T. Herman, S. Bondrup-Neilsen, J.H.M. Willison, and N.W.P Munro, eds, *Ecosystem Monitoring and Protected Areas.* Wolfville, NS: SAMPAA, 299–303.

KEY WORDS/CONCEPTS

acceptable change
carrying capacity
direct management strategies
ecological carrying capacity
ecological integrity
ecosystem management
encounter norms
indirect management strategies
Limits of Acceptable Change (LAC)
physical carrying capacity
recreation conflict
Recreation Opportunity Spectrum (ROS)
social carrying capacity
Visitor Activity Management Process (VAMP)
Visitor Experience and Resource Protection Framework (VERP)
Visitor Impact Management (VIM)
wilderness

STUDY QUESTIONS

1. Discuss the confusion that sometimes occurs between visual impacts and ecological impacts (ecological integrity).
2. Compare 'design carrying capacity', 'ecological carrying capacity', and 'social carrying capacity'.
3. 'Ecological carrying capacity has not been seriously applied in parks and protected areas.' Discuss.
4. Compare 'indirect management actions' with 'direct management actions'. Which approaches do you think would work best in a national park and why?
5. Compare ROS and carrying capacity.
6. Compare VAMP and carrying capacity.
7. Compare LAC and carrying capacity.
8. Compare VIM and carrying capacity.
9. Compare VERP and carrying capacity.
10. Compare VAMP and ROS.
11. Examine the national park zoning system, and compare this with the approach to zoning developed within ROS.
12. A significant component of LAC is the inclusion of stakeholders as part of the planning process. Discuss strengths and weaknesses of this approach. What other forms of public participation are possible? Discuss strengths and weaknesses of each.
13. Examine a park management plan of your choice, and outline the extent to which the plan uses visitor management concepts described in this chapter.
14. Critique the visitor management strategies described in the plan.

CHAPTER 8

The Role of Interpretation

Glen T. Hvenegaard, John Shultis, and James R. Butler

> I'll interpret the rocks, learn the language of flood, storm and the avalanche. I'll acquaint myself with the glaciers and wild gardens, and get as near the heart of the world as I can.
>
> *John Muir, 1871* (cited in Mackintosh, 2000: 1)

INTRODUCTION

Interpretation is a communication process designed to reveal meanings and relationships of cultural and natural heritage to the public, through first-hand involvement with an object, artifact, landscape, or site (Interpretation Canada, 2008). Interpretation has been fundamental to the original concept of parks and protected areas, at first led by energetic volunteers, and later supported by administrators and implemented with professionally trained staff. Today, interpretation, in its many forms—and the related communication techniques of environmental education and tour guiding—is employed in the vast majority of Western protected areas systems. The provision of interpretation provides important benefits, not only to park visitors, but to the natural environment, park agencies, and society in general. The purpose of this chapter is to examine the definitions, principles, history, theoretical foundations, effectiveness, and practice of interpretation, with a focus on Canadian protected areas at the federal level, as municipal, provincial, and territorial interpretation legislation, policy, and practices vary considerably throughout Canada.

DEFINITION AND PURPOSE OF INTERPRETATION IN PROTECTED AREAS

Tilden (1977: 8) defined interpretation as 'an educational activity which aims to reveal meanings and relationships through the use of original objects, by firsthand experience, and by illustrative media, rather than simply to communicate factual information'. *Interpretation* differs from *information* in two fundamental ways. First, interpretation depends on information, but seeks to reveal meanings based on that information, so that 'visitors increase knowledge and deepen understanding.' Second, the chief aim of inter-

FIGURE 8.1 Visitor scanning for wildlife, Firth River, Vuntut National Park. 'Interpretation should fill visitors with a greater sense of wonder and curiosity.' *Photo: P. Dearden.*

pretation is not instruction, but provocation (ibid.). Such provocation works to develop appreciation, respect, and a sense of responsibility to those protected places being interpreted (Canadian Environmental Advisory Council, 1991; Parks Canada, 1998a).

Thus, interpretation is not mere transfer of information to others. Effective interpretation fills visitors with a greater sense of wonder and curiosity. It leaves the visitor both better informed and with a desire to know more. Finally, interpretation provides challenges or opportunities to act on this new sense of respect, benefiting the ecological integrity of national parks and the surrounding environments. A number of similar definitions exist (e.g., Sharpe, 1982; Knudson et al., 2003), but typically embrace the following attributes of interpretive services:

1. They are *on-site*, emphasizing *first-hand experience* with the natural environment (e.g., they introduce, then encourage visitors to spend time outdoors for direct interactions with the park's features, as distinct from a museum, which functions as the destination itself).
2. They provide an *informal* form of education (i.e., interpretation does not employ a classroom-based approach).
3. They deal with a *voluntary, non-captive audience*: visitors participate by free choice during their leisure time.
4. They satisfy visitors' normal *expectation of gratification* (i.e., visitors want to be rewarded or to have a need or want satisfied).

5. They are *inspirational* and *motivational* in nature; they do not merely present factual information.
6. Their goals are *expansions of knowledge, shifts in attitude, and alterations in behaviour* of visitors; visitors should increase their understanding of, and their appreciation and respect for, the park environment.
7. They create experiences based on the *constructed values* of natural and cultural features; individuals and societies continuously generate and reassess the meaning of various resources, including protected areas. Interpretation facilitates the understanding, appreciation, and protection of the park's intrinsic and constructed values.

The purpose of park interpretation is shaped by park legislation and policy as well as by societal attitudes and values and visitor behaviour. Canadian legislation notes that national parks are created for the 'benefit, education and enjoyment' of the people of Canada (Canada, 1990). Existing national parks policy further outlines the goals, methods, content, and target audiences of interpretation. Specifically, the national parks are to provide programs to encourage and assist Canadians in 'understanding, appreciating, enjoying and protecting their national parks' (Parks Canada, 1994: 37).

Interpretation provides essential facts about an area and its facilities, and helps the visitor understand, appreciate, and enjoy the park's natural and cultural features. By doing so, interpretation helps to minimize uncertainty and to maximize opportunity for visitors. A well-rounded interpretation program serves to awaken public awareness of park purposes and policies and strives to develop a stronger concern for preservation. The degree to which a visitor enjoys and values his or her experience in a park depends largely on the individual's perception of that area's environment. For this reason, interpretive approaches should be designed to enhance the visitor's perception of these landscapes and ultimately to be a positive influence on the interactions between visitors and the ecosystem.

Interpretation, Environmental Education, and Tour Guiding

Interpretation, environmental education, and tour guiding all can occur in protected areas, and all are important in park operations. Traditionally, interpreters were Parks Canada staff, while environmental educators and tour guides were hired by other government agencies, commercial operators, and NGOs. As messages created by external groups are difficult to control by Parks Canada (or other park agencies), greater efforts at partnering with these external groups have been attempted, to ensure that the messages provided match the messages desired by park agencies. In addition, Parks Canada has recently moved towards increasing its environmental education capacity in an attempt to better engage with Canadians outside park boundaries (Parks Canada Agency, 2006a). Additional information on the move to off-site education and interpretation by Parks Canada will be provided later in this chapter.

While an extended discussion of the theoretical and professional links between interpretation, environmental education, and tour guiding is beyond the scope of this chapter, it is important to note the common objectives that these educational endeavours sometimes share. *Environmental education* programs are developed principally for pri-

FIGURE 8.2 An interpreter interacts with his audience on Bonaventure Island, Quebec. *Photo: G. Hvenegaard.*

mary and secondary school students: *personal services* include teacher training and events led by volunteers and park staff, while *non-personal services* include pre- and post-trip resource kits, brochures on self-guided field studies, exhibits developed specifically for children, and CD programs and videos to be used in the classroom before field trips. In its broadest perspective, environmental education is aimed at producing people who are knowledgeable about the whole environment and its associated problems, aware of how to help solve these problems, skilled at helping others do the same, and motivated to work towards their solution (Grant and Littlejohn, 2004).

Tour guiding normally involves private companies engaged in commercial operations. Most often used in pre-booked packages, tour companies hire trained guides to lead these tours, and in doing so provide a conduit between each site visited on the tour and the tourists. Tour guides based in or nearby protected areas also can be hired locally by both guided and independent recreationists and by tourists for day trips and overnight excursions in protected areas. The tour guides schedule and lead these trips, and go far beyond merely providing information to their clients: they mediate between differing cultures, act as mentors to clients, develop narratives of place and time, and choose from a breadth of information to inform clients (Cohen, 1985; Dahles, 2002; Reisinger and Steiner, 2006). Thus, in addition to interpretation, these two related sources of communication have important roles to play in influencing visitor behaviour and individual and societal attitudes towards protected areas.

FIGURE 8.3 Living interpretation at one of Canada's most famous sites, the fortress of Louisbourg, Nova Scotia. *Photo: G. Hvenegaard.*

The Links between Interpretation and Ecotourism

There are implicit links between interpretation and ecotourism; the latter is often touted as the best means to influence park use by tourists. While the definition of ecotourism is still hotly debated, most agree that three concepts are fundamental to ecotourism: (1) a focus on nature-based activities; (2) an educational component that informs and inspires tourists; and (3) a conservation ethic, so that the visitors' activities do not harm the cultural and ecological conditions of the area (Weaver, 2002; Diamantis, 2004). Ecotourists' desire to learn about the natural world emphasizes the need for effective interpretation by ecotourism operators. Ecotourism usually refers to non-consumptive activities, such as birding, whale watching, nature photography, and botanical study. As such, ecotourism is reliant on natural features in relatively undisturbed sites, most often in parks and protected areas (Hvenegaard, 1994; Diamantis, 2004). Interpretation, whether delivered by park staff (usually in the public sector) or tour guides (normally in the private sector), is viewed as the best means of maximizing the positive and minimizing the negative impacts of park visitors (see Bramwell and Lane, 1993).

The role of the tour guide in providing interpretation and education to ecotourists is well documented (Ballantyne and Hughes, 2001; Reisinger and Steiner, 2006). For example, Cohen (1985) suggests four dimensions of tour guiding exist: instrumental (organization and management of tours), interactional (facilitating encounters with people and places), social interaction (group development and cohesion), and communicative (selecting and disseminating information to the group). As well as the obvious link to the latter dimension, the interactional dimension can include minimizing impacts on people and place. Weiler and Davis (1993) also suggest that tour guides must take responsibility for managing the group's environmental footprint and—when necessary—changing the environmental attitudes and behaviours of tourists (see also Christie and Mason, 2003).

Of course, many researchers are unconvinced of the appropriateness of ecotourism and question whether ecotourism is different from 'normal' tourism. For example, Zell (1992: 31) states that 'Tourism creates more tourism, the location becomes well known and thus desirable creating demand, more supply and ultimately destruction of the original reason for going there.' On the other hand, ecotourism has the potential to facilitate change in visitors' knowledge and beliefs (Orams, 1996), but, as Price (2003) suggests, significant obstacles need to be overcome.

A few examples illustrate interpretation's various roles. First, interpretation can help to reduce environmental impacts by highlighting appropriate activities in sensitive areas. Interpreters and tour guides often describe species and habitats that are of concern, due to disturbance or overuse. Second, interpretation can be used to alleviate social impacts by involving local communities and redirecting traffic. For example, in response to high levels of use at Point Pelee National Park, 'Operation Spreadout' seeks to inform visitors of other nearby attractions. Third, by providing additional opportunities for visitors, interpretation can improve local economic impacts by increasing the number of days that visitors stay in an area. As well, interpretation can emphasize opportunities to contribute to conservation. Ecotourists tend to donate more money to conservation than general tourists, but they need additional interpretation to encourage them to do so (Hvenegaard and Dearden, 1998). Finally, interpretation responds to the ecotourists' need for awareness and understanding of a park's natural features. Visitors to Grasslands National Park in southern Saskatchewan, for example, are primarily motivated to learn more about the environment (Saleh and Karwacki, 1996). As Ham and Weiler (2002) suggest, interpretation has the potential to promote sustainability, in the context of satisfying visitors, altering behaviour, and promoting a conservation ethic (see also Orams, 1996).

ORIGINS AND HISTORY OF INTERPRETATION IN PROTECTED AREAS

Soon after the creation of federal national park systems in Canada (1911) and the United States (1916), the fundamental role of interpretation in national parks was established. In his initial annual report, the US Park Service's new director, Stephen T. Mather, emphasized that 'one of the chief functions of the national parks and monuments is to serve educational purposes' (cited in Carlsbad Caverns National Park, 2004: 1).

Mather was first introduced to interpretation in private resorts in the Lake Tahoe area. After a visit to Yosemite, in 1919, Mather attended a popular program provided by university professors Miller and Bryant at the Fallen Leaf Lodge (Weaver, 1982). This new approach to education was supported by a Sacramento patron who had enthused about interpretive guides he had encountered in Europe, and by the California Nature Study League and the State Fish and Game Commission. Upon observing the success of this program, Mather recognized that such programs could increase public support for national parks, and could therefore increase government support and funding for park creation and park management. The following summer, the two professors became the first park naturalists employed in the American park system. Private interpretive programs had already begun within the national parks in a more informal manner, and Mather's 1919 annual report specifically mentioned eight instances of campfire education already underway that summer (Shankland, 1970).

Thus began a tradition that would, in spite of wavering commitments and administrative cutbacks, continue within most national park systems in the developed world to the present time (see Box 8.1). The abolition of interpretation in British Columbia's provincial parks in 2001 provides a worst-case scenario of budget cuts and the increasing commercialization of protected areas. As of early 2008, British Columbia has yet to directly reinstate interpretation in its provincial parks, although limited funding through the Conservation Corps includes some interpretation in provincial parks in the summer season.

Shifts in Emphasis in Park Interpretation

Since the 1920s, interpretation in parks has become considerably more sophisticated, in terms of technological applications, systems planning, and interagency co-ordination. Important shifts in interpretive focus over this period may be divided into four phases, each of which reflects key roles within park management. In Phase One, beginning with the earliest interpretive programs, park interpretation was concerned with acquainting visitors with *features* in the park, often those that were most dramatic,

BOX 8.1 History of Park Interpretation in North America

1784 First natural history museum to utilize interpretive techniques opens in Philadelphia, with Charles Wilson Peale exhibiting wildlife collections from the American West.

1869 First park interpretive book, *The Yosemite Guidebook*, is published by California State Geologist J.D. Whitney.

1870s John Muir leads groups on interpretive hikes into Yosemite backcountry. Muir later becomes an influential spokesperson for wilderness preservation.

1887 Scottish caretaker and guide, David Galletly, conducts visitors through the lower Hot Spring cave, Banff. These are the first formal interpretive walks conducted by an interpreter in a Canadian national park.

1889 Enos Mills, the founder of nature guiding, formalizes and teaches principles of nature guiding in Rocky Mountain National Park, Colorado. He later wrote *Adventures of a Nature Guide and Essays in Interpretation.*

1895 First park interpretive museum, and first museum in any national park, is established at Banff.

1904 First park interpretive trail is established at Yosemite; Lieutenant Pipes of the Army Medical Corps establishes a trail with labelled trees and other plants.

1905 C.H. Deutschman, who discovered Nakimu Caves in Glacier National Park, British Columbia, begins to conduct visitors through the cave system.

1911 Evening campfire programs and tours of park features are well established in several Canadian and US national parks, but all are conducted by concessions.

1914 First Canadian national park interpretive publications appear in Banff.

1915 Esther Burnell Estes becomes first licensed woman interpreter in the US.

1918 US establishes its first park museum in Mesa Verde, Colorado, with exhibits and lectures given; the next museum opens in 1921, in Yosemite.

1919 Nature guiding becomes popular in Rocky Mountain resorts in US. Steven Mather, Director of US Parks Service, has the idea of institutionalizing interpretation in the US national parks system.

1920 First US Park Service interpretive programs begin with government-employed interpreters in Yosemite and Yellowstone.

1929 First seasonal interpretive programs begin in the Rocky Mountain national parks of Canada, with the appointment of J. Hamilton Laing.

1931 Grey Owl (Archie Belaney) employed as interpreter by Parks Canada at Riding Mountain, Manitoba; later transferred to Prince Albert, Saskatchewan.

1944 Early interpretive events conducted in Banff; wildlife warden, Hubert Green, feeds aspen cuttings to beavers of Vermilion Lakes before 25–30 tourists nightly while discussing beaver life history.

1954 Interpretive programs begin in provincial parks of Ontario.

1958 First co-ordinated interpretive service established in Ottawa for Canada's national park system.

1964 First permanent naturalists located in Canadian Rocky Mountain national parks.

1969 First Canadian Wildlife Service interpretation centre opens at Wye Marsh, near Midland, Ontario.
1985 A national Canadian assembly on national parks and protected areas, formed to mark the centennial of Canada's national parks, encourages the development of more interpretive programs.
1988 & 1994 Governmental downsizing results in drastic budget cuts for interpretive staff and services in Canada's national parks.
1990s Scattered attempts to 'privatize' interpretive services, particularly in provincial parks.
1991 The Canadian Environmental Advisory Council (1991) report advises Parks Canada to dedicate more resources to support interpretation and education programs.
1998 In response to budget cuts and perceived decreasing quality and quantity of interpretation in national parks, Parks Canada creates the *Action Plan for the Renewal of Heritage Presentations.* 'Outreach' communication is also touted for urban and youth audiences.
2000 Panel on the Ecological Integrity of Canada's National Parks recommends that interpretation have ecological integrity as its core purpose and that interpretive funding in national parks be doubled.
2001 'Engaging Canadians', a new approach for external communications for Parks Canada, is created. It incorporates three goals: to raise awareness of protected areas by Canadians; to foster understanding and enjoyment of parks and thus influence their attitudes towards parks; and to strengthen emotional connections to Canadian heritage places by creating more opportunities for residents to become directly involved in parks.
2001 BC Parks is dismantled as a separate provincial ministry in British Columbia, and interpretation in BC's provincial parks is cut.
2005 The 2005 session of the Minister's Round Table on Parks Canada focuses on facilitating memorable visitor experiences, including an increased emphasis on reaching new Canadians, youth, ethnic minorities, and urban residents in off-site outreach communications (e.g., environmental education). Parks Canada responds in part by creating an External Relations and Visitor Experience Directorate to address these new priorities.

majestic, and exceptional. The emphasis was on providing explanations for these phenomena, often as examples of the wonders of God's creation. This reflected the focus of creating parks that included the most spectacular and monumental natural features in each nation in order to generate national pride in these relatively young countries (Shultis, 1995; Runte, 1997).

In Phase Two, during an era of higher environmental awareness, beginning in the early 1960s, interpretation began to stress ecological interrelationships and the broader landscape, even when these were less dramatic than high-profile features like hot springs or waterfalls. Management issues such as crowding and environmental impacts of outdoor recreation also received greater attention (Hendee and Dawson, 2002). Communication, however, was focused only on natural and cultural features existing within the park boundary.

In Phase Three, interpretation began to foster a broader ecological consciousness among park visitors and the Western public at large. This involved a shift from an internal focus to a greater external awareness of the ecosystems surrounding the park. This phase was influenced greatly by the photograph of Earth from outer space in 1969 and paralleled the heightened interest in ecology in the 1970s. During this period, environmental education emerged as an important and critical interdisciplinary subject integrated into school curricula systems throughout North America (Rasmussen, 2000).

This new 'ecological mission' of park interpretation, with its shift of emphasis from the park in isolation to ecological perspectives beyond the limits of the park boundary, paralleled the development of several new policy and management perspectives occurring in national parks as part of a redefinition of their role in society as critical and important ecological landscapes. The new ecological interpretation affirmed the perspective that national parks, which had come to be revered as benchmarks of the natural environment (to be compared and contrasted to landscapes where renewable resource activities such as logging are undertaken), could not survive independent of the surrounding landscapes. A federal vehicle was needed to reach the public in a campaign to support a national environmental strategy, and national park visitors were recognized as a more highly educated segment of the population, with greater receptivity to environmental education and a disproportionately higher influence on decision-making. Relatively healthy national park environments could assist in building a new philosophy and ethical system among visitors, using interpretation as a key communication tool. Such a reorientation is critical to the long-term protection of the ecological integrity of national parks and the environment in general. This federal responsibility was expressed by Parks Canada (1997: 1): 'In a world of rapid change, our parks, historic sites, and marine conservation areas are seen as models of environmental stewardship and as an important legacy to be preserved for future generations. They represent one of the most positive, tangible and enduring demonstrations of the federal government's commitment to the environment.'

A Phase Four, beginning at the turn of the twenty-first century, seems to be developing. While the focus on ecological issues inside and outside of park boundaries from Phase Three is still in evidence, several park agencies around the world are beginning to rethink their traditional reliance on on-site interpretation in protected areas. For example, both the National Parks Service in the United States and the Parks Canada Agency have expressed concern with focusing on park visitors only, especially when many park systems are experiencing stagnating or declining visitation. Both agencies have identified a need to reach out to non-visitors, particularly the so-called 'x-box generation' (i.e., youth), 'new' citizens (e.g., recent immigrants), ethnic minorities, and urban residents (National Park Service, 2003; Parks Canada Agency, 2006b). Particular concern towards the low interest in and use of protected areas by contemporary youth has recently been demonstrated (e.g., Pergrams and Zaradic, 2006), with the suggestion that today's children have '*nature deficit disorder*' from their focus on computer-based leisure activities, suburban sprawl, ever busier schedules for children, and an increased focus on academic performance (Louv, 2005). Educational programs are now defined much more broadly, and include not only interpretation, but environmental education and, to a lesser extent, tour guiding; the shift from on-site to off-site locations for heritage communications is a significant shift. On-site interpretation will continue, but an increased emphasis will be placed on off-site communications strategies.

In much the same way as Parks Canada reconfigured its organizational structure to emphasize the primary management directive of ecological integrity in the 1990s, it has made similar (though less extensive) changes to emphasize this new communication strategy. Based on the 'Engaging Canadians' strategy created in 2001 and feedback from the Minister's Round Table on Parks Canada in 2005 (Parks Canada Agency, Performance, Audit and Review Group, 2005; Parks Canada Agency, 2006b), an External Relations and Visitor Experience Directorate has been created at the head office to allow the agency to strengthen its educational efforts:

> Engaging Canadians in heritage conservation and celebration is key to creating awareness of the necessity to preserve diverse cultural and natural resources. As part of this effort to reach out to Canadians, Parks Canada is moving ahead with plans to reach urban audiences, new Canadians and Canadian youth. Parks Canada is exploring the addition of visitor facilities in urban areas to act as multimedia windows into Canada's special natural and cultural places. School programs and Parks Canada curriculum material will be strengthened and Web-based learning materials will be broadened. (Parks Canada Agency, 2006b: 15)

The increased significance of heritage communications in Parks Canada is also reflected in the list of program activities for Parks Canada: after the first two objectives of 'establishing heritage places' and 'conserving heritage resources', the third and fourth programs are 'promoting public appreciation and understanding' and 'enhancing visitor experiences'. These four program activities are identified as the 'core programs' of the agency (Parks Canada Agency, 2006b: 17), and are used by Parks Canada to base its reporting to Parliament. Approximately $70 million (of a total budget of approximately $500 million) has been targeted for promoting public appreciation and understanding from 2006 to 2011. According to Parks Canada (ibid., 58–9), these public information and education programs 'must go beyond the mere communication of information to become rich learning experiences that resonate with Canadians to empower them to take action as stewards and to encourage them to visit and personally experience Canada's special heritage places'; to make such a connection, they 'must go beyond the visitors to its parks and sites, and reach out to Canadians—in their homes, their schools, their communities and their place of work'.

THEORETICAL FOUNDATIONS OF INTERPRETATION

Early professional interpreters based their planning and evaluation of interpretation—primarily personal forms of interpretation—largely on intuition and feedback from peers. As academic researchers began to study interpretation, theories from a variety of

FIGURE 8.4 Self-guiding trail booklets are one way to inform a large number of visitors about a park. *Photo: G. Hvenegaard.*

disciplines were used in an attempt to provide a conceptual framework for interpretation beyond that of Tilden's principles (Ballantyne and Uzzel, 1999) and to ensure that interpretive research efforts would be considered an area of academic research equal to other, more established research areas in academia.

As interpretive research is still in its infancy, a host of theoretical approaches may be drawn on to explain the impact of interpretation; in addition, commonalities exist between theoretical approaches to interpretation, environmental education, and tour guiding. But as Larson (2004: 70) notes, 'interpretation is disadvantaged by not having a coherent professional language based on shared theory and understanding.' Without a foundational conceptual and theoretical base, it is difficult for interpreters and researchers to generate the necessary support from potential funders, politicians, administrators, and managers. For the purposes of this section, the primary theories and epistemological stances normally used only in interpretation research will be reviewed, taken largely from the disciplines of social psychology and education.

The theories used most frequently to explain and predict the potential impact of interpretation are taken from social psychology. The theories of *reasoned action* and *planned behaviour* suggest that behaviour is best explained and predicted through individuals' *intentions* to perform a specific behaviour, which are linked with people's behavioural beliefs (attitudes about the consequences of the behaviour), normative beliefs (perceived social support for the behaviour and their motivation to comply),

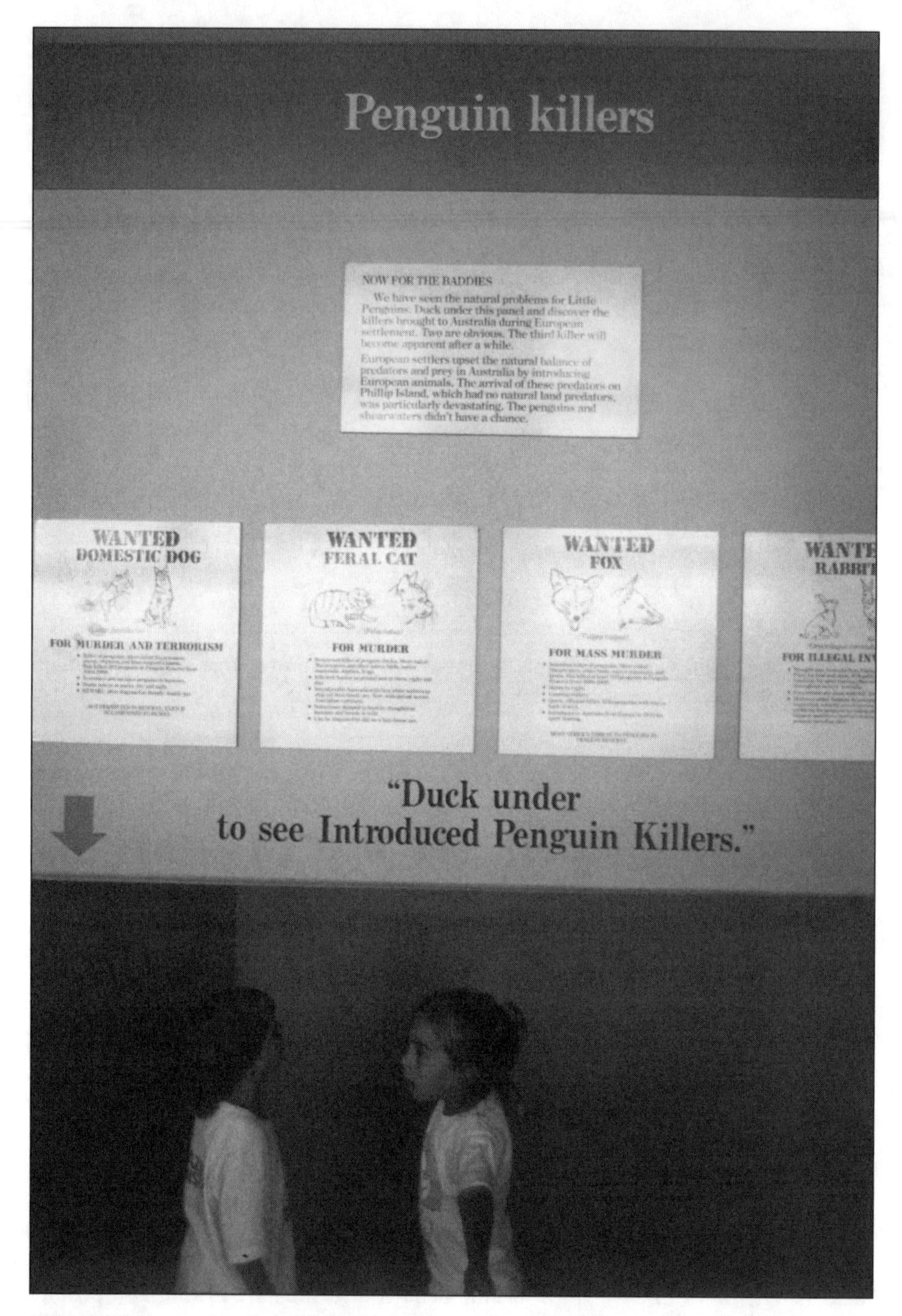

FIGURE 8.5 A critical aspect of interpretation is to orient the message and delivery to the audience. Here we see signs designed to appeal to a younger audience. *Photo: G. Hvenegaard.*

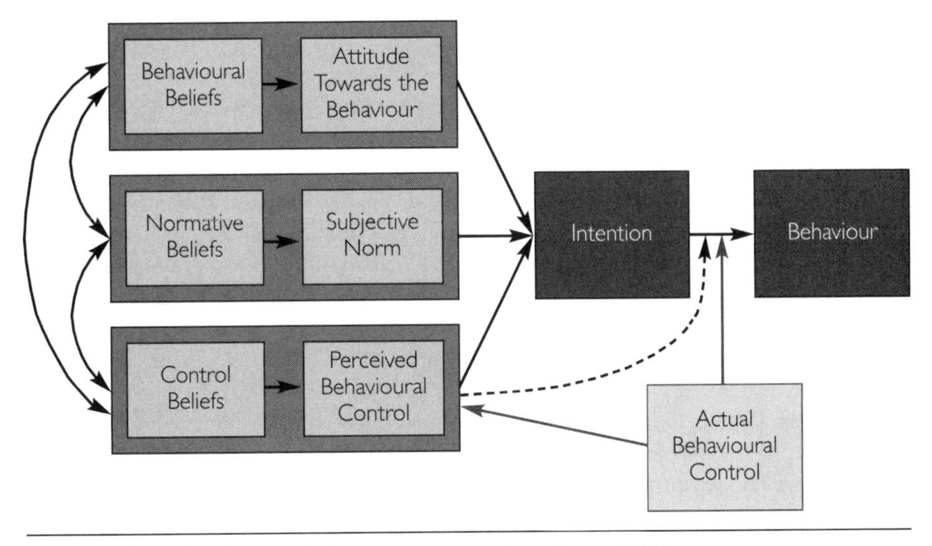

FIGURE 8.6 The theory of planned behaviour (from Ajzen, 2006).

and control beliefs (perceived ability and difficulty of undertaking the behaviour (Figure 8.6; Ajzen, 1991; Fishbein and Manfredo, 1992). Thus, in order to change behaviour, one or more of the above variables must be altered. For example, changing a behavioural belief through an interpretive program on the impact of feeding animals might note that habituation and death of the animal could result from such behaviour, leading individuals to 'rethink' their behavioural or normative beliefs for this behaviour (see Box 8.2 for another example).

Theories from persuasion research are also frequently used in interpretation research (Knudson et al., 2003; Marion and Reid, 2007). Two routes to persuasive communication have been identified in the *elaboration likelihood model* (Petty et al., 1992). Information provided in protected areas (e.g., 'Feeding the animals is illegal') attempts to target the peripheral route to persuasion, where an individual's behaviour is directed more by the source of the message than by the cognitive change from the message itself. Tour guides' modelling of appropriate behaviour is another example of the peripheral route to persuasion. Interpretation and environmental education normally attempt to access the central route to persuasion, which is characterized by individuals attending to the message and then internalizing the message through active cognition. The information in the message is compared with existing beliefs, attitudes, and values, and—if successful—new beliefs, attitudes, values, or behaviours are created. For example, if visitors are informed that feeding wildlife will lead to habituation, illness, and death for a number of animals, they will process this information, relate it to their existing beliefs, attitudes, and values, and may adjust their behaviour accordingly.

Other positivistic approaches to assessing interpretation include decision-making, mindfulness, and moral development. *Decision-making* theory examines how individuals behave, suggesting it is a result of assessing consequences and benefits of various

BOX 8.2 Application of the Theory of Reasoned Action

Bright et al. (1993) used the theory of *reasoned action* to explore public acceptance of the US National Park Service's controlled burn policy. Based on an initial sample of visitors to Yellowstone National Park, behavioural intentions (to support the policy) were strongly correlated with attitudes and subjective norms. Further, there was a correlation between attitudes and normative beliefs. Several key outcomes of a controlled burn policy differentiated visitors who did or did not support the policy. Positive attitudes were affected by the outcomes of improving conditions for wildlife and allowing natural events to occur. Negative attitudes were influenced by effects on private property, destroying scenery, and allowing fires to get out of control.

Based on these key outcomes, researchers prepared belief-targeted messages to change support for the controlled burn policy. Researchers conducted three major tests about the theory of reasoned action. First, the change in attitudes and subjective norms about the policy were good predictors about support for the policy, although attitudes were slightly better predictors. Second, the change in beliefs and outcomes was positively correlated with the change in attitude towards the burn policy (but not for those who did not originally support the policy). Third, those beliefs targeted in the messages were more strongly endorsed over time, but again, not for those who did not originally support the policy. The researchers conclude that, while the theory can be used as a tool to understand relationships among belief change, attitude change, and behavioural intentions, the effectiveness of belief-targeted messages in changing beliefs is less clear and requires further study.

courses of action (Marion and Reid, 2007). In this scenario, visitors would assess the pros and cons of feeding wildlife and of not feeding wildlife. Moscardo (1999) suggests that visitors must be made *mindful* through having control of behaviour, receiving personally relevant interpretation, and having a variety of novel interpretive options available. *Moral development* theory suggests that people progress through six stages in three levels of moral development, ranging from pre-conventional (where fear of punishment largely dictates behaviour) to conventional (where the opinions of others largely determine behaviour) to post-conventional (consideration for fairness, justice, and equality largely influence behaviour) (Kohlberg, 1976). Moral development theory suggests that, depending on the audience and stage of moral development (e.g., children, students, seniors), different interpretive messages may be tailored to each moral development stage.

However, much like the lack of clear, predictable links between beliefs, attitudes, intensions, and behaviour in the context of research using the theories of reasoned action and planned behaviour, questions have been raised as to the link between decision-making, moral reasoning, and behaviour (Marnburg, 2001). Humans are often stubborn, illogical, and irrational, but each of the theories noted above uses a 'rational actor model', where human decision-making and behaviour is conceived as completely rational and thus predictable. However, the rational actor model has come under increasing criticism for ignoring many instances where behaviour does not follow the precepts of 'rational actors'; the foundational assumptions of this model have also been questioned (e.g., Green and Shapiro, 1996; Pellikaan and van der Veen, 2002; Mueller, 2004). The assumption that people have equal access to information, review all alternatives impartially, and act according to rational cognitive processes may not always be true; people behave in certain ways for many different reasons, not all of them rational.

Notwithstanding the important contributions quantitative studies have made in this field, a growing number of researchers have highlighted the potential utility of incorporating qualitative research to study interpretation. Social scientists in many disciplines have turned to so-called 'interpretive' approaches to science (Denzin and Lincoln, 2005). The traditional positivist, reductionist approaches to studying behavioural issues focus on quantifying human behaviour and experience to enable identification of constituent variables and use them to predict behaviour of wider populations. The interpretive perspective uses reflexive, qualitative approaches in research to create a deeper understanding of experiential dimensions of human behaviour and acknowledge the social context of behaviour. Normally, a relativist (as opposed to realist) epistemological stance is adopted (Denzin and Lincoln, 2005). For example, if realists consider resources to have innate properties, the role of interpreters would be to highlight these properties to visitors. On the other hand, relativists suggest that meanings are socially constructed by humans, and change through time and space and across cultures. Thus, if there is no one 'truth' to interpret, the interpreters' job is to 'deconstruct' the meanings attached to natural and cultural features and thus bring to light their hidden assumptions and contexts.

The use of postmodernism and social constructivism to frame communication between interpreters/visitors and environmental educators/non-visitors acknowledges the two-way process of communication and the constructed nature of reality for both 'sender' and 'receiver'. Traditional *reductionist* approaches typically view communication as a one-way process and ignore situational contexts in the making of meaning. *Constructivist* approaches 'extend the focus from the exhibition or experience itself to include the visitor who interprets, understands, and imposes meaning on the [interpretation], often within a social context' (Ballantyne, 1998: 84). Meaning is not simply contained within the protected area or specific feature being interpreted; meaning is constantly and actively being shaped and reshaped by both senders and receivers, based on individual and societal contexts (Archer and Wearing, 2003; Reisinger and Steiner, 2006). For example, many agencies around the world have recently re-evaluated interpretive themes and language regarding the history and role of Aboriginal peoples in areas later designated as protected areas.

Knapp and colleagues (e.g., Knapp and Yang, 2002; Knapp and Benton, 2004) have taken the lead in using qualitative approaches towards studying interpretation (see also Archer and Wearing, 2003; Deacon, 2004; Reisinger and Steiner, 2006). We view the combination of realist/positivist and relativist/interpretive approaches to the study of interpretation, environmental education, and tour guiding as a sign of the increasing epistemological and theoretical sophistication in recent research in interpretation. It reflects similar trends throughout the social sciences, and suggests a growing awareness of the utility of theory among interpretation researchers. However, it must also be noted that the long-standing split between interpretation professionals and academic professors is still in evidence, despite the efforts by both groups to bridge this gap: professionals are sometimes less convinced of the need for research and theoretical frameworks (Larson, 2004).

INTERPRETATION IN PRACTICE

Interpretation is both a science and an art (Tilden, 1977). As a science, it relies on proven learning principles in psychology, sociology, communications, and education; it also requires familiarity with the area's visitors and their motivations. In addition, interpretation must be based on a thorough understanding of the natural and social sciences, in areas as diverse as geology, botany, history, paleontology, psychology, anthropology, geography, folklore, and zoology.

As an art, interpretation responds to the unpredictable interactions with the park audience. The process and rationale for selecting a given concept may be based in the sciences, but the effective communication of that concept requires art. Elements such as drama, visual design, and music improve visual and verbal communications. Complex relationships may be introduced in an entertaining fashion to arouse the visitor's interest; more refined treatments of specific concepts may then follow. Because interpretation involves meanings and relationships, it strives for a holistic approach rather than merely presenting isolated facts (Knudson et al., 2003).

In the past, some interpreters and park managers have mistakenly assumed that all visitors desire interpretation. However, Stewart et al. (1998) suggest there are at least four types of potential users of interpretive sources: (1) 'seekers' actively search for interpretation; (2) 'stumblers' normally interact with interpretation by accident rather than design; (3) 'shadowers' are chaperoned by other people through interpretation; and (4) 'shunners' actively avoid interpretation. Each type of interpretive user poses unique challenges and opportunities to interpretive and park planners and staff. Another challenge is presented by the diverse range of interpretive needs, based on the variety of visitor ages, backgrounds, experiences, and personalities, acknowledged some time ago by Tilden (1977). This may require a variety of interpretive programming strategies to reach a variety of audiences on a variety of levels. Knowing the characteristics of park visitors requires thorough research. Such research is currently limited in Canada; Parks Canada and other park agencies need to increase their capacity for social science research, beginning by focusing on segmenting park visitors and non-visitors (see, e.g., Danchuk et al., 2000).

Important communication principles have been summarized by Dick et al. (1974), Sharpe (1982), Knudson et al. (2003), and Marion and Reid (2007). While more complicated than can be presented here, communication and the tendency to change opinion will be improved by understanding and integrating three components of the interpretive experience: the communicator, the message, and the receiver. First, communication is improved if the communicator is: well-informed, intelligent, and trustworthy; is well-liked; states at least one opinion with which the receiver agrees; gets and holds the attention of the receiver; and uses credible sources. In one example, Manfredo and Bright (1991) found that users of the Boundary Waters Canoe Area Wilderness were more persuaded by information packets if they were perceived to be credible.

Second, messages are more effective if they use good grammar; are clear, concise, and consistent; gain the attention of the receiver; are understandable; relate to the interests of the receiver; have a moderate, versus substantial, difference in opinion; cater to diverse interests; provide an appropriate means of action for the receiver; and, for complicated messages, state conclusions clearly. Messages with a request to change behaviour are more likely to be successful if based on ecological rather than social reasons (Marion and Reid, 2007). For example, closing an area due to ecological impacts from visitor use is normally much more strongly supported than closing an area due to lack of funds or overcrowding. Overall, the experience should be rewarding and fun for the receiver. To provide one example, Morgan and Gramann (1989) found that the effectiveness of interpretive programs (in this case, reflecting improved attitudes towards snakes) was not related to the amount of information. The level of involvement, especially a combination of exposure, modelling, and direct contact, was much more important in improving attitudes about snakes (see also Guy et al., 1990). Similarly Bitgood and Patterson (1993) found that visitors spent more time reading if the exhibits minimized the number of words used in the labels, enlarged letter size, avoided distractions, and were relevant to and placed near the objects being discussed.

Third, acceptance of the interpretive message is more likely if the receiver exposes him/herself to communications of interest; is made to feel at ease; is motivated by group membership; and continues discussion on the issue with like-minded people. For example, Manfredo and Bright (1991) found that less-experienced wilderness canoeists (i.e., with less confidence about the area or with canoeing in general) were more persuaded by information than more experienced canoeists.

Orams (1996) outlines a model for interpretation programs that emphasizes the affective (i.e., attitudes, feelings, emotions, and value systems) and cognitive domains (i.e., knowledge and perception). Including the former recognizes that knowledge alone does not change attitudes and values; they need to be addressed directly. Related to the latter, an interpretive program can develop cognitive dissonance or a dynamic disequilibrium by 'throwing people off balance'. Returning to the example of feeding wildlife, if people want wildlife populations to thrive but learn that such feeding behaviour can cause harm, their behaviour is dissonant with their goals. Such psychological discomfort creates a 'teachable moment' (Forestell, 1991), causing people to ask 'how', 'why', and 'when' types of questions about their behaviour. To induce behaviour change to achieve management goals, interpretive programs should follow these attempts to change knowledge and attitudes by providing a motivation to act (e.g., to resolve their cognitive dissonance) and opportunity to act (appropriate to the skill level of visitors). This should be followed by evaluation and feedback.

FORMS OF INTERPRETATION

In selecting a form of interpretation, interpreters should consider a variety of factors (Sharpe et al., 1994). First, interpreters should consider visitors' need for orientation, use of other facilities, group size, language barriers, and varying backgrounds. Second, related to the natural environment, interpreters should consider relevant interpretive themes, appropriate timing of events, vulnerability of park features, and potential safety hazards. Finally, interpreters should consider how interpretive efforts further the park agency's goals, in the context of investment, maintenance, and replacement costs. Table 8.1 provides a list of organizations and publications that examine many of these issues. Interpretation takes two basic forms: (1) personal interpretation (e.g., guided hikes) involves direct contact between the interpreter and the visitor, (2) and non-personal interpretation connects visitors with inanimate interpretive media (e.g., signs).

Personal Interpretation

Informational services tell visitors where specific facilities and opportunities are located and how to make use of them, but do not involve interpretive programs, as described below. *Scheduled services*, using people as interpreters in this era of high-tech gadgetry, remain one of the most popular and effective forms of interpretive events. These events take place at a predetermined time and are advertised accordingly. Many examples are described below.

Guided tours, led by interpreters, encourage interactions between visitors and the natural environment. *Audiovisual presentations* are effective ways to convey abstract messages and provide excellent substitute experiences. *Prop talks*, using artifacts as focal points of a talk, can also provide valuable first-hand involvement. *Dramatic presentations* require more elaborate production but can prove highly effective and entertaining. *Point duty* involves stationing an interpreter at a prominent feature or gathering place during

TABLE 8.1 Interpretation Organizations and Publications

Organization	Publication	Website
Interpretation Canada: An Association for Heritage Interpretation	*Interpscan*	www.interpcan.ca/new/
National Association for Interpretation	*Journal of Interpretation Research, The Interpreter*	www.interpnet.com/
Association for Heritage Interpretation	*Interpretation*	www.heritage-interpretation.org.uk/
Association for Experiential Education	*Journal of Experiential Education*	www.aee.org/
Canadian Network for Environmental Education and Communication	*Canadian Journal of Environmental Education*	cjee.lakeheadu.ca/
North American Association for Environmental Education	*Environmental Communicator*	naaee.org/
Heldref Publications: Helen Dwight Reid Educational Foundation	*Journal of Environmental Education*	www.heldref.org/jenve.php

periods of high visitation. *Travelling point duty* or *roving duty* is similar, but the interpreter moves through an area, informally interpreting sites to people who are encountered. *Impromptu events* are not formally scheduled but are well planned; they are impromptu only for the visitor. This might involve, for example, the sudden presence of an interpreter on a beach with table, aquarium, net, bioscope, and sample jars to offer visitors a close look at sand particles and aquatic invertebrates.

Living interpretation demonstrates a historical lifestyle that is different from that of the visitors. Living interpreters in period costumes and authentic settings carry out day-to-day activities, showing visitors how people actually lived, often with technical information or authentic products as an additional feature. *Extension programs* are taken into communities or schools or communicated through various media, with the intention of expanding the audience for an interpretive message.

Non-Personal Interpretation

Visitor centres provide visitors with essential information about an area and its values, special features, opportunities for the visitor, and overall role in the park system. Visitor centres can also provide important facilities for visitors and staff. *Exhibits* at the visitor centre or in a park near key natural features may include kiosks, dioramas, artifacts, reconstructions, and models. Exhibits should be versatile, so they can respond to changes in information and season. *Signs* interpret natural or cultural features in the immediate vicinity; readers can decide what to read and how fast to read it. Self-guided *interpretive trails* use signs or brochures to help guide visitors to interesting features that might otherwise be overlooked or not fully appreciated (Sharpe, 1982; Knudson et al., 2003). *Publications* provide more detailed information, and can be taken home as souvenirs and referred to many times after a visit. *Websites* can be referenced off-site, allowing the visitors to be better informed about unique park features, opportunities, and services prior to their visits.

Recent research suggests that park managers should be aware of these many forms of interpretation. The effectiveness of interpretation seems to be increased through the provision of several types of interpretative media, as various types of visitors with different motivations will either ignore interpretation altogether or will access only certain types of interpretation in certain locations (Porter and Howard, 2002; Manning, 2003). The question of the effectiveness of interpretation is a critical one, and is discussed in greater detail below.

INTERPRETATION AS A MANAGEMENT TOOL

As Sharpe (1982: 9) notes, one of the main benefits of park interpretation programs is to 'motivate the public to take action to protect their environment'. Managers use interpretation to reduce negative visitor impacts on the environment, minimize destructive behaviour, decrease enforcement problems, and guide visitors towards designated and selected locations (Hendee and Dawson, 2002). Interpretation can also minimize public safety incidents by emphasizing common public safety concerns in the park. As Boxes 8.3 and 8.4 suggest, park agencies should view interpretation as one of several approaches needed to ensure protection of the park environment, including protective facility design, proper legal designation, and a commitment to enforcement. In some cases, interpretation has unintended side effects. For example, interpretation can emphasize that rare species have been severely impacted by off-trail use; however, managers should be aware that this can result both in increased understanding of the issues and in increased searching, and subsequent off-trail use, for those same rare species (Cialdini, 1996).

Interpretation can promote public understanding of an agency's goals and objectives. Interpretation can demonstrate, by example, the agency's philosophies about environmental protection and the intrinsic value of its natural and cultural heritage. This will hopefully improve its public image and solidify public support for the agency's goals. The park agency may also choose to highlight, through interpretation, the values of and issues facing the agency.

Is Interpretation Effective?

The number of studies assessing the effectiveness of interpretation has increased over the last decade, driven in part by a focus on accountability in the public sector and budget cuts in most protected area systems (Dearden and Dempsey, 2004; Shultis, 2005).

BOX 8.3 Examples of Interpretation-Effectiveness Studies

Interpretation and Resource Protection

In Kananaskis Country, Alberta, one year after starting an innovative poster campaign illustrating commonly picked flowers (e.g., 'Wanted ALIVE not dead'), the number of visitors reprimanded by park staff for picking flowers decreased by 50 per cent (Wolfe, 1997).

In the 1970s, staff at Dinosaur Provincial Park, Alberta noted low compliance from visitors to restricted-area signs. Within days after organizing interpretive hikes into interesting areas outside of the restricted area, the number of visitors observed within the restricted areas decreased nearly 90 per cent (Wolfe, 1997).

Interpretation and Public Safety

Near the Bonneville Lock and Dam, along the Columbia River, at least one person had drowned for five years in a row. The drownings resulted from improper anchoring. Sturgeon anglers were using anchor ropes that were too short in the deep, fast-moving water; the bows of the boats were simply pulled under the water. In response, interpreters developed new safe-anchoring posters and passed them out to anglers. In addition, interpreters placed large signs at boat ramps, developed an interactive video, and made contacts with supply stores. In the five years after the 'Safe Anchoring in Current' program was initiated, no visitors drowned (Barry, 1993).

After noting an increase in bears frequenting campsites, staff at Peter Lougheed Provincial Park, Alberta, initiated a 'Bear Paw Program'. They placed bear paw-shaped cards (with information from the bear's perspective) on the tables of campers who left their campsites in an unsuitable manner. Within one year, the number of problem campsites had decreased significantly (Wolfe, 1997).

Interpretation and Long-term Knowledge

Many studies show that interpretive programs increase short-term knowledge, but how long is that knowledge retained? Knapp and Yang (2002) interviewed park visitors on the bats of Indiana one year after they had participated in a one-hour evening interpretive program on the subject in Hoosier National Forest. Based on visitor responses, three factors affecting attention paid to the program included novelty, personal significance, and speaker qualities. Long-term memory was influenced by program activities, prior knowledge/misconceptions, and visual imagery.

Interpretation Impacts on Knowledge, Attitudes, and Behaviour

Does knowledge influence attitudes and, subsequently, behaviour? Tubb (2003) conducted pre- and post-visit interviews with respondents at the High Moorland Visitor Centre in Dartmoor National Park, England. First, post-visit respondents answered knowledge questions 4–13 per cent more correctly than pre-visit respondents. Second, among 11 attitude statements, post-visit respondents had significantly stronger protection attitudes than pre-visit respondents for only two attitudes (role of park agency and feeding of animals). Third, regarding behavioural change, post-visit respondents were more aware of actions they could take to protect the park than pre-visit respondents, but no significant difference was found in their intentions to act in a way to reduce their environmental impact.

Who Is Likely to Use or Read Interpretive Information?

On Fraser Island, Australia, all registering visitors receive a 'Be Dingo-Smart' interpretive brochure about how to have safe experiences around dingoes. According to Porter and Howard (2002), only 59 per cent of respondents recalled receiving the brochure (lower for younger respondents), and half stated they had read the brochure (lower for individuals and groups of friends and for younger respondents). Wildlife watchers and hikers were more likely to read on-site interpretation. Knowledge about dingoes was lower for younger respondents and for respondents travelling with friends.

In addition, a greater number of academic researchers, primarily from Australia and the United States, are conducting research in this area. The majority of researchers use a positivistic, empirical approach to assessing interpretive effectiveness, with quantitative research approaches using visitor surveys the norm in this research area.

However, a small but increasing number of scholars have questioned this traditional conceptualization of interpretation and interpretation research (Staiff et al., 2002). Tubb (2003) notes that broader research in education and communication points to one main finding: human behaviour is extremely complex and resistant to change, and educational theories and processes have only limited success in effecting attitudinal or behaviour change. Moreover, the link between beliefs, attitudes, intentions, and behaviour—as posited by theories of reasoned action and planned behaviour—has led to important advances in understanding human behaviour, but predicting human behaviour using these models has seen limited success (Conner and Armitage, 1998; Godin and Kok, 1996). Given the difficulty of researchers in education, psychology, and communications to be able to predict or meaningfully effect attitudinal or behavioural change, the ability of one short-term interpretive program during recreationists' or tourists' leisure time in a non-captive setting (Orams, 1997; Tubb, 2002; Knapp and Yang, 2002) to create significant attitudinal or behavioural change may be an unfair proposition.

Perhaps the most difficult issue in measuring the effectiveness of interpretation is deciding what interpretation is supposed to achieve (Knapp and Benton, 2004). As previously noted, interpretation has always had very lofty goals: 'Through interpretation, understanding; through understanding, appreciation; through appreciation, protection' (Tilden, 1977: 38). Tilden's objectives of 'provocation' and 'revelation' are also difficult to assess (Beckmann, 1999). Interpretation is both a management tool and a leisure experience, which also adds to the difficulty in defining 'effectiveness'. The link between information, interpretation, environmental education, and tour guiding also provides challenges, as all of these activities are related yet each may have differing goals and objectives.

Marion and Reid (2007) identify four categories of research assessing interpretive effectiveness: knowledge gain, behavioural change, redistributing visitors, and changes in resource conditions (see also Manning, 2003). Interpretation is conceptualized as both a management tool and a communication/education strategy. No research has tackled the 'protection' part of the interpretive equation; it is extremely difficult to design research that could assess the impact of specific interpretive programs on long-term visitor behaviours that would serve to 'protect' parks (e.g., volunteering, employment/career choices, public input into decision-making, and membership in environmental groups). Similar research assessing the potential long-term impact of early-life outdoor experiences on environmental attitudes and behaviour has also proven to be methodologically challenging (e.g., Bixler et al., 2002; Hahn and Kellert, 2002; Ewert et al., 2005).

In terms of information and knowledge gained from interpretation, the research suggests that interpretation often provides statistically significant increases in knowledge (reviewed in Manning, 2003; Marion and Reid, 2007). These studies typically assess short-term information or knowledge gains, normally using a quasi-experimental, pre- and post-test soon after the interpretive intervention. The research on impacts on behaviour are less impressive, suggesting that interpretation programs do not always significantly affect visitor behaviour; informational sanctions often worked as well as or better than interpretive programs, and non-personal forms of interpretation (e.g., signs) often were as effective as personal forms of interpretation (e.g., presentations) (see Manning, 2003; Marion and Reid, 2007). This type of research is sometimes limited by the use of self-reported behaviour rather than actual behaviour (or a combination thereof) to assess visitor behaviour. Even though redistributing visitors to low-use areas is not always considered an appropriate management strategy for minimizing visitor impacts (Leung and Marion, 2000), a limited amount of research has shown that different types of information and interpretation can disperse use in protected areas. Finally, limited research suggests that changes in resource conditions (including depreciative behaviour such as littering) can be lowered through the use of interpretation and information (Wells and Smith, 2000; Manning, 2003; Marion and Reid, 2007).

In sum, an increasing body of knowledge using a variety of theoretical and research approaches to study the effectiveness of interpretation is available. Increasingly, qualitative research is being used to assess the social construction of meaning created by interpreters and visitors. The nascent nature of much of this research in interpretation, the variety of definitions of 'effectiveness' and goals of interpretation used, the assessment of various types of heritage communication (i.e., information, personal and non-personal forms of interpretation, environmental education and tour guiding each can be referred to as 'interpretation'), and the variety of locations used (e.g., frontcountry versus backcountry; parks versus museums) all make it difficult to come to any firm conclusions on the effectiveness of interpretation. It does seem apparent, however, that interpretation in various forms and locations in protected areas can be effective in creating short-term changes in visitor attitudes, knowledge, and (to a lesser extent) behaviour. The longer-term implications, or how these experiences combine with other contextual information (e.g., media, other life experiences) to 'make meaning' for visitors participating in interpretation in protected areas is as yet *terra incognita* (Ewert et al., 2005). These findings for interpretation are equivalent to findings in the voluminous environmental education literature: research suggests that one-time environmental education programs may have short-term impacts on knowledge and attitudes, but behavioural changes are much more difficult to generate. For example, Bogner (1998: 28) concluded that, to be effective, environmental education programs needed to incorporate 'a combination of first-hand experience, participatory interaction, adequate preparation and subsequent reinforcement'. It seems likely that similar requirements would be necessary for interpretation to become more effective in the long-term.

INTERPRETIVE PLANNING

Without planning, interpretive programs can easily become ineffective or redundant. Several planning models are available for interpretation. The model by Sharpe et al. (1994) involves seven logical steps. The interpretive planner should work with other management specialists to determine objectives, take inventory, analyze data, synthesize alternatives, develop the plan, implement the plan, and evaluate and revise the plan. Throughout the planning process, planners should consider each objective of interpretation and make appropriate connections among them.

Another interpretation planning model by McArthur (1998) requires three key stages for successful interpretation. The first stage is to define a target audience, gaining an understanding of the group's demographics, motivations, expectations, and satisfactions. The second stage is to determine the content and structure of the interpretive message. This is considered within the hierarchy of interpretive themes, concepts, and messages. The third step is to select a technique, understanding its inherent advantages and disadvantages.

The latter model is applicable to interpretive efforts in Canada's national parks. First, visitors can be grouped according to experience, use history, trip behaviour, motivations,

BOX 8.4 Interpretation to Manage Dolphin Watching in Australia

At a resort in Tangalooma, Australia, about 40 km east of Brisbane, tourists have the opportunity to hand-feed wild bottlenose dolphins. Prior to 1993, resort staff running the dolphin feeding program provided only informal instructions as part of their management strategy to control interactions between dolphins and tourists. Some of the rules included: only supervised entry allowed into the feeding area, limited amount of fish fed to dolphins, no touching or swimming with dolphins, contact time minimized, people with colds or flu not allowed to feed dolphins, and no photography in or near the water. One staff member made sure that tourists wanting to feed dolphins disinfected their hands before touching the animals, and another staff member helped people feed the dolphins.

In 1994, a structured interpretive program was implemented with three main features. First, a Dolphin Education Centre provided information and served as the location where tourists could collect a limited number of free tokens, which allowed visitors to feed dolphins later that night. Second, people with tokens were given an educational briefing about the dolphins and the rules regarding feeding. Third, a public address system allowed staff to talk to feeders and observers during the feeding sessions (focusing on the behaviour and ecology of dolphins, and related environmental responsibilities of visitors).

Researchers measured the effectiveness of interpretation on visitor behaviour. They observed and videotaped feeding sessions before and after the new interpretive program to count inappropriate behaviours and cautions (Orams and Hill, 1998). Standardized to 100 feeding events, the number of touches in the control group dropped from 6.7 in 1993 to 1.2 in 1994. As well, the number of staff cautions dropped from 2.6 to 1.2, and the number of other inappropriate behaviours dropped from 3.2 to 1.1.

At the same time, researchers measured the impact on the tourists (Orams, 1997) by asking questions about their experience immediately after and 2–3 months after the feeding. Although high levels of enjoyment exist for both groups, the interpretive program increased the level of enjoyment for 1994 visitors, especially their desire for more information. Knowledge levels of 1994 visitors were significantly higher than those of 1993 visitors. The same was true for increases in environmentally conscious attitudes, but intentions to participate in environmentally friendly behaviour were not significantly different. However, the 1994 visitors increased their actual environmentally friendly behaviour; they removed beach litter, made donations, and became involved in environmental issues.

attitudes towards interpretation and park management, or demographic characteristics. Visitor types are best analyzed at the park level, and planners often use techniques from the Visitor Activity Management Process (Nilsen and Tayler, 1998). This process requires an understanding of the demographic, attitudinal, and activity characteristics of each visitor group to guide visitor management. This information also can assist interpreters in planning, marketing, and evaluating interpretive activities.

Second, the content of interpretive messages is guided by national policy and adapted to individual parks. Nationally, the purpose of interpretation is to encourage visitors to appreciate, understand, and support:

- Canada's system of nationally significant heritage places;
- the essence of each heritage place and how it is significant to the country and relevant to individuals;
- the need to protect heritage resources (Parks Canada 1998b); although not directly stated, the protection of ecological integrity should be the primary purpose for interpretation (Parks Canada Agency 2000).

Thus, across the system, interpretation focuses on messages about the national park system, ensuring ecological integrity, and issues of national significance. Individual parks then highlight national and park-specific issues as they relate to these messages. For example, each park could interpret the status of completing the national parks system, how the park is representative of its natural region, and why the park is of national significance. Similarly, each park could interpret species that are endangered at the national level and that are found in the park. Messages should complement current site-level management practices to promote ecological integrity. Interpretive messages should also be placed in a regional or national context, and relate to broader environmental issues (Parks Canada, 1994), such as global warming, acid rain, recycling, and water quality.

Finally, each park chooses the mode of delivering each interpretive message, as discussed earlier. In times of cutbacks, choosing an interpretive technique is heavily influenced by funding and staffing. However, subject to these constraints, an interpretive technique should be chosen for its ability to reach the target audience, effectively interpret the target message, and fulfill the broader goals of interpretation.

CONCLUSION

Interpretation in North America has an amazing history of inspirational, dedicated, and professional staff. Interpretation in protected areas fulfills several important goals derived from legislation, policy, societal attitudes and values, and visitor behaviour. Significant and tangible benefits result from park interpretive efforts; these accrue to the visitor, park environment, park agency, and ecosystems beyond the park boundaries.

Unfortunately, the full importance of interpretation is not often reflected in a park's staffing and budgetary priorities. Evidence for this comes in many forms. For example, as budgets rise or fall, interpreters are often the last to be hired and the first to be fired. Many full-time interpreters have been replaced by seasonal employees (Parks Canada Agency, 2000). Budgets have been severely cut, forcing staff to rely on the least costly, and in many cases least effective, interpretive methods, when such programs survive at all. In recent years, attempts have been made by several provincial governments to privatize programs in a global trend to convert park operations from public domain to profit-motivated privatization. The BC government's short-sighted decision to cut publicly funded interpretation in 2001 provides an extreme example of these trends.

In many parks, interpretation is under-utilized as a park management service. There are at least three reasons for this. First, few managers—who control the park budgets—actually possess a background in interpretation or communications, leading to a poor understanding of interpretation's potential and role in the ecological and visitor management mandates of parks. Second, and perhaps most importantly, interpretation is generally poorly integrated into the park planning and management efforts. This results, for example, in interpretive staff located in buildings that offer little interaction with other park operational staff. Moreover, it is often unclear how interpretation and environmental education relate to individual park management plans or system planning policies. Third, managers are rarely familiar with published research on the effectiveness of interpretation. Without this knowledge, such managers rarely consider interpretation as an important management tool, nor do they actively promote interpretation in setting budgetary priorities. Programs are also poorly supported because of a lack of useful information on park user demographics and preferences; interpretive content is often narrowly or inappropriately focused; unimaginative communication techniques are often used; and evaluation is generally lacking (Sharpe, 1982; Nyberg, 1984; Wolfe, 1997; McArthur, 1998; Knudson et al., 2003).

As a further constraint on the expanse and support of interpretive programs, the unique and intrinsic features of individual parks are poorly celebrated from a systems perspective or poorly aligned to the seasonal specialties of the calendar year. Environmental education opportunities with local schools seem more reluctantly accommodated than encouraged and facilitated. Few parks effectively promote their relevance beyond the park boundaries and training opportunities are limited and rarely encouraged. However, at the federal level, at least in Parks Canada and the National Park Service (United States), this situation appears to be changing; indeed, outreach programs are being emphasized much more than on-site interpretation, especially among 'new' (recent immigrants), urban, and young residents (e.g., National Parks Service, 2003; Parks Canada Agency, 2006b). There is a concern that these populations are not engaging in protected areas to the same extent as the 'traditional' visitors, which may lead to decreased support for these areas. Parks Canada is responding to these concerns by increasing the profile of 'heritage communication' through its 'Engaging Canadians' campaign and its creation of the External Relations and Visitor Experience Directorate at the head office level (Parks Canada Agency, 2006b).

Effective interpretation is essential to the successful management and operation of park and protected area systems. Protected areas cannot survive as islands. Their survival is closely tied to people's attitudes, beliefs, and values. Public support, at both the political and community levels, is necessary to help an area succeed in meeting its conservation and preservation goals (Parks Canada Agency, Performance, Audit and Review Group, 2005).

An important goal of any protected area is to give local residents and visitors information to increase their awareness and understanding of the area's natural values and to relate these experiences to modern life. Achieving this goal will result in informed people who have a deeper appreciation for their area's natural and cultural heritage, and who transfer these values and experience into their daily lives. Attitudes towards the environment are learned, not inborn.

Protecting parks and wilderness areas is, in many ways, comparable to a library's acquiring important works to ensure the availability of the literature of the past and present. Acquisition and protection are indeed important, but the books have to be read and understood for their true worth to be realized. While the librarian (or park manager) may conserve the volumes, the visitor must also be shown how to read them. Most visitors to parks and wilderness settings today lack the experience to adequately 'read' such places. Their visits are comparable to passing through a corridor of valuable books, most of which are unreadable. Interpretation resolves this dilemma. It guides the visitor to discover the wonders contained within these volumes, to 'experience this sense of wonder', as Rachel Carson (1965) described it, to be moved emotionally and mindfully along the path of discovery, to be motivated to understand more, and to experience in the volumes of nature's wealth and complexity an upwelling of pride in our heritage and an inner sense of belonging and richness in our personal lives. While this is an extremely tall order, interpretation—and the related techniques of environmental education and tour guiding—appear to be the best approaches we have for making such substantial changes at the individual and societal level.

REFERENCES

Ajzen, I. 1991. 'The theory of planned behavior', *Organizational Behavior and Human Decision Processes* 50: 179–211.

———. 2006. 'The theory of planned behavior'. At: <www.people.umass.edu/aizen/tpb.diag.html>.

Archer, D., and S. Wearing. 2003. 'Self-, space, and interpretive experience: The interactionism of environmental interpretation', *Journal of Interpretation Research* 8, 1: 7–23.

Ballantyne, R. 1998. 'Interpreting "visions": Addressing environmental education goals through interpretation', in D.L. Uzzell and R. Ballantyne, eds, *Contemporary Issues in Heritage and Environmental Interpretation*. London: The Stationery Office, 79–99.

——— and K. Hughes. 2001. 'Interpretation in ecotourism settings: Investigating tour guides' perceptions of their role, responsibilities and training needs', *Journal of Tourism Studies* 12, 2: 2–9.

——— and D. Uzzell. 1999. 'International trends in heritage and environmental interpretation: Future directions for Australian research and practice', *Journal of Interpretation Research* 4, 1: 59–75.

Barry, J.P. 1993. 'Anchoring safely in current: An example of using interpretation to solve a management problem', *Legacy* 4, 4: 22.

Beckmann, E.A. 1999. 'Evaluating visitors' reactions to interpretation in Australian national parks', *Journal of Interpretation Research* 4, 1: 5–19.

Bitgood, S.C., and D.D. Patterson. 1993. 'The effects of gallery changes on visitor reading and object viewing time', *Environment and Behaviour* 25: 761–81.

Bixler, R.D., M.F. Floyd, and W.E. Hammitt. 2002. 'Environmental socialization: Quantitative tests of the childhood play hypothesis', *Environment and Behavior* 34, 6: 795–818.

Bogner, F. 1998. 'The influence of short-term outdoor ecology education on long-term variables of environmental perspective', *Journal of Environmental Education* 29, 4: 17–30.

Bramwell, B., and B. Lane. 1993. 'Interpretation and sustainable tourism: The potential and the pitfalls', *Journal of Sustainable Tourism* 1, 2: 71–80.

Bright, A., M.J. Manfredo, M. Fishbein, and A. Bath. 1993. 'Application of the theory of reasoned action to the National Park Service's control burn policy', *Journal of Leisure Research* 25, 3: 263–80.

Canada. 1990. *National Parks Act.* Ottawa: Minister of Supply and Services Canada.

Canadian Environmental Advisory Council. 1991. *A Protected Areas Vision for Canada.* Ottawa: Minister of Supply and Services Canada.

Carlsbad Caverns National Park. 2004. *The National Park Service Role in American Education. Chihuahuan Desert Lab Manual: Umbrella – Project 0, Supplement 0.7.* At: <www.nps.gov/archive/cave/cdl/supp/p0s7.pdf>.

Carson, R. 1965. *The Sense of Wonder.* New York: Harper & Row.

Christie, M.F., and P.A. Mason. 2003. 'Transformative tour guiding: Training tour guides to be critically reflective practitioners', *Journal of Ecotourism* 2, 1: 1–16.

Cialdini, R.B. 1996. 'Activating and aligning two kinds of norms in persuasive communications', *Journal of Interpretation Research* 1, 1: 3–10.

Cohen, E. 1985. 'The tourist guide: The origins, structure and dynamics of a role', *Annals of Tourism Research* 12: 5–29.

Conner, M., and C. Armitage Jr. 1998. 'Extending the theory of planned behavior: A review and avenues for further research', *Journal of Applied Social Psychology* 28, 1429–64.

Dahles, H. 2002. 'The politics of tour guiding: Image management in Indonesia', *Annals of Tourism Research* 29, 3: 783–800.

Danchuk, G., R. Weaver, and D. Dugas. 2000. 'Social science at Parks Canada', *Proceedings of the Travel and Tourism Research Association Conference*, 17–19 Sept. Whitehorse, Yukon, 45–51.

Deacon, H. 2004. 'Intangible heritage in conservation management planning: The case of Robben Island', *International Journal of Heritage Studies* 10, 3: 309–19.

Dearden, P., and J. Dempsey. 2004. 'Protected areas in Canada: Decade of change', *Canadian Geographer* 48: 225–39.

Denzin, N.K., and Y.S. Lincoln. 2005. *The Sage Handbook of Qualitative Research*, 3rd edn. Thousand Oaks, Calif.: Sage.

Diamantis, D. 2004. *Ecotourism: Management and Assessment.* London: Thomson.

Dick, R.E., D.T. McKee, and J.A. Wagar. 1974. 'A summary and annotated bibliography of communication principles', *Journal of Environmental Education* 5, 4: 1–13.

Ewert, A., G. Place, and J. Sibthorp. 2005. 'Early-life outdoor experiences and an individual's environmental attitudes', *Leisure Sciences* 27: 225–39.

Fishbein, M., and M.J. Manfredo. 1992. 'A theory of behavior change', in M.J. Manfredo, ed., *Influencing Human Behavior: Theory and Applications in Recreation, Tourism and Natural Resources Management.* Champaign, Ill.: Sagamore, 29–50.

Forestell, P.H. 1991. 'Marine education and ocean tourism: Replacing parasitism with symbiosis', in M.L. Miller and J. Auyong, eds, *Proceedings of the 1990 Congress on Coastal and Marine Tourism.* Newport, Ore.: National Coastal Resources Research & Development Institute, 35–9.

Godin, G., and G. Kok. 1996. 'The theory of planned behavior: A review of its applications to health-related behaviors', *American Journal of Health Promotion* 11: 87–98.

Green, D., and I. Shapiro. 1996. *Pathologies of Rational Choice Theory: A Critique of Applications in Political Science.* New Haven: Yale University Press.

Grant, T., and G. Littlejohn. 2004. *Teaching Green: The Middle Years.* Toronto: Green Teacher.

Guy, B.S., W.W. Curtis, and J.C. Crotts. 1990. 'Environmental learning of first-time travellers', *Annals of Tourism Research* 17: 419–31.

Hahn, P.H., and S.R. Kellert. 2002. *Children and Nature: Psychological, Sociocultural, and Evolutionary Investigations.* Cambridge, Mass.: MIT Press.

Ham, S.H., and B. Weiler. 2002. 'Interpretation as the centrepiece of sustainable wildlife tourism', in R. Harris, T. Griffin, and P. Williams, eds, *Sustainable Tourism: A Global Perspective.* Amsterdam: Elsevier Butterworth-Heinemann, 35–44.

Hendee, J., and C. Dawson. 2002. *Wilderness Management: Stewardship and Protection of Resources and Values*, 3rd edn, Golden, Colo.: Fulcrum Publishing.

Hvenegaard, G.T. 1994. 'Ecotourism: A status report and conceptual framework', *Journal of Tourism Studies* 5, 2: 24–35.

——— and P. Dearden. 1998. 'Ecotourism versus tourism in a Thai national park', *Annals of Tourism Research* 25: 700–20.

Interpretation Canada. 2008. Interpretation Canada Homepage. At: <www.interpcan.ca/new/>.

Knapp, D., and G.M. Benton. 2004. 'Elements to successful interpretation: A multiple case study of five national parks', *Journal of Interpretation Research* 9, 2: 9–25.

——— and L.-L.Yang. 2002. 'A phenomenological analysis of long-term recollections of an interpretive program', *Journal of Interpretation Research* 7, 2: 7–17.

Knudson, D.M., T.T. Cable, and L. Beck. 2003. *Interpretation of Cultural and Natural Resources.* State College, Penn.: Venture.

Kohlberg, L. 1976. 'Moral stage and moralization: The cognitive-developmental approach', in T. Lickona, ed., *Moral Development and Behaviour.* New York: Holt, Rinehart and Winston, 31–53.

Larson, D.L. 2004. 'Research: A voice of our own', *Journal of Interpretation Research* 9, 1: 69–71.

Leung, Y.F., and J.L. Marion. 2000. 'Recreation impacts and management in wilderness: A state-of-knowledge review', in S.F. McCool, W.T. Borrie, and J. O'Loughlin, eds, *Wilderness Science in a Time of Change Conference Proceedings.* RMRS-P-15-VOL-5. Ogden, Utah: USDA Forest Service, Rocky Mountain Research Station, 23–48.

Louv, R. 2005. *Last Child in the Woods: Saving Our Children from Nature-Deficit Disorder.* Chapel Hill, NC: Algonquin Books of Chapel Hill.

McArthur, S. 1998. 'Introducing the undercapitalized world of interpretation', in K. Lindberg, M.E. Wood, and D. Engeldrum, eds, *Ecotourism: A Guide for Planners and Managers*, vol. 2. North Bennington, Vt: Ecotourism Society, 63–85.

Mackintosh, B. 2000. *Interpretation in the National Park Service: A Historical Perspective.* US National Park Service. At: <www.cr.nps.gov/history/online_books/mackintosh2/origins_before_nps.htm>.

Manfredo, M.J., and A.D. Bright. 1991. 'A model for assessing the effects of communication on recreationists', *Journal of Leisure Research* 23, 1: 1–20.

Manning, R. 2003. 'Emerging principles for using information/education in wilderness management', *International Journal of Wilderness* 9, 1: 20–7, 12.

Marnburg, E. 2001. 'The questionable use of moral development theory in studies of business ethics: Discussion and empirical findings', *Journal of Business Ethics* 32: 275–83.

Marion, J.L., and S.E. Reid. 2007. 'Minimizing visitor impacts to protected areas: The efficacy of low impact education programmes', *Journal of Sustainable Tourism* 15, 1: 5–27.

Morgan, J.M., and J.H. Gramann. 1989. 'Predicting effectiveness of wildlife education programs: A study of students' attitudes and knowledge toward snakes', *Wildlife Society Bulletin* 17: 501–9.

Moscardo, G. 1999. *Making Visitors Mindful: Principles for Creating Sustainable Visitor Experiences through Effective Communication.* Champaign, Ill.: Sagamore.

Mueller, D.C. 2004. 'Models of man: Neoclassical, behavioural and evolutionary', *Politics, Philosophy and Economics* 3, 1: 59–76.

National Park Service, US Department of the Interior. 2003. *Renewing Our Education Mission: Report to the National Leadership Council.* At: <www.nature.nps.gov/LearningCenters/new/renewmission_jun03.pdf>.

Nilsen, P., and G. Tayler. 1998. 'A comparative analysis of protected area planning and management frameworks', in S.F. McCool and D.N Cole, comps, *Proceedings—Limits of Acceptable*

Change and Related Planning Processes: Progress and Future Directions; 1997 May 20–22; Missoula, MT. Gen. Tech. Rep. INT-GTR-371. Ogden, Utah: US Department of Agriculture, Forest Service, Rocky Mountain Research Station, 49–57.

Nyberg, K.L. 1984. 'Some radical comments on interpretation: A little heresy is good for the soul', in G.E. Machlis and D.R. Field, eds, *On Interpretation: Sociology for Interpreters of Natural and Cultural History*. Corvallis: Oregon State University Press, 151–6.

Orams, M.B. 1996. 'Using interpretation to manage nature-based tourism', *Journal of Sustainable Tourism* 4, 2: 81–94.

———. 1997. 'The effectiveness of environmental education: Can we turn tourists into "greenies"?', *Progress in Tourism and Hospitality Research* 3: 295–306.

——— and G.J.E. Hill. 1998. 'Controlling the ecotourist in a wild dolphin feeding program: Is education the answer?', *Journal of Environmental Education* 29, 3: 33–8.

Parks Canada. 1994. *Guiding Principles and Operational Policies*. Ottawa: Minister of Supply and Services Canada.

———. 1997. *National Parks System Plan*, 3rd edn, Ottawa: Minister of Supply and Services Canada.

———. 1998a. *State of the Parks, 1997 Report*. Ottawa: Minister of Supply and Services Canada.

———. 1998b. *The Role of Heritage Presentation in Achieving Ecological Integrity*. Ottawa: Minister of Supply and Services Canada.

Parks Canada Agency. 2000. *Unimpaired for Future Generations? Protecting Ecological Integrity with Canada's National Parks*, vol. 2: *Setting a New Direction for Canada's National Parks*. Ottawa: Report of the Panel on the Ecological Integrity of Canada's National Parks.

———. 2006a. *Performance Report for the Period Ending March 31, 2006*. At: <www.pc.gc.ca/docs/pc/rpts/rmr-dpr/archives/2005-06/par-par_e.pdf>.

———. 2006b. *Corporate Plan 2006/07–2010/11*. At: < www.pc.gc.ca/docs/pc/plans/plan2006-2007/cp_0607-E.pdf >.

———, Performance, Audit and Review Group. 2005. *National Performance and Evaluation Framework for Engaging Canadians: External Communications at Parks Canada*. At: <www.pc.gc.ca/docs/pc/rpts/rve-par/pdf/audit_engaging_cdn_framework-e.pdf>.

Pellikaan, H., and R.J. van der Veen. 2002. *Environmental Dilemmas and Policy Design*. New York: Cambridge University Press.

Pergams, O.R.W., and P.A. Zaradic. 2006. 'Is love of nature in the US becoming love of electronic media? 16-year downtrend in national park visits explained by watching movies, playing video games, Internet use and oil prices', *Journal of Environmental Management* 80: 387–93.

Petty, R.E., S. McMichael, and L.A. Brannon. 1992. 'The elaboration likelihood model of persuasion: Applications in recreation and tourism', in M.J. Manfredo, ed., *Influencing Human Behavior: Theory and Applications in Recreation, Tourism and Natural Resources Management*. Champaign, Ill.: Sagamore, 77–101.

Porter, A.L., and J.L. Howard. 2002. 'Warning visitors about the potential dangers of dingos on Fraser Island, Queensland, Australia', *Journal of Interpretive Research* 7, 2: 51–63.

Price, G.G. 2003. 'Ecotourism and the development of environmental literacy in Australia', *E-Review of Tourism Research* 1, 3: 72–5.

Rasmussen, K. 2000. 'Environmental education evolves: Developing citizens, furthering education reform', *Education Update* 42, 1: 1–4. At: <www.ascd.org/ed_topics/eu200001_rasmussen. html>.

Reisinger, Y., and C. Steiner. 2006. 'Reconceptualizing interpretation: The role of tour guides in authentic tourism', *Current Issues in Tourism* 9, 6: 481–98.

Runte, A. 1997. *National Parks: The American Experience*, 3rd edn, Lincoln: University of Nebraska Press.

Saleh, F., and J. Karwacki. 1996. 'Revisiting the ecotourist: The case of Grasslands National Park', *Journal of Sustainable Tourism* 4, 2: 61–80.

Shankland, R. 1970. *Steve Mather of the National Parks.* New York: Alfred A. Knopf.

Sharpe, G.W. 1982. 'An overview of interpretation', in G.W. Sharpe, ed., *Interpreting the Environment,* 2nd edn, New York: John Wiley & Sons, 3–26.

———, C.H. Odegaard, and W.F. Sharpe. 1994. *A Comprehensive Introduction to Park Management,* 2nd edn, Champaign, Ill.: Sagamore.

Shultis, J.D. 1995. 'Improving the wilderness: Common factors in creating national parks and equivalent reserves during the nineteenth century', *Forest and Conservation History* 39, 3: 121–9.

———. 2005. 'The effects of neo-conservatism on park science, management and administration: Examples and a discussion', *George Wright Forum* 22, 2: 51–8.

Staiff, R., R. Bushnell, and P. Kennedy. 2002. 'Interpretation in national parks: Some critical questions', *Journal of Sustainable Tourism* 10, 2: 97–113.

Stewart, E.J., B.M. Hayward, and P.J. Devlin. 1998. 'The "place" of interpretation: A new approach to the evaluation of interpretation', *Tourism Management* 19, 3: 257–66.

Tilden, F. 1977. *Interpreting Our Heritage,* 3rd edn. Chapel Hill: University of North Carolina Press.

Tubb, K.N. 2003. 'An evaluation of the effectiveness of interpretation within Dartmoor National Park in reaching the goals of sustainable tourism development', *Journal of Sustainable Tourism* 11, 6: 476–98.

Weaver, D.B. 2002. 'The evolving concept of ecotourism and its potential impacts', *International Journal of Sustainable Development* 5, 3: 251–64.

Weaver, H.E. 1982. 'Origins of interpretation', in G.W. Sharpe, ed, *Interpreting the Environment,* 2nd edn, New York: John Wiley & Sons, 28–51.

Weiler, B., and D. Davis. 1993. 'An exploratory investigation into the roles of the nature-based tour leader', *Tourism Management* 14, 2: 91–8.

Wells, M., and L. Smith. 2000. *The Effectiveness of Non-Personal Media Used in Interpretation and Informal Education: An Annotated Bibliography.* Fort Collins, Colo.: National Association for Interpretation.

Wolfe, R. 1997. 'Interpretive education: An under-rated element of park management?', *Research Links* 5, 3: 11–12.

Zell, L. 1992. 'Ecotourism of the future: The vicarious experience', in B. Weiler, ed., *Ecotourism: Incorporating the Global Classroom.* Canberra, Australia: Bureau of Tourism Research, 30–5.

KEY WORDS/CONCEPTS

communication
decision-making theory
ecotourism
elaboration likelihood model
environmental education
extrinsic activity
first-hand experience
informal education
interpretation
interpretation effectiveness research
interpretive planning
intrinsic value
mindfulness
moral development theory
non-personal interpretation
personal interpretation
rational actor model
reductionist versus constructivist approaches
theories of reasoned action and planned behaviour
tour guiding
visitor motivation

STUDY QUESTIONS

1. List the major attributes of interpretation. How is providing interpretation different from providing information?
2. List the major benefits of interpretation for park managers and visitors.
3. Discuss the similarities and differences among interpretation, environmental education, and tour guiding.
4. Discuss the role of interpretation in achieving the objectives of ecotourism.
5. Discuss the historical phases of park interpretation in Canada and the US.
6. What are the major theoretical foundations for interpretation?
7. Discuss why interpretation is both a science and an art.
8. Describe four types of potential users of interpretive services.
9. Describe how each of the following can make interpretation more effective: the communicator; the message; the receiver.
10. Describe the potential benefits and costs of personal and non-personal forms of interpretation.
11. Why is it difficult to measure the effectiveness of interpretation? What can researchers hope to learn about effectiveness?
12. Why is interpretive planning important?
13. Discuss the impact of interpretation on the experience of dolphin watching in Australia.
14. 'In many parks interpretation is under-utilized.' Discuss.
15. Select a park in your area. Review and critique the range of interpretive services provided.
16. Join a guided walk provided by interpreters in a park setting. Discuss the techniques used by the interpreter. Comment on what worked and did not work.

PART IV

Putting It Together

> Enter Glacier National Park and you enter the homeland of the grizzly bear. We are uninvited guests here, intruders, the bear our reluctant host. If he chooses, now and then, to chase someone up a tree, or all the way to hospital, that is the bear's prerogative. Those who prefer, quite reasonably, not to take such chances, should stick to Disneyland in all its many forms and guises.
>
> *Edward Abbey*, The Journey Home

This section applies the conservation theory and social science theory developed in the previous sections to the management of parks and protected areas. In this section of the book, we extend earlier discussions by examining the impact of theory on: policy and legislation (Chapter 9); the struggle to use policy and legislation to effect decisions on the ground (Chapter 10); the application of park management principles in a unique region—the North (Chapter 11); the role of private-sector ecotourism in park management (Chapter 12); and the role of parks within larger landscapes (Chapter 13).

The section begins with a description of how national parks are managed in Canada, reviewing the legislation, policies, and procedures developed by Parks Canada (Chapter 9). Of particular interest here is the approach to systems planning, ecological integrity, and visitor management. Regarding systems planning and ecological integrity, the influence of conservation theory and application (Chapters 4 and 5) should be evident. Regarding visitor management, the influence of social science theory and application (Chapters 6, 7, and 8) should also be apparent.

National park legislation, policy, and procedures have been developed in order to provide consistency in approach and direction for park managers. However, the realities of applying these directives on the ground are not straightforward. To illustrate, we have provided a case study of Banff National Park (Chapter 10), the best-known but perhaps most challenging park in the national system, and pose the very serious question: If we can't get it right here, what chance do we have anywhere else?

Banff is well known to the Canadian public and to many international visitors, as are most parks found in southern Canada. Many of these parks in southern Canada were created as a result of intense interest on the part of people living in the more heavily populated southern parts of the country, especially those in closer proximity to these natural areas and so more aware of the various ecological and recreational values these more accessible parks might provide. However, Canada north of 60 degrees latitude is different. The landscape is different, but so, too, are the cultures, the history, and the politics. Chapter 11 explores these differences and describes the unique challenges for park management in this very important region of the country.

Until recently, most thinking about park management in the North American context has focused almost entirely on the responsibility of government (federal, provincial, regional, or local) and the role of non-governmental organizations (NGOs) in the creation and management of parks and protected areas. In Chapter 12, we look at the role of the private sector, most strongly evidenced in the emergence of ecotourism. In the past, the tourism industry has been viewed as one of the major stressors on park environments, as described in Chapters 9 and 10, but not all tourism is the same—ecotourism is viewed today as a more sustainable form of tourism providing a new vehicle for involving the private sector in a variety of ways that will support and improve park conservation. This is particularly compelling at a time when financial resources for parks have declined while many park systems have expanded.

The final chapter in this section (Chapter 13) looks at the management of parks from the perspective of the park ecosystem or landscape in which parks are found. Parks are not large enough to protect all species and natural processes. Even the very large parks are impacted by external forces, such as acid rain and global warming. Furthermore, many species travel across park boundaries and are affected by conditions found outside parks. The ecosystem approach recognizes that humans at some level of activity are a natural component of most ecosystems and cannot be ignored in the pursuit of ecological integrity. This includes the interaction of park visitors within park environments, as well as the interaction of parks with adjacent communities.

CHAPTER 9

Managing the National Parks

Pamela Wright and Rick Rollins

INTRODUCTION

Millions of people explore the national parks of Canada each year. They clamber to lookouts, gaze from belvederes, and rest at scenic pull-offs that highlight special features. Other visitors probe deeper within national parks, perhaps travelling for days by canoe, horseback, or on foot with a backpack. Visitors may encounter park staff, providing information and interpretation to visitors or enforcing park regulations. Sometimes, commercial guides or outfitters lead groups into national parks, offering their clients a unique experience that an experienced guide can bring. Others appreciate Canada's national parks from afar through photographs and movies and through the Internet. Although some will never visit a national park directly, they may still develop a sense of appreciation and understanding of their natural beauty and cultural significance. National parks are icons of wildness and contribute to building our sense of Canada identity.

From migrating caribou herds, to pods of orca whales, to the cultural richness of the mortuary poles at Ninstints in Gwaii Haanas National Park Reserve, to a refuge for remnant short-grass prairie, the value of national parks, and the foundation for the visitor experience, is in their diverse and rich ecological and cultural resources. Our history of park management has focused primarily on internal issues within the boundaries of the park. We are concerned, for example, about the amount, type, and nature of recreation and tourism that occur in national parks. We are concerned that the numbers of park visitors, their behaviour, and the services and facilities visitors require or demand will erode the very qualities of a natural setting the park is intended to protect. For example, the behaviour of bears, elk, and other wildlife may change in response to campers leaving garbage, taking photos, or establishing trails and campsites in places that interfere with wildlife feeding or other range requirements. In some parks, townsites have developed over the years to support a booming nature-based tourism industry, but the impact these developments have on park ecosystems is now becoming better understood.

Parks, as the caribou and whales illustrate, will never be big enough to be managed just within their borders—instead, managing a park requires recognizing that it's not just what goes on within the park boundary or on the neighbouring ranchland or forest block that matters. Indeed, the realities of climate change force us to acknowledge that the ecological integrity of parks requires us to understand and respond to stressors hundreds and thousands of kilometres away. Thus, concerns about the impacts of visitor use in parks are exacerbated by the growing realization that parks are threatened by human activity in adjacent lands: hunting, farming, ranching, logging, petroleum development, and urban growth.

Hence, the job of managing the national parks is a complex and complicated endeavour. The issues can be summarized into the following kinds of considerations, discussed in this chapter and elsewhere in the book:

- What is the purpose of national parks? What values are guiding park management decisions? (See Chapters 1 and 2.)
- Where should national parks be located, and how many are needed? How large must they be, and how are boundaries resolved? (See Chapters 3, 4, and 14.)
- How are natural resources managed in parks? (See Chapters 5, 10, and 11.)
- How is visitor use to be managed in parks? (See Chapters 6, 7, and 8.)
- How are decisions made about management issues? (See Chapters 1, 2, 11, and 15.)
- How will park managers work collaboratively with adjacent communities and landowners and resolve conflicts when they arise? What relationships will be established with Aboriginal communities? (See Chapter 12, 13 and 14.)

This chapter will provide an overview of the direction provided by legislation and by policy and then focus on some of the primary management concerns described above. Further details on many of these subjects are provided in other chapters.

DIRECTION PROVIDED BY LEGISLATION

Legislation and policy form two of the primary tools that guide management. Legislation is in the form of acts or laws, is approved by Parliament, must be followed by the government and citizens, and is enforced by the court system. Legislation is generally broad in scope and vague in details. It has often been described as the tool that enables policies and practices and that gives guidance on what can be done but not usually the details on how it is done.

Canada National Parks Act

Several key pieces of legislation guide the management of national parks, but at the core of these is the Canada National Parks Act. Prior to 1930, each national park was established by an individual Act and management was subject to whatever stipulations were contained within that legislation. The National Parks Act was first established in 1930 with a major set of amendments made in 1988. With the passing of the 1930 Act, a comprehensive set of rules for the management of every national park was established. In October 2000, a new Canada National Parks Act was approved in Parliament. In this chapter, references are to this consolidated Act unless otherwise noted.

FIGURE 9.1 Horse-drawn carriages stand ready to whisk visitors from the Banff railway station to hotels such as the Banff Springs (1913). Because it owned both the hotel and the railway station, the Canadian Pacific Railway held considerable influence over park visitors. Other hotels expressed resentment over the CPR's influence during this period, particularly when they learned the CPR was advising tourists that the Banff Springs was the only hotel in town.

Purpose of Parks

The Canada National Parks Act states that 'The national parks of Canada are hereby dedicated to the people of Canada for their benefit, education and enjoyment, subject to this Act and the regulations, and the parks shall be maintained and made use of so as to leave them unimpaired for the enjoyment of future generations' (2000, c. 32, s. 4[1]). This statement contains several noteworthy points. That parks are dedicated to the public good is clear, but what is also clear is the intent to maintain the ecological values of these areas. The use of the dedication 'for future generations' and of the desired status—'unimpaired'—provided the initial impetus for management and protection of the ecological values within national parks. Early conceptualizations of appropriate use and enjoyment and what would impair national parks were not based on the current numbers of people that visit parks, the competing demands for use of the parks, the increasing stresses and pressures placed on parks from the outside, or our understanding of the complex ecological systems that support and are supported by parks. As a result, there has been much debate about and difficulty encountered in regard to reconciling human use within national parks. Many have characterized this debate as a 'dual-use mandate' with the competing purposes of recreation and conservation. Indeed, much of the history of our use of parks and their management (see, for example, the issues around ski hills and golf courses in national parks) has focused on these issues. However, as our understanding of the impacts of recreation and tourism in parks has increased, and as pressures on conservation have increased from land management around parks, legislation and policy have sought to clarify this potentially troubling issue. Subsequent changes in policy and legislation have sought to contemporize our understandings of these situations and to further clarify the purpose.

Publication of the 1979 policy statement for the agency, the National Park Act Amendments of 1988, the subsequent revised *Guiding Principles and Operational Policies* of 1994, and the Parks Canada Agency Act (1988) have successively indicated that ecological integrity for national parks (and commemorative integrity for national historic sites) should be Parks Canada's first consideration and is to be regarded as a prerequisite to use.

Further clarification of the purpose of parks and consolidation of ideas from the *Guiding Principles and Operational Policies* and the Parks Canada Agency Act (discussed in more detail in other sections of this chapter) are made in the 2000 Canada National Parks Act. Section 8(2) states that 'maintenance or restoration of ecological integrity, through the protection of natural resources and natural processes, shall be the first priority of the Minister when considering all aspects of the management of parks.' This wording is intended to be broader still than that included in the 1988 amendments, where it could be questioned that the imperatives of ecological integrity were intended to be applied only to park zoning and visitor use in a management plan. To resolve confusion regarding terms and to resolve ongoing debates concerning the actual definition of ecological integrity, the Act defines ecological integrity as follows: 'ecological integrity means, with respect to a park, a condition that is determined to be characteristic of its natural region and likely to persist, including abiotic components and the composition and abundance of native species and biological communities, rates of change and supporting processes.'

The result of these evolving purpose statements is an agency mandate that is clear about intent and priority. The focus resulting from this clarified purpose is an increased emphasis on ecosystem-based management and on building scientific understanding and collaborative relationships regionally. Ecosystem-based management is an approach to management that integrates 'scientific knowledge of ecological relationships within a complex sociopolitical and values framework toward the general goal of protecting native ecosystem integrity over the long term' (Grumbine, 1994).

As the purpose of national parks has evolved, so, too, have the definitions of what activities and pursuits constitute 'benefit, education, and enjoyment'. In the beginning, the definition was very broad, and embraced a wide variety of activities including extractive activities such as hunting. As attitudes changed, the range of activities started to be limited to outdoor recreation activities, and extractive uses were largely omitted. Later still, only those outdoor activities with a minimum or non-consumptive impact on the environment were allowed. Clearly, one of the continuing and primary challenges for management within park boundaries is determining where, what type, and how much use should be allowed in order to provide for 'use without abuse' of the national parks. Indeed, much of the criticism that the agency has received over the past few decades (see, e.g., Searle, 2000; Banff–Bow Valley Task Force, 1996; Bella, 1987) has focused on this issue. Recent discourse, however—in the corporate plan discussed later in this chapter and in various internal communications—is increasingly giving direction to what is internally referred to as a more balanced approach. Within this context, the three pillars of national parks include protecting the natural and cultural resources of the park, providing innovative educational opportunities, and facilitating memorable experiences where visitors can connect and enjoy this truly unique landscape. This shifting emphasis is aimed at ensuring that visitor experiences, awareness, and education are recognized as integral to achieving the mandate of the agency. Indeed, without creating ambassadors for parks through meaningful visitor experiences, Parks Canada won't have the support it needs to make tough management decisions. Critics are likely to note, however, that this shift in emphasis towards visitor experience may tip the scales again in favour of use over ecological integrity.

Direction Provided by Policy

Policies are statements of intent for management and are usually much more detailed and explicit than legislation. Although policy direction, such as the interim policy on the recreational activity of geocaching, wherein recreationists navigate to various caches to collect clues, should be followed by the bureaucracy, it is not directly enforceable in the courts unless it is incorporated into a legislated document such as a park management plan. Any statement expressed in the Canada National Parks Act can be enforced in the courts, whereas policy cannot. Policy guidelines provide extremely important direction for the day-to-day planning and management of national parks. The national parks policy (Canadian Heritage, 1994) describes the rationale for national parks and the philosophy for managing them. The first comprehensive policy document was passed in 1964, with the latest, most comprehensive revisions published in 1994. The purpose of national parks, as outlined in the legislation, is further elaborated in these policy documents:

> Protecting ecological integrity and ensuring commemorative integrity take precedence in acquiring, managing, and administering heritage places and programs. In every application of policy, this guiding principle is paramount. The integrity of natural and cultural heritage is maintained by striving to ensure that management decisions affecting these special places are made on sound cultural resource management and ecosystem-management practices. (Canadian Heritage, 1994)

The guiding principles and specific policies provide guidance with respect to the national park system and establishment of national parks, management planning, ecosystem-based management, public understanding, appreciation and enjoyment of national parks, historical activities and infrastructure, and land tenure and residency. The implementation and challenges of these policies are discussed in other chapters. The statement of guiding principles provides an overview of the direction for management activities. Important concepts include:

- the paramountcy of protecting ecological integrity and commemorative integrity;
- managing on an ecosystem basis, with parks seen as part of larger ecosystems and not as islands;

- providing leadership and demonstrating through example environmental and heritage ethics and practices;
- protecting new parks and historic sites based on a systems approach, informed by science and on a co-operative basis;
- recognizing that education through a variety of approaches is the key to longer-term success of the protection of park values;
- basing management decisions on the best available knowledge;
- providing appropriate, basic, and essential services for park visitors within the objectives of maintaining ecological integrity;
- providing opportunities to build public understanding and make sound decisions through public involvement, co-operation, and collaboration with a full range of levels of government and interest groups;
- ensuring the accountability of Parks Canada for adherence to these principles through state of the parks reporting.

Although Parks Canada published its comprehensive policies in the 1994 document, new policy direction, largely in the form of ministerial statements, has been of primary importance. These policy directions often come in the form of individual issues and are harder to track. Some of these include: the revenue policy (Parks Canada, 1998a); a principle of 'no net environmental impact' for parks containing communities (e.g., Banff, Jasper, and Yoho), such that the ecological or development footprint of the community may not increase, along with other policies designed to regulate use and development (Canadian Heritage, 1998); National Park Ski Area Management Guidelines released in 2006 (Parks Canada Agency, 2006a; see Box 9.1); a moratorium on commercial development within parks but outside the boundaries of designated communities such as Lake Louise, Banff, Jasper, and Field (ibid.); and interim policies on a range of other issues including the emerging recreational activity of geocaching.

Strategic direction for parks has come at the national level through the creation of a systems plan for the completion of the national park system, a comprehensive policy document, agency corporate plans, park management plans, park-level business plans, and, more recently, accountability measures such as a national reporting system. These elements are discussed in the subsequent sections of this chapter.

LOCATION AND NUMBER OF NATIONAL PARKS

As allocation decisions on land use throughout Canada become more restricted, making decisions about where new national parks are located can be quite controversial, and the process of identifying suitable candidate areas that are supported by the public becomes increasingly challenging. Establishment of Gwaii Haanas National Park Reserve and Haida Heritage Site on South Moresby Island, BC, illustrates this point. The involvement of several levels of government and stakeholder groups (Dearden, 1987) played a part in the final decisions about the park. Conservation groups wanted to see a large national park established to provide maximum protection to significant ecological resources, while the tourist industry urged for creation of a national park to serve

BOX 9.1 Highlights of a New Park Policy for Ski Area Management

Downhill skiing has a long history in Canada's national parks, starting with the first commercial ski facility in Banff in 1934. Although a cornerstone of winter tourism in the mountain parks, the pressures that this activity has placed on alpine and subalpine environments has resulted in legislation prohibiting the development of any new commercial ski areas inside national parks and policy and guidelines to regulate the activity. In 2006, new Ski Area Guidelines were published, updating the existing direction. These guidelines note that Parks Canada's primary goal for the management of ski areas is to achieve certainty of long-term land use that:

- ensures ecological integrity will be maintained or restored;
- contributes to facilitating memorable national park visitor experiences and educational opportunities; and
- provides ski area operators with clear parameters for business planning in support of an economically healthy operation.

New development for ski areas is to be considered through the development of long-range plans developed under the conditions outlined in these guidelines. Some of the principles that will guide these plans include:

- Inside the existing developed area, new development can be considered where potential ecological impacts can be mitigated.
- Outside the existing developed area, new development can be considered if there is a substantial environmental gain within or adjacent to the leasehold.
- Ski areas will contribute to a unique, memorable national park experience.
- Ski areas will promote public appreciation and understanding of the heritage values of the park and world heritage site and local conservation initiatives.
- Ski areas will be leaders in the application of environmental management, stewardship, and best practices.

as a focus for tourism. The Haida Nation, whose traditional territory the area is within, wanted the area protected, but their concerns did not exactly mirror those of the tourism industry or the conservation community; they value the area for its spiritual significance to their culture. Opposed to this notion was the forest industry, which wanted to harvest the valuable old-growth temperate rain forest within the area. The provincial government was reluctant to give up any territory to the federal government as a national park and would have preferred some other solution. The debate over Gwaii Haanas was fiercely fought among polarized interests; protests were staged over harvest activities, and arrests were made.

The issue was resolved with the signing of a memorandum of understanding in 1987 between the federal government and the government of British Columbia. This was followed by the Canada–British Columbia South Moresby Agreement, signed in 1988, which provided compensation to the forest industry and other industries as well as funds for a regional economic initiative, based on tourism and small business development in the Queen Charlotte Islands (Environment Canada, 1991). Perhaps the most significant aspect of the Agreement was an 'agreement to disagree' between Canada and the Haida Nation in which the Haida and the federal government agreed to manage the area as a national park while continuing to disagree over ultimate ownership of the land.

The solution will endure as one of the most unique in national park creation worldwide and a turning point in the process of establishing new national parks. Gwaii Haanas signalled the increasing role and importance that First Nations now have in park creation and the need for the development of a common vision and agreement from all parties on the protection of ecological integrity.

As the Gwaii Haanas case illustrates, the creation of national parks is not the act of government alone—at times, in fact, the lead has been taken by others. Prominent among the motivational forces for the expansion of the current national park system was the Endangered Spaces Campaign launched by the non-governmental community in 1989. The World Wildlife Fund and the Canadian Parks and Wilderness Society launched this campaign to rally public support and push public policy towards representing each of the country's 486 ecoregions (a much finer-scale classification system than the Parks Canada system of 39 regions) with a protected area (Hummel, 1989). The choice of targets for representation is one of the principal critiques and contributions made by the non-governmental community. A common critique is that the Parks Canada approach based on 39 natural regions results in regions that are generally far too large and diverse for a single national park to represent adequately; as a consequence, most ecoregions still need more adequate representation. Many of the methods used to identify representative areas predated modern principles of conservation biology. In reality, the adequacy of representation of the protected areas systems depends on the scale of analysis, the strength of the candidates, the degree to which candidate areas are unimpaired, the willingness of the levels of government to protect designated areas, and the legislative and management tools used for protection. Although Parks Canada has stuck to its system of representation of 39 regions, the critiques on ecological complexity and the need for richer analysis of representation have informed both the national park system and many of the provincial park systems.

The influence of the non-governmental community in new park establishment goals is profound. In 1992, the Endangered Spaces goal became public policy with the signature by the Tri-Council of Environment, Parks and Wildlife Ministers (federal, provincial, and territorial ministers) to the 'Statement of Commitment to Complete Canada's Network of Protected Areas'. This statement committed the governments to completing the terrestrial protected areas network by 2000.

Systems Planning

Every potential national park is established in the context of complexity of ownership and jurisdiction, existing land use, and competing values and visions for the land. As a result, it has become imperative to define some sort of process and rationale for establishing new national parks.

Canada's early national and provincial parks were set aside as opportunities arose. These lands were preserved for their scenic beauty, revenue potential from tourism, wildlife, or other wonders of nature. Growth in the number of parks for the first half of the century was not part of a system plan and certainly not explicitly linked to protecting biodiversity. The initial ideas of systems planning suggested that each province or regional area should have a protected area. Gradually, more ecological rationale has crept in. The creation of a systems planning framework was an effort to develop a rational basis for establishing national parks. In 1971, a national parks systems plan was approved as a basis for deciding where national parks were needed (National and Historic Parks Branch, 1971; Environment Canada, 1990). These natural regions were based generally on broad physiographic characteristics and defined as:

> natural landscapes and/or environments of Canada which may be separated from other such landscapes and environments by surface features which are readily observable, discernible, and understandable by the layman as well as by scientists and others more familiar with the natural features of Canada. (National and Historic Sites Branch, 1971: 3)

Although the development of this ecologically based system plan is noteworthy among protected area systems worldwide, the foundation of the system in readily understandable 'natural regions' has been critiqued as ecologically simplistic.

In the 1980s, this natural region approach was updated to give it a broader ecological foundation that focused on ecoregions, with the notion that establishing a representative protected area within each region would capture the typical range of variability of landforms, vegetation, and wildlife and therefore help conserve the native biodiversity of the region. The plan classified the land mass of Canada into 39 natural regions, each with its own characteristic vegetation patterns, landforms, climate, and wildlife. Parks Canada's progress towards its commitment to represent these natural regions has been significant, with the creation of a number of new national parks, particularly in the Canadian Arctic, and as a result of land claim settlements. Parallel system plans have been made for national historic sites (Parks Canada Agency, 2000) and for national marine conservation areas.

Progress towards completion of this system of representation proceeds very slowly (Figure 9.2). The complexities of new park creation are compounded in the parts of the country where private land dominates; consequently, in these areas new park acquisition is a careful and slow process of relationship building, fear of loss of traditional use and access, and often high financial costs for land acquisition. These situations can result in establishment of small parcels of national park intermingled with private lands (e.g., Gulf Islands National Park and Grasslands National Park).

The goal of Parks Canada Agency to achieve representation of 34 of 39 terrestrial regions and eight of 29 marine regions by 2008, along with expansion of three existing parks and land acquisition for three other national parks, has not been met. The *Parks Canada Agency Performance Report* issued for 2005–6 (Parks Canada Agency, 2006b) noted, however, that one new national park reserve was designated in Labrador

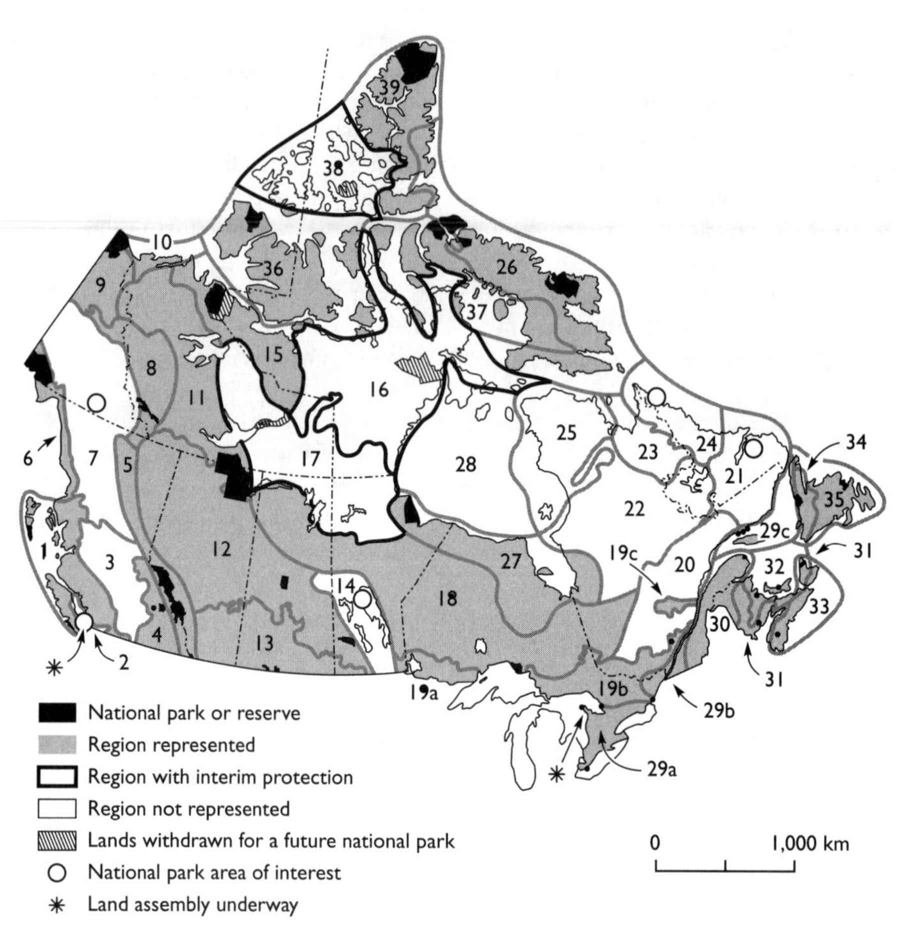

1 Pacific Coast Mountains
2 Strait of Georgia Lowlands
3 Interior Dry Plateau
4 Columbia Mountains
5 Rocky Mountains
6 Northern Coast Mountains
7 Northern Interior Plateaux and Mountains
8 Mackenzie Mountains
9 Northern Yukon
10 Mackenzie Delta
11 Northern Boreal Plains
12 Southern Boreal Plains and Plateaux
13 Prairie Grasslands
14 Manitoba Lowlands
15 Tundra Hills
16 Central Tundra
17 Northwestern Boreal Uplands
18 Central Boreal Uplands
19a West Great Lakes–St Lawrence Precambrian Region
19b Central Great Lakes–St Lawrence Precambrian Region
19c East Great Lakes–St Lawrence Precambrian Region
20 Laurentian Boreal Highlands
21 East Coast Boreal Region
22 Boreal Lake Plateau
23 Whale River
24 Northern Labrador Mountains
25 Ungava Tundra Plateau
26 Northern Davis Region
27 Hudson–James Bay Lowlands
28 Southampton Plain
29a West St Lawrence Lowland
29b Central St Lawrence Lowland
29c East St Lawrence Lowland
30 Notre Dame–Megantic Mountains
31 Maritime Acadian Highlands
32 Maritime Plain
33 Atlantic Coast Uplands
34 Western Newfoundland Highlands
35 Eastern Newfoundland Atlantic Region
36 Western Arctic Lowlands
37 Eastern Arctic Lowlands
38 Western High Arctic
39 Eastern High Arctic

FIGURE 9.2 Terrestrial system plan for Parks Canada, showing degree of representation.

(Torngat Mountains National Park Reserve), that an agreement in principle had been achieved for a proposed marine conservation area, and that some progress had been made on five other candidate national parks and three other national marine conservation areas. Similarly, park expansions and acquisitions are proceeding slowly with progress made on one of each.

On the historic sites side, the systems plan and the targets for new site designation or commemoration are somewhat different. Nineteen new designations were made in 2004 (compared to a goal of designating sites for an average of 27 new places, persons, and events per year) and 50 commemorative plaques placed in the 2005–6 year (compared to a goal of 30 commemorative plaques placed annually).

Establishing a New Park

Identifying, selecting, and establishing new national parks has proven to be a complex exercise, although the normal sequence of events can be summarized in five steps:

1. Identify representative natural areas within the natural regions.
2. Select potential park areas, known as 'Natural Areas of Canadian Significance'.
3. Assess park feasibility.
4. Negotiate a new park agreement.
5. Establish a new national park in legislation.

In the past, establishing a new national park in legislation, thus allowing the full authority of Parks Canada to manage the park, has been a difficult process taking up to 20 years. The Canada National Parks Act contains a revised park establishment process that allows a new national park to be added quickly to the legislation—within months rather than years.

New national parks must be established with the co-operation of the provinces and territories, which must, under current legislation, make formal transfer of the land to federal jurisdiction. Provincial governments have been reluctant to surrender lands to the federal government, thereby slowing the process of creating new national parks. Increasingly in southern parts of Canada, new park creation requires private land acquisition on a willing-seller, willing-buyer basis. Although voluntary in nature, new park acquisition from private lands can bring with it its own suite of problems, from patchwork park ownership, to property price inflation, to conflict between neighbours, to access problems across private land.

Aboriginal peoples also have a key role to play in negotiating new parks within traditional territories: the recent establishment of parks such as Aulavik, Sirmilik, and Tuktut Nogait in the North are testimony not only to the increasingly important role of Aboriginal peoples but also to the new and creative ways that must be sought for park establishment. National parks established through land claim agreements, while perhaps increasing the overall management complexity, may indeed be examples of park creation that allows for parks to be managed in a way that takes into account the greater park ecosystem. For example, in Ivvavik National Park in Yukon, established through the Inuvialuit Final Agreement, the local hunter and trapper committees, the north-slope

Wildlife Management Advisory Committee, and other mandated boards and councils all play a role in regional integration of the park.

Forgoing Options

Competing resource values such as forestry and mining on provincial Crown land and urbanization on private land mean that options for new park creation are being foreclosed. The ultimate location, size, shape, and condition of the new parks have a significant effect on the ability to maintain ecological integrity, and, consequently, the negotiation for new parks is exceptionally important. The Mealy Mountain area of Labrador was proposed since the 1970s as a candidate national park, representing the East Coast Boreal Region. The area is part of an Innu land claim and the Innu Nation supports the establishment of this park. Only recently, however, was the feasibility study on the candidate site launched. The feasibility study continued with a second round of public consultations in March 2006, which focused on a proposed boundary for the national park and a management framework for the continuation of traditional land uses by local Labradorians. An interim report on the park proposal is being drafted based on the results of the consultations.

During the initial phases of the feasibility study, phase III of the Trans-Labrador highway was slated to traverse the proposed national park. If the road was built through the proposed area prior to the completion of the feasibility study, resource users would have gained access and legal rights to the land, compromising the park values and making more difficult the land claim negotiations with the Innu Nation (Parks Canada Agency, 2000). The Canadian Environmental Advisory Council (1991) has described the problems inherent to a lack of co-ordination among different stakeholders and agencies:

> Establishing protected areas in isolation from regional planning and decision-making processes is not an effective way to ensure the maintenance of their long-term ecological integrity. Past experience has shown that surrounding communities, landowners and commercial developers systematically encircle and encroach on protected areas. The result is often the loss of protected areas values and demands for inappropriate uses of these resources.

In March of 2004, the Newfoundland and Labrador government decided to construct the highway outside the heart of the proposed Mealy Mountains National Park, potentially avoiding significant impacts to the park.

Clearly, a more co-operative and inclusive approach by Parks Canada is required to achieve the protected areas goals. The 1996 Auditor General of Canada's report on national parks noted, however, that 'By simply waiting for other governments and local communities to adopt favorable positions, Parks Canada is reducing the likelihood of achieving representation in several natural regions and maintaining ecological integrity' (Auditor General of Canada, 1996: 18). Both the Auditor General's report and the Ecological Integrity Panel report noted that by the nature of the process it uses, Parks Canada may be encouraging other jurisdictions to adopt a defensive position at the outset. The challenge is to facilitate the development of a common vision with the relevant governments or communities within which a new national park could be created *before* the identification of a candidate area. Based on this common vision, stronger interim protection measures (such as withdrawal from mining claims) meant to ensure park conservation values are not lost during the slow process of negotiations.

Towards a National Protected Areas Strategy

National and provincial parks, wildlife management areas, heritage rivers, conservation easements, wilderness areas, marine conservation areas, and special management areas established under Aboriginal land claims, along with a host of other conservation tools, make up some of the components of a system of protected areas. But is it really a 'system'? Are the designation and management of these areas co-ordinated to ensure that, collectively, they meet the nation's needs and its international obligations for the conservation of biodiversity, wilderness, ecological integrity, sacred lands and waters, and for recreation? To achieve the national conservation objectives, Canada needs a comprehensive national protected areas systems plan that folds in the myriad layers of conservation goals within a co-operative implementation plan. A true protected areas systems approach includes:

- representative core areas such as national or provincial parks in each ecoregion, designed to play a key role in maintaining ecological integrity;
- protection of special natural and cultural features and landscapes;
- protection of wildlife habitat and species populations throughout the country;
- protection of rare and endangered species throughout their ranges;
- maintenance of ecological connectivity between protected areas;
- management of human use outside of protected areas in ways that help conserve biodiversity as well as ecosystem functions. (Parks Canada Agency, 2000)

Achievement of a national protected areas strategy requires close federal–provincial–territorial co-operation and collaboration. As the purpose and mandate for national parks have evolved, so, too, have many of the provincial and territorial park systems. The goals and resources devoted to building protected area systems at the provincial and territorial levels vary considerably, both between jurisdictions and over time, and this variation seems more affected by the politics of the government of the day (see Chapter 3). In 1997, Ontario, through a comprehensive land-use strategy, focused effort on new park establishment (Lands for Life) in the boreal heart of the province covering an area of 46 million ha. This initiative led to the creation of many new protected areas. Strengthening this is renewed provincial legislation (Provincial Parks and Conservation Reserves Act, 2006) with a stated purpose of managing 'these areas to ensure that ecological integrity is maintained'. In the West, British Columbia went through an impressive and unprecedented park expansion initiative in the 1990s, more than doubling the amount of the province designated as protected. This expansion was accomplished through an extensive multi-stakeholder consultation process coupled with goals for representation based on conservation biology theory. However, designation of protected areas is only the first step to achieving goals. Although a renewed focus on conservation, discussion, and training on ecological integrity initially accompanied the protected area system expansion in BC, successive personnel and resource cuts and bureaucratic reorganizations have left their mark on the organization. In addition, current provincial government emphasis for the park system is on use—a 'parks for people' philosophy. While encouraging more people to visit and appreciate provincial protected areas is a noteworthy goal, the political focus is on fostering provincial economic development and tourism promotion (see, for example, the BC Parks 2006 Fixed-Roof Accommodation Policy), potentially subverting the conservation objectives. These two provincial cases illustrate not only the challenges of developing a comprehensive protected areas strategy for Canada, but also the need to have the purpose and mandate of parks, the system and method of park establishment, and the management of the parks closely linked.

HOW PARKS ARE MANAGED

Legislation and policy form the foundation for management in the national parks, with that foundation translated into park-specific direction through the development of plans. Parks Canada currently divides planning activities into three tiers at the park level: strategic, implementation, and work planning. While there are many types of plans in each tier, the main focus here will be on the park management plan at the strategic tier, the business plan at the implementation tier level, and specific tactical plans at the third tier.

BOX 9.2 Other Legislation and Constitutional Guidance

The National Parks Act and the associated Parks Canada Agency Act are not the only legislative guidance for the management of national parks. Several other important pieces of legislation have direct implications for specific management actions such as those associated with the enforcement of rules and regulations referred to in the section on park regulations.

Canadian Environmental Assessment Act (CEAA)

CEAA is federal legislation that requires environmental screening, assessment, and review of projects involving federal land or federal dollars. The environmental assessment process is a key mechanism for examining use and development decisions within national parks, and Parks Canada has associated directives guiding the procedures used for environmental assessments. Although a strong tool, one common criticism is that, in practice, the environmental assessment of a project is, de facto, the final review, resulting only in recommendations on ways to mitigate the effects of proposed projects. In a review from 1998 to 1999, only six of 962 Parks Canada projects registered with CEAA were rejected through the environmental assessment process. A new policy directive (Parks Canada, 1998a) now requires that projects not undergo environmental assessment until the project is proven to be in compliance with Parks Canada legislation, policies, and directives. This directive, if fully implemented, may go a long way towards strengthening the utility of the CEAA tool for park management.

Marine Legislation

While Canada's system of terrestrial national parks is one of the oldest and most entrenched in the world, progress towards establishing a marine protected areas system has lagged substantially behind. In the past decade some progress has been made towards the establishment of legislative guidance for marine protected areas, including the Canada Oceans Act passed in 1996. Critical to full implementation of a marine protected areas strategy is the National Marine Conservation Area Act proclaimed in 2006.

Species at Risk

Unlike many other countries, including the United States, Canada lacks a major legislative tool, an Endangered Species Act, to protect biodiversity. In 2000 federal legislation for the protection of threatened and endangered species was proposed, and the Species at Risk Act (SARA) was passed in 2003. Conservation groups, however, have criticized SARA for inadequately addressing the needs of threatened and endangered species because politicians, not natural scientists, have the final word on which species are at risk and because the law does not apply across all jurisdictions and ownerships. The legislation, however, does require Parks Canada to take action to protect species.

Constitutional Guidance

Aboriginal rights, land claims, and treaties that are enabled under the Constitution provide guidance—with authority that overarches legislation and policy—regarding the establishment and management of national parks with respect to Aboriginal peoples. As Aboriginal rights are clarified, comprehensive land claim agreements are achieved, and specific treaty claims are negotiated and met, a variety of new challenges and new opportunities arise for national park management. These issues, discussed in more detail in Chapter 12, affect where and how national parks are established and how they are managed.

Park Management Plans

The park management plan is the primary tool for directing management at the level of the individual park. Management plans are required by law and must be tabled in the House of Commons within five years of park establishment. The Canada National Parks Act (s. 11[1]) further clarifies the requirements:

> The Minister shall, within five years after a park is established, prepare a management plan for the park containing a long-term ecological vision for the park, a set of ecological integrity objectives and indicators and provisions for resource protection and restoration, zoning, visitor use, public awareness and performance evaluation, which shall be tabled in each House of Parliament.

Similarly, the Parks Canada Agency Act legislates management plans for national historic sites and the Canada National Marine Conservation Area Act legislates management plans for NMCAs, including more prescriptive requirements for consultation than found for national parks.

In addition to requiring the development and filing of management plans on a timely basis, legislation also requires that management plans must be reviewed every five years with consultation, and tabled in Parliament. This requirement for review has led to a number of innovative developments discussed later in the chapter. Two other clauses in the Canada National Parks Act provide further legislative direction for park management.

Detailed management planning guidelines have been prepared by Parks Canada Agency to ensure consistency and standardization in plan preparation, including everything from required elements in the plans to guidance on the level and types of public participation for plan development. Both the policy statements and the 2000 Act state that maintenance of ecological integrity must be the first consideration in management planning. This increased focus on ecosystem-based management for ecological integrity is resulting in a fundamental shift in park management planning. In the past, most plans could best be described as human use and development plans with the relationship of visitor opportunities and allowable developments and activities tied only loosely to the underlying ecological resource.

Newer plans and associated guidelines are calling for a much more explicit focus and priority on ecological integrity and, within that foundation, for a vision and objectives for human use. Most plans now include an emphasis on establishing long-term processes (such as advisory committees) for working with partners and stakeholders to manage issues that cannot be managed in isolation. Related to this is an emphasis in northern plans on integrating local and traditional knowledge with scientific knowledge in recognition of increased participation by Aboriginal peoples in protected areas management in the North.

Parks Canada has been criticized in the past for the development of so many different types of plans and strategies that making a logical linkage between plans is exceptionally difficult (Charron, 1999). This wealth of plans emerged often based on good rationale. Since the early 1980s, with an increased focus on ecosystem-based management and ecological integrity, a series of plans or strategic documents, including ecosystem conservation plans, ecological integrity statements, and vegetation and aquatic management plans, began to be developed. Similarly, at various times there has been an increased focus on human use—from marketing, to communications, to community outreach, to visitor activities management. Often, all of these different issues and concerns have been seen to merit separate plans. Not only has the overwhelming number of plans resulted in continuous planning and a lack of action, but accountability and sustained public involvement become difficult. Parks Canada now faces the challenge of consolidating planning to ensure not only ecological and commemorative integrity, but also meaningful visitor experiences and collaborative and co-operative approaches to planning and management. Revised management planning guidelines are beginning to focus on this change in emphasis, but it remains to be seen if the next generation of park management plans reflects this change.

Public involvement and participation in park management planning have been required for many years, but the nature of that involvement and the forums for participation have greatly evolved. The national parks are entering a new era of more inclusive forms of decision-making, not mere consultation, in park management planning. This evolution away from traditional consultation processes, such as written submissions and public hearings, to more co-operative or consensus processes, such as round tables and other multi-stakeholder processes, is both opportune and problematic. Clearly, a more participatory democracy in which citizens are more invested in the management of national parks and in which managers are more responsive to the knowledge and concerns of the citizenry is ideal. That said, this ideal is not without challenges because the interests of a more empowered and involved local citizenry must be tempered with respect for the interests of the broader Canadian public. These civics-based approaches acknowledge the range of values and interests held by different parties that will be necessary to manage parks not as islands but as core protected areas connected to working landscapes. However, some of the early attempts by Parks Canada to engage in these more participatory processes have been critiqued as entering naively into processes in which they are unclear of the impacts in terms of the ability of the agency to uphold legislation and policy and to form lasting relationships. Recognizing the skills shortfall among employees in this area, a series of new training workshops have recently been developed (see, e.g., Chinook Institute, 2007) and are to be delivered in a series of parks. The training focuses on a spectrum of involvement that ranges from information, to influence, to involvement, to collaboration.

FIGURE 9.3 In front of the Wikkaninish interpretation centre in Pacific Rim National Park Reserve a full-scale wedding is taking place. Is this an appropriate activity to be encouraging in national parks? *Photo: P. Dearden.*

Park Zoning

Through the planning process, one of the principal techniques for organizing management activities and objectives within parks is the park zoning system (Box 9.3). The zones are an attempt to reflect both the resilience of different areas within the park and, subsequently, the resulting different intensities of visitor use. Zone I (Special Preservation) and Zone II (Wilderness) are intended to provide the highest level of protection, although the only specific guideline is that motorized access will not be allowed

BOX 9.3 Zoning System for National Parks

Zone I: Special Preservation

- Contain or support unique, threatened, or endangered natural or cultural features, or among the best examples of features that represent a natural region.
- Motorized access and circulation will not be permitted.

Zone II: Wilderness

- Extensive areas that are good representations of a natural region.
- Perpetuation of ecosystems with minimal interference is key.
- Require few if any rudimentary services and facilities.
- Visitors will experience remoteness and solitude.
- Visitor activities will not conflict with maintaining the wilderness.
- Motorized access and circulation will not be permitted.
- A variety of direct and indirect management strategies will be used to manage visitors.

Zone III: Natural Environment

- Provide outdoor recreation opportunities requiring minimal services and facilities of a rustic nature.
- Motorized access may be allowed, but will be controlled.
- Public transit will be preferred.

Zone IV: Outdoor Recreation

- Limited areas that are capable of accomplishing a broad range of opportunities.
- Activities, services, and facilities will impact ecological integrity to the smallest extent possible.
- Defining feature is direct access by motorized vehicle.

Zone V: Park Services

- Communities in existing national parks that contain a concentration of visitor services and support facilities.
- Major park operation and administration functions may also be found.

Source: Adapted from Parks Canada (1994b).

in these two zones. Zones III–V allow for more facilities and services, presumably to deal with higher numbers of visitors. The intent of the zoning system is to emphasize Zones I and II, such that each national park consists mainly of these two zones. In addition, the 2000 Act specified that Zone II wilderness boundaries would be protected under legislation, whereas the designation of other zones is subject to national park policy.

The Parks Canada zonation system has some parallels to visitor management tools developed for other organizations, such as the Recreation Opportunity Spectrum (ROS) and the Limits of Acceptable Change (LAC) (see Chapter 7). Like the ROS, the park zoning system is a broad tool for classifying land into a range of capacities and use intensities; however, the focus for park zoning is based more explicitly on a blend of conservation and use, whereas ROS focuses primarily on land classification for desired recreational experiences, based on providing diversity in ease of access, facilities provided, rules and regulations, and numbers of encounters. Although sharing similar foundations, the LAC process is a more encompassing visitor management framework with explicit public consultation and participation components linked to develop a joint plan for the allocation of types and levels of use; this process also involves an embedded monitoring framework to assess achievement or change over time. Although there are many similarities between park zoning, ROS, and LAC, some parks, such as Yoho National Park, have found it useful to employ a variety of these tools to produce both a more ecologically based and a more visitor experience-based system for management.

Management of Natural Resources

While one of the major vehicles for protecting park resources is the zoning system, a number of policies guide planning and management. Managing for ecological integrity within a framework of ecosystem management requires a number of specific management guidelines, as indicated in the listing of relevant policies in Box 9.4. Increasingly, parks are taking a more active management approach to achieving ecosystem management objectives, using a mix of tools that includes species reintroduction, vegetative thinning, prescribed burning, and restoration (see Chapter 5). To assist this, a good park management plan should clearly state the goal to be achieved; the ecological constraints, conditions, and benchmarks of this goal; and the measurable objectives to be monitored.

Visitor Management in National Parks

Although external stressors from climate change perhaps represent the largest threat to conservation within national park boundaries, visitor use and infrastructure development are still potential threats to ecological integrity. Yet, the provision of appropriate visitor activities is an increasingly important function of national parks. Box 9.5 lists a number of policies developed to guide visitor management in national parks. Chapter 7 provides more discussion of appropriate activities for national parks and how proposed new activities are assessed.

Park Regulations and Enforcement

Regulations for national parks are made by cabinet and include such regulatory powers as: the powers to make detailed rules governing the protection of flora and wild animals; public safety; management of fishing; public works; traffic; domestic animals;

BOX 9.4 Selected Resource Management Policies

- Ecosystem management provides the conceptual and strategic basis for the protection of park ecosystems.
- Decision-making associated with the protection of park ecosystems will be scientifically based on internationally accepted principles and concepts of conservation biology.
- An integrated database will be developed for each national park, along with research and environmental monitoring. Data requirements will regularly extend beyond park boundaries.
- Human activities within a national park that threaten ecological integrity will not be permitted.
- Sport hunting will not be permitted.
- Sport fishing may be permitted.
- Fish stocking will be discontinued.
- National parks will make efforts to prevent the introduction of exotic plants and animals.
- National park ecosystems will be managed with minimal interference to natural processes.
- Manipulation of naturally occurring processes such as fire, insects, and disease may take place if no alternative exists, or if:
  - there will be serious adverse effects on neighboring lands, or
  - major park facilities, public health, or safety will be threatened, or
  - management of certain natural features or cultural resources cannot be achieved otherwise.
- Integrated management agreements will be made with adjacent landowners and land management agencies.

Source: Adapted from Parks Canada (1994b).

control of fires; firearm discharge and licensing. While these regulations are applied through planning and management activities, section 2 of the National Parks Act states that a park warden is designated under section 18 'for the preservation and maintenance of the public peace in parks, and for those purposes park wardens are peace officers within the meaning of the *Criminal Code*.' Penalties and fines associated with violations have been increasing over time in order to serve as effective deterrents or punishments of infractions. Although the Act designates park wardens as peace officers capable of enforcement, the issue of whether park wardens should carry firearms has been the subject of significant debate not only externally, but also internally within the warden service. An Occupational Health and Safety Officer decision in 2000 found that wardens, who are currently unarmed, are not properly equipped to protect themselves while performing the law enforcement component of their jobs. After numerous reviews, appeals, and reversals, in May 2007 the Parks Canada CEO, Alan Latourelle, removed park wardens from all law enforcement duties and announced that the RCMP

BOX 9.5 Selected Visitor Management Policies

- Not every kind of use requested by the public will be provided.
- Provincial, territorial, municipal, and private agencies will be encouraged to provide complementary opportunities that respect shared ecosystems.
- The practice of ecotourism will constitute an important mutual linkage with other land management agencies and private interests.
- Each park may develop a variety of outdoor recreation opportunities, but these must conform to the park zoning plan.
- A minimum of built facilities will be permitted.
- An integrated visitor activities database will be developed and kept up to date for each national park.
- Risk control measures will consider the experience needs of the visitor and promote visitor self-reliance accordingly.
- Parks Canada will use a variety of direct and indirect management strategies for managing public use. Direct strategies include rationing use, restricting activities, and law enforcement. Indirect strategies include facility design, information dispersal, and cost-recovery mechanisms.
- Any built facilities in a national park will be designed so that the scale, site, accessibility, and function are in harmony with the setting.
- Non-motorized means of transportation will be favoured.
- Aircraft will not be allowed in national parks unless reasonable travel alternatives are not available.
- Trails and roads may be constructed if their primary function is to serve park purposes.
- New commercial skiing areas will not be permitted in national parks.
- Through interpretation and public education, both within and outside national parks, Parks Canada will provide the public with interesting and enjoyable opportunities to observe and discover each park's natural, cultural, and historical features.
- Interpretation and education will be used to inform visitors of park management issues and practices.
- Interpretation and education will be used to relate park themes to broader environmental issues, to provide the public with knowledge and skills to make environmentally responsible decisions.

Source: Adapted from Parks Canada (1994b).

will perform law enforcement in Canadian national parks until further reviews are undertaken. The issues of this debate are numerous, however; from a conservation perspective, critics, including the National Park Warden Association and the Canadian Parks and Wilderness Society, are concerned that the solution will be costly and ineffective. A previous decision ceding enforcement authority to the RCMP resulted in a significantly reduced number of charges laid.

FIGURE 9.4 The Icefields Centre in Jasper National Park. The Centre is a public-private partnership between Parks Canada and Brewster Tours. See Edward Abbey's *Desert Solitaire* (1968: ch. 5) for some comments on Coke machines in parks! *Photo: P. Dearden.*

Business Plans

The main planning tool at the implementation level is the business plan. These business plans set out the financial accountabilities and short-term actions of Parks Canada Agency and provide a potential tracking mechanism for monitoring. At the corporate (Agency) level, preparation of business plans was added to the required direction for management (Parks Canada Agency Act). At the park or field unit (a grouping of similar parks and historic sites) level, business plans are prepared but are not mandated by legislation. The business plan, at the field unit level, covers five years and combines planning for national parks and national historic sites. Just as the business plan is the primary document in regard to implementation, the park management plan, as legislatively required, is the paramount accountability mechanism. From a managerial perspective, the business plans are the key accountability mechanism between the field unit superintendents and the Parks Canada CEO, and they spell out both the specific actions that parks will take to implement the park management plans during that time period as well as the financial requirements, including revenues, to implement them.

Tactical Plans

Tactical plans represent the detailed set of tasks and actions that are prescribed for a particular functional area within a park, such as an urban outreach plan or a fire management plan. These plans typically contain detailed actions over a fairly short time horizon (1–3 years). The tactical planning function is an important part of daily operations and is the final stage where direction—from legislation and policy through systems and management planning—is translated into action. Although these plans are typically not prescribed or required, guidance for them can be found in policy and guidelines. Important aspects often addressed through specific tactical plans, such as fire management and restoration plans and visitor use management, are discussed in other chapters in this text.

ADMINISTRATION: HOW MANAGEMENT IS STRUCTURED

The purpose of this chapter has been to examine national park legislation, systems plans and policies, and management direction for parks. Legislation and policy provide direction for Parks Canada in managing the national parks and provide for the Canadian public a framework for debating the purpose and rationale for various management actions. To fully understand how national parks are managed, we must, however, also understand the management culture: the bureaucracy of the Parks Canada Agency. At the millennium, reviews and critiques of Parks Canada indicated that the organizational culture of the Agency was the single biggest factor and barrier in affecting management (Parks Canada Agency, 2000a; Searle, 2000). Such critiques from external scientists, conservation organizations, and other advocacy organizations are often difficult for the corporate culture to receive. However, the organization has made a concerted effort to improve management structure and process as well as staff involvement, morale, and enthusiasm. This, along with increased financial resources, has renewed and re-energized the organization.

A More Corporate Approach to Management: The Parks Canada Agency

Since the 1980s, Parks Canada underwent a series of rapid reorganizations, moves, renaming, and downsizing. In 1993, Parks was shifted from its older home in the Department of the Environment to reside in the Department of Canadian Heritage. In the spring of 1999, Parks Canada became a separate operating agency—a unique structure within the government. While the Parks Canada Agency now reports to the Minister of the Environment, it has greater autonomy, particularly with respect to fiscal management and revenue generation. The creation of Agency status provides an opportunity to attempt new creative management structures as well as a chance to renew the organization. Now, the organization is focused on renewal of management functions and processes, increasing staff, and dedicating new resources to critical core functions. Although the new Agency status provides opportunities to pursue more creative fiscal strategies, critics are ever vigilant to ensure that cost recovery and revenue generation are not pursued at the expense of the core of the mandate.

Jurisdiction

The National Parks Act gives clear authority over the Parks Canada Agency to the Minister of Environment. The Chief Executive Officer of the Agency, the executive board, field unit and park superintendents, and park wardens are cited in either the National Parks Act or the Parks Canada Agency Act as being involved in some way with the routine responsibilities of managing each park, regulating activities, leases, licences, camping permits, and so on. Although Parks Canada operates as a separate agency, the CEO still reports directly to the minister and cabinet, and the minister's office has considerable authority and responsibility with respect to policy and legislation. The National Parks Act and policy provide general direction and authority, yet most of the day-to-day management decisions in the park are guided by directives and regulations passed pursuant to the Act. Within the Agency, the field unit superintendents are directly accountable to the CEO for decisions and actions at the park level.

Although Parks Canada has sole jurisdiction over national parks, park managers are by no means completely autonomous. Parks Canada must also take direction from central agencies, including the Treasury Board, on such areas as program expenditures, personnel policies, and the management of real estate. In addition, an increasing number of parks have been created through formal co-operative management agreements (e.g., Gwaaii Haanas National Park and Haida Heritage Site) with First Nations—as a result, park managers may share responsibility for some or all parts of park management activities with duly appointed representatives or councils. For example, wildlife management within many of the newer northern national parks involves local or regional wildlife councils. Similar arrangements exist for a host of resource issues in an increasing number of parks.

The Organizational Structure

The Parks Canada Agency has responsibility for national parks, national historic sites, national marine conservation areas, and a range of other designations, including historic canals and Canadian heritage rivers. The area of the Parks Canada Agency with jurisdiction over national parks is currently made up of an executive board, service centres, and field units (Figure 9.5). The executive board is headed by the CEO along with a number of executive positions including the directors general and other senior administrators. Policy and direction established by this executive are aimed to reflect the national parks mandate, be responsive to field staff concerns, and meet the needs of politicians.

In the past, regional offices were spread across the country to provide technical and professional support on a range of issues, from planning and data management to environmental assessment assistance. In the mid-1990s the organizational structure of the Agency was flattened as part of a staff reduction and cost-saving initiative. One principal effect of this initiative was the removal of the regional offices. In their place are service centres with centralized professional and technical staff. Although some reduction in numbers of professional services staff did occur in the beginning, the primary result of this realignment was the removal of the regional directors, who by way of directives and other organizational authorities could affect park management and give direction to superintendents in areas of research, planning, heritage program-

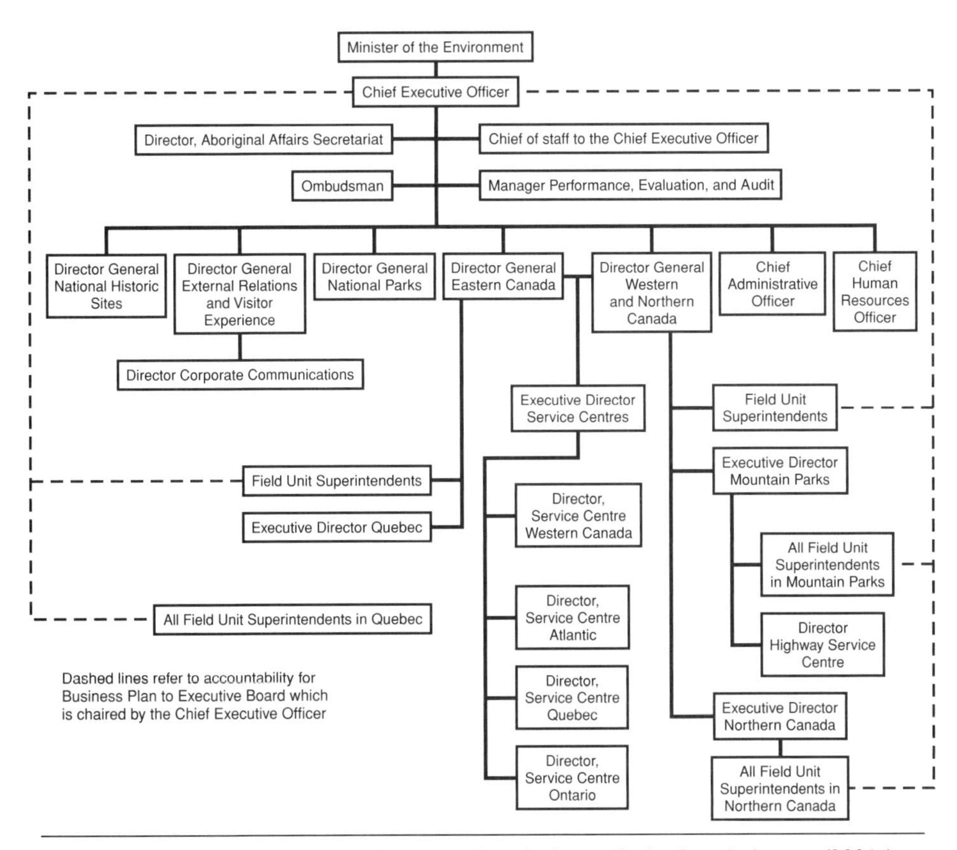

FIGURE 9.5 Organization chart for Parks Canada. *Source*: Parks Canada Agency (2006c).

ming, and resource management. Now, with the service centres, the authority to give direction has been replaced with a more advisory structure that is set in motion at the request of the field unit superintendents.

The core of the Agency operations occurs at the field unit level managed by a field unit superintendent. These field units are of varying size and spatial area and can be made up of a combination of national park(s), national historic site(s), and national historic canal(s). Depending on the size and location, some parks also have a park superintendent whose specific focus is the national park. This new and less hierarchical Agency organizational structure has been criticized as resulting in highly politically responsive but balkanized units with a focus so broad that no one purpose is well served.

The structure of management at the national park level is less standard and varies to some degree from park to park and from region to region. Job positions or functions common to most units include:

- park superintendent
- chief park warden and warden staff
- park ecologist(s) and resource conservation staff
- heritage interpretation/communication specialists and staff
- business management staff, including operations, maintenance, personnel, finance
- park planner (often a shared position within a field unit or service centre)
- cultural resource manager, historians, and archaeologists.

At the park level, the job requirements are many and varied and the staffing levels low. Increasingly, park staff are being called on to work outside of park boundaries, for example, by participating in or leading initiatives for the restoration of species at risk across the broader landscape or by consulting with neighbours about land management practices that affect the park. This broader approach to conservation management places even more demands on staff time.

A Daily Profile

One of the emerging management challenges within national parks is that for each staff category or position, protecting ecological integrity should be the top priority. For example, the park wardens, ecologists, and resource conservation staff provide expertise and guidance in ecological issues and management. Wardens involved in enforcement and resource conservation are instrumental in ensuring that park visitors comply with the requirements and laws that protect ecological integrity, from conducting environmental assessments to apprehending poachers. Interpretation and outreach staff should raise awareness and knowledge about the role of the park within the greater ecosystem and encourage action by park visitors and partners. Maintenance and cleaning staff affect management directly through their choice and use of environmentally safe cleaning products and indirectly by demonstrating to the public the relationship between environmental awareness and individual actions. Box 9.6 outlines the major types of positions found in national parks and the responsibilities associated with each position.

Financial Management

Traditionally, Parks Canada, like other government departments, was given an annual allocation from Treasury on the discretion of Parliament. These allocations were based on work plans and multi-year operational plans prepared within Parks Canada and across regions. Budget plans are generally divided into operating plans—the ongoing costs associated with operating the park, such as salaries and standardized expenses—and capital projects such as the construction or refurbishment of visitor centres and

BOX 9.6 A Day-to-Day Profile

Field Unit Superintendent

Field unit superintendents are responsible for managing a mix of national parks and national historic sites within a geographic area. They spend much of their time out of the park at meetings and in consultations with various community groups, industry associations, and various levels of government. Among their responsibilities, depending on the specific location, are:

- annual budget planning;
- reviewing seasonal personnel requests for warden service;
- attending negotiations and meeting with the park Aboriginal liaison officer and local First Nation about traditional use and co-operative management of resources;
- responding to phone inquiry from headquarters regarding monthly revenue targets from park fees;
- meeting with the chief park warden and operations staff regarding public safety issues and vandalism;
- scheduling a meeting with provincial and industry representatives to discuss a proposed forest management plan;
- attending an evening meeting in the adjacent town regarding community concerns over community access and boat use.

Park Warden

Depending on their skills and the needs of the park, wardens have a variety of responsibilities, including public safety, operations, communicating with the public, resource management activities, and conducting environmental assessments. Among these activities might be:

- patrolling in the frontcountry area of the park, including checking self-pay fee station in campground;
- responding to emergency call regarding aggressive elk in campground, which could involve arranging transport for injured person and incident report assessment;
- follow-up in the campground, talking to visitors about appropriate human use during calving season;
- planning for next day's vegetation inventory fieldwork.

Heritage Interpreter

Heritage interpreter, park naturalist, and outreach co-ordinator are among the many names for park staff with a primary responsibility for communication with the public. Historically, interpreters have been involved primarily with traditional park interpretive programs; however, today they are becoming more involved in communicating ecological messages to a range of audiences both inside and outside the park. The daily tasks of a heritage interpreter might include:

- making a presentation to regional real estate board members on corridors for wildlife, green space, and property values;
- working with team members and schoolteachers to develop educational outreach materials for urban schools;
- meet with wardens to revise ecological and public safety messages for boating in the park;
- preparing a public service announcement on safety during elk calving season for the local radio station;
- talking to a university graduate student regarding threatened and endangered species in the park.

Resource Conservation Staff

Resource conservation staff, such as ecologists, are often the primary co-ordinators of research, monitoring, and restoration efforts conducted by everyone from wardens to volunteer groups to heritage interpreters. Their responsibilities can include:

- reviewing methods and writing contracts for university study of impacts of recreation on grizzly bears;
- providing technical advice on environmental assessment being prepared by wardens on the proposed filming of a TV commercial in the park;
- negotiating with transport companies for compensation from the aquatic impacts of an oil spill on the highway bisecting the park;
- arranging a meeting with the superintendent and chief park warden to discuss the ecological impacts of park staff use of ATVs for trail maintenance;
- chairing an evening meeting of the Model Forest Association to discuss co-operative woodlot management and biodiversity issues.

new trails. The passage of a Parks Canada Revenue Policy and the creation of the Parks Canada Agency launched the organization into a new era of financial management.

The Parks Canada revenue policy introduced in the mid-1990s means that most national parks are involved in revenue generation with respect to charging fees for various products and services of both commercial and personal benefit. The purposes of the revenue policy are varied, although, clearly, the intent is to provide funding, through traditional governmental appropriations, for the core mandate of the Agency and to ensure that public goods are paid for from the public purse. Guiding principles within the revenue policy state that revenue initiatives will ensure long-term sustainability of ecological integrity and commemorative integrity; will be consistent with market demand for services; and will deliver services in a cost-efficient manner. Services operated or provided to benefit only select individuals who choose to consume or use those services (private goods) are to have fees charged for them.

> The fundamental principle guiding Parks Canada's revenue policy is that tax dollars pay for the cost of establishing and protecting national parks and national historic sites; those who use them, will pay for the additional personal or commercial benefits that they receive. Services providing both a public good and personal benefit, such as heritage presentation programs in parks and sites, will be financed through a combination of tax-based appropriations and fees. (Parks Canada Revenue Policy, 1998)

Revenue generation targets and plans are developed on an annual basis for each field unit and approved by the Agency executive as part of annual business planning approvals. The portion of individual park budgets that these targets represent varies considerably from year to year and from park to park. Revenue targets are not meant to directly offset expenditures within parks to make the parks self-sufficient, as this would violate the intent of government to provide funding for the core mandate of parks. Instead, revenue resources are distributed by complicated funding schemes to help fund new park acquisitions and to subsidize parks with lower visitation levels. If targets are exceeded, the excess amount is reallocated by the executive board. If a revenue target is not met, the field unit must reduce its expenditures by an equivalent amount.

From its inception, the revenue policy of Parks Canada has been controversial both inside and outside of the agency. Criticisms from the public and opposition politicians cite concerns that fees will be accessible only to the privileged in Canadian society and not to the disadvantaged. Another objection is that charging entrance or public goods fees in parks, in addition to the general appropriations that ultimately come from income tax, amounts to double taxation (see, e.g., Bella, 1987). Within the Agency, employees have cited concerns that revenue generation in some parks is driving activities or levels of activities that are in conflict with the maintenance of ecological integrity. Mechanisms are needed to ensure that revenue targets are appropriately set and that these targets neither prevent people from visiting parks nor generate a motivation and reward system for managing in contravention of the primary purpose of national parks.

The Ups and Downs of Budgets: When Building a Fence Can Serve Ecological Integrity

The funding scenario for the organization also has dramatically changed in the last decade. In the early 1990s a unique five-year funding program (ending in 1996–7), the Green Plan, provided an envelope of money that supported the creation of new parks and sites as well as for many ecological initiatives related to inventory, monitoring, research, and education. Despite this funding boost, by 1998–9, Parks Canada saw an annual reduction of $104 million or 25 per cent from 1994–5. This funding decline was further exacerbated by the fact that a number of new parks or park reserves, new marine parks, and new park study areas had been created over the same period. The result of decrease in funding was a major reorganization with associated staffing declines, an increased focus on cost-recovery policy, and reduced subsidies to users that receive specific services. Entrance fees to parks were increased and new fees introduced. Today, funding for Parks Canada for all operations hovers around $600 million, with estimated revenues of about $100 million. A 1999 study by the World Conservation Monitoring Centre of global protected areas budgets and staffing (James, 1999) showed that, while Canada is one of the world leaders in the amount of land in protected area status, it had significantly less staff (13 staff/1,000 km^2 in Canada compared to an average of 27 staff/1,000 km^2 in developed countries) and budget ($1,017/km^2 in Canada compared to an average of $2,058/km^2 in developed countries) than many other developed, and some developing, nations. With the creation of many new parks still on the agenda and an increasingly complex management job, staffing and budgetary requirements were clearly inadequate. Since this study was completed significant progress has been made: the staff/area ratio is now approximately 18.4/1,000 km^2, and the budget/area ratio has more than doubled, to $2,673/1,000km^2.

Issues of funding don't just revolve around how much money there is; more importantly, they focus on how that money is allocated. Given the complexity of the task at hand, in 2000, the Ecological Integrity Panel (Parks Canada Agency, 2000) noted that merely to address the primary ecological integrity recommendations that they were able to cost would require an additional $330 million over five years. However, the Panel went on to note that increased revenues, without a clear plan and a focus on accountability, would not solve the problem. For several years after the Panel report, additional monies to address the recommendations and other critical concerns trickled only sporadically from Treasury. While disheartening, the lag was not used as a delay tactic for action. A first phase—comprehensive ecological integrity training for all employees—was implemented, a number of strategic new positions (e.g., director general for ecological integrity) were created, and the Agency focused on getting its internal house, the organizational culture, in order. Although funding remains a challenge, recent programs have led to the release of significant funds for ecological integrity issues.

Through the management and business planning process—along with special funding calls—parks are forced to produce clear plans of action with measurable goals for how funding will be spent. The old adage—'Put your money where your mouth is'—proves true when we take a deeper look at the projects the new resources have gone to support. One of the biggest fears of critics is that any new financial resources for ecological integrity would quickly be siphoned off to build infrastructure and more development. So, when is a fence not an infrastructural development but an ecological management tool? In Grasslands National Park, located in the mixed prairie ecosystems of southern Saskatchewan, traditional grazers such as bison and, more recently, cattle perform vital ecological functions. Grazers reduce thatch, promote plant growth, reduce the spread of some exotic weeds, and create structural diversity of grass species and heights that are relied on by nesting songbirds, burrowing owls, swift foxes, and black-tailed prairie dogs. As a relatively recent national park created amid a mix of private ranchland and communal grazing land, Grasslands is a patchwork of blocks of national park across a larger landscape. In this instance, good fences do indeed make good neighbours—not just because they define ownerships but also because they lead to ecological restoration. Grasslands National Park has been busily building kilometres of fencing around sections of the park for ecological integrity. Good science, combined with good fences and friendly neighbours, led to the reintroduction of bison into a large block of the park, and in smaller sections where it wasn't feasible to reintroduce bison, experimental and managed grazing of cattle is ongoing. Maintaining and restoring remnant native prairie is something that everyone in the Grasslands vicinity can agree on.

PROVIDING FEEDBACK TO MANAGEMENT: THE EVALUATION AND REPORTING FUNCTION

Clear legislation and policy are not sufficient without strong accountability mechanisms. Accountability mechanisms from the parliamentary level to the personnel level provide the means of ensuring that laws, policies, and plans are being followed and are effective. In reality, the purpose and utility of such measures vary, depending on the mechanism, on whom it holds accountable, and at what level. Accountability mechanisms for Parks Canada are aimed either to ensure compliance with the laws, policies, and plans or to measure the effectiveness of the laws, policies, and plans and actions. Some mechanisms serve both purposes.

At the parliamentary level, accountability mechanisms have traditionally consisted of those associated with budgetary processes and the tabling of park management plans. To date, the latter has been generally more of a formality than a useful accountability mechanism. Perhaps one of the most significant elements of the 1988 amendments to the National Parks Act was the introduction of a new, system-wide accountability mechanism. The Act calls for a report to Parliament 'on the state of the parks and progress towards establishing new parks' (s. 5.1.5). These State of the Parks reports, released roughly every two years since 1990, provide an overview of the status towards completion of the national park system, the ecological integrity of national parks, commemorative integrity of national historic sites administered by Parks Canada, and the status of national marine conservation areas (Parks Canada Agency, 2005c).

These State of the Parks reports have been increasing in quality over the years, with a better information base gradually being built to inform them. Strengthening the science and knowledge base that informs the reports is a need recognized both inside and outside of the Agency, and recent actions, including hiring a chief of social science, work on regional and park-level ecological indicators, and development of visitor experience performance measures, represent steps towards improved reporting.

Within the Agency, the major accountability mechanisms in addition to the State of the Parks reports are the systems plan, corporate plan, and annual performance report (Parks Canada Agency, 2006b, 2006c) at the Agency level and the park management plan and business plan at the field unit level. All of these mechanisms can be used to hold managers and the executive accountable for actions. The Auditor General for Canada performs an important role in conducting periodic reviews, which include close analysis of these documents, and has the power to hold the Agency to account federally. Since the release of the Ecological Integrity Panel report, the Parks Canada Agency has also instituted an interesting mechanism for public reporting and accountability. The Minister's Roundtable on Parks Canada (Parks Canada Agency, 2005a) is one means of vetting progress and performance with a broad group of stakeholders. Although attendees at these roundtables are by invitation only, the performance reports used as information items for the meeting and the Agency's responses to the meeting (Parks Canada Agency, 2005a) are public documents available on-line.

The State of the Parks reports, written at the national level, have provided the only mechanism to evaluate or report on park-specific actions, but this happens in aggregate data form supplemented by a few individual case studies. The institution of a formalized park-based evaluation and reporting mechanism—an individual state-of-the-park report that would be conducted at least prior to management plan review—was recommended as a mechanism to improve this shortfall (Parks Canada Agency, 2000) and has since been adopted. These park-level reports are to be done in advance of and to facilitate scoping for management plan reviews and for public awareness and consultation (see, e.g., Banff National Park, 2003).

Not to be neglected in a discussion of accountability mechanisms is the increasingly important role that the public at large has in affecting Parks Canada policy and actions. Some parks are experimenting with various ways of conducting more enhanced reviews of progress and accomplishments. Banff National Park, for example, has taken a leadership role by presenting in a public forum an annual report of its actions and progress towards achieving its objectives, and this example, coupled with a park-based evaluation report, could serve as a model for improving evaluation and reporting. Grasslands National Park has used a slightly different format by assembling an external expert panel to review and comment on management progress prior to management plan review. Results from the expert panel are to be shared publicly through the park's stakeholder advisory committee. Public accountability mechanisms are not always formalized, but through public participation, litigation, lobbying, and campaigning, citizens are holding the government, the minister, and the Agency accountable for its actions. Likewise, external specialists and critics of national park management (see, e.g., Banff–Bow Valley Task Force, 1996; Bella, 1987; Parks Canada Agency, 2000; Searle, 2000) have played, and continue to play, a valuable role in calling attention to critical issues in park management. As noted throughout this chapter, many of these contributions have resulted in fundamental shifts in Agency direction. In a much broader context, public responsibility and personal accountability for actions in keeping with the purpose of national parks are central to protecting park values.

CONCLUSION

That Canadians support the concept of national parks and of protecting wilderness and natural settings is beyond doubt. Consistently, national polls and the media more generally state the public concern regarding the value of parks and the threats to parks. Ensuring the maintenance and restoration of ecological and commemorative integrity while providing memorable, educational experiences for visitors can be done simultaneously.

This chapter discussed the process and challenges associated with new park establishment and the progress that has been made—and remains to be made—in this regard. With the recognition of growing stresses on parks from management within national parks and land management outside of parks, the notion that parks, once established, will be self-managing has been dispelled. The human footprint is everywhere, from the clogging of the Bow Valley wildlife corridor with highways, towns, and developments to the generation of airborne pollutants in cities that are deposited thousands of miles away in the pristine high alpine lakes of a national park or in the fatty tissue of a marine mammal within Saguenay Fjord National Marine Conservation Area. In the Canada we face today, the notion that 'nature will heal itself' if we protect the right areas has been dispelled as romantic and fraught with problems. The stresses placed on parks are too many and too great for us to be sure that their integrity will persist without active management (see Chapter 5).

Park management, like many other resource management activities, is an adaptive and evolving process. Although developing a plan and implementing it are important, the critical step is evaluation and reporting. Providing an assessment of what has gone right, what needs improvement, and how fast management is progressing towards its goals is critical and should happen at each level of the organization. The renewed focus and emphasis on monitoring, evaluation, and reporting—both to improve management and to increase public accountability—is a sign of positive change.

ACKNOWLEDGEMENTS

This chapter is revised from the second edition and supported by contributions from Chapter 4 (Paul Eagles) and Chapter 5 (Rick Rollins) in the first edition of this book, by the collective writings of the Ecological Integrity Panel, and by suggestions from two Parks Canada Agency planners, Graham Dodds and Kevin Lunn. However, any errors, omissions, or mistakes in interpretation are our own.

REFERENCES

Auditor General of Canada. 1996. *Report of the Auditor General of Canada to the House of Commons*, Chapter 31: 'Canadian Heritage–Parks Canada: Preserving Canada's National Heritage'. Ottawa.

BC Parks. 2006. Fixed-Roof Accommodation Policy. At: <www.env.gov.bc.ca/bcparks/fixed_roof>.

Banff–Bow Valley Task Force. 1996. *Banff–Bow Valley: At the Crossroads*. Technical Report of the Banff–Bow Valley Task Force, prepared for the Honourable Sheila Copps, Minister of Canadian Heritage. Ottawa.

Banff National Park. 2003. *State of the Park Report 2003*.

Bella, L. 1987. *Parks for Profit*. Montreal: Harvest House.

Canada National Parks Act, 2000, c. 32

Canadian Environmental Advisory Council. 1991. *A Protected Areas Vision for Canada*. Ottawa: Minister of Supply and Services.

Canadian Heritage. 1994. *Parks Canada: Guiding Principles and Operational Policies*. Ottawa: Minister of Supply and Services.

Charron, L. 1999. *An Analysis of Planning Processes of Parks Canada*, prepared for the Panel on the Ecological Integrity of Canada's National Parks.

Chinook Institute. 2007. 'Skills for working together in the management of protected heritage areas: Planning and getting organized'. Canmore, Alta: Chinook Institute for Community Stewardship. At: <www.chinookinstitute.org>.

Dearden, P. 1987. 'Mobilizing public support for environment: The case of South Moresby Island, British Columbia', in *Need-to-Know: Effective Communication for Environmental Groups*, Proceedings of the 1987 Annual Joint Meeting of the Public Advisory Committees to the Environmental Council of Alberta. Edmonton: Environment Council of Alberta, 62–75.

Environment Canada. 1990. *National Parks Systems Plan*. Ottawa: Supply and Services Canada.

———. 1991. *State of the Parks: 1990 Report*. Ottawa: Supply and Services Canada.

Grumbine, R.E. 1994. 'What is ecosystem management', *Conservation Biology* 8, 1: 27–38.

Hummell, M. 1989. *Endangered Spaces*. Toronto: Key Porter.

James, A. 1999. 'Institutional constraints to protected area funding', *Parks* 9, 2: 15–26.

National and Historic Sites Branch. 1971. *National Parks Systems Planning Manual*. Ottawa: Information Canada.

Parks Canada. 1994a. 'Parks Canada guide to management planning', internal document. Ottawa.

———. 1994b. *Guiding Principles and Operational Policies*. Ottawa: Parks Canada.

———. 1996. 'Revenue policy', internal document. Ottawa.

———. 1998a. 'Management directive 2.4.2, impact assessment', internal document. Ottawa.

———. 1998b. *State of the Parks, 1997*. Ottawa: Minister of Public Works and Government Services.

Parks Canada Agency. 2000a. *Unimpaired for Future Generations? Protecting Ecological Integrity with Canada's National Parks*. Vol. 2: *Setting a New Direction for Canada's National Parks*. Report of the Panel on the Ecological Integrity of Canada's National Parks. Ottawa.

———. 2000b. *Parks Canada Guide to Management Planning*. Ottawa: Minister of Public Works and Government Services Canada.

———. 2000c. *National Historic Sites of Canada System Plan*. Ottawa: Minister of Public Works and Government Services Canada.

———. 2005a. *Response of the Minister of the Environment to Recommendations Made at the Third Minister's Round Table on Parks Canada (2005)*. Ottawa: Minister of Public Works and Government Services Canada.

———. 2005b. *Action on the Ground: Ecological Integrity in Canada's National Parks*. Ottawa: Minister of Public Works and Government Services Canada.

———. 2005c. *State of Protected Heritage Areas (for the period ending March 31, 2005)*. Ottawa: Minister of Public Works and Government Services Canada.

———. 2006a. *Ski Area Management Guidelines*. Ottawa: Minister of Public Works and Government Services Canada.

———. 2006b. *Parks Canada Agency Performance Report (for the period ending March 31, 2006)*. Ottawa: Minister of Public Works and Government Services Canada.

———. 2006c. *Parks Canada Agency Corporate Plan 2006/07–2010/11*. Ottawa: Minister of Public Works and Government Services Canada.

Parks Canada Agency Act 1998, c. 31.

Provincial Parks and Conservation Reserves Act, 2006, c. 12.

Searle, R. 2000. *Phantom Parks*. Toronto: Key Porter.

KEY WORDS/CONCEPTS

business plan
Endangered Spaces Campaign
field unit superintendent
financial management
heritage interpreter
legislation
park management plan
park warden
policy
regulations
resource conservation staff
revenue policy
State of the Parks report
systems plan
tactical plan
zoning system

STUDY QUESTIONS

1. Describe the evolution of 'appropriate activities' in national parks.
2. Distinguish between legislation, policy, and regulations.
3. During the winter of 2001, a debate surfaced regarding the carrying of firearms for law enforcement purposes by national park wardens. Some people in the agency supported this idea, but others were opposed. Develop a position on this issue by examining relevant sections of national park legislation, policies, and regulations. Examine also any media accounts of the debate.
4. What is the significance of species-at-risk legislation for Parks Canada.
5. Discuss the significance of the Endangered Spaces Campaign.
6. Critique the systems planning approach used by Parks Canada (list positive and negative features).
7. Examine a recent national park management plan for evidence of ecological integrity, ecosystem management, and appropriate activities.
8. Discuss the pros and cons of a more corporate approach to management.
9. What is romantic or naive about the belief that if we protect the right areas, nature will take care of the rest?
10. What is the purpose of the zoning system developed by Parks Canada. Compare this with the type of zoning described in ROS (Chapter7).
11. What circumstances described in national park policy allow for the 'manipulation' of naturally occurring processes? Discuss your views supporting and opposing this policy.
12. Discuss advantages and disadvantages of the policies for allowing motor vehicle access into national parks.
13. Explain the value of State of the Parks reports.
14. Compare the roles of park warden, heritage interpreter, and field unit superintendent.

CHAPTER 10

Case Study: Banff and Bow Valley

Joe Pavelka and Rick Rollins

INTRODUCTION

Banff is Canada's oldest and most visited national park. The park is relatively easy to reach, with an international airport in nearby Calgary, and access into the heart of the park by car or bus on the Trans-Canada Highway, or by rail on the Canadian Pacific Railway (CPR). Over 4 million people visit the park each year, most arriving by car or bus travelling along the Trans-Canada Highway that runs through the park. Entering the park, one is struck by the spectacular mountain scenery and resplendent forest environment extending up the slopes of the mountains (Figure 10.1). From the highway, it is possible at times to see a variety of wildlife, including black bear, grizzly bear, and bighorn sheep.

The highway runs through the Bow River Valley and into the town of Banff, the usual first stop for most park visitors. The town of Banff, a community of over 5,000 people, rests in the heart of the Bow River Valley, and provides a world-class tourism resort destination. Park visitors may choose from a variety of types of accommodation, including hotels, motels, lodges, chalets, condominiums, private cabins, hostels, and campgrounds. Many recreational activities are available, including hiking, mountaineering, angling, canoeing, river rafting, cycling, horseback riding, downhill skiing, and viewing wildlife. Commercial guiding companies located in Banff promote these activities and lead guided outings for park visitors. Within the town, visitors can golf, swim in hot springs, or shop in the various stores and arcades. Visitor amenities include restaurants, pubs, clothing stores, jewellery stores, grocery stores, liquor stores, banks, commercial outfitters, and tourist information centres (Figure 10.2). The town of Banff also caters to the business traveller, providing conference and convention facilities in luxury hotels.

The tourism experience provided in Banff National Park is very much in keeping with the original vision for the park as a tourism destination, an island of civilization in a sea of wilderness, as outlined in Chapter 2. However, this romantic notion appears to mask the reality of changes that have occurred in this wilderness as a result of human development and activity that has been allowed and encouraged, until recently, in the park. Most of this development has occurred in the montane region of the park, that is, below the tree line—the Bow River Valley, a natural place to locate

FIGURE 10.1 View from Sulphur Mountain showing the town of Banff and illustrating its critical position within the montane zone, and the constricted nature of wildlife corridors. *Photo: Guy Swinnerton.*

a town, a railway, and a highway. However, the montane zone is small, comprising just 3 per cent of the park, but is critical for wildlife, including elk, moose, and grizzly, whose numbers have been declining in the park as a consequence. Further, one of the most significant causes of wildlife mortality in the park occurs when wildlife cross roads and railroad lines. Human activity has caused other changes in wildlife populations. Wolves, for example, were eradicated in the park and have only recently been reintroduced. Black bears, normally shy of humans, have become habituated to humans, as a consequence of access to human garbage and being fed by tourists. Popular sport fishing areas were stocked to cater to park visitors, and in some cases with non-native species that threatened indigenous stocks.

The forest lands in the valleys and on mountain slopes have also been altered by human intervention, in the form of fire suppression. In the past, all fires were put out to prevent 'damage' to the park, out of a desire to protect scenery, wildlife, park facilities, and park visitors. We now understand that fire is an important natural process in this environment, needed to maintain the diversity of forest vegetation and wildlife. Decades of fire suppression have had the effect of creating an atypical forest, one that is overstocked with old-growth coniferous forest and understocked with deciduous forest, including aspen and other vegetation characteristic of early forest succession following a fire. This, in turn, has contributed to the decline in moose and other wildlife in the park, and also to the spread of the mountain pine beetle (see Chapter 5).

In recent years our understanding of natural processes in parks has matured while the impact of human activities on natural systems has grown substantially (Chapters 4 and 5), and these factors have influenced revisions to legislation and policies to stress ecological integrity in the management of national parks (Chapters 2 and 9). For Banff, a new focus on ecological integrity, in an environment significantly altered by tourism and commercial development, poses huge challenges. To propose new management practices reflecting ecological integrity would almost certainly curtail growth and development of the business community in Banff and in the smaller centres at Lake Louise and Sunshine Village. Also, many park visitors are reluctant to see changes that might restrain traditional recreational pursuits in the park. However, to ignore the evidence of human impact outlined above would lead to further compromising of the natural values of the park. Despite the best science, policy, and legislative tools at its disposal, Banff continues to struggle with the consistent application of a balanced management approach between environmental protection and human uses (Swinnerton, 2002). Hence, the purpose of this chapter is to describe how Parks Canada and the community of residents living in the park have attempted to address these developments. As Banff represents one of the more extreme situations of human-induced landscape changes in a protected area, this case study aims to inform the kinds of challenges that can occur in any protected area, by outlining the events that contributed to the current situation and examining recent attempts to mitigate the problems that have occurred.

FIGURE 10.2 Park visitors and a street scene in Banff. *Photo: Guy Swinnerton.*

Balancing long-term environmental protection and human use is particularly daunting for Banff, given the history of the park, the unique and complex human-use pressures facing the park (Lovelock, 2000), and the role it plays as Canada's crown jewel of national parks (Draper, 2000). Banff is different—an anomaly among other national parks within Canada—for several reasons:

- Banff National Park is Canada's first national park.
- Banff is Canada's most visited national park, a world-class tourism destination.
- Banff includes strong summer and winter tourism products and corresponding year-round visitation.
- Banff is bisected by the Trans-Canada Highway and CPR Railway, with heavy year-round traffic on both.
- Banff is situated adjacent to the town of Canmore, with a population of 17,500 (Canmore, 2007), and is only 90 minutes from Calgary (a city of one million). The Calgary International Airport provides access for approximately 4.6 million visitors travelling through the park (Golder Associates, 2004: 1).
- The town of Banff includes a population of approximately 5,600 permanent residents (Biosphere Institute of the Bow Valley, 2002) and exists as a separate corporate entity from Parks Canada. Banff National Park is one of the few national parks in Canada to contain a community of permanent residents.
- Banff National Park includes the hamlet of Lake Louise, with its significant resort and ski facilities.
- Banff includes more commercial activity within its boundaries than the four largest national parks in the United States combined (Dunkel, 2003).

A BRIEF HISTORICAL PERSPECTIVE

Banff National Park came into being for reasons that had little to do with long-term environmental protection or the maintenance of ecological integrity (Lovelock, 2000; Bella, 1987), but much has changed since it was declared a park reserve in 1885 (Chapter 2). In recent years, Parks Canada has attempted to reconcile developments in Banff with contemporary thinking about park management as reflected in the current National Parks Act, National Parks Policy, and other policies that are described in Chapter 9. The 1996 report of the Banff–Bow Valley Task Force, described later in this chapter, provided an unprecedented review and direction in establishing the protection of ecological integrity as its central priority. This priority was clear in the 1998 Banff National Park Management Plan (Parks Canada, 2004).

The central theme guiding all aspects of management and planning since the turn of this century has been to protect ecological integrity. At the same time, pressure to expand tourism development in the park continues, but mainly for frontcountry use; backcountry use in the park actually has declined in recent years (Parks Canada, 2006). The major issue for backcountry management has been the existence of three downhill ski areas, the management of which is described later in the chapter. Other aspects of backcountry management in Banff are less problematic and, for the most part, have been addressed by the Parks Canada zoning system (Chapter 9) and the application of various visitor management strategies outlined in Chapters 6 and 7. The focus of this case study is the management of the more heavily impacted frontcountry areas of the park in the Bow Valley corridor (Figure 10.3)

Swinnerton (2002), citing Belland and Zinkan (1998), presents the history of the four mountain parks (Banff, Jasper, Yoho, and Kootenay national parks) as four major periods of development, from 1885 to the 1990s. The first period, from 1885 to the 1950s, is characterized by the promotion of the parks for public enjoyment, tourism, and economic development. The preservation of the environment was important for its support of human uses and goals.

The second period, the 1960s and 1970s, also focused on tourism growth with greater involvement from Parks Canada. However, the increased amount of tourism at the time and, most notably, the rise of winter tourism precipitated concern regarding the scale of development within the park. Towards the latter portion of this period, Parks Canada policy began to address specifically ecological and historical integrity.

The third period is the 1980s. Significant in this period was the 1981 Four Mountain Parks Planning Program, which provided, for the first time, a long-term planning framework for these parks. Subsequently, in 1988, a management plan for Banff National Park was approved, focusing on the primacy of ecological integrity. Other developments of the 1980s included the Canadian Rocky Mountain Parks World Heritage Site designation (1984).

The final period—the 1990s—is highlighted by three key events: (1) the incorporation of the town of Banff in 1990; (2) the 1996 Banff–Bow Valley Task Force report; and (3) the 1998 Banff National Park Management Plan. This final period significantly clarified the role of the park and shifted the management agenda towards greater environmental protection, though this shift was not always supported by all involved stakeholders (Swinnerton, 2002).

Another way to approach the history of Banff National Park is to examine it as a tug-of-war between environmental protection and human use. This type of struggle is not so different from mountain communities throughout the world or unlike virtually any other nature-based tourism community (Godde et al., 2000). Banff National Park's unique features make it a living laboratory of the interplay between environmental protection and human use.

Recently created national parks in Canada have benefited from a contemporary vision of parks, usually expressed as a balance between environmental protection and human use. This strategy, of course, was not in place at the creation of Banff National Park; its original mandate was to be tourism and economic development (Box 10.1). The original declaration establishing what is now Banff National Park clearly presents the park mandate as a place for human recreation and enjoyment, while the retention of its natural condition was to serve the tourism purpose. To illustrate, the 1886 culling of predator species such as wolves and the development of a tree nursery were undertaken to enhance the aesthetic appeal of the area for tourists (Sanford, 1999). Little attention was given to maintenance of ecological integrity. The prospect of allowing fires to burn to maintain natural ecological processes likely was not considered. In these

BOX 10.1 The Early Vision of Banff National Park

1885: Banff Springs Reserve—It is hereby ordered that whereas near the station of Banff on the CPR there have been discovered several hot springs which promise to be of great sanitary advantage to the public, and in order that proper control of the lands surrounding these springs may be vested in the crown, the said lands . . . be . . . hereby reserved from sale or squatting.

1887: Rocky Mountains Park—The said tract of land hereby is reserved and set apart as a public and pleasure ground for the benefit, advantage and enjoyment of the people of Canada . . . preservation from injury or spoliation, of all timber and mineral deposits, natural curiosities or wonders . . . and their retention in their natural condition . . . .

Source: Parks Canada (2003: 3).

early days, wilderness was not viewed as a scarce commodity. Wilderness was abundant, particularly in the Canadian West.

Firmly adding to its position as a tourism destination was the 1888 opening of Banff Springs Hotel, catering to an elite global clientele. Other parks, such as Elk Island (1903) and Wood Buffalo (1908), had been designated, but none of these struggled with the unique and intense human pressures of Banff National Park, and none included a community of permanent residents within its boundaries. By 1911, automobile access to the park had been established. Over time, vehicle access to the park changed the nature of park visitors from that of a more predictable train-reliant, long-stay visitor of two weeks or more, to a much higher volume of short-stay, automobile-reliant overnight visitors. Visitation to the park tripled from 71,500 in 1912 to 217,000 by 1929, with a full 80 per cent of them arriving by automobile (Hart, 2003). Automobile access was the primary motivation behind the development of Kootenay and Yoho national parks as part of an attempt to create a Grand Circle Tour of automobile-driving tourism (Hart, 2003).

When J.B. Harkin became the first Commissioner of Dominion Parks in 1911, he provided a much-needed direction to park development and management (Sanford, 1999). The 1930 National Parks Act crafted by Harkin represents a skilful attempt to improve the balance between environmental protection and human use. Harkin understood, better than most at that time, that for parks to receive protection they must possess more than intrinsic value; they must be of use to and valued by the public, and nowhere was the match more evident than in Banff (ibid.).

With a growing community of permanent residents within Banff, park management was pressured to ensure the economic and social sustainability of the park. The economy of Banff relied heavily on a seasonal tourism industry. However, the Great Depression of the 1930s impacted tourism travel, and the nature of park visitation was changing as a result of automobile access. It is not surprising that Banff residents lobbied for greater access to markets (Hart, 2003). Thus, during the Depression era and into the 1940s several make-work projects, such as improvements to park amenities and the soon-famous Banff–Jasper highway, were undertaken more to support local residents than out of need. Many of these projects would influence the patterns of human use that developed (ibid.).

The 1940s and 1950s witnessed sharp increases in tourism travel to Banff, with annual visitation at just under a half-million as compared with the 219,000 in 1929 (Draper, 2002). While other parks in Canada had to contend with similar issues of growth brought on by the automobile and increased leisure travel, few, if any, had the added challenge of winter or four-season use. This explains, in part, the origins of the ski industry in Banff. Skiing had began at Skoki Lodge in 1922, but skiing did not become accessible to a travelling public until the 1940s and 1950s (Hart, 2003). Banff residents were generally supportive of skiing because it added to tourism and made Banff a more economically stable community year-round. However, skiing added to the pressures on the park from human use by increasing visitation, as well as adding to the increased reliance on tourism by Banff residents.

In 1962 the Trans-Canada Highway finally was paved west through to Banff and just prior, in 1959, powered ski lifts at Lake Louise were introduced, making summer and winter recreation in Banff National Park much more accessible (ibid.). Today, it is difficult to grasp the enormous significance of these two developments that opened Banff National Park to the new driving public, the booming middle class of the time. The Trans-Canada Highway meant that virtually anyone could visit the park for a two-week vacation, an evening, or just for the day.

The introduction of powered lifts for skiing greatly expanded the winter use of the park. Prior to the introduction of power lifts, skiers had to use the laborious technique of side-stepping up slopes, which meant they could manage only two or three runs in a day. Powered lifts meant that most people of average fitness could access the hills. Winter recreation spurred further development and encouraged increased visitation and population growth within the park, and was seen as a viable economic development strategy for the community (ibid.). Trans-Canada Highway improvements and powered lifts influenced an increase in visitation from 500,000 in 1950 to more than 3.5 million in the early 1980s (Draper, 2000).

The struggle between environmental protection and human use took a new turn during the 1960s and 1970s. Despite the growth of tourism, especially summer tourism, Banff residents continued in their efforts to place Banff on the world stage of tourism. Towards that goal a concerted bid for the 1968 Winter Olympics was initiated (Hart, 2003). At the same time, a growing environmental sensibility gave pause to the bid and arguably forced Parks Canada at length to put at end to it (Sanford, 1999). However, the Olympic bid process highlighted the ambiguous role of Banff National Park. On one hand, it was a tourist destination on which a permanent community of residents relied; on the other hand, as a national park Banff was viewed as a place where ecological integrity measures should be asserted.

The 1980s brought about several important changes that ushered the park into the new millennium. The four mountain parks (including Banff) received UNESCO World Heritage Site status in 1984, further adding to an imperative to establish strong environ-

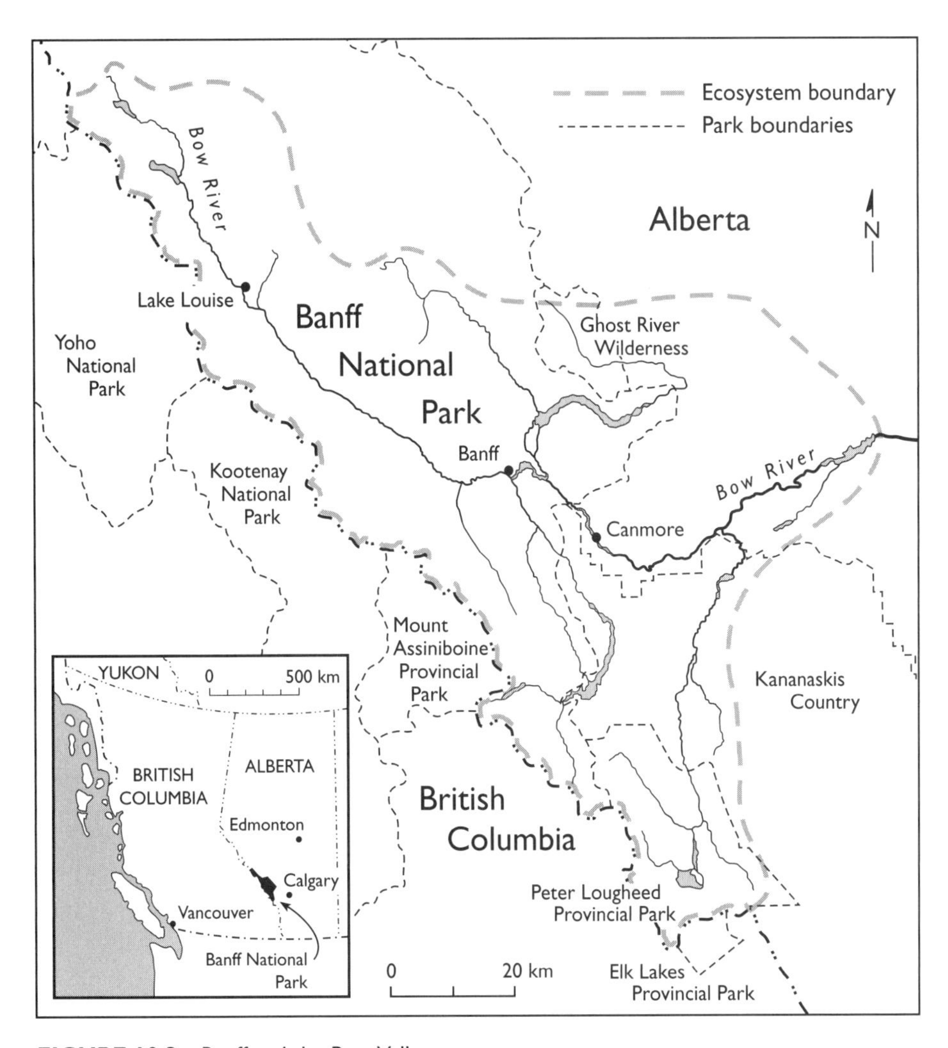

FIGURE 10.3 Banff and the Bow Valley.

mental protection (Draper, 2000). In 1985 the first Four Mountain Parks Management Plan was presented, recognizing a need to address the ambiguity of the past and address challenges of growing human use. In 1988, amendments to the National Parks Act were put forth that strengthened the environmental protection mandate and formally introduced the concept of ecological integrity (Swinnerton, 2002). During the 1980s Banff National Park experienced strong increases in visitation. The town of Banff continued to prosper as an international tourist destination operating throughout the year with a resident population of close to 5,000 (Hart, 2003). In 1990 the town was incorporated with its own governance separate from but highly intertwined with Parks Canada.

During the early 1990s it became apparent that 'the ecological integrity of the Bow Valley could be permanently impaired on account of the continued increase in visitation and infrastructure development' (Swinnerton, 2002: 243). This awareness resulted in the initiation of the Banff–Bow Valley Task Force in 1994. The Banff–Bow Valley Task Force conducted its business in round table format and involved an unprecedented level of public consultation (see Figure 10.4) and numerous studies (ibid.). The final technical report, released in 1996, warned that the region's ecological integrity was under such pressure that its very identity as a tourism destination was at risk. The final report included 500 recommendations addressing management and planning issues, many of which were controversial, such as capping the resident population, capping commercial growth, and closing many existing facilities (see Box 10.2). The establishment of the Banff–Bow Valley Task Force is noted as one of the most important events in the history of Banff National Park, and its report provided the framework for the development of the 1997 Banff National Park Management Plan and many subsequent management decisions (ibid.).

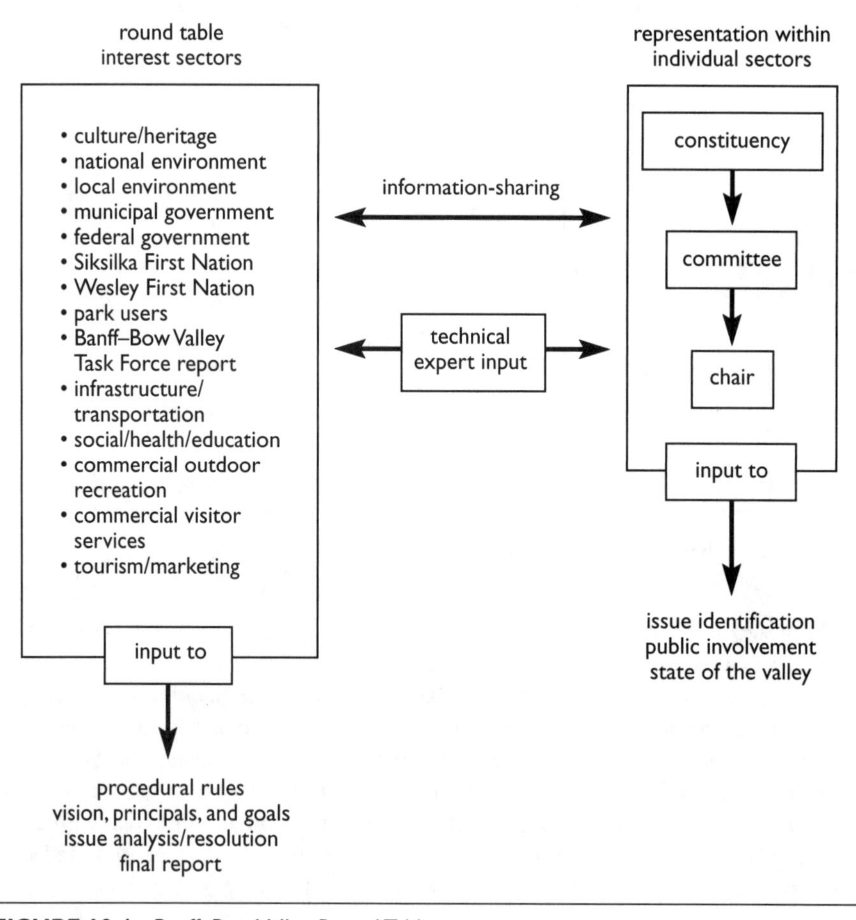

FIGURE 10.4 Banff–Bow Valley Round Table.

BOX 10.2 Key Conclusions of the Banff–Bow Valley Task Force

1. While Parks Canada has clear and comprehensive legislation and policies, Banff National Park suffers from inconsistent application of the National Parks Act and Parks Canada's Policy.
2. Despite the fact that ecological integrity is the primary focus of the National Parks Act and Parks Canada's Policy, we found that ecological integrity has been, and continues to be, increasingly compromised.
3. While scientific evidence supports conclusion #2 above, a significant percentage of the population find it difficult, based on what they see around them, to understand the ecological impacts that have occurred.
4. The current rates of growth in visitor numbers and development, if allowed to continue, will cause serious and irreversible harm to Banff National Park's ecological integrity. Stricter limits to growth than those already in place must be imposed if Banff is to continue as a national park.
5. More effective methods of managing and limiting human use in Banff National Park are required.
6. To maintain natural landscapes and processes, disturbances such as fire and flooding must be restored to appropriate levels in Banff National Park.
7. There are existing anomalies in the Park, such as the Trans-Canada Highway, the Canadian Pacific Railway, and the Minnewanka dam.
8. We are proposing the refocusing and upgrading of the role of tourism. Tourism in Banff National Park will, to a greater extent, reflect the values of the Park, and contribute to the achievement of ecological integrity and the quality of the visitor experience.
9. We acknowledge that mountain tourism in Alberta will continue to expand. Any new, related facilities will have to be located outside national park boundaries.
10. Current growth in the number of residents, and in the infrastructure they require, is inconsistent with the principles of the national park. Revisions to the General Municipal Plan for the Town of Banff must address these inconsistencies and the needs for limits to growth.
11. Public skepticism and lack of trust in the decision-making process have led to the polarization of opinion. There must be an overhaul of the development review process.
12. Visitors must be better informed about the importance of the Park's natural and cultural heritage, the role of protected areas, and the challenges that the Park will face in the third millennium.
13. Improvements in Parks Canada's management are central to the successful future of Banff National Park. This should begin with a comprehensive revision to the Banff National Park Management Plan that will reflect the recommendations of this report.
14. We believe that current funding will be inadequate to implement the recommendations of the Task Force and maintain normal operations in Banff National Park.

Source: Banff–Bow Valley Task Force (1996), from Swinnerton (2002).

RECENT DEVELOPMENTS

The purpose of this section is to highlight some recent developments since the Banff–Bow Valley Task Force presented its final report in 1996. The focus will be on the *Banff Community Plan* of 1998 and its ongoing revisions to the present.

Banff Community Plan

The presence of a community of permanent residents in the park is a feature that clearly distinguishes Banff National Park from other national parks in Canada. Since the park's inception, the challenge of managing a park that includes a townsite has been difficult (Hart, 2003). Along with the fragmentation of the park's montane region by the Trans-Canada Highway, the town of Banff represents a significant focus of human use. The town is home to a thriving year-round community of residents, and it hosts an influx each year of approximately 1.5 million visitors (Parks Canada, 2006). The Banff–Bow Valley Task Force provided considerable direction to the *Banff Community Plan* by capping the permanent population at 10,000, eliminating certain developments, requiring the town to develop bylaws to support existing heritage assets, and providing appropriate levels of services for its residents (Banff–Bow Valley Task Force, 1996).

The *Banff Community Plan* was initiated in June of 1995 with a goal of addressing the basic questions: Where are we now? Where are we going? How do we get there? Following comprehensive consultations with resident and stakeholder groups and after numerous amendments, a final report was approved in 1998. The *Banff Community Plan* (1998) addressed several aspects of town management and growth:

- growth and development;
- environmental protection;
- heritage preservation and development quality;
- resident land use;
- commercial land use;
- community, institutional, and recreational land use;
- tourism and economy;
- transportation and parking;
- utilities and infrastructure;
- municipal finance;
- community and visitor services;
- special policy areas;
- regional integration.

  (Banff, 1998)

Some of the more contentious areas of policy found in the plan were: limits to population growth; limits to commercial growth; determination of appropriate uses; and the 'need-to-reside' policy (discussed later in the chapter) (Draper, 2000). In many ways the *Banff Community Plan* represents a unique and forward-looking direction for a community desperately challenged to balance environmental (and heritage) protection

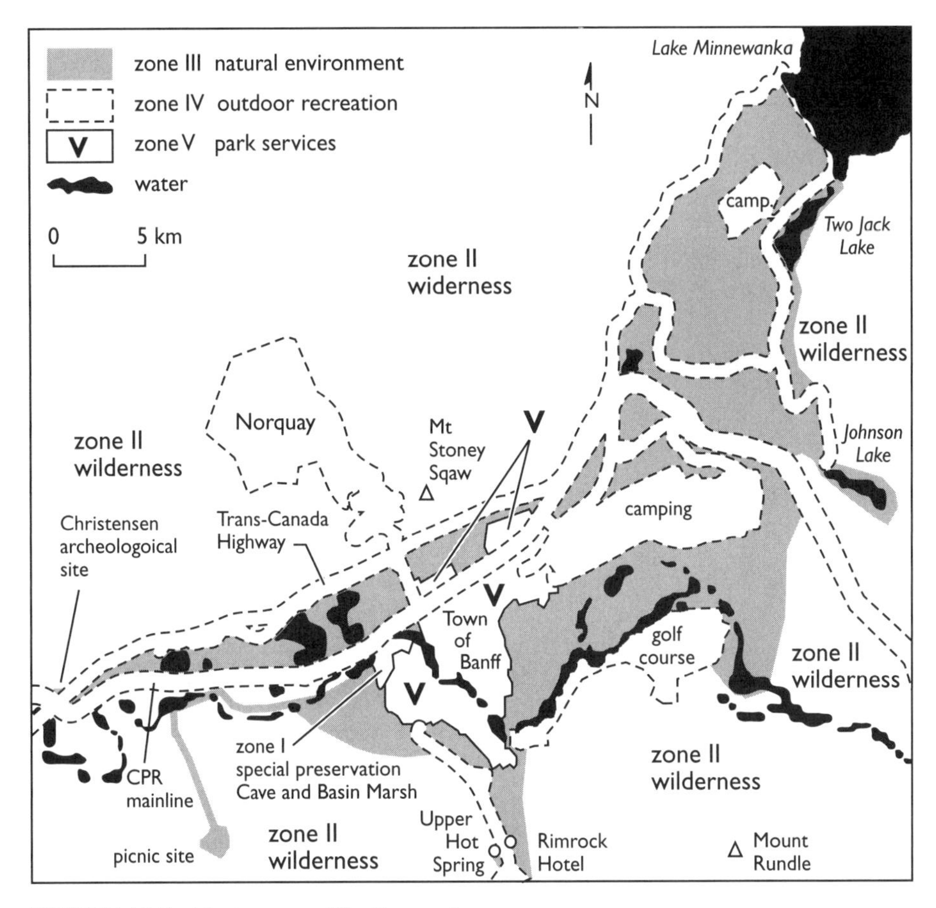

FIGURE 10.5 Zoning map of Banff townsite.

with numerous and intense human uses. A similar community plan for Lake Louise received approval in 2001 (Parks Canada, 2006a).

In 2005 the town of Banff began the process of revising the *Banff Community Plan.* Some preliminary themes, emerging from community consultations in 2006, are presented in Box 10.3. These themes call for the town to focus on balance—enhancing efforts to retain authenticity through heritage and related tourism, while remaining mindful of the needs of its residents. The emerging themes presented thus far highlight many of the issues of human use that Banff National Park has had to contend with since its inception. The revised *Banff Community Plan* was approved by Banff Town Council in October of 2007 and awaits final approval from Parks Canada.

The 'Need-to-Reside' Policy

The 'need-to-reside' policy is a feature of the town of Banff that distinguishes it from other such communities (Banff, 2007). This policy restricts residency in Banff generally

BOX 10.3 Common Themes Emerging in the Revised Banff Community Plan

1. Provide an attractive town with a positive and safe social atmosphere.
2. Provide the necessary infrastructure and services to residents and visitors.
3. Develop a unique and desirable 'signature' for Banff as a destination, showcasing natural and heritage values.
4. Enhance alternative transportation opportunities and work to resolve conflicts between pedestrians and automobiles.
5. Deal with major issues of promotion, regional partnerships, and communication.
6. Develop a clear vision and direction for the town that recognizes it as a town within a national park, and the need to balance community and tourism.
7. Preserve the history and heritage of Banff while remaining an internationally competitive tourism destination.
8. Provide goods, services, and housing for residents that are affordable and appropriate to their needs.
9. Move towards being a sustainable tourism community and better understand the implications of key aspects of growth management.
10. Embrace and lead in the area of environmental standards.
11. Enhance the spirit of vitality of Banff for permanent and non-permanent residents.
12. Provide education and learning opportunities for employees and have the community recognize the important role they play.

to people who work or own a business in the park and to their partners. The policy was first met with resistance, primarily because of concerns that the policy would lead to decreased property values (Kuczaj, 2002), but over time its most positive feature is to have closed the door on second-home ownership (part-time residency). Second-home ownership has drastically altered the character of Canmore, Banff's neighbour community located just east of the park boundary. For Banff residents, this policy has been difficult to accept, but in the long run it is perhaps one of the most crucial steps needed to maintain some semblance of control over future growth.

Residents of Banff form one constituency of people who influence park policy. Other groups influencing park policy include park visitors, independent experts, park staff, politicians, and the general public (Eagles, 2002). The values and opinions of each group shape the type of influence each group asserts. Research has examined the views of many of these groups, and new research in the Bow Valley suggests that community residents are not a homogeneous group, but consist of several distinct types that need to be understood (Box 10.4).

Heritage Tourism Strategy

The Banff Heritage Tourism Corporation came about as a result of the Banff–Bow Valley Task Force report and the Banff National Park Management Plan. The purpose of the corporation is to serve the objective of a 'place for people' (Parks Canada, 2004).

BOX 10.4 Amenity Migration in the Bow Valley

Amenity migration—the movement of people for reasons of pleasure rather than for economic reasons (Chipeniuk, 2004)—is becoming a critical issue for Banff National Park. Amenity migration appears in many forms, from the low-budget outdoor adventurer who will stay as long as the cost of living remains low, to entrepreneurs seeking new destination opportunities, to new residents who seek an idyllic lifestyle, through to second-home owners (part-time residents) who seek an accessible retreat—among many other 'groups'. As each group arrives it impacts and attempts to manipulate the destination to suit its own needs. The natural environment is a big part of the attraction to the town, and each group lays claim to it, while sharing the environment with short-stay tourists. However, as infrastructure and capital investment in the destination increase, some resident groups leave and others arrive. Those who leave generally do so because of the increased cost of living or because the destination has lost its initial character. It can be said that the destination evolves in part because of the recreation needs of the local residents.

Recent research into amenity migration in the Bow Valley has focused on residents of Banff and Canmore (Pavelka, 2007). Results indicate that nature-based tourism communities such as the Bow Valley are more complex than was thought at first.

People who choose to live in the Bow Valley do so because of what the area offers. Most of the time that will involve the natural attributes related to the mountains and mountain recreation, but it also includes urban-type amenities such as restaurants, bars, cafés, and even paved pathways. Some of the highlights of the research include:

1. Most of the people who reside in the Bow Valley do so to experience the mountains in some way, and most do so by trying to balance meaningful work with a mountain recreation lifestyle.
2. Most are very satisfied with the recreation life that living in the Bow Valley offers and find very few barriers to accessing recreation opportunities.
3. Some serious challenges exist to living and recreating in a community such as the Bow Valley, including:
   - Everyday cost-of-living concerns that translate into poor housing options and issues related to working too much (two jobs, etc.) and, therefore, not being able to undertake desired activities.
   - Everyday cost-of-living concerns that force individuals to take jobs/work that they really do not like.
   - Everyday cost-of-living concerns that result in not enough money to actually do the activities desired—buy a ski pass, etc.
   - Career opportunity concerns—the realization that many people have post-secondary education that is incompatible with employment and career opportunities in a nature-based tourism community. This represents a nagging concern that they someday will be forced to leave the area for a 'real' career.
   - Education concerns, such as in order to further their careers they must leave the Bow Valley—perhaps not to return.

Not everyone experiences these challenges in the same way. Five different groups of amenity migrants are found in the Bow Valley:

1. Those seeking a mountain-recreation lifestyle and who rely on the Bow Valley for a livelihood and *make it work* through the sacrificing of career, home ownership, or work life. These people are seriously committed to living a life of mountain recreation and have grappled with its trade-offs.
2. Those seeking a mountain-recreation lifestyle and who rely on the Bow Valley for a livelihood and *cannot make it work*, as concerns of cost of living, home ownership, career, and education result in leaving the Bow Valley. Though the primary reason for leaving the area is the cost of living, it may be exacerbated by a perception that the area has lost its small town, wilderness, or remote character. Many of those will leave to discover other, smaller, nature-based tourism communities with a lower cost of living.
3. Those who seek a career and/or a mountain-recreation lifestyle, who have found careers compatible with the Bow Valley (parks, senior tourism, etc.) and who have carved out relatively normal lives (they have children who play hockey, etc.). The key is to find a job that provides a good standard of living, and those who do so are more likely to stay in the area.
4. Those who are seeking hospitality careers (not senior positions) and strong social connections, but who are not drawn to a mountain-recreation lifestyle. This group may include career bartenders and hotel staff. This group tends to stay (as they say, 'until the party is over', which may be many years).
5. Those second-home owners and/or Canmore residents who commute to work elsewhere and thereby do not rely on the Bow Valley for a livelihood (as distinct from those who can make it work and stay). Those who do not rely on the Bow Valley for their livelihood often have high-paying careers elsewhere.

Each of these five groups varies in its demographic makeup and brings its own recreation demands on either the landscape or the local resources: from informal trails for mountain biking to upscale dining and furniture retail outlets. The recreation demands of all groups impact the natural and built environment and alter it so that it will become more or less attractive to others in the future who either seek the same or other uses of the environment.

Underlying these findings is the question of who has a greater impact and alters more the nature-based tourism destination? Is it tourism, and the short-stay tourist, who in most cases remains within a tourist route or area? Or is it the amenity migrant, who comes to live in the region and expresses a different range of needs?

The Banff Heritage Tourism Corporation came about in 1999. It is a volunteer-driven organization responsible for the Heritage Tourism Strategy, which it implements in a partnership format with other stakeholders. The significance of the Heritage Tourism Strategy is that it provides direction for tourism for a community within a national park. The strategy must be compatible with a host of documents, plans, bylaws, and guidelines, including the Banff–Bow Valley report, the Banff National Park Management Plan, and the *Banff Community Plan*. Further, the strategy requires acceptance from the various tourism-related stakeholders. Tourism, by its commercial nature, is driven by the free market, while parks and protected areas are driven by notions of public good. The Heritage Tourism Strategy is intended to bridge the two, within the policy and legislative frameworks that have preceded it. In addition, the strategy frames the type of tourism that may be appropriate for an urban setting located within a national park. The vision for the Heritage Tourism Strategy is that every visitor experience and be inspired by the authentic heritage of Banff National Park (Banff Heritage Tourism Corporation, 2004). It is framed by four objectives:

1. to make all visitors to Banff and residents of the town of Banff aware that they are in a national park and World Heritage Site, by fostering appreciation and understanding of the nature, history, and culture of the park, the town, and surrounding areas;
2. to encourage, develop, and promote opportunities, products, and services consistent with heritage and environmental values;
3. to encourage environmental stewardship initiatives on which sustainable heritage tourism depends;
4. to strengthen employee orientation, training, and accreditation programming as it relates to sharing heritage understanding with visitors.

In addition to these initiatives, the Heritage Tourism Corporation provides 'Banff's Best' training to thousands of seasonal employees—1,700 in 2005 and 2006 alone (Heritage Canada, 2006: 1–18), and is active in the delivery of a heritage tourism awards program.

FIGURE 10.6 Wildlife overpass. *Photo: Guy Swinnerton.*

Updates to the Banff National Park Management Plan: Human Use Strategy and the Lands Adjacent to Banff Report

The Banff National Park Management Plan asserted the primacy of ecological integrity in all aspects of short- and long-term management. Human use takes various forms in the park: transportation along the Trans-Canada Highway; tourist activity and development; and resident activity. This human activity is widely acknowledged to be the park's most important issue, generating the most controversy and debate. The emphasis on ecological integrity represents a departure from the earlier emphasis on tourism development, and, not surprisingly, many objected to a strategy that was seen to not support human uses, such as tourism business (Gartner, 2002) and recreation uses (LeRoy and Cooper, 2000). Park management did not wish to eliminate human uses from its boundaries, but saw clearly that the role of human uses within the park needed to be addressed. Since 2000, Parks Canada has endeavoured to provide clarity to human use within the overarching framework guided by the goal of ecological integrity. For the purposes of our discussion, the focus will be on the 'Lands Adjacent to Banff' report and the development of 'ski area guidelines' by Parks Canada, as well as an integrated management approach.

The 'lands adjacent to Banff' are those areas of high use generally located adjacent to the town of Banff. They represent some of the areas most intensely used by tourists, and include the Upper Hot Springs, the Cave and Basin, and Mount Norquay, as well as the myriad formal and informal trails used by permanent and seasonal residents. The 'Lands Adjacent to Banff' report was to provide specific and detailed direction for the management of these areas, in accordance with Banff National Park Management Plan. In 2002, Parks Canada began consultations with a wide range of stakeholder groups on the project. This project required considerable clarification surrounding the role of human uses in the park because of the detailed recommendations regarding amenities, trail closures, and the like that it was expected to bring forth. At the onset of the project, Parks Canada (2005) stated the goals of the human use strategy as:

- to improve frontcountry facilities to reflect the World Heritage Site status of the park;
- to reduce ecological impacts associated with human use;
- to maintain the range of visitor opportunities within the backcountry;
- to preserve truly wild areas within the park.

In 2004, the Banff National Park Management Plan was amended to include the following human use strategy (Parks Canada, 2004: 42):

> Human use management is the direction and guidance of people, their numbers, their behaviour, permissible activities and the necessary infrastructure. The objective of human use management is to allow people to enjoy a national park without damaging its ecological integrity.

Throughout a four-year period (2002–6) the 'Lands Adjacent to Banff' process addressed a variety of concerns, including: encouraging human use in appropriate areas; human and animal safety; decreasing the proliferation of informal trails; the need for improved information and interpretation; user conflicts; and quality of experience for a diverse range of users. The report provided recommendations to enhance the visitor experience at each of 15 locations, and it was included as an amendment to the human use strategy of the Park Management Plan in July of 2007.

Banff is required to develop ski area guidelines in order to rationalize the development associated with the three major ski areas (Lake Louise, Sunshine Village, and Mount Norquay). On 8 December 2006, National Park Ski Area Guidelines were announced by Rona Ambrose, the Minister of the Environment. In regard to Banff, it was recognized that the 2000 ski area guidelines for Banff were out of date. New guidelines for the park were hoped to be completed by 2002, and in the interim all development plans were put on hold. Draft guidelines have been drawn up, but consultations remain ongoing, and the 2002 deadline has not been met (Parks Canada, 2006b). The fact that the ski area guidelines have not been amended into the Banff Management Plan (2004) is a testament to the challenge of incorporating a vibrant ski industry within the boundaries of a national park.

FUTURE CONSIDERATIONS

Most visitors to Banff National Park like what they see. However, for some more informed observers, the upscale form of tourism found in Banff townsite is a major concern, as are the ecological impacts the town has created. Banff can be seen as the quintessential symbol of our struggle to find our place in the natural world, to address the question: 'What does nature mean to us?' Banff is a living laboratory of contemporary trade-offs between efforts to preserve a special part of the natural world and desire to address other human wants and needs, including daily subsistence, commerce, convenience, and vanity. Many other tourism communities with similar levels of commerce and traffic have given into global forces and radically changed. Banff does not have that option; it has a legal responsibility to manage change within the structure of the National Parks Act. Banff also represents our need to control change in a world where there is often little sense of control. The future of Banff National Park rests as much with controlling key processes as it does with understanding larger currents of change that are more difficult to control.

Following the entrenchment of the primacy of ecological integrity in national parks during the 1990s, the new millennium has been marked by efforts to clarify human use within a park that receives close to 5 million visitors each year. This is not an easy task, but some progress has been achieved—such as the amendment of the Banff Management Plan to include the human use strategy, including the 'Lands Adjacent to Banff'. The positioning of human use within a strong ecological integrity framework is likely to be the most important issue facing the park into the future.

Some are more pessimistic about the future of Banff, arguing that significant improvement in ecological integrity is unlikely given the persistence of roads, railways, and residential communities in the park. Part of this sentiment is captured in the concept of the 'right to roam', discussed in Box 10.5. Others make the argument that to significantly change human use in Banff would damage public support for national parks, since Banff, in its present form, remains a prominent image that sustains public interest in the national park concept—'there remains a reluctance to accept the reality that aesthetics or general appearance of a landscape is a very limited indicator of its biological health and ecological integrity' (Swinnerton, 2002: 259).

Other issues face Banff National Park. Climate change will impact all aspects of the park if projections are accurate (Scott and Jones, 2005; Luckman and Kavanagh, 2000), as will pine beetle infestation (McFarlane et al., 2006). Other issues include water (Schindler, 2000) and drought conditions that have significantly affected the famed Hot Springs since 1998 (Mason, 2004). The highway and railway have contributed to habitat fragmentation and human-caused animal mortality (Nielsen et al., 2004; Alexander et al., 2005). These issues highlight the central theme of this chapter, the past and present struggle in Banff between the protection of the environment and human uses. Over a relatively short time period, the right to roam through, within, and at its outskirts has contributed to the evolution of Banff from 'an island of civilization within a sea of wilderness' to 'an island of wilderness within a sea of civilization'. The question for the future remains: How will our cultural values and subsequent legal frameworks address the critical questions of trade-offs to come?

BOX 10.5 The Right to Roam

Williams (2001) makes a strong case for park managers to rethink the way we try to understand human uses in our parks. He claims that despite all the science-based tools at our disposal that are intended to change behaviours and even to rationalize quotas, they have resulted in marginal success at best. Williams goes on to explain that people will do what they want to do; they will find a way, even in the face of science-based arguments, because of what he calls a 'right to roam'. The right to roam is culturally rooted and politically enabled—culture and politics have always been worthy adversaries for science and they are today. The right to roam can be seen as our sense of entitlement in accessing the land in all its forms. In Banff, the right to roam manifests itself in a variety of ways: travelling through the park on the Trans-Canada Highway; accessing various types of tourism, the mass tourism within the park (Weaver, 2001); roaming the backcountry for days at a time; the right to live, work, earn a living, and start a business in a national park. This rationale can be extended outside of the park gates eastward to the town of Canmore, where the right to roam is manifested in the right to own recreational property that is adjacent to a national park. How has Banff National Park addressed the right to roam, and what may be the bigger questions for the future?

REFERENCES

Alexander, S.M., N. Water, and P.C. Paquet. 2005. 'Traffic volume and highway permeability for a mammalian community in the Canadian Rocky Mountains', *Canadian Geographer* 49, 4: 321–31.

Banff, Town of. 1998. *Banff Community Plan.*

———. 2007. 'Doing Business in Banff'. At: <www.banff.ca/business/planning-development/growth-management.htm>.

Banff–Bow Valley Task Force. 1996. *Banff–Bow Valley: At the Crossroads.* Technical Report of the Banff–Bow Valley Task Force. Edited by R. Page, S. Bayley, J.D. Cook, J.E. Green, and J.R.B. Ritchie. Prepared for the Honourable Sheila Copps, Minister of Canadian Heritage. Ottawa: Minister of Supply and Services Canada.

Banff Heritage Tourism Corporation. 2004. 'Banff–Bow Valley Heritage Tourism Strategy'. At: <www.banffheritagetourism.com/index.htm>.

Banff National Park. 2006. Data Management Human Use. 2005 Backcountry Summary, 1.

Barlow, L. 2007. 'Handing out condoms plus a bit of education', *Banff Crag & Canyon*, 13 Feb.

Bella, L. 1987. *Parks for Profit.* Montreal: Harvest House.

Biosphere Institute of the Bow Valley. 2002. 'Selected eco-facts for business planning in the Bow Valley'. At: <www.biosphereinstitute.org/docs/Eco-Facts-Business-Business-Planning.pdf>.

Canmore, Town of. 2007. 'Doing business in Canmore: Population and demographics'. At: <www.canmore.ca/business/population-and-demographics/population-demographics.html>.

Chipeniuk, R. 2004. 'Planning for amenity migration in Canada: Current capacities of interior British Columbia mountain communities', *Mountain Research and Development* 24, 4: 327–35.

Draper, D. 2002. 'Toward sustainable mountain communities: Balancing tourism development and environmental protection in Banff and Banff National Park, Canada', *Journal of Human Environment* 29, 7: 408–15.

Dunkel, T. 2003. 'Deep Banff', *National Geographic Traveler* 20, 5: 70.

Eagles, P.F.J. 2002. 'Environmental management', in P. Dearden and R. Rollins, eds, *Parks and Protected Areas in Canada: Planning and Management*, 2nd edn. Toronto: Oxford University Press, 265–94.

Gartner, B. 2002. 'Small business in mountain parks', *Canadian Federation of Independent Business Research Bulletin* (Dec.).

Golder Associates. 2004. *Screening Report for the Trans-Canada Highway Twinning Project Phase 3B Banff National Park.* Submitted to Parks Canada Agency, Nov.

Godde, P.M., M.F. Price, and F.M. Zimmermann. 2000. 'Tourism and development in mountain regions', in Godde, Price, and Zimmermann, eds, *Tourism and Development in Mountain Regions: Moving Forward into the New Millennium.* Wallingford, UK: CABI Publishing, 1–27.

Hart, E.J. 2003. *The Battle for Banff: Exploring the Heritage of the Banff Bow Valley.* Banff: EJH Literary Enterprises.

Heritage Canada. 1996. 'Copps releases Banff–Bow Valley Report', press release, 7 Oct.

———. 2006. *Banff National Park of Canada Park Management: A Year in Review.* At: <www.pc.gc.ca/pn-np/banff/plan/plan7-06a-3_E.asp>.

Kuczaj, Sonia. 2002. 'Need-to-reside discriminatory says association', *Banff Crag & Canyon*, 20 Nov.

LeRoy, Sylvia, and Barry Cooper. 2000. 'Off limits: How radical environmentalists are stealing Canada's national parks', *Public Policy Sources* (Fraser Institute): 3–58.

Lovelock, Brent. 2002. 'Why it's good to be bad: The role of conflict in contributing towards sustainable tourism in protected areas', *Journal of Sustainable Tourism* 10, 1: 5–30.

Lukman, Brian, and Trudy Kavanagh. 2000. 'Impact of climate fluctuations on mountain environments in the Canadian Rockies', *Ambio* 29, 7: 371–81.

McFarland, B.L., R.C.G. Stumpf-Allen, and D.O. Watson. 2006. 'Public perceptions of natural disturbance in Canada's national parks: The case study of the mountain pine beetle', *Biological Conservation* 130, 3: 340–8.

Mason, Chris. 2004. 'Hot springs run dry', *Canadian Geographic* 124, 5: 28.

Nielsen, Scott, Stephen Herrero, Mark Boyce, Richard Mace, Byron Benn, Michael Gibeau, and Scott Jevons. 2004. 'Modeling the spatial distribution of human-caused grizzly bear mortalities in the Central Rockies ecosystem of Canada', *Biological Conservation* 120: 101–13.

Parks Canada. 2003. *Mountain Parks Heritage Interpretation Association Training Manual, Park Management 2003.* Ottawa: Parks Canada.

———. 2004. Banff National Park Management Plan, 'Introduction to amended management plan'. At: <www.pc.gc.ca/pn-np/ab/banff/docs/plan1a_E.sp>.

———. 2005. *Lands Adjacent to the Town of Banff—Final Report*, 4 May.

———. 2006a. *Park Management A Year in Review 2005/2006 Summary Report*, Oct. At: <www.pc.gc.ca/pn-np/ab/banff/plan/plan7-06a-3_E.asp>.

———. 2006b. 'Park Management Planning, Ski Area Guidelines'. At: <www.pc.gc.ca/pn-np/ab/banff/plan/plan5_E.asp>.

Sanford, R.W. 1999. *The Best of Banff Heritage Orientation Program.* Banff, Alta: Banff–Bow Valley Heritage Tourism Council and Mountain Parks Heritage Tourism Association.

Schindler, David. 2000. 'Aquatic problems caused by human activities in Banff National Park, Alberta, Canada', *Ambio* 29, 7: 401–7.

Scott, D., and B. Jones. 2005. *Climate Change & Banff National Park: Implications for Tourism and Recreation.* Report prepared for the Town of Banff. Waterloo, Ont.: University of Waterloo.

Swinnerton, G. 2002. 'Banff and the Bow Valley', in P. Dearden and R. Rollins, eds, *Parks and Protected Areas in Canada: Planning and Management*, 2nd edn. Toronto: Oxford University Press, 240–64.

Weaver, David B. 2001. 'Ecotourism as mass tourism: Contradiction of reality?', *Cornell Hotel and Restaurant Administration Quarterly* 42, 2: 104–12.

Williams, D.R. 2001. 'Sustainability and public access to nature: Contesting the right to roam', *Journal of Sustainable Tourism* 9, 5: 361–7.

KEY WORDS/CONCEPTS

amenity migration
automobile tourism
Banff–Bow Valley Task Force
Banff Community Plan
Banff Management Plan
ecological integrity
Four Mountain Parks Management Plan
Heritage Tourism Strategy
Lands Adjacent to Banff Report
need-to-reside policy
powered ski lifts
ski area guidelines
townsite
Trans-Canada Highway

STUDY QUESTIONS

1. In what ways is Banff National Park different from other national parks in Canada?
2. What types of human use challenges does Banff National Park face presently?
3. What have been the role and impact of the Trans-Canada Highway on the evolution of the park?
4. Given the history of Banff National Park, what directions could have been taken that were not?
5. In what ways is the town of Banff different from other nature-based tourism communities in Canada?
6. What is the significance of the Heritage Tourism Strategy?
7. How can Banff National Park integrate the principles of ecological integrity in light of the high levels of human use, including winter activity?
8. What is meant by the right to roam? What are the cultural values in Canada that make up our right to roam?
9. Given its realities, has Banff National Park achieved a balance of environmental protection and human use?
10. Explain the history of Banff National Park in light of trade-offs between environmental protection and human use.
11. What might the next decade of challenges look like for Banff National Park?

CHAPTER 11

Northern Parks and Protected Areas

R. Harvey Lemelin and Margaret E. Johnston

INTRODUCTION

The Canadian North is a vast territory that covers nearly 80 per cent of the land and water of Canada (Bone, 2003). Yet, the inhabitants of the North make up only about 3 per cent of the country's population (www.ainc-inac.gc.ca/pr/info/info102_e.html). The focus of this chapter is on the area north of 60 degrees, a primarily political and perceptual delineation that takes the sixtieth parallel as the southern boundary of the region and reflects the division between the provinces and the territories from Manitoba west. Almost all of the Canadian Arctic and some of the Subarctic are within this delineation. The North is comprised largely of ocean, islands, tundra, and the boreal forest; however, mountainous zones, including the highest peak Mt Caubvick/D'Iberville (1652 m) east of the Rockies in Torngat Mountains National Park Reserve, are also found throughout northern Canada. Key features of the North include the existence of permafrost (permanently frozen ground), low average annual temperatures, comparatively low biodiversity, and ecosystems characterized by large mammals.

The original human population developed a way of life that followed the movement and abundance of these large mammals; this traditional life is still followed to greater or lesser degrees across the North by the Aboriginal groups that inhabit these lands: the Inuit, the Dene, and the Métis. Aboriginal people comprise a much larger proportion of the population in Canada's North than in southern Canada. In 2001, Aboriginal people, that is, people who self-identified as Indian (i.e., treaty or registered Indian), Métis, or Inuit, comprised nearly one-quarter (22.8 per cent) of the Yukon population, 50.1 per cent of the population of the Northwest Territories, and 85 per cent of the population in Nunavut. These percentages are significantly higher than the 4.4 per cent of Aboriginal people in the total Canadian population, yet in the Territorial North Aboriginal peoples made up nearly 52 per cent of the total population (Bone, 2008: 481).

From a very early period, beginning in the sixteenth century, non-Aboriginal people began to venture to the North for economic reasons: exploration for a trading route to the Far East; the development and expansion of the fur trade; mineral exploration and exploitation. Today, a number of mixed communities in the North function as centres of government and administration, education, and service. These are not large by southern standards: in the 2006 census, Whitehorse, Yukon, had a population of 22,898; Yellowknife, NWT, had 18,700; and Iqaluit, Nunavut, had 6,184 (Statistics Canada, 2007). In addition to these three main centres and territorial capitals, there are many small communities; these are largely Aboriginal and often isolated (e.g., Old Crow, Gjoa Haven, Pangnirtung). Many of these communities are accessed only by air or winter road. In fact, Whitehorse is the only territorial capital that has year-round road access; Yellowknife's access is periodically interrupted during freeze-up and breakup, when the ferry across the Mackenzie River cannot run, while Iqaluit can only be accessed by air year-round and by water during the warmer summer months. These geographical and socio-cultural features mean that northern Canada faces unique resource management needs and governance strategies.

In terms of resource management and self-government, seven of the 14 First Nations in the Yukon Territory have already settled self-government agreements with the federal government. In the NWT, claims settled to date include the Inuvialuit Final Agreement in 1984, the Gwich'in Agreement in 1992, and the Sahtu Dene and Métis Agreement in 1994. The Treaty 11 Dogrib Claim is currently in negotiation. In Nunavut, the land claim agreement was settled in 1993 and the territory of Nunavut was established in 1999. The most recent agreement, between the Inuit of northern Labrador and the federal and provincial governments, which created a homeland called Nunatsiavut, also provided the opportunity to create one of Canada's newest national parks, Torngat Mountains. In this agreement the Inuit share the management of the park with Parks Canada and Inuit from Nunavik in northern Quebec (Kobalenko, 2007).

Though Yukon has a devolution agreement with the federal government, the Northwest Territories and Nunavut still do not have the control over natural resources and access to resource revenue that the provinces have. This inequity is increasingly obvious, as both territories are experiencing mining booms based on an abundance of minerals and very strong international demand. Further issues in the Canadian North include Arctic sovereignty, globalization, the biomagnification of pollutants, and the physical, social, and economic repercussions of climate change (see Ford, 2005).

This chapter provides an overview of protected areas and parks 'north of 60' and located in Yukon, the Northwest Territories, Nunavut, Nunavik (Quebec), and Labrador (the northern, mainland portion of the province of Newfoundland and Labrador). The chapter describes the institutional and policy approaches used by the various organizations with responsibility or involvement in northern parks and protected areas, and outlines the protection significance of northern landscapes, wildlife, culture, and ecosystems. The chapter then discusses three influences of importance in the present and for the future management of protected areas in northern Canada.

PROTECTED AREAS IN NORTHERN CANADA

Protected areas in northern Canada are diverse in size, focus, resources, and context, ranging from small territorial parks to the largest protected area in the country, and the challenges for management of these protected areas are diverse. Yet, opportunities also exist in protected area management related to Aboriginal self-government, collaborative management strategies, the use of traditional ecological knowledge (TEK) and *Inuit*

Qaujimajatuqangit (IQ). The involvement of Aboriginal peoples is central to the establishment and/or the management of many of these protected areas. Further, protected areas in the North are, to a certain extent, tourist destinations that offer economic opportunities for communities and individuals.

Protected areas in the North include national parks (a federal term nationally and a provincial term in Quebec), national park reserves, territorial parks, national wildlife areas and migratory bird sanctuaries, World Heritage Sites, wetlands of international importance, and marine protected areas. Though national historic sites, heritage rivers, and other conservation initiatives are relevant and are mentioned, it is not possible to cover them in any detail here.

Protected area management in the North occurs through numerous and often overlapping systems of protected areas legislation and policies. Territorial governments manage protected areas through departments: in Nunavut, the Department of the Environment; in Northwest Territories, the Department of Environment and Natural Resources; and, in Yukon, the Department of the Environment, Parks Branch Yukon. In Quebec, the Ministère du Développement durable de l'Environnement et des Parcs du Québec has this role. At the federal level, Parks Canada, the Canadian Wildlife Service, Environment Canada, and Indian and Northern Affairs have responsibilities for protected areas. International organizations, such as UNESCO's World Heritage Committee, and environmental not-for-profit organizations, such as Ducks Unlimited, also are active in the North. Aboriginal involvement stems from land claims and co-management arrangements, as well as proximity and traditional use.

As of 2007, there are 12 national parks, three national park reserves, 16 migratory bird sanctuaries, three national wildlife areas, seven wetlands of international importance, three World Heritage Sites, and 49 territorial parks in northern Canada (see Table 11.1). These sites protect approximately 80,000 km^2 of terrestrial wildlife habitat across the North and 14,000 km^2 of marine wildlife habitat, all of which is in Nunavut (New Parks North, 2005: 5).

Parks Canada and Northern Parks

The first northern national park was Wood Buffalo, located in the Northwest Territories and northern Alberta. Created in 1922 as a sanctuary for the last remaining wild buffalo herd on the continent, it was considered by some 'to be the most farsighted and important conservation measure ever taken by the Canadian government' (Griffith, 1987: 26). It would be several decades before any other national park would be established in the North. These next parks, Nahanni, Kluane, and Auyuittuq, were first established in 1972 as national park reserves. The development of national park reserves is perhaps, for the North, the most significant component of the National Parks Act, a change effected through amendments to the statute in 1972. The three national park reserves were to become full national parks upon settlement of comprehensive land claims. As Dearden and Langdon explain in Chapter 14 of this text:

> The 'reserve' designation allowed Parks Canada to treat and manage the areas in question as national parkland, but did not extinguish any Aboriginal rights or title to the areas. Importantly, this designation does not prejudice the ability of Aboriginal peoples to select parkland in the course of land claim negotiations.

TABLE 11.1 Northern Parks and Protected Areas

Jurisdiction	Year of Designation	Total Area km^2	Additional Designation*	IUCN Category
Newfoundland & Labrador				
Torngat Mountains National Park Reserve	2005	9,700 km^2		II
NWT				
Aulavik National Park	1992	12,200 km^2		II
Hannah Bay Migratory Bird Sanctuary	1939	295 km^2		IV
Nahanni National Park Reserve	1972	4,765 km^2	World Heritage Site (1978)	
Tuktut Nogait National Park	1996	18,181km^2		II
Wood Buffalo National Park	1922	44,807 km^2	World Heritage Site (1983) Ramsar Site (1982)	II
33 territorial parks				
Nunavut				
Auyuittuq National Park	1993	19,384 km^2		II
Igaliqtuuq National Wildlife Area		5,928 km^2		IV
Nirjutiqawik National Wildlife Area		1,780 km^2		IV
Queen Maud Gulf Migratory Bird Sancutary	1961	62,782 km^2	Ramsar Site (1982)	IV
Quttinirpaaq National Park	1986	37,775 km^2		II
Simirlik National Park	1999	22,200 km^2		II
Thelon Wildlife Sanctuary (shared with NWT)	1927	56,000 km^2		Ib
Ukkusiksalik National Park	2003	20,500 km^2		II
13 territorial parks				
Yukon				
Herschel Island Territorial Park		113 km^2		II
Ivvavik National Park	1984	16,000 km^2		II
Kluane National Park Reserve	1972	22,000 km^2	World Heritage Site (1979)	II
Old Crow Flats Special Management Area	1982	7,742 km^2	Ramsar Site (1982)	IV
Tombstone Territorial Park		2,113 km^2		II
Vuntut National Park	1993	4,345 km^2		II
Cold River Springs Territorial Park		16 km^2		
Fishing Branch Territorial Park		7,000 km^2		

*World Heritage Sites are places of international historical and cultural significance, as determined by the Convention on World Heritage Sites established through UNESCO. Ramsar Sites are wetlands of international significance designated as such under the 1971 Ramsar Convention on Wetlands, which was negotiated in Ramsar, Iran, and to which most nations are signatory.

In 1979, Parks Canada's policy further integrated this more contextual approach by embracing the concept of joint management by government and Aboriginal peoples; the establishment in 1984 of Ivvavik National Park in northern Yukon is an example of this new process. The establishment in 1995 of Vuntut National Park, which borders on Ivvavik to the north, further demonstrated how different goals can be achieved through a national park designation. The local community in Old Crow, the Vuntut Gwitchin, viewed the park designation as a means of protecting traditional lands for harvesting and other cultural values, while Parks Canada saw the park designation as adding to the representation of a natural region: both perspectives were accommodated (Njootli, 1994). In 1994, Parks Canada policy was revised to include a more comprehensive approach to working with Aboriginal peoples, recognizing the value of local knowledge systems for the management of protected areas (Doberstein and Devin, 2004). This policy modification recognizes section 35 of the 1982 Constitution Act, which affirms Aboriginal and treaty rights, and the creation in 1999 of the Parks Canada Aboriginal Affairs Secretariat institutionalizes the importance of the Aboriginal context in protected areas management.

Northern Canada, as stated earlier, has some of Canada's oldest protected areas, and it is also the location of some of Canada's newest national parks, including Simirlik (Nunavut, 2001), Ukkusiksalik (Nunavut, 2003), and Torngat Mountains National Park Reserve (Labrador, 2005). Of the 12 national parks and national park reserves in northern Canada, three are in Yukon, four are in the NWT, four are in Nunavut, and one is in Newfoundland and Labrador. Parks are also being proposed for several locations:

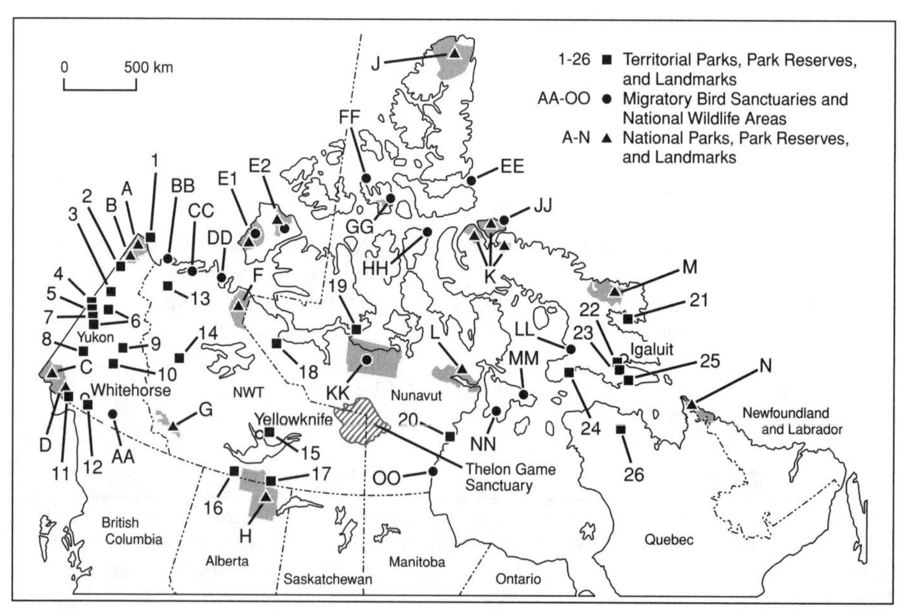

FIGURE 11.1 Parks and protected areas in Canada's North.

the East Arm of Great Slave Lake, Lutsel K'e, NWT (see Box 11.1); the Wolf Lake and Jennings Lake area of the Yukon and BC; Northern Bathurst Island; and Utkuhiksalik in Nunavut. Further protection through expansion was successful when 1,841 km^2 of Sahtu lands were added to Tuktut Nogait National Park in 2005. In August 2007, Prime Minister Stephen Harper announced the addition of 5,400 km^2 to the Nahanni National Park Reserve and World Heritage Site. This latest announcement, combined with the interim land withdrawal of 23,000 km^2 by the Dehcho First Nations in 2003, will provide further protection of the greater Nahanni ecosystem (CBC News, 2007).

National Wildlife Areas, Migratory Bird Sanctuaries, and Wetlands of International Significance: The Canadian Wildlife Service

The Canadian Wildlife Service (CWS), an agency of Environment Canada, is the steward of Canada's second largest land area after Parks Canada, yet the public appears to know little of this agency or its role in protected area management in northern Canada. The CWS currently manages three national wildlife areas (NWAs) and 16 migratory bird sanctuaries (MBSs). It also oversees the management of seven wetlands of international significance in northern Canada. Early MBSs, such as Akimiski Island and Hannah Bay

BOX 11.1 Lutsel K'e Dene First Nation and Parks Canada

After 25 years of discussion, the Lutsel K'e Dene First Nation and Parks Canada announced in the fall of 2007 the withdrawal of 26,350 km^2 of land in the NWT, which is a significant step towards establishing a new national park in the Northwest Territories along the East Arm of Great Slave Lake (www.parcscanada.pch.gc.ca/agen/bulletin/num3-vol1/page4_E.asp). Much of the proposed new national park lies in the Mackenzie watershed, and will protect Thaydene Nene, the Lutsel K'e Dene name for their land, a region home to migrating caribou herds, the barren ground grizzly bear, wood bison, and internationally significant populations of resident and migratory bird species from mining developments (www.borealcanada.ca/news_e.cfm?p_id=323). A portion of the area was proposed for national park designation in the 1970s and land was withdrawn; however, local concerns about these developments and a lack of consultation resulted in a veto of the proposed park by the Chief and Council of the Lutsel K'e Dene, delaying the implementation of the park (Griffith, 1987).

Unlike earlier proposals (see ibid.), the 2007 agreement between Parks Canada and the Lutsel K'e Dene First Nation commits the parties to work co-operatively towards establishing Thaydene Nene as a national park of Canada. Over the next several years, the parties will undertake a feasibility study and negotiate a model for the Dene to participate fully in all aspects of park operations and management. Capacity-creation, respect of culture and traditions and of the environment, equity generation through employment, tourism and other opportunities, and empowerment through education will be key components of the model.

(Nunavut), were created in the late 1930s and early 1940s, predating all national parks created in northern Canada except for Wood Buffalo. Other CWS sites, such as Nirjutiqavvik (Coburg Island) in Nunavut and the Nisutlin Delta in Yukon, are relatively more recent NWAs.

Seven of these wetland sites have also been designated under the Ramsar Convention on Wetlands, signed in Ramsar, Iran, in 1971. The Ramsar List of Wetlands of International Importance is a multilateral environmental agreement providing a framework for national action and international co-operation for the conservation and wise use of wetlands and their resources. There are presently 147 signatory countries ('contracting parties') to the Convention, with 1,524 Ramsar-designated wetland sites, totalling 129.2 million hectares. Canada has 36 such sites and seven of these, including the Old Crow Flats Special Management Area (Yukon) and the Queen Maud Gulf MBS (Nunavut) (see Box 11.2), are located in northern Canada.

An NWA or an MBS is designated legally under the Wildlife Area Regulation and Migratory Bird Sanctuary Regulations, and then is listed in the *Canada Gazette*. Additional protection is also provided through the Canadian Wildlife Act, the Species at Risk Act, the Migratory Birds Convention Act, and the Canadian Environmental Protection Act. For a site to be considered for designation as a national wildlife area, it must contain 'nationally significant' habitat for migratory birds, support wildlife or ecosystems at risk, or represent a rare or unusual wildlife habitat. A migratory bird sanctuary must regularly support at least 1 per cent of a population of a migratory bird species or subspecies. Research and management, including wildlife studies, maintaining and improving wildlife habitat, periodic inspections, and enforcement of hunting prohibitions and regulations, are also requirements in these protected areas (www.mb.ec.gc.ca/nature/whp/sanctuaries/dc01s00.en.html).

Highlighting the role that Aboriginal peoples play in managing protected areas in northern Canada, the Dene in the NWT are examining the creation of Ts'ude niline Tu eyeta (The Ramparts) and Edéhzhíe as NWAs. The Canadian Wildlife Service is currently working with the Dehcho and Dogrib First Nations under the NWT Protected Area Strategy to have the Edéhzhie area in the NWT—a 2.5-million-hectare (25,000 km^2) plateau west of Great Slave Lake, which is recognized by First Nations as a cultural and spiritual gathering place—legally designated as 'Aboriginal Cultural Landscapes' (New Parks North, 2005).

World Heritage Sites

A UNESCO World Heritage Site is a site of exceptional natural and/or cultural heritage (e.g., unique ecosystems, monuments) that has been nominated and confirmed for inclusion on a list maintained by the international World Heritage Programme administered by the UNESCO World Heritage Committee. The program is embodied in an international treaty called the Convention concerning the Protection of the World Cultural and Natural Heritage, adopted by UNESCO in 1972. Each World Heritage Site is the property of the country on whose territory the site is located, and is often managed by a national agency such as Parks Canada.

Canada's first World Heritage Site, Nahanni National Park Reserve, was designated in 1978, and the 'St Elias complex' (Kluane/Wrangell–St Elias/Glacier Bay/

BOX 11.2 Queen Maud Gulf Migratory Bird Sanctuary

At 63,655 km^2 the Queen Maud Gulf is the largest migratory bird sanctuary in Canada, the second largest Ramsar Site in the world, and Canada's largest protected area. The Queen Maud Gulf MBS lies between the northern coast of the mainland and the southwestern corner of Victoria Island in Nunavut. At its western end lies Cambridge Bay, to the east is the Simpson Strait, and to the north, Victoria Strait. The area has been used by Inuit and, before them, the Paleo-Eskimos for centuries. The area is of special historical significance to Euro-Canadians since it is near the Queen Maud Gulf where the ill-fated expedition to find the Northwest Passage, led by Sir John Franklin, perished on King William Island in 1847. Fifty years later, the area was navigated by Norwegian explorer Roald Amundsen and named in honour of Maud of Wales, the Queen of Norway (www.mb.ec.gc.ca/nature/whp/ramsar/df02s03.en.html).

The Queen Maud Lowlands shelter the largest variety of geese of any nesting area in North America, including Ross's geese, lesser snow geese, Canada geese, white-fronted geese, Atlantic brant, and black brant, as well as other waterfowl. The area is also an important staging area for caribou and muskoxen. Protection of the waterfowl was the main reason for its establishment as a migratory bird sanctuary in 1961 under the Migratory Bird Sanctuary Act. The area was later declared a Ramsar Wetland of International Importance in 1982 and, subsequently, an Important Bird Area. The IBA program is an international conservation initiative co-ordinated by BirdLife International. Research in the Queen Maud Gulf is often conducted from the Karrak Lake research facility (ibid.).

Following the creation of Nunavut, the area has been subject to co-management agreements under the Nunavut land claim agreement and the Migratory Bird Sanctuary Act. A habitat survey of the sanctuary was recently completed, and the completion of a management plan is expected (ibid.). Though the area is relatively remote, potential threats from proposed mining projects and shipping in the Coronation Gulf are of concern. The Queen Maud Gulf region has been listed as a potential national marine conservation area for Canada's Arctic marine environment, under the National Marine Conservation Areas System Plan. The Queen Maud Gulf MBS and Thelon Wildlife Sanctuary together are being considered for World Heritage designation as a single protected area.

Tatshenshini–Alsek transboundary area, 1979), and Wood Buffalo National Park (1983) soon followed. Although many other important sites throughout the North have been identified as potential UNESCO World Heritage Sites, issues of accessibility and cost associated with these destinations made many of these virtually forgotten by UNESCO (www.thelon.com/sanctuary.htm).

However, growing concerns regarding proposed development projects (i.e., mining, hydroelectric generation) and climate change in the early twenty-first century revived interest in northern World Heritage Sites. Since then, several areas were submitted to UNESCO for the tentative list for World Heritage Sites (e.g., the Klondike, in BC/Yukon, and Quttinirpaaq National Park in Nunavut). Additional sites being considered include Aulavik National Park (NWT), Auyuittuq National Park (Nunavut), the fossil forest on Axel Heiberg Island (Nunavut), the North Magnetic Pole (Nunavut), the Northwest Passage, Pingo Canadian Landmark (NWT), Sirmilik National Park (Nunavut), the Thelon Wildlife Sanctuary (NWT/Nunavut), and Tuktut Nogait National Park (NWT). Inspired by the St Elias complex World Heritage Site, two similar proposals have arisen: Ivvavik/Vuntut national parks, Qikiqtaruk Territorial Park, and the Arctic National Wildlife Refuge (see Box 11.3), and the Queen Maud Gulf MBS/Thelon Wildlife Sanctuary. While other sites in northern Canada contain large wildlife populations, the diversity and abundance of wildlife found in the Thelon Wildlife Sanctuary and the Queen Maud Gulf MBS are of particular importance to UNESCO because no other World Heritage Site exists in the Canadian tundra biogeographical region. In addition, these lands protect a substantial portion of the region, which is equivalent in size to the St Elias complex (ibid.).

Territorial Parks

All three territories (NWT, Nunavut, Yukon) and the province of Quebec have departments or agencies assigned to the management and protection of territorial and provincial parks (there are no Newfoundland and Labrador provincial parks located above the sixtieth parallel in Labrador). The Yukon Department of the Environment, Parks Branch, manages four territorial parks, the smallest being Coal River Springs at 16 km^2

BOX 11.3 The Ivvavik and Vuntut National Parks, Qikiqtaruk Territorial Park, and the Arctic National Wildlife Refuge World Heritage Site and Transboundary Area

Inspired by the St Elias complex in Alaska, British Columbia, and Yukon, the land of the Inuvialuit and Vuntut Gwitchin, partly protected by two national parks (Ivvavik and Vuntut), and Qikiqtaruk (Herschel Island) Territorial Park in the Yukon, along with the Arctic National Wildlife Refuge (nominated in the early 1980s by the United States government), have been listed on UNESCO's tentative list for World Heritage Sites. The areas under Canadian control make up over 14,500 km^2 of wilderness on the Yukon coastal plain, in the Richardson Mountains, and on a portion of the Old Crow Flats wetlands, and would combine with the Arctic National Wildlife Refuge in Alaska (76,000 km^2) to create a massive and significant protected area. Few sites on the World Heritage list have such a spectrum of habitats, including mountains, coastal plains, wetlands, and boreal forests, and such faunal diversity, including the Porcupine caribou herd (www.unep-wcmc.org/protected_areas/world_heritage/BorealFo.pdf). Of prime importance to the Vuntut Gwitchin of Old Crow is protecting the calving grounds of the Porcupine caribou herd in the coastal plain of the Arctic National Wildlife Refuge. The caribou herd crosses the traditional territory of the Vuntut Gwitchin twice a year during the migration from the calving grounds to the wintering grounds, providing a twice-yearly hunt on which the community depends: this is 'the centre of Vuntut Gwitchin culture and life' (Vuntut Gwitchin First Nation, 1998: 10).

and the largest, Fishing Branch Ni'iinlii'njik, encompasses 7,000 km^2. The two other parks are Qikiqtaruk (Herschel Island) and Tombstone (see Box 11.4). The Department of Industry, Tourism and Investment, NWT, manages 33 territorial parks. These parks range from day-use areas (Northwest Territories/Yukon Border) to large territorial park reserves (Gwich'in). The management of these parks is guided by the NWT Protected Area Strategy (1999), which describes an overall framework and set of collaborative criteria to guide the work of identifying and establishing protected areas in the NWT (New Parks North, 2005). Nunavut's 13 territorial parks, managed by the Department of the Environment, are quite diverse and include campgrounds (e.g., Tamaarvik), community parks (e.g., Kugluk/Bloody Falls), conservation parks (e.g., Thelon Wildlife Sanctuary), destination parks (e.g., Katannilik), and historic parks (e.g., the Northwest Passage). In Quebec, where national (i.e., provincial) parks are guided by the 'parks north of 52 strategy', only one park, Pingualuit National Park created in 2004, currently exists north of 60. The main features of this park (1,133.9 km^2) are a meteoric crater and the Riviere aux Feuilles caribou herd. Two sites, Monts-de-Puvirnittuq and Assinios, are also being considered for designation as parks in Quebec.

Marine Protected Areas

Though issues of climate change and sovereignty continue to dominate northern politics, and despite the existence of nine of Parks Canada's marine natural regions in the North (Beaufort Sea, Northern Arctic, Viscount Melville Sound, Queen Maud Gulf, Lancaster Sound, Eastern Baffin Island Shelf, Fox Basin, Davis and Hudson Straits, and Hudson Bay), no marine protected areas (MPAs) have been designated or established in northern Canada. This should come as no surprise, considering that Canada's attempts at creating marine protected areas have gone slowly (see Chapter 15). It should be noted, however, that a number of national parks, including Auyuittuq on Baffin Island (Eastern Baffin Island Shelf zone) and the Nirjutiqawik NWA, both in Nunavut, can be described as marine park initiatives, though these initiatives 'can be seen largely as add-ons to terrestrial park programs with minimal marine preserva-

BOX 11.4 Tombstone Territorial Park, Yukon

Tombstone (2,164 km^2) officially became a territorial park in the fall of 2004 and is designated a Natural Environment Park, a park established to protect the unique cultural and natural heritage of this subarctic environment, in accordance with the Tr'ondëk Hwëch'in First Nation Final Agreement (1998) (New Parks North, 2005).

One of the most interesting opportunities for the Aboriginal people (the Tr'ondëk Hwëch'in) involved with the management of Tombstone Territorial Park will be in the area of cultural interpretation. Since the area is particularly rich in Aboriginal artifacts and archaeological sites, under the management plan all of the artifacts and objects belong to the Tr'ondëk Hwëch'in people. Furthermore, guidelines and protocols are being developed to provide interpretation of these sites and the design of a new visitor interpretive centre (National Aboriginal Forestry Association and Wildlands League, 2003).

tion mandates' (Dearden, 2002: 359). There are only two proposed marine protected areas for this region, the Beaufort Sea Beluga Management Plan, and the Bowhead Whale MWA of Igaliqtuuq on the east coast of Baffin Island. Despite this, the community of Clyde River is proposing the creation of a marine wildlife area at Isabella Bay for the protection of the bowhead whale and its habitat. Negotiations with the Canadian Wildlife Service are ongoing. Nonetheless, the designation of such a protected area is highly unlikely considering the controversy surrounding the harvest of bowhead whales (see Vlessides, 1998) and the difficulty the CWS has encountered in establishing new national wildlife areas and migratory bird sanctuaries, let alone a marine wildlife area (Dearden, 2002).

Other Approaches

To some Aboriginal communities, the creation of national parks, while desired, entails time-consuming and expensive processes. Often, interim measures are required to ensure that the land and its resources remain protected. The establishment of the Sahoyúé-ʔehdacho as a national historical site in 1997 is an example of such a measure. This area, actually two peninsulas on Great Bear Lake, is of importance to the cultural heritage of the Shtugot'in (the Dene of Great Bear Lake). Though national historic site designation provides official recognition and commemoration, it does not provide protection from industrial development. Current strategies being contemplated include protection under NWT's Protected Area Strategy. However, the preferred management option for Sahoyúé-ʔehdacho is a protected area made up of both Crown and Sahtu lands, jointly managed by Parks Canada and the Dene of that area (New Parks North, 2005). In March 2007 an agreement was signed by Parks Canada, the Deline First Nation, and the Deline Land Corporation to work towards permanent protection and co-operative management of the site, with the federal government funding initial development and ongoing operational costs. The site is significant because of its size (5,587 km^2), and, also, it is the first national historic site in Canada with lands acquired by Parks Canada on the basis of consultation with Aboriginal peoples.

Wildlife sanctuaries—areas set aside for the protection of specific wildlife species, often game species—also have been established in northern Canada. Two of the most significant are Ddhaw Ghro (Yukon) and the Thelon Wildlife Sanctuary (Nunavut/NWT) (see Box 11.5). In Yukon, the Ddhaw Ghro Habitat Protection Area (1,595 km^2), formerly known as McArthur Wildlife Sanctuary, supports populations of rare Fannin and Dall sheep. Ddhaw Ghro was identified as a special management area under the final agreements of the Nacho Nyak Dun and Selkirk First Nations, and will be designated a habitat protection area (ibid.).

CLIMATE CHANGE AND NORTHERN PARKS AND PROTECTED AREAS

The Arctic is often depicted as the globe's barometer measuring the scale and speed of climate change. Therefore, environmental changes threatening the Arctic environment are actively monitored through various national and international networks (Fenge, 2005). Research by the Arctic Climate Impact Assessment network and other interna-

BOX 11.5 The Thelon Wildlife Sanctuary

At 56,000 km^2, the Thelon Wildlife Sanctuary, straddling the border between Nunavut and the NWT, is one of Canada's largest protected areas, second only to the Queen Maud Gulf MBS. First established as the Thelon Game Sanctuary in 1927, only a few years after the establishment of Wood Buffalo National Park, some of the lands were withdrawn from the sanctuary in 1930 and then again in 1956 to accommodate mining and other interests in the southwest portion. Studies carried out by the International Biological Program, under the auspices of the International Council of Scientific Unions (ICSU), the International Union of Biological Sciences (IUBS), and the National Academy of Sciences–National Research Council in the 1960s and 1970s highlighted the outstanding geological features of the area, including the world's largest drumlin field and a tremendous abundance of migratory waterfowl, shorebirds, and mammals (e.g., muskoxen, caribou). The Thelon Wildlife Sanctuary was identified as a 'Biological Site of Universal Importance'.

The draft management plan, co-ordinated by the government of Nunavut in keeping with its obligations under the Nunavut land claim agreement, and supported by Nunavut Thunngavik Inc., the Nunavut Wildlife Management Board, the Kivalliq Inuit Association, and the community of Baker Lake, proposed special management areas extending beyond the current sanctuary boundaries to more fully protect the calving grounds of the Beverly caribou herd and to allow Aboriginal subsistence hunting. The Department of Indian Affairs and Northern Development approved the Thelon Wildlife Sanctuary Management Plan in 2005 (www.nunavutparks.com/on_the_land/thelon_wildlife_about.cfm).

tional bodies has shown that the Arctic as a whole has undergone the greatest warming on earth in recent decades, with annual temperatures now averaging 2° to 3°C higher than in the 1950s (Ford, 2005). The polar amplification of global warming is well known and relates to changes in albedo (reflection of sunlight) that form a positive feedback component of the warming process. Snow and ice are highly reflective, so as more snow and ice melts, more of the land or water surface can absorb sunlight, causing more melting of snow and ice. Sea ice, glaciers, permafrost, snow cover, and peatlands are all sensitive indicators of change, susceptible to subtle climatic variations and ocean temperatures. These factors play a key role in many global processes, such as global atmosphere and ocean circulation, and involve potentially important sources and sinks of trace gases. It is in this marine environment that climate change 'is most visibly altering ecological relationships and the cultures and economies of the region' (Fenge, 2005: 25).

Other outcomes of climate change include increased ice instability, increased thawing of the permafrost, shifts in vegetation zones, changes in animal species' diversity, ranges, and distribution, increasing storm exposure for communities and shoreline erosion, and increased levels of ultraviolet radiation reaching people, plants, and animals (AICA, 2004). As environmental change in the Arctic appears to be increasingly evident, with major implications for wildlife, resources, and even sovereignty, greater international attention will be focused on this region of the world. This might mean increased interest in protected areas north of 60, and assuredly it will mean an increase in resource management challenges for those involved in managing protected areas and parks as the environment changes in concert with climate. Some of these changes may be dramatic and fast, and others will be slower. Regardless of the pace and scale of change, managers will need to be prepared for how these changes affect landscapes, wildlife, and people (for further information on climate change and protected areas, see Lemieux and Scott, 2005).

NORTHERN TOURISM AND PROTECTED AREAS

A wide variety of tourism activity takes place across this broad and diverse region. Though some cultural attractions are visited, much tourism involves wildlife and landscape. Wildlife attractions include numerous bird species, ungulates, and charismatic mega-fauna (e.g., whales, polar bears); landscape attractions include glaciers, icebergs, fjords, mountains, coasts, islands, sea ice, tundra, and forest. Throughout the Arctic and Subarctic, tourism opportunities can be divided into two basic categories: consumptive activities such as trophy hunting (e.g., polar bears in Nunavut) and fishing outings; and non-consumptive activities such as expedition-style and destination cruising, photographic safaris, wildlife viewing, northern lights tourism, skiing and snow-machine riding, canoeing, kayaking and rafting, dogsledding, and cultural and historical tourism. In much of northern Canada roads are sparse and air travel is limited and quite expensive. Yukon is the exception, since it can be accessed by several major roads, including the well-travelled Alaska Highway, in itself an attraction for tourists.

Non-resident travellers to the Northwest Territories numbered 61,282 in 2004–5, with a total annual estimated spending of over $100 million (www.iti.gov.nt.ca/parks/tourism/research_and_statistics.htm). In 2002, Yukon received about 32,000 tourists, and Nunavut had 12,000 visitors (Pagnan, 2003). Tourism is often sought as a new form of income in the mixed-economy communities looking for incoming cash flows and in regions where other forms of industry have declined. A number of studies suggest there is local support for tourism growth, provided it meets particular requirements related to community interests. Other research (e.g., Smith, 1997, in Baker Lake) shows that traditional activities can be conducted in conjunction with tourism activities.

Parks and protected areas are important destinations for many northern travellers and tourism often figures in the considerations surrounding park establishment. For example, Cruises North, a three-year-old Inuit-held company, runs 10 summer cruises, many of these visiting national and territorial parks. Other existing operators include Bathurst Inlet Lodge, Pond Inlet, Baffin Island, Toonoonik Sahoonik Tours, Silah Lodge, Tununiq Travel and Adventures, and Whitewolf Adventure Expeditions. Some, such as Nahanni River Adventures, rely heavily on protected areas as destinations for river-based trips, while other protected areas, such as Fishing Branch Territorial Park in the Yukon, provide wildlife viewing opportunities. Wright and McVetty (2000) describe the need to foster community economic and social well-being in the development of a national park on Banks Island in the Northwest Territories. Discussions with park managers indicate that visitation patterns to Canadian national parks in the North show some variability—some parks have experienced declines (e.g., Aulavik, Auyuittuq Reserve, Quttinirpaaq); others, such as Ivvavik, have seen modest growth; and at least one, Sirmilik National Park, has had a significant increase in visitors (Parks Canada, Communications Manager, Iqaluit Field Unit, communication with R.H. Lemelin, March 2007).

Although local peoples and governments expect that tourism will benefit communities and regions, promote conservation, and encourage cultural exchange, the potential negative impacts associated with tourism are numerous (e.g., disruption to subsistence activities, loss of autonomy), and some are more problematic in small, remote settlements than in larger, road-accessible locations (Johnston, 2006; Stewart et al., 2005). The ongoing challenges that face northern tourism development (e.g., distance to market, economic leakage, industry structure) and the additional threats, such as environmental change, could compromise the ecological integrity of protected area attractions and destinations. An example where climate change may actually be benefiting tourism (albeit for the short term) is the idea that time is running out for people to see certain mega-fauna and landforms. Here, operators seeking to capitalize on climate change in the North and the plight of polar bears, for example, market their outings as 'last-chance tourism' or 'polar bears on thin ice'. Indeed, the thawing of Canada's Arctic has contributed to a mini-boom in tourism in Nunavut and Nunavik (the homeland of the Inuit of northern Quebec) as curious travellers rush to see the North before it changes and animals disappear (Buhasz, 2007). This adds further chal-

FIGURE 11.2 Despite their remote location, parks such as Sirmilik, at the northern tip of Baffin Island, are experiencing increasing visitation levels. *Photo: © Parks Canada/L. Narraway/ 13.01.03.10 (205).*

lenges to existing management approaches, for as accessibility increases, wildlife and environmental vulnerability also increases.

Wildlife-Human Interactions

Because of low visitation numbers and remoteness, some northern parks, such as Vuntut National Park, have the 'best' situation for maintaining ecological integrity, yet new stressors such as climate change, invasive species, and biomagnification may change this advantage. Polar bears have become the flagship species of climate change, and as people clamour to the North to view these animals, the likelihood of polar bear–human encounters in protected areas and in northern communities increases. Incidents of human–wildlife encounters in the North, specifically polar bear–human encounters, have been reported by researchers (Clark, 2003; Dyck, 2006; Stenhouse et al., 1988) and journalists (Kobalenko, 2007; Struzik, 2004). Although no one has been killed by a polar bear in Canadian national parks, the number of close calls has been growing (Struzik, 2004). For example, in 2002, paddlers camped in Ivvavik National Park, Yukon, were visited by a polar bear (ibid.). Although no one was hurt in this encounter, the situation indicates how important safety and training are in these regions. Examples of polar bear–human encounters in the past five years include three people who were attacked and injured by polar bears in two separate incidents involving researchers and tourists (Clark, 2003). In another incident, a hiker in Auyuittuq National Park, Nunavut, was slightly injured by a polar bear in 2000 (Clark, 2003). This was the first recorded 'injury by a polar bear in a national park' (Lunn et al., 2002: 50). In 2001, a polar bear injured two canoeists in Katannilik Territorial Park Reserve in Nunavut.

In Ukkusiksalik National Park an estimated 130 polar bears are thought to be living and using the areas near Sila Lodge. The polar bears are attracting wildlife viewers to this national park, yet no one really knows how many bears are in the park or how park management will deal with the increasing potential for polar bear–human confrontations brought about through tourism. Similar observations were made by Kobalenko (2007) during an excursion into Torngat Mountains National Park Reserve. One of the challenges for northern parks is providing visitors with adequate protection from polar bears (i.e., firearms). At present, only Aboriginal people, some researchers (with valid certification), and park wardens can legally carry firearms (Struzik, 2004). To some managers and Aboriginal peoples this indicates that federal rules and regulations have little, if any, contextual relevance in the Canadian North. That said, the territorial government of Nunavut has implemented a number of proactive policies and regulations (e.g., visitor safety education, providing guides) aimed at wildlife management that should assist with this type of situation.

CONCLUSIONS

Confronted by a shrinking wilderness and declining resources, early managers and policy-makers responded by protecting selected areas in the North through various mechanisms, including national parks, migratory bird sanctuaries, national wildlife areas, wildlife sanctuaries, and territorial parks. In essence, these early management strategies were guided by conservation philosophies that promoted the protection of charismatic

FIGURE 11.3 Polar bear watching has become one of the most popular tourist activities in the North, with some operators advertising it as a 'last chance to see' the bears before their numbers decline as a result of global warming. *Photo: Keith Levit Photography/World of Stock.*

mega-fauna (e.g., wood buffalo) and game species (e.g., waterfowl). Unfortunately, by setting aside these areas and protecting them through enforcement, these agencies also created a 'fortress mentality' that often alienated the Aboriginal peoples who depended on these resources. Despite the imposition of these protected areas, Aboriginal peoples in northern Canada continued their traditional activities and, as demonstrated in the case of Lutsel K'e, sometimes resisted the creation of new national parks.

Concerns regarding proposed development projects and the pursuit of self-government (see Chapter 14) in the late twentieth century transformed both Aboriginal politics and the management of protected areas in Canada. Since then, a number of changes to Parks Canada and other management agencies have occurred, including the national park reserve system established in 1972; collaborative management approaches; the acknowledgment of traditional harvesting rights (i.e., s. 35 of the Constitution Act, 1982); the recognition of TEK and IQ in parks management in Canada; and a general shift away from preservation without people, to preservation with people. These changes have made national parks and protected areas, in some cases, more acceptable to Aboriginal peoples.

In essence, the earlier commitments to establish new national parks and protected areas in the North, formerly based on the traditional models of species protection and the 'fortress mentality', have been overtaken by rapid socio-political evolution and Aboriginal empowerment (Sadler, 1989). Thus, new types of northern protected areas such as Aboriginal cultural landscapes, and transboundary protected areas, more closely linked to the traditional landscape perspectives and an ethic of conservation based on sustainable development rather than wilderness preservation, have emerged (ibid.). As we have seen, these new management strategies and philosophies have already provided the opportunity for new approaches to preservation, interpretation, and tourism.

However, challenges—both social and ecological—exist since the size, remoteness, and relative ecological integrity of northern parks and protected areas no longer provide assurances against global changes and anthropogenic activities. For example, increasing demands for minerals and fossil fuels (e.g., oil and gas exploration and development in the tar sands of northern Alberta, at the southern edge of Wood Buffalo National Park) and ongoing consideration for hydroelectric generation (e.g., Nahanni National Park) create new challenges to park managers (www.pc.gc.ca/docs/rspm-whsr/rapports-reports/r11_e.asp). Even the northward expansion of timber harvesting and agricultural development (e.g., ranching in northern Alberta), as well as backcountry recreation and tourism facilitated by increasing access to remote regions, creates new challenges as human–wildlife encounters increase in these regions (Wiersma et al., 2005). As we discussed in regard to Queen Maud Gulf, issues of sovereignty in the North, especially as related to the Northwest Passage, may also increase anthropogenic impacts in northern protected areas.

Last-chance tourism demonstrates that the most significant environmental pressure being placed on protected areas in the North may very well be climate change (Lemieux and Scott, 2005; Scott et al., 2004). Some parks, such as Wood Buffalo National Park, are already experiencing a reduction in annual precipitation, thereby affecting water quality and quantity, ecosystem health, the potential for invasive species, and an increasing likelihood of fire cycles (Wiersma et al., 2005). Many of these stressors cannot be managed by protected areas alone (see Chapter 13), and will require national and even international collaboration. However, the North and its people have always proven to be quite resilient and adaptable—some have called this process 'creative learning from great challenges'. This is perhaps why Nunavut representatives have even stated that some good, including the preservation of some species (i.e., muskoxen in the Thelon Wildlife Sanctuary), has come out of some of these protected area strategies. This perspective may have resulted in the following approach to protected areas:

> Nunavut's Territorial Parks, Heritage Rivers, Historic Sites, and Conservation Areas are places that help us remember the way our ancestors lived thousands of years ago; they are places that provide habitat for caribou, polar bear, and other wildlife; and they are places we like to use for dog-sledding, boating or camping. They are places where we can be happy on the land, and places that we feel are truly special. . . . Our parks are also one of the main reasons people come to visit Nunavut. And, while visitors are here, they rely on Nunavummiut for guiding and outfitting; they buy local arts and crafts and participate in home-stay programs—which helps to build our economy, and provide employment opportunities. (www.nunavutparks.com/on_the_land/index.cfm)

This approach promotes the concepts of respect, equity, and empowerment, while also providing an opportunity to guide future management strategies and challenges in the Canadian North through the benefit of hindsight and attention to foresight.

REFERENCES

Arctic Climate Impact Assessment (ACIA). 2004. 'Impacts of a warming climate: Arctic climate impact assessment. Summary'. At: <www.acia.uaf.edu/PDFs/Testimony.pdf>.

Bone, R. 2003. *The Geography of the Canadian North: Issues and Challenges*, 2nd edn. Toronto: Oxford University Press.

———. 2008. *The Regional Geography of Canada*, 4th edn. Toronto: Oxford University Press.

Buhasz, L. 2007. 'Last-chance tourism', *Globe and Mail*, 17 Feb.

Canadian Boreal Initiative. 2006. 'Canadian Boreal Initiative applauds agreement for a new northern National Park between Lutsel K'e Dene First Nation and Parks Canada'. At: <www.borealcanada.ca/news_e.cfm?p_id=323>.

Canadian Broadcasting Corporation. 2007. 'Harper announces expansion of NWT park'. At: <www.cbc.ca/canada/north/story/2007/08/08/nwt-nahanni.html>.

Clark, D. 2003. 'Polar bear–human interactions in Canadian national parks, 1986–2000', *Ursus* 14, 1: 65–71.

Dearden, P. 2002. 'Marine parks', in P. Dearden and R. Rollins, eds, *Parks and Protected Areas in Canada: Planning and Management*, 2nd edn. Toronto: Oxford University Press, 354–78.

Doberstein, B., and S. Devin. 2004. 'Traditional ecological knowledge in parks management: A Canadian perspective', *Environments* 32, 2. At: <proquest.umi.com/pqdlink?index=4&did=701262681&SrchMode=3&sid=1&Fmt>.

Dyck, M.G. 2006. 'Characteristics of polar bears killed in defense of life and property in Nunavut, Canada, 1970–2000', *Ursus* 17: 52–62.

Environment Canada. n.d. 'Migratory bird sanctuaries'. At: <www.mb.ec.gc.ca/nature/whp/sanctuaries/dc01s00.en.html>.

———. n.d. 'Queen Maud Gulf Migratory Bird Sanctuary, Nunavut—Ramsar Site'. At: <www.mb.ec.gc.ca/nature/whp/ramsar/df02s03.en.html>.

Fenge, T. 2005. 'Arctic alarm', *Alternatives Journal* 31: 25–6.

Ford, J. 2005. 'Living with climate change in the Arctic', *World Watch* (Sept.–Oct.): 18–21.

Griffith, R. 1987. 'Northern park development: The case of Snowdrift', *Alternatives* 14, 1: 26–7.

Indian and Northern Affairs Canada. 2004. 'The North, March 2000'. At: <www.ainc-inac.gc.ca/pr/info/info102_e.html>.

Johnston, M.E. 2006. 'Impacts of global environmental change on tourism in the polar regions', in S. Gössling and M. Hall, eds, *Tourism and Global Environmental Change*. New York: Routledge, 37–53.

Kobalenko, J. 2007. 'Between Nanuk and the cold grey sea', *Canadian Geographic* (May–June): 38–52.

Lemieux, C., and D. Scott. 2005. 'Climate change, biodiversity conservation and protected areas planning in Canada', *Canadian Geographer* 49, 4: 384–99.

Lunn, N.J., et al. 2002. 'Polar bear management in Canada 1997–2000', in N.J. Lunn, S. Schliebe, and E.W. Born, eds, *Polar Bears: Proceedings of the 13th Working Meeting of the IUCN/SSC Polar Bear Specialist Group, 23–28 June 2001, Nuuk, Greenland*. Occasional Paper of the IUCN Species Survival Commission. Gland, Switzerland: IUCN, 41–52.

National Aboriginal Forestry Association and Wildlands League. 2003. *Honouring the Promise—Aboriginal Values in Protected Areas in Canada*. Ottawa: NAFA/Wildlands League.

New Parks North. 2005. 'An annual progress report on natural and cultural heritage initiatives in northern Canada', newsletter 14 (Mar.). At: <www.newparksnorth.org/npn2005_e.pdf>.

Njootli, S. 1994. 'Two perspectives, one park: Vuntut National Park', in J. Peepre and R. Jickling, eds, *Northern Protected Areas and Wilderness. Proceedings of a Forum on Northern Protected Areas and Wilderness, November, 1993.* Whitehorse, Yukon: Canadian Parks and Wilderness Society and Yukon College, 132–6.

Northwest Territories, Industry, Tourism and Investment. n.d.. 'Tourism and Parks: Tourism research and statistics. At: <www.iti.gov.nt.ca/parks/tourism/research_and_statistics.htm>.

Nunavut Parks. n.d.. 'On the land: Thelon Wildlife Sanctuary, Thelon Heritage River'. At: <www.nunavutparks.com/on_the_land/thelon_wildlife_about.cfm> and <www.nunavutparks.com/on_the_land/index.cfm>.

Pagnan, J. 2003. 'Climate change impacts on Arctic tourism—A preliminary review', in *Climate Change and Tourism.* Proceedings of the 1st International Conference on Climate Change and Tourism, Djerba, Tunisia, April 2003. Madrid: World Tourism Organization.

Parks Canada. n.d. *Periodic Report on the Application of the World Heritage Convention,* Section II, 'Report on the state of conservation of Wood Buffalo National Park'. At: <www.pc.gc.ca/docs/rspm-whsr/rapports-reports/r11_e.asp>.

———. 2007. 'Information bulletin: Canada's 42nd national park now formally established'. At: <www.pc.gc.ca/agen/bulletin/torngat_e.asp>.

———. 2007. 'New heritage park to open in Metepenagiag'. At: <www.parcscanada.pch.gc.ca/agen/bulletin/num3-vol1/page4_E.asp>.

Peepre, J., and P. Dearden. 2002. 'The role of Aboriginal peoples in parks and protected areas in Canada: Planning and management', in P. Dearden and R. Rollins, eds, *Parks and Protected Areas in Canada: Planning and Management,* 2nd edn. Toronto: Oxford University Press, 323–53.

Sadler, B. 1989. 'National parks, wilderness preservation, and Native peoples in northern Canada', *Natural Resources Journal* 29: 185–204.

Scott, D., G. McBoyle, and M. Schwarzentrube. 2004. 'Climate change and the distribution of climatic resources for tourism in North America', *Climate Research* 27: 105–17.

Smith, V.L. 1997. 'The Inuit as hosts: Heritage and wilderness tourism in Nunavut', in M.F. Price and V.L. Smith, eds, *People and Tourism in Fragile Environments.* Toronto: John Wiley & Sons, 33–50.

Statistics Canada. 2007. *2006 Community Profiles.* At: <www12.statcan.ca/english/census06/data/profiles/community/Index.cfm?Lang=E>.

Stenhouse, G.B., L.J. Lee, and K.G. Poole. 1988. 'Some characteristics of polar bears killed during conflicts with humans in the Northwest Territories', *Arctic* 41: 275–8.

Stewart, E.J., D. Draper, and M.E. Johnston. 2005. 'A review of tourism research in the polar regions', *Arctic* 58, 4: 383–94.

Struzik, E. 2004. 'Difficult births: Creating new parks', *Edmonton Journal,* 7 Nov.

Thelon Game Sanctuary. n.d. 'The Thelon Game Sanctuary'. At: <www.thelon.com/sanctuary.htm>.

United Nations Education, Scientific, and Cultural Organization (UNESCO). n.d. 'World Heritage Sites: Brief history'. At: <www.whc.unesco.org/en/169/>.

Vlessides, M. 1998. 'Licence to whale', *Canadian Geographic* 118, 1: 24–34.

Wiersma, Y.F., T.J. Beechey, B.M. Oosenbrug, and J.C. Meikle. 2005. *Protected Areas in Northern Canada: Designing for Ecological Integrity.* Gatineau, Que.: Canadian Council on Ecological Areas.

World Conservation Union. 2004. *Proceedings of the World Heritage Boreal Zone Workshop Held in St. Petersburg, Russia, 10-13 October 2003.* At: <www.unep-wcmc.org/protected_areas/world_heritage/BorealFo.pdf>.

KEY WORDS/CONCEPTS

Aboriginal cultural landscape
Aboroginal rights
Arctic sovereignty
collaborative management
fortress mentality
Inuit Qaujimajatuqangit (IQ)
last-chance tourism
migratory bird sanctuary (MBS)
national wildlife area (NWA)
north of 60 degrees
Nunatsiavat
Nunavut
self-government
territorial park
Torngat Mountains National Park Reserve
traditional ecological knowledge (TEK)
transboundary protected area
World Heritage Site

STUDY QUESTIONS

1. What is unique about the natural environment of the North, and how does this influence the way national parks are managed?
2. What is unique about the social/cultural environment of the North, and how does this influence the way national parks are managed?
3. Why are other types of protected areas, besides national parks, found in the North? Is this a good development or not? Discuss.
4. What is unique about Wood Buffalo National Park?
5. What are some of the consequences of climate change on northern parks?
6. In what ways do northern parks contribute to marine conservation?
7. What is significant about Thelon Wildlife Sanctuary?
8. What are some of the challenges of tourism management in northern parks?
9. How did 'fortress mentality' as applied to protected area management contribute to the alienation of Aboriginal people in the North?
10. What is meant by a shift from conservation without people, to conservation with people?

CHAPTER 12

Tourism, Ecotourism, and Protected Areas

Rick Rollins, Paul Eagles, and Philip Dearden

INTRODUCTION

The many different roles that protected areas play have been discussed in Chapter 1. Public use of parks and protected areas is a long-standing and fundamental component of the Canadian experience. Governments create parks when public demand is effectively expressed. Governments provide financial and managerial resources each year according to public demand. Therefore, both the creation and the management of parks are based on the presence of a mobilized and supportive constituency in society. Typically, this constituency is created by citizens who visit parks, enjoy their recreation, and then lobby for government action. Both Canadians and international visitors value the opportunity to spend time in wild places and a significant nature-based tourism industry has developed in Canada, with much of this tourism activity focused on people visiting parks and protected areas. However, management must balance the benefits and disbenefits of nature-based tourism on natural environments, on host communities, and on the experience of some park visitors (see Chapter 10).

Views on the role of park tourism vary according to the perspective of the observer (Table 12 1). All motivations are inherently self-centred, with each person wanting to maximize his/her own benefits. These benefits can be viewed quite differently depending on the role of the person, and one person may hold divergent views at different times according to the multiple or changing roles he or she plays. For example, one person could be a local resident, a municipal politician, and also a park visitor. Another person could be a foreign citizen, a scientist, and a tourist. These roles influence their opinions on what should be the most important goals for park tourism.

The large number of views on the goals for park tourism leads to a challenging management environment. Many of the ideas are complementary, but some are contradictory. For example, the private tourism industry may view park tourism primarily as a means to creating profit, whereas ecotourists may attach more significance to tourism as a vehicle for promoting conservation of natural and cultural features. Park managers must constantly work on obtaining as much consensus as possible in regard to major planning and management decisions on park tourism.

TABLE 12.1 Stakeholders' Views of Tourism in Parks and Protected Areas

Stakeholder Group	Goals for Park Tourism
Politicians	Satisfy constituents Gain public approval Establish a legacy
Park managers	Gain income Build political alliances Develop heritage appreciation
Visitors and tourists	Gain widespread access to the park Enhance personal experiences, which include: cognitive objectives (e.g., learning about nature and wildlife) affective concepts (e.g., gaining peace of mind) psychomotor desires (e.g., getting exercise) Gain health benefits Enhance social relations with family and friends
Aboriginal people	Source of jobs and income Ability to teach others about culture
Potential visitors	Possible vacation destination
Scientists	Place for scientific research Increase status in society
Tourism industry	Create profit Develop products that match visitor demand Ability to use entrepreneurial abilities
Non-governmental organizations	Fulfillment of the goals of the members Ecological conservation Community development
General public	Contribute to improved quality of life Promote the conservation of natural and cultural heritage Sustain and commemorate cultural identity Provide education opportunities to members of society
Local communities	Securing additional income Source of employment Enhance respect for local traditions, cultural values, local environment Access to better services Enhancement of self-esteem

Source: Adapted from Eagles et al. (2002).

This chapter describes some of the issues associated with protected area tourism, the emerging role of ecotourism, and how park agencies in Canada appear to be responding to this form of tourism. This discussion differs from what is found in other chapters, where the roles of public-sector agencies like Parks Canada (see Chapters 6, 7, 8, and 9) are emphasized. In contrast, this chapter focuses mainly on the role of the private, profit-making sector in providing visitor experiences in parks and protected areas. The private, non-profit sector is often very important in park tourism in Canada, but is not discussed in this chapter.

Sometimes, to distinguish between tourist and local visitor use of protected areas is important. Both tourists and local visitors engage in recreational activities in parks. Originally, all national park visitors were tourists, because the national parks were so remote that overnight stays were required to access them (Chapter 2). This is now less likely the case as urban development near parks brings the human population much closer to them and, therefore, provides many more local visitors. Therefore, much of Chapters 6 and 7, in regard to understanding the social science and visitor management aspects of parks, is directly relevant to this chapter. It is instructive here to make a clear distinction between 'a tourist' and a 'local visitor'. The World Tourism Organization (1994) provides the following definitions:

> A *tourist* is a person travelling to and staying in a place outside his or her usual environment for not more than one consecutive year for leisure, business, and other purposes.
>
> A *local visitor* is a person visiting a park or protected area and living within an 80-km radius of the park.

In other words, a tourist who visits a park or protected area is someone who travels a greater distance to access it than does a local visitor. A local visitor typically lives close enough not to require overnight accommodation away from home, although may wish to camp or stay in other overnight facilities. This distinction is important because local residents may receive benefits or negative impacts that differ from what a non-local tourist might receive, as described in the following.

THE COSTS AND BENEFITS OF NATURE-BASED TOURISM

Nature-based tourism has the potential to deliver positive and negative outcomes (impacts) that often are described along three dimensions: economic impacts, social impacts, and environmental impacts.

Economic Impacts of Tourism

Tourism is one of the fastest-growing industries in the world, and much of this activity occurs in or near natural areas. Internationally, tourism employment for 2006 was estimated at 234,305,000, and this is estimated to grow to 279,374,000 by 2016 (World Travel and Tourism Council, 2006). In Canada, this has translated into 104 million days of visitor use in national, provincial, and territorial parks (Table 12.2), and one study estimates the annual direct impact of park tourism in Canada at between $11 billion and $17 billion (Eagles et al., 2000). These figures are certainly underestimates as many parks do not count visitation in low-use periods and many do not count visitation at all.

TABLE 12.2 Park Visitation in Canada

Provinces, Territories, and National Agencies	Visitor Days in 2006
Newfoundland and Labrador	188,144
Nova Scotia	53,068
Prince Edward Island	33,760
New Brunswick	659,306
Quebec	3,859,956
Ontario Provincial Parks	10,096,074
Ontario Conservation Areas	5,000,000
Ontario Park Commissions	16,000,000
Manitoba	5,222,130
Saskatchewan	2,537,685
Alberta	8,484,305
British Columbia	18,302,732
Northwest Territories and Nunavut	32,429
Yukon	80,000
National Parks	24,415,164
National Historic Parks	9,227,377
National Wildlife Areas	96,980
Total	**104,289,110**

This park visitation leads to visitor expenditures in and near parks. The money ripples through a local economy in the form of spending on accommodation, transportation, restaurants, gift shops, and other amenities. In addition, taxes collected from this tourism activity contribute to improvements in public services, such as schools, hospitals, roads, recreation facilities, and other forms of community infrastructure. Eagles et al. (2002) suggest the following economic benefits of park tourism:

- an increase in jobs for local residents;
- an increase in incomes;
- new tourism enterprises;
- stimulation and diversification of the local economy;
- local manufacture of goods;
- new markets and foreign exchange;
- improved living standards;
- local tax revenues;
- new skills learned by employees;
- increased funding for protected areas and local communities.

One example of the economic benefits of park tourism is illustrated in a study of birdwatching in Point Pelee National Park (Hvenegaard et al., 1989), which revealed that 48 per cent of visitors were from the United States and that these visitors spent $3.2 million annually in the local area. These expenditures included the purchase of food, accommodations, and souvenirs. Local businesses reported hiring additional staff and extending hours of operation during peak birdwatching season. Often, initial spending by tourists in parks is 'multiplied' in a community through additional rounds of spending on goods and services (Lindberg, 2001).

However, these economic benefits are not without problems (Johnson and Leahy, 2004), as outlined below:

- Some tourism-related employment is seasonal and low-paying.
- During peak tourism season, prices for some products may be increased, negatively impacting on community residents, unless they are the people selling these products. In this case, they benefit from higher prices.
- Some tourism operations are owned and operated by people not living in the region, so there is a leakage of tourism revenues away from local communities. This occurs when local expenditures flow out of the community through the employment of non-resident seasonal workers and the purchase of products from outside the local area.
- Sometimes tourism displaces other sources of income. It is common for nature conservation supporters to demand the reduction in the exploitation of local resources.
- Increases in property values as a result of amenity migration to tourism destinations can make housing less affordable for new residents moving into the area. This is occurring in Canmore, Alberta, where tourism and amenity migration have been spurred in part by the close proximity of Banff National Park (Chapter 10).

Tourism in rural areas may provide new economic opportunity when there is a decline in resource-based industries such as forestry, mining, commercial fishing, or agriculture (Box 12.1). However, some of this tourism employment can be unstable, and it is vulnerable to some of the same and different factors that influence other resource-based industries. For example, tourism in British Columbia has experienced periodic downturns associated with external factors such as fuel prices, terrorism (9/11), SARS, and strengthening of the Canadian dollar. All resource-based industries suffer from upswings and downturns that occur due to external factors beyond the control of the local managers.

Social Impacts of Tourism

Many of the social benefits associated with tourism development are related to economic growth and diversification and the provision of additional infrastructure (improved roads, schools, recreation facilities, etc). A number of other social benefits include community pride, cultural appreciation between tourists and local residents, and maintenance and appreciation of local culture (Johnson and Leahy, 2004; Lankford and Howard, 1994; Rollins, 1997; Reid et al., 2000).

On the negative side, high visitor numbers may stress local services, such as causing congestion at stores, banks, service stations, beaches, and other locations. A 'power shift' may also occur when local control over the host community erodes in response to demands from tourists and when the tourism industry responds to their needs. Some tourists may feel less constrained while on holiday and may be rude, condescending, rowdy, or obnoxious to local residents and other tourists. Sometimes visitors offend

BOX 12.1 Rural Tourism on Vancouver Island

A recent study on Vancouver Island (Vaugeois and Rollins, 2007) examined tourism employment to determine if tourism attracts people who have been displaced by downturns in other sectors of the economy. Until the 1990s much of the rural population of Vancouver Island (outside of Victoria) were employed in resource-based industries such as forestry, commercial fisheries, and mining. Recent declines in these industries created economic havoc in some rural communities, such as Lake Cowichan, Port Hardy, and Gold River. However, other communities, such as Tofino, Ucluelet, and Chemainus, made a successful transition into a tourism economy and were better able to withstand the economic decline that occurred elsewhere.

To better understand the nature of this tourism development, a study examined people employed in rural tourism operations. Results indicated that mobility from resource-based industries accounted for 14 per cent of recent tourism employment, but 20 per cent had moved from other careers and 19 per cent were recent college/university graduates. The most important reasons for seeking tourism employment were related to a more desirable lifestyle (to work in pleasant surroundings, to do interesting work, to work with people, to work in a job related to one's lifestyle, etc). However, job satisfaction varied somewhat, with many people not satisfied with: level of benefits provided (36 per cent satisfied); opportunities for advancement (41 per cent satisfied); training opportunities (40 per cent satisfied), and income received (51 per cent satisfied).

These findings indicate that the rural tourism industry on Vancouver Island, which depends largely on attractive scenery and parks, is still experiencing growing pains, with some new tourism employees not fully satisfied with their employment. When the tourism industry has to compete with other industries for good employees, as was the case with the strong Canadian economy in 2006–7, employee satisfaction is critical to retaining good employees who can sustain the tourism industry.

without meaning to offend because they are unaware of local customs or sensitivities. Parks often are located near remote, rural communities where lifestyles may not be as affluent as those of their visitors. This 'demonstration effect' may create stress within the rural community when its people aspire to a lifestyle difficult to obtain (Weaver, 2001). Other tourism-related concerns include:

- The presence of affluent tourists sometimes gives rise to increases in crime and the sex trade.
- The trivialization or commodification of local cultures can occur when tour operators organize displays of local culture that have been altered in ways thought to be more interesting to tour groups, rather than presenting authentic displays of local culture.
- Growth in tourism may result in displacement of local residents through spiralling property values induced by the tourism industry.
- Traditional community decision-making structures may be stressed if some people in a community receive economic benefits and others do not.

Keogh (1989) argues that social impacts of tourism vary from place to place and are influenced by three factors: visitor behaviour, destination characteristics, and characteristics of the resident population. *Visitor behaviour* can be positive or negative on host communities as a function of visitor attitudes, expectations, and preferences as well as the host communities' ability to accept and adapt. Other visitor characteristics, such as visitor numbers and length of stay, can influence the way tourism is perceived in a host community.

Important *destination characteristics* that influence the social impact of tourism include the character of the natural setting, the nature of political development, and the level of development of the host community. For example, the community of Bamfield, located on the northern terminus of the West Coast Trail segment of Pacific Rim National Park, is a tiny community of 300 permanent residents. The community has not been incorporated and so has little control over land-use planning and tourism development, which has stagnated in recent years, unlike the nearby communities of Tofino and Ucluelet, or other gateway communities like Canmore, Wells, and Radium Hotsprings in Alberta.

A third factor influencing social impacts of tourism is the *character of local communities and cultures.* Some communities have strong local cultures, as evidenced in several First Nations communities found throughout the more remote areas of Canada. Not all of these communities welcome tourists, who often are attracted by nearby parks, out of concern for how tourists might impact their cultural traditions. Another example occurs with amenity migration to desirable rural communities. These newer residents may be reluctant to see the growth of urbanization that accompanies some forms of tourism growth. This is illustrated in Box 12.2. However, other cultures welcome tourists, since they provide a valuable economic and cultural opportunity for advancement.

It is important to note that parks and the associated tourism industry can stimulate the creation of local communities near parks. There are many examples in Canada and elsewhere where jobs in parks and tourism encourage some people to move close to park entrances, creating gateway communities designed around and servicing park activities. In Canada, well-established communities have grown up over time in concert with park development. In Ontario, the town of Huntsville on the west side of Algonquin Provincial Park and the town of Whitney on the east side are classic examples of service communities that developed along with the park over its 115-year history.

Environmental Impacts of Tourism

One of the major reasons for the creation of most parks is environmental conservation. Parks are established when sufficient numbers of citizens demand that governments do so and spend money for their management. Since environmental protection is a major objective, much concern and management effort is placed on ensuring that visitor use supports conservation and has as little negative environmental impact as possible.

The negative environmental impacts of tourists visiting parks have received much attention, particularly with the emergence of ecological integrity priorities in national parks and other park agencies (Chapter 5). These impacts include damage to trails, vegetation, soils, and wildlife populations (see Box 12.3). Some of this damage is created with very little use, from activities like hiking, camping, canoeing, and horseback travel.

BOX 12.2 Social Impacts of Tourism on Saltspring Island, British Columbia

Saltspring Island is located in the Gulf Island region between Vancouver and Victoria. The island experienced a great deal of amenity migration, due in part to the many parks, as well as to desirable island ambience. Nearby, the newly created Gulf Islands National Park Reserve may increase this growth.

To understand how local residents feel about the recent growth in tourism, a community survey (Saltspring Island Chamber of Commerce, 2004) revealed many perceived benefits and concerns. The benefits of tourism receiving highest support were:

- Tourism provides employment (94 per cent agree).
- It provides recognition and marketing opportunities for local artists and craftspeople (91 per cent agree).
- It results in more shops and restaurants (84 per cent agree).
- It contributes directly and indirectly to the healthy economy of the community (83 per cent agree).

As well as these benefits, a number of concerns of tourism were identified:

- Tourism creates problems with car parking (93 per cent agree).
- It creates problems with crowding in town (75 per cent agree).
- It contributes to air pollution and over-consumption of water (74 per cent agree).
- It results in more litter and garbage (61 per cent agree).
- It reduces privacy in the community (35 per cent agree).
- It creates crowding on roads (31 per cent agree).

When asked to reflect on the possible benefits and concerns associated with tourism development, respondents stated they mostly supported tourism, with 83 per cent indicating they were 'somewhat' or 'strongly' supportive. These results suggest that Saltspring Island residents attach more importance to the benefits of tourism than to their concerns about tourism. However, if the future growth of tourism is not managed properly, this could lead to an erosion of community support for tourism.

Other impacts on some park environments are created by built structures (hostels, golf courses, swimming pools, ski developments, etc.). Negative environmental impacts within parks can be exacerbated if the tourism industry demands higher visitor numbers without ensuring the proper management actions are undertaken to service visitor needs and demands. A thorough review of the environmental impacts of natural area tourism is provided by Newsome et al. (2002).

In addition to tourism-induced environmental impacts within parks, the tourism industry creates a variety of environmental impacts within surrounding communities. These can include increases in garbage, demands on sewage and water systems, and

FIGURE 12.1 Popular sites can sustain considerable damage from visitors, especially in fragile habitats such as this grassy headland in Cape Breton National Park that provides excellent viewing of the whales that swim below. In response, Parks Canada has 'hardened' the site with a wooden boardwalk, but some people feel that this construction is too intrusive. What do you think? *Photo: P. Dearden.*

impacts on vegetation through construction of new roads and tourism facilities. Tourism-related impacts on wildlife can be significant, through altered habitat, hunting, wildlife viewing, or feeding of wildlife; these impacts are extensively reviewed by Newsome et al. (2005). Tourism impacts are significant also in marine communities, and include dredging, sewage, loss of beach vegetation from driving vehicles on beaches, and the inappropriate locations of marinas and beach resorts (Jackson and Leahy, 2004; Yasue and Dearden, 2006), as well as the impacts of marine recreation such as diving (Dearden et al., 2007). In addition, consumptive activities such as fishing and shellfish-harvesting contribute to loss of marine diversity and even so-called non-consumptive uses such as watching whales can have significant impacts (Duffus and Dearden, 1993; Parsons et al., 2006).

Many negative impacts can be reduced or eliminated with effective park management. A large body of literature outlines policies and plans that can be implemented for this purpose (Eagles et al., 2002).

Most park agencies have interpretive programs and public communication strategies aimed at influencing behaviour to reduce negative environmental impacts and to increase support for conservation measures through increases in the knowledge of visitors to the area (Chapter 8). However, the park agency messages are not always supported by the tourism industry, which plays a major role in shaping the behaviours of their clients who visit parks. The tourism industry has not always been consistent in

BOX 12.3 Recreation Ecology

Recreation ecology is a field of study that examines the impacts of recreational activities on the environment. It is the scientific underpinning to assessments of ecological carrying capacity (see Chapter 6) and a key source of information for assessing the impacts of ecotourism. Two recent Canadian examples of recreation ecology are discussed below.

Duchesne et al. (2000) assessed the impact of ecotourist visits during winter on woodland caribou (*Rangifer tarandus caribou*) time budgets in the Charlevoix Biosphere Reserve, Quebec. The authors compared the behaviour of caribou during and after ecotourist visits with their behaviour during days without visits. In the presence of ecotourists, caribou increased the time spent being vigilant and standing, mostly at the expense of time spent resting and foraging. After visits, caribou tended to rest more than during control days. Caribou reduced the time spent foraging during ecotourist visits as the number of observers increased. The impact of ecotourists appeared to decrease as winter progressed. Visits were short (average 39.3 minutes) and caribou never left their winter quarters because of human presence. However, caribou abandoned their wintering area twice in response to wolf presence. Although winter is a difficult period of the year for caribou, the authors suggest that with proper precautions caribou in Charlevoix can tolerate ecotourist visits. This study may reveal the principle that some animals are able to adapt to the presence of people when they learn that those people are not a danger.

A second study (Dyck and Baydack, 2004) examined the viewing of polar bears from tundra vehicles, an activity offered at Churchill, Manitoba, since the early 1980s. This form of wildlife viewing provides a unique and safe way for tourists to observe and learn about polar bears. Researchers studied vigilance behaviour (a scanning of the immediate vicinity and beyond) of resting polar bears to evaluate the impacts on the bears from tundra vehicle activity. They recorded the numbers of head-ups, vigilance bout length, and between-bout intervals for polar bears. In general, the frequency of head-ups increased, and the between-bout intervals decreased for male bears, when vehicles were present. Female bears' behaviour was opposite to males' behaviour. The vigilance bout lengths did not differ significantly between vehicle presence and absence. Vigilance behaviour of male bears was not magnified with increasing numbers of vehicles; therefore, the authors suggest that the threshold is one vehicle, but low-impact polar bear viewing is possible. Lemelin and Smale (2006) looked at the satisfactions of polar bear tourists and found that the only significant influence on visitor satisfaction

FIGURE 12.2 Mother polar bear with cubs. *Photo: iStockphoto.*

was the number of bears seen. This puts quite an onus on the tours to locate the animals, but suggests that the number of vehicles involved needs to be limited.

Both cases suggest that the presence of ecotourists affects the focal species involved. Scientists and park managers then have to determine whether that presence is a *significant* impact on the animal. One might assume that all such impacts are negative. However, this might not be so and there may also be differences in how animals react to tourists based on their age and sex, as described above with polar bears, and also according to time of year. For example, grizzly bears have become an important ecotourism resource on the coast of BC, especially where they congregate to feed on large numbers of returning salmon. Owen and Gilbert (2005) studied the interactions among the bears and tourists, and reached a surprising conclusion. Under natural conditions the rivers and marine food are dominated by large male bears. However, the large males are the most averse to tourist presence, and hence the presence of tourists often provided some protection for the females and their cubs to enjoy the bounty, which otherwise they might not get. In other words, ecotourism might actually benefit the species overall by reducing the competition among bears for food supplies.

Deciding whether a particular impact on wildlife is negative or positive has a large subjective element. Decision-makers must decide whether the negative impacts are counterbalanced by positive benefits. For example, the minor impacts on polar bears described above could be balanced by the massive increase in societal knowledge and concern about polar bears that came about through viewing the bears. Or, the minor impacts on caribou could be balanced by the income from tourism fees that pays the salaries of park wardens, who then ensure that the caribou are not killed by poachers. Tourists may also be involved in collecting data on wildlife populations that are useful for conservation, as described by Theberge and Dearden (2006) in reference to whale sharks.

delivering tourism experiences in parks that are in full compliance with the objectives of park managers in protecting park resources. In some countries, the messages from private tourism enterprises are the only conservation messages given to tourists. This occurs when the park management does not have sufficient financial capability to deliver education programs.

Tourism Sustainability

Assessments of tourism sustainability are especially important in relation to protected areas and impacts on surrounding communities. One of the first critical assessments of tourism impacts was Doxey's (1975) analysis of how tourism is perceived by local residents. He argued that an emerging tourism industry introduced to an existing community that has no experience with tourism may evolve through different stages over time, as follows:

- *Euphoria.* In this stage, the anticipated benefits of tourism are promoted, and tourists are welcomed. Little control of tourism development occurs.

- *Apathy.* In this stage, the tourists are no longer a novelty for community residents; they are part of the social landscape. Some of the problems that tourists bring to a community are more apparent, but these are balanced somewhat by the benefits of tourism.
- *Annoyance.* As the tourism industry grows and becomes more conspicuous in terms of numbers of tourists, tourist behaviour becomes an increasing concern for local residents. Rather than controlling or managing growth, more infrastructure is created, further alienating local residents.
- *Antagonism.* Sometimes, in the absence of tourism planning, annoyance spills into overt hostility or aggression towards tourists and the tourism industry.

The Saltspring Island case study (Box 12.2) describes a community that seems to have moved beyond the euphoria level in the Doxey model. It is important to note that this progression need not always occur, if appropriate measures are taken by the community, the tourism industry, or the park (assuming a park is the major draw of tourism to the area).

The Doxey model has contributed to the development of other models, one of the best known being Butler's (1980) model of a tourism area life cycle (Figure 12.3). This model builds on Doxey's reasoning by suggesting that tourism destinations change over time, sometimes reaching a critical point where stagnation occurs. Stagnation may be a result of a variety of factors: a degraded natural environment; an unsupportive local community; or a tourism product that has lost its market. Over time, a lack of response

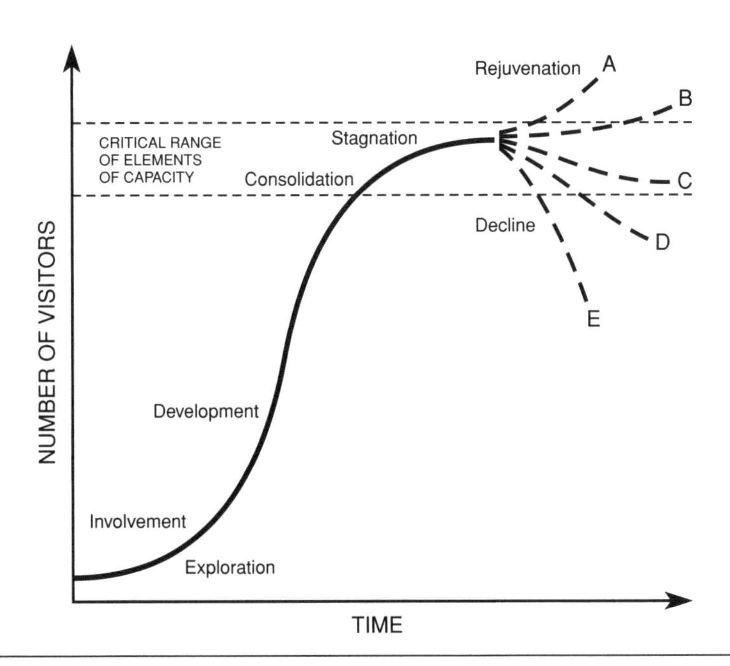

FIGURE 12.3 The Butler model of tourism life cycle.

to stagnation can lead to the decline of the local tourist industry, as tourists move on to other more preferred locations in the globally competitive industry.

When considering nature-based tourism, the tourist area life cycle model can be unforgiving—declines in resource quality are difficult to reverse, and a declining resource base is likely to impact the associated tourism industry. The whale-watching industry illustrates this point. If whale watching contributes to declines in whale populations, the related tourism industries also will experience declines. This thinking led Duffus and Dearden (1990) to introduce a new factor into the tourism destination model—the changing character of the type of visitor attracted to an area as the tourism destination changes over time (Figure 12.4). 'As the awareness of the site and associated activity grow, a less ambitious user will dominate the group. There will be a concomitant demand for more facility development, more mediation, and increased pressure on both the social system and the ecosystem of the host area.' This model incorporates the notion of spe-

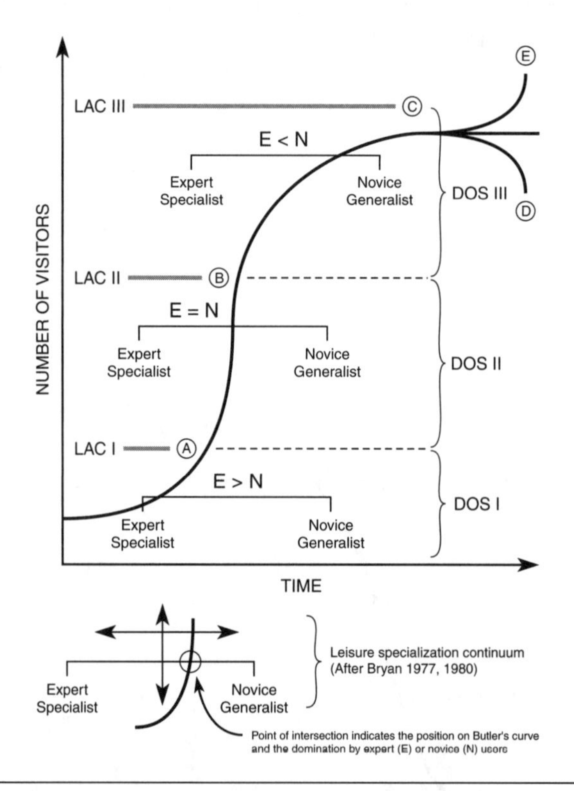

FIGURE 12.4 The Duffus–Dearden model. As the number of visitors rises over time there is a change in the proportions of experts (E) and novices (N) at the site and the site ultimately becomes dominated by novices. This may not be good for sustainability and Limits of Acceptable Change (LAC) should be set to reflect site objectives. This may also lead to different zones reflecting different management objectives, in this case Dive Opportunity Spectrum (DOS) zones discussed in more detail in Dearden et al. (2006).

cialization discussed in Chapter 6, and suggests that specialists are more likely to support the protection of natural values over increases in the built environment. However, the whale-watching industry, in its zest to grow and be profitable, may be introducing far more 'generalists' into the mix, driving the industry to provide spectacular close encounters with whales, thereby compromising marine conservation (see Box 12.4).

The Duffus–Dearden model suggests the use of an LAC approach (Chapter 7) to set limits on the growth of nature-based tourism that will achieve maximum benefits for

BOX 12.4 Changing Values on Acceptable Levels of Impact

The history of Pinery Provincial Park in Ontario reveals shifting values on the importance of various impacts of tourism over a 50-year time span. Pinery had an initial rapid development, a major reassessment of priorities, and then a substantial policy change designed to bring recreation impact more into line with an emerging ecological value set.

Pinery Provincial Park, created in 1957, contains a spectacular landscape of forested sand dunes on the Lake Huron shore. It is large, for a southern Ontario park, at 2,532 hectares. Its location in southwestern Ontario means it is accessible to a huge market from nearby cities in Ontario and Michigan.

The initial ethic for this park was one of rapid tourism development. No overall plan was used; the development was directed by the whims of the park managers of the time. Within 15 years of its creation, the park was fully developed and the use had peaked at 1.7 million visitor days of use per year. As the concept of ecology emerged in the 1960s, many people started to question the negative impacts of the development and of heavy recreational use on the sand dune forests. By 1970 the Parks Branch of the Department of Lands and Forests decided that the carrying capacity problems identified at Pinery necessitated a full planning review. This resulted in 1971 in the publication of the first-ever Master Plan for an Ontario provincial park (Department of Lands and Forests, 1971).

This 1971 plan called for major changes in the operation of the park. Prominent new policies included:

- reduction in tourism volume;
- removal of inappropriate facilities;
- prohibition of all-terrain vehicles;
- building of new tourism facilities more compatible with environmental resources;
- enhanced education programs;
- restoration of damaged ecosystems;
- field research on targeted species and ecosystems;
- strict enforcement of new rules governing visitor behaviour.

Figure 12.5 shows the changes in tourism volume over time, as expressed in visitor days of activity. From 1.7 million in 1971, the volume was reduced by 1975 to 400,000 visitor days of recreation per year, a figure maintained for the next 20 years. This plan was very successful in achieving its objective of reducing visitor use to a level that was more appropriate to the ecological carrying capacity of this sensitive landscape.

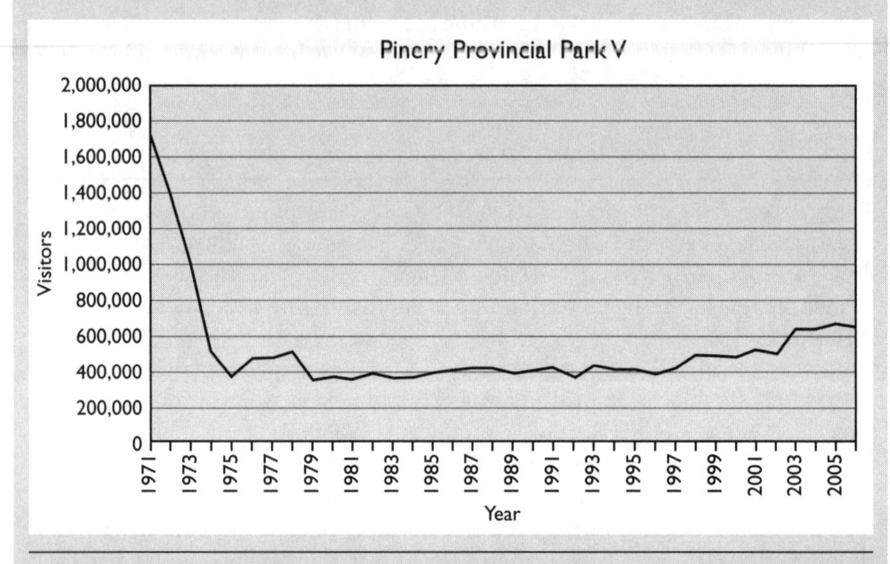

FIGURE 12.5 Visitor Use of Pinery Provincial Park

The initial Pinery development occurred between 1954 and 1969. When the major management re-evaluation took place in 1971, environmental impact assessment legislation did not exist in Ontario, the first recreation management program at an Ontario university had not yet produced graduates, the teaching of ecology had just started at universities, and land-use planning had recently become a university program. This Pinery case study shows how changes in society's attitudes towards recreation, ecology, and park management worked together.

When problems were recognized, new policies and plans were quickly introduced to deal with them. As values changed in society, the park management priorities changed accordingly. This illustrates how a park agency must develop new procedures, in this case a whole new management planning policy, in response to perceived value shifts. In the long run, it shows how changing societal values in regard to the acceptability of certain activities are reflected in park management. All park management ultimately is concerned with social values.

conservation and local communities while minimizing negative impacts, and this has been adopted by many researchers (e.g., Higham, 1998; Sorice et al., 2006; Roman et al., 2007) looking at wildlife tourism. The same concepts apply equally well to protected area tourism. The limits set to meet the goals of the protected area can be both physical (e.g., what level of biophysical impact is desirable?) and social (what level of social impact is desirable?). Key indicators are established to monitor trends and standards established to assess disturbance in relationship to goals. Management interventions (Chapter 7) can reduce negative impacts where necessary, and can also stimulate positive impacts that nature-based tourism can provide. Vaske et al. (2003) and Manning (2007) offer useful overviews and examples. The Pinery Provincial Park in Ontario provides an example of how value changes over time lead to different interpretations of what is acceptable in terms of tourism development and environmental protection.

Summary of Tourism Impact Discussion

For years, managers of parks and protected areas attempted to ensure that the changes sometimes caused by the tourism industry are balanced appropriately with resource conservation. The tools used by the managers include determining levels of park use, stopping inappropriate visitor behaviours in parks, and managing demands for tourism facilities and services within parks (Eagles et al., 2002). The Pinery Provincial Park case study reveals many management options used by park planners. Tourism planners have also been concerned about sustaining the tourism industry in ways that avoid stagnation

BOX 12.5 Whale Watching on Vancouver Island

The Duffus-Dearden model was explored in a study that examined the whale watching in three locations on Vancouver Island (Malcolm, 2003). One site, Telegraph Cove, is a remote community in the northeast region of the island near a whale conservation area known as Robson Bight–Michael Bigg Ecological Reserve. Interviews with whale watchers revealed a high number of 'intermediate' and 'advanced' whale watchers (72 per cent), as determined by the amount of previous whale watching, the importance attached to whale watching, and the amount of personal reading and research about whales. These more specialized whale watchers had stronger attitudes about conservation in general and whale conservation in particular.

The second site, Tofino, near Pacific Rim National Park, had fewer advanced and intermediate specialists (46 per cent). The third site, Victoria, near Gulf Islands National Park Reserve and the proposed Gulf Islands National Marine Conservation Area, had the highest proportion of novice participants (70 per cent). The higher number of whale-watching companies in Victoria caters to a much larger market of whale watchers who are mainly novices lacking a high commitment to whale conservation. If whale watching continues to grow in Victoria, in response to market demand from novices for these spectacular viewing opportunities, the small and endangered population of orca whales may continue to decline, leading also to a decline in the whale-watching industry (although a number of other factors, such as toxins found in some marine areas, influence the viability of whale populations). Alternatively, if the whale-watching industry is able to provide powerful interpretive messages, these novices may become more like whale-watching specialists with stronger commitment to whale conservation.

and decline. Nature tourism is particularly dependent on the quality of natural settings, often within a park or protected area. Part of the sustainability equation is rooted in the support of local communities for tourism. Lack of community support for tourism can undermine the tourism product. In most countries, support for parks and protected areas depends in part on a positive economic benefit occurring for local communities. If local communities are asked to forgo traditional economies such as logging, hunting, or agriculture when a protected area is first created, a viable local tourism economy that can offset these losses may create support for park conservation (Chapter 17).

As the linkages between local communities, park conservation, and nature-based tourism have become better understood, new ways of thinking about more sustainable tourism practices, including 'ecotourism' and related concepts, have developed. The next section provides an overview of these different tourist types.

TYPES OF TOURISM

Although tourism has always been a heterogeneous activity with different types of tourists visiting various types of attractions, only more recently have these differences been categorized. A basic distinction to make is between mass tourism and alternative tourism (Figure 12.6). Mass tourism generally involves large numbers of people visiting staged settings, such as resorts, that may have little relationship to the actual environment within which they are set. The attraction base often is related to the 'four S's' of tourism: sea, sun, sand, and shopping. In contrast, alternative tourism usually

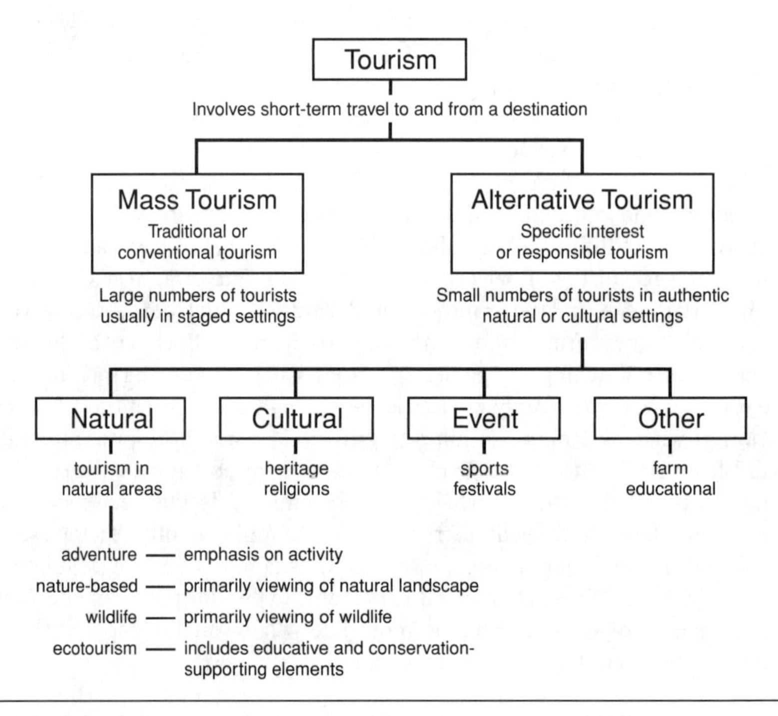

FIGURE 12.6 Different types of tourism (after Newsome et al., 2002).

involves smaller numbers of tourists visiting more authentic settings with an emphasis on attractions that are consistent with local natural, social, and community values.

Given this distinction, tourism in protected areas in Canada has elements of both mass and alternative tourism. Banff, with its large numbers of visitors, extensive infrastructure, international shopping attractions, and significant impacts, fits within the mass tourism category and generates conflicts with some park goals (Chapter 10). In fact, whether some visitors' touring is at all nature-based is a serious question. Nature-based tourism takes place in natural settings where the settings are an integral part of the attraction. As such, nature-based tourism is a form of alternative tourism. Most tourism to protected areas is nature-based, although there are good opportunities in Canada to further develop other forms of alternative tourism in parks, such as cultural tourism in First Nations communities.

Within nature-based tourism occur some important differences. For example, adventure tourism takes place *in* the environment, but it is not *about* the environment. Adventure tourism has a sharp focus on personal accomplishment through an activity, such as rock climbing or whitewater rafting, rather than on fostering a better understanding of the natural setting or local cultures. Wildlife tourism is a specific form of nature-based tourism that is focused on wildlife, and sometimes on only one species of wildlife. This form of nature-based tourism is growing rapidly with the increasing interest in viewing birds, whales, sharks, bears, and the like. Geotourism, the viewing of unique geological phenomena, is also becoming popular.

Ecotourism is a more stringent form of nature-based tourism and is tourism *for* the environment. Not only does the activity seek to minimize negative impacts, but ecotourism attempts to benefit the natural environment. This benefit can take many forms. It may involve alleviating poverty in local communities so that locals no longer poach in parks. This is a very common benefit in lesser-developed countries (Chapter 17). Benefits may also involve volunteers collecting data on an aspect of the environment that assist future conservation efforts, or building trails that minimize erosion, or undertaking ecological restoration activities or similar kinds of activities. It often provides the funds that are necessary to manage the natural resource.

Ecotourism has another very important component: teaching/learning. This component ensures that visitors have a greater understanding of natural ecosystems when they finish their activity than when they started it. Ecotourism should change the way people behave in their everyday settings (Chapter 8). However, this is one of the most difficult aspects of ecotourism to test, as behavioural change at home may not occur until some time after the tourism activity has taken place, and tourists may be hard to locate for questioning.

Sustainable tourism is 'tourism . . . in a form [that] can maintain its viability for an indefinite period of time' (Butler, 1993). Sustainable tourism development is 'the management of all resources in such a way that we can fulfill economic, social, and aesthetic needs while maintaining ecological processes, biological diversity, and life support systems' (Murphy, 1994). Although all tourism should be sustainable, the requirements outlined above mean that ecotourism has greater potential to contribute to sustainability than other forms of tourism, and should be a central focus for protected areas. Unfortunately, considerable confusion persists regarding the term 'ecotourism' (Wight, 1993; Fennell, 1999). In some countries, all tourist activity that doesn't obviously fall under the 'mass' category is thought to be 'ecotourism' (Dearden, 1992). Furthermore, the 'eco-' prefix sometimes is used as a marketing tool to attract more customers and charge higher prices to facilities that have little relationship to true ecotourism. These misuses have led many people to distrust the word 'ecotourism', despite the appropriateness of the concept for aiding nature conservation and informing park management.

A MARKET FOR AN ECOTOURISM INDUSTRY

The ecotourism concept provides direction for private-sector tourism companies to manage their operations in ways that sustain the environment, local communities, and the tourism industry. However, these definitions do not identify whether a market exists that would support this form of tourism. Recent polls show that Canadians have a strong interest in conserving the environment, but this does not necessarily translate into demand for ecotourism. Early studies revealed substantial differences between Canadian ecotourists and the average Canadian tourists (Kretchman and Eagles, 1990; Eagles, 1992; Ballantine and Eagles, 1994; Eagles and Cascagnette, 1995). Canadian ecotourists were more likely to place emphasis on the natural attraction and less on the social aspects of the destination. They ranked wilderness, water, mountains, parks, and rural areas much more highly than average tourists. They showed strong interest in specific elements of the natural environment, such as birds, mammals, trees, and wildflowers. Their social motives were concentrated on learning, physical activity, and photography. More recently, Randall (2000) examined commercial kayak guiding in the Broken Group Islands segment of Pacific Rim National Park, through interviews conducted with kayakers travelling with a kayaking guide. Survey findings indicated strong support for the learning/educational domain and the environmental/sociocultural domain of ecotourism. The nature-based domain of ecotourism was not examined, as this appeared to be self-evident in the study.

Randall's study also examined the growth of paid commercial backcountry wilderness trips, such as kayaking in the Broken Group Islands. One aspect of park tourism that has grown significantly over the last decade is guided eco- and adventure tours in the parks. Whether hiking the West Coast Trail in Pacific Rim or sailing through the majesty of Gwaii Haanas or rafting the Firth River into the Arctic Ocean through Vuntut National Park, many visitors are choosing guided trips. In fact, so great is the demand that Parks Canada has established limits in many parks on the number of operators who can offer trips and the number of visitors who can take such commercial trips. The trips raise many questions. Randall, for example, found that conflict was generated between groups on commercial trips and other park visitors. The commercial trippers often feel that they have a right to camp in certain sites that supersede the rights of other visitors. Commercial trip groups also tend to be larger than non-commercial groups and may violate social carrying capacity norms more frequently (Chapter 6).

Other questions relate to the trip experience. Sax (1980), for example, suggested in his classic treatise, *Mountains without Handrails*, that a principal role of national parks is to retain areas where we can test ourselves in a manner that we would not be able to do in an everyday environment, as discussed in Chapter 1. When visitors sign up for a guided trip, part of that responsibility is taken by the trip guide. The guide decides where to camp, the route through rapids, how to avoid getting hurt, and how to deal with inclement weather. Unless visitors are encouraged to make these decisions themselves, they are not truly engaging the challenges of wilderness travel. On the other hand, such guided trips also provide access into areas that many people might otherwise not be able to see. They also provide a safety net, ensure that certain safety standards are met, and generate park income. Organized tour groups can be attractive to park managers because such groups are much easier to direct than free, independent travellers. Their operational licences can include high standards of behaviour, facilities, equipment, and programs. One again, the key issue here is the capability of the managers in properly directing this activity, not so much the activity per se.

The apparent growth of the ecotourism industry suggests that some companies are tapping at least some of the ecotourism dimensions outlined in this chapter, but other companies may not be aware of all aspects of ecotourism or choose not to stress all components. The Canadian Tourism Commission (1999) attempted to communicate some of this ecotourism thinking directly to commercial tourism operators through a *Catalogue of Exemplary Practices in Adventure Travel and Ecotourism.* Key components of this guide are outlined in Box 12.6

One example of an ecotourism company that appears to illustrate many ecotourism principles is Stubbs Island Whale Watching Company, located in Telegraph Cove on Vancouver Island (www.stubbs-island.com/english/ethics/ethics.html). This company had a pivotal role in the formation of Robson Bight–Michael Biggs Ecological Reserve, in stewardship programs, and in developing guidelines for responsible wildlife viewing. The company assists in the monitoring of orca and humpback populations in the area, contributes to orca recovery plans, provides a whale and marine interpretive centre, provides educational programs, and sponsors research and development of wildlife viewing guidelines. The company strives to reduce its ecological footprint through the use of biodegradable cleaning products, organic fair trade coffee, and the use of non-disposable items and compostable plastic bags. In June 2007, the head naturalist was selected as one of the top three candidates for the educator category of the 'Me to We' awards, focused on encouraging people to lead lives that are socially conscious and responsible (www.metowe.org/about/me-to-we-philosophy.html).

HOW PARKS RESPOND TO ECOTOURISM

Many national and provincial parks in Canada have an international reputation as outstanding areas for nature-based travel. The key principles of ecotourism described above appear to be in close congruence with Parks Canada policy (Parks Canada, 1994), as outlined in Chapter 9:

- *Protecting ecological integrity and ensuring commemorative integrity take precedence in acquiring, managing, and administering heritage places and programs* (ibid., 16). Ecotourism concepts support ecological integrity in national parks, in that visits to pristine natural settings are central to the appeal of ecotourism.
- *It is recognized that these places are not islands, but are part of larger ecosystems and cultural landscapes* (ibid.). One tenet of ecotourism supports the economy and culture

of local communities. This serves to sustain the tourism product and creates support in local communities for national parks.

- *Leadership is established by example, by demonstrating and advocating environmental and heritage ethics and practices, and by assisting and co-operating with others* (ibid.). Commercial ecotourism guides are in a unique position to demonstrate and advocate environmental and ethical practices consistent with this policy. In fact, some of the longer ecotourism experiences, extending over several days, may provide a better opportunity to instill appropriate behaviours compared to the relatively short interaction that park visitors might otherwise have with park staff or with interpretive media.
- *The long-term success of efforts to commemorate, protect, and present Canada's natural and cultural heritage depends on the ability of all Canadians to understand and appreciate this heritage* (ibid., 17). The presentation of heritage themes and messages

BOX 12.6 Outline of the *Catalogue of Exemplary Practices in Adventure Travel and Ecotourism*

The Canadian Tourism Commission developed guidelines for nature-based tourism companies that embrace many of the ecotourism concepts described in this chapter, and provided examples to illustrate each of the following:

- Resource protection and sustainability
  - be sensitive to the environment
  - conserve and manage energy, water, waste, and transportation
  - develop policies for purchasers and suppliers
  - minimize impacts on wildlife
  - minimize impacts on natural environments
  - guide visitor behaviour
  - support regional conservation efforts
- Social and community contribution
  - consult with and involve local people and groups
  - employ local people
  - purchase local goods and services
  - share with or contribute other benefits to local communities
  - adapt to local conditions over time
  - minimize impact on and be sensitive to communities
  - enable guests to experience local communities and culture
- Product and delivery
  - implement quality assurance measures
  - consider the way the company operates in the field
  - develop high service standards
  - define codes and standards of behaviour
  - communicate with visitors through interpretation and guiding

is normally achieved through on-site interpretation and through outreach. While these are important initiatives by Parks Canada, the agency ensure these messages are communicated by collaborating through private-sector ecotourism operators.

- *Management decisions are based on the best available knowledge, supported by a wide range of research, including a commitment to integrated scientific monitoring . . . local knowledge is also of value to Parks Canada in managing heritage areas* (ibid., 18). Ecotourism guides are often intimately familiar with the places they visit and tend to return frequently to the same place. They are in a unique situation to provide local knowledge in a timely way that complements and enhances scientific monitoring of ecological conditions (Chapter 5).
- *Opportunities will be provided to visitors that enhance public understanding, appreciation, enjoyment, and protection of the national heritage and which are appropriate to the purpose of each park and historic site* (ibid.). Ecotourism operators have the knowledge and experience to help create national park experiences that many Canadians and foreign visitors might otherwise not have the skill, experience, or understanding to pursue.

This analysis illustrates a close congruence between published ecotourism concepts and Parks Canada policy. Some would argue that private-sector tourism, motivated in part by a bottom line of profit, must be viewed with suspicion (Wight, 1993; Fuller, 1977). However, one would not have much of an industry if the operators did not make money. On the other hand, it could be argued that public agencies should be viewed with suspicion—for example, some agencies may not fully appreciate the potential of

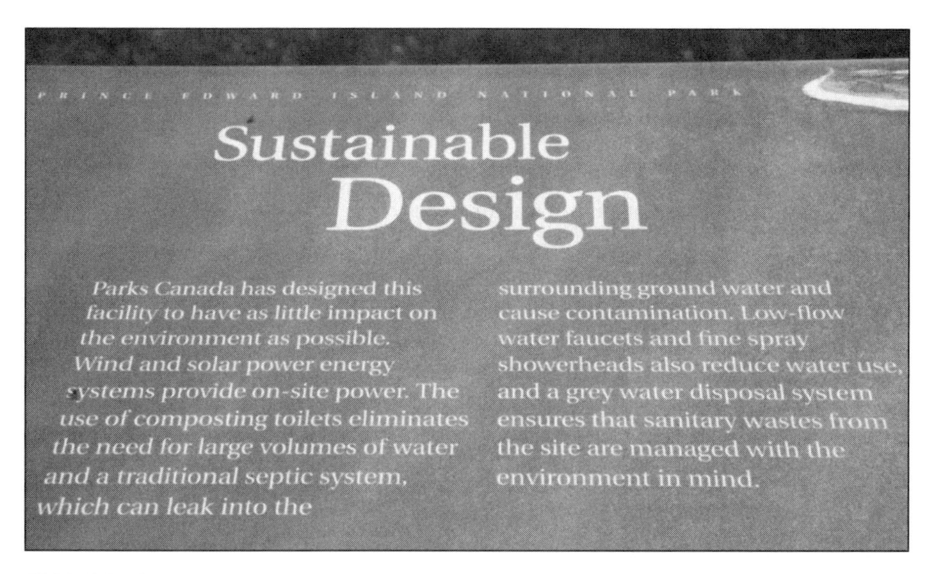

FIGURE 12.7 Parks Canada has invested in trying to reduce the impacts of visitor facilities. In this case, at Prince Edward Island National Park, the washroom facilities and beach changing rooms have been designed with sustainability in mind. *Photo: P. Dearden.*

the private sector to contribute appropriately to park management. To address this latter argument we present a case study (Box 12.7) of commercial operator standards for Dive Charters within Pacific Rim National Park.

Boyd (1995) undertook a study on the perceptions of experts such as policy-makers, park superintendents, and academics on what tourism sustainability in national parks in Canada might look like. Interesting differences arose among the groups. Park superintendents, for example, scored all forms of tourism, including resort and mass tourism, much higher on the sustainability scale than did academics. Nonetheless, certain types of tourism, such as remote, mountain, and ecotourism, were recognized as being potentially more sustainable by all groups when compared with other types of tourism.

BOX 12.7 Commercial Dive Charter Standards for Pacific Rim National Park

The following outlines the key provisions of the standards developed by Pacific Rim National Park for commercial dive charter companies applying for a business licence to operate within the park. These regulations indicate close correspondence with Parks Canada policies and ecotourism principles in a number of factors:

- Providing a safe enjoyable experience in a natural setting (a national park).
- Minimizing environmental impacts (e.g., from anchors, not feeding fish, not collecting natural objects).
- Minimizing impacts on cultural resources within the park (e.g., diving near wrecks).
- Complying with a voluntary 'no-take' fishing policy. This is a good example of using the private sector to advance a park management initiative.
- Monitoring condition of dive sites. This illustrates the use of local knowledge through collaboration with ecotourism dive operators, to assist in the assessment of ecological integrity.
- Providing visitor use data. This is a way of assessing possible environmental impacts and monitoring the quality of visitor experience. Both of these are important aspects of Parks Canada's mandate. This collaboration produces important park management data and is consistent with one aspect of ecotourism (Orams, 1995) that stresses more active involvement of ecotourism in conserving park resources.
- Sharing dive log information. Operators are encouraged, but not required, to share their dive log information and contribute to the database for the dive area.

This analysis also suggests a number of key opportunities that are not addressed in these standards:

- There is no mention of interpretation or education, to make divers more aware of ecological processes or the role of Parks Canada in protecting marine resources in national parks.
- There is no reference to the use of local providers, hiring local staff, etc., to support local communities.

This discussion suggests a number of ways that Parks Canada has embraced or could embrace ecotourism. This assessment takes the position that ecotourism provides an opportunity for parks to increase capacity and to deliver services and benefits that might not otherwise be fully realized. This is particularly important at a time when public-sector budgets for parks management have declined while in most jurisdictions in Canada the number of parks has increased. Further, there is growing awareness within park agencies in Canada of the need to address ecosystem management and to engage with communities peripheral to park boundaries to better address ecological integrity (Chapter 13). While this principle is well understood, it is difficult at times to find resources for park managers to move in this direction. Ecotourism is one avenue for connecting local communities with park management operations and for supporting ecosystem management.

In 2007 Parks Canada launched a new tourism policy, entitled 'External Relations and Visitor Experience'. This involved creating a new branch of the agency, hiring new staff, and creating a new approach to the provision of visitor experiences in national parks and national historic parks. The goal of the program is on 'enhancing the visitor experience' and on 'turning many one-time visitors into lifelong stewards and ambassadors of Parks Canada's natural and historic sites' (Parks Canada, 2007). This is an ambitious objective because it requires a personal commitment of visitors after the on-site park experience. The agency places more emphasis 'on nature tourism, adventure tourism, eco-tourism, authentic heritage experiences, and first-hand learning' (ibid.). The program has two major aspects. The first deals with information before the visit. To this goal, the agency 'will work to provide travellers with access to an imaginative website, toll-free telephone service, readily available trip planning materials, convenient reservation methods, and fair pricing' (ibid.). The second deals with the visitor experience during the visit. For this goal, the agency will provide 'warm hospitality, top-notch facilities, onsite programs and activities, as well as knowledgeable and friendly staff' (ibid.). To measure attainment of the goals, the agency will conduct on-site surveys and post-trip assessments.

CONCLUSION

This chapter describes the emergence of ecotourism as a form of private-sector activity that contributes to ecological integrity, positive visitor experiences in parks, and the sustainability of local communities. Although all tourism should be sustainable, ecotourism focuses on nature experiences, learning, and community development. As such, ecotourism can be seen as an important component of ecosystem management, whereby ecotourism operators take some of the initiative in connecting park agencies, local communities, and park visitors in ways that focus on sustaining parks and the broader ecosystems where parks are embedded.

While many tourism companies use the ecotourism label, not all companies practise these ecotourism principles. This divergence can be explained by purposely misleading promotional materials or by an ignorance of ecotourism practices. To this end, some argue for accreditation of ecotourism companies, in much the same way as the lumber industry uses sustainable forest practices as a form of accreditation to communicate better practices to consumers. Park agencies have a role in setting standards and procedures when granting business licences for tourism companies operating within parks and protected areas. They can, therefore, in a way create their own certification scheme to ensure that appropriate ecotourism becomes the dominant mode of park tourism, which enriches visitor experiences while protecting park resources.

REFERENCES

Ballantine, J.L., and P.F.J. Eagles. 1994. 'Defining Canadian ecotourists', *Journal of Sustainable Tourism* 2, 1: 1–5.

Blamey, R.K. 2001. 'Principles of ecotourism', in D. Weaver, ed., *Encyclopedia of Ecotourism.* Wallingford, UK: CABI, 4–22.

Boyd, S.W. 1995. 'Sustainability and Canada's national parks: Suitability for planning, policy and management', Ph.D. thesis, University of Western Ontario.

Butler, R.W. 1980. 'The concept of a tourist-area cycle of evolution and implications for management', *Canadian Geographer* 24: 5–12.

———. 1993. 'Tourism—an evolutionary perspective', in J.G. Nelson, R. Butler, and G. Wall, eds, *Tourism and Sustainable Development: Monitoring, Planning, Managing.* Waterloo, Ont.: Department of Geography Publication Series, Heritage Resources Centre, University of Waterloo, 27–44.

——— and S.W. Boyd. 2003. 'Tourism and parks—a long but uneasy relationship', in R.W. Butler and S.W. Boyd, eds, *Tourism and National Parks: Issues and Implications.* Toronto: John Wiley, 3–11.

Canadian Tourism Commission. 1999. *Catalogue of Exemplary Practices in Adventure Travel and Ecotourism.* Ottawa.

Dearden, P. 1992. 'Tourism and development in Southeast Asia: Some challenges for the future', in A. Pongsapich, M.C. Howard, and J. Amyot, eds, *Regional Development and Change in Southeast Asia in the 1990s.* Bangkok: Social Research Institute, Chulalongkorn University, 215–29.

———, M. Bennett, and R. Rollins. 2006. 'Dive specialization in Phuket: Implications for reef conservation', *Environmental Conservation* 33: 353–63.

———, ———, and ———. 2007. 'Perceptions of diving impacts and implications for reef conservation', *Coastal Management* 35: 305–17.

Department of Lands and Forests. 1971. *Pinery Provincial Park Master Plan.* Toronto: Queen's Printer.

Diamantis, D. 2004. *Ecotourism: Management and Assessment.* London: Thomson.

Doxey, G.V. 1975. 'A causation theory of visitor–resident irritants: methodology and research inference', paper presented at San Diego, TTRA Conference.

Duchesne, M., S. de Cote, and C. Barrette. 2000. 'Responses of woodland caribou to winter ecotourism in the Charlevoix Biosphere Reserve, Canada', *Biological Conservation* 96: 311–17.

Duffus, D.A., and P. Dearden. 1990. 'Non-consumptive wildlife-oriented recreation: A conceptual framework', *Biological Conservation* 53, 3: 213–31.

——— and ———. 1993. 'Killer whales, science and protected area management in British Columbia', *George Wright Forum* 9: 79–87.

Dyck, M.G., and R.K. Baydack. 2004. 'Vigilance behaviour of polar bears (*Ursus maritimus*) in the context of wildlife-viewing activities at Churchill, Manitoba, Canada', *Biological Conservation* 116: 343–50.

Eagles, P.F.J. 1992. 'The travel motivations of Canadian ecotourists', *Journal of Travel Research* 31, 2: 3–7.

——— and J.W. Cascagnette. 1995. 'Canadian ecotourists: Who are they?', *Tourism Recreation Research* 20, 1: 22–8.

——— and S. McCool. 2002. *Tourism in National Parks and Protected Areas: Planning and Management.* Wallingford, UK: CABI.

———, ———, and C. Haynes. 2002. *Sustainable Tourism in Protected Areas: Guidelines for Planning and Management.* Gland, Switzerland: United Nations Environment Programme, World Tourism Organization, and World Conservation Union.

———, D. McLean, and M.J. Stabler. 2000. 'Estimating the tourism volume and value in parks and protected areas in Canada and the USA', *George Wright Forum* 17, 3: 62–76.

Epler Wood, M., F. Gatz, and K. Lindberg. 1991. 'The Ecotourism Society: An action agenda', in: J. Kusler, ed., *Ecotourism and Resource Conservation.* Selected papers from the Second International Symposium: Ecotourism and Resource Conservation. Madison, Wis.: Omnipress, 75–9.

Fennell, D. 1999. *Ecotourism: An Introduction.* New York: Routledge.

———. 2001. 'A content analysis of ecotourism definitions', *Current Issues in Tourism* 4, 5: 403–21.

Fuller, W.A. 1977. *Tragedy in Our National Parks.* Toronto: National and Provincial Parks Society of Canada.

Higham, J.E.S. 1998. 'Tourists and albatrosses: The dynamics of tourism at the Northern Royal Albatross Colony, Taiaroa Head, New Zealand', *Tourism Management* 19: 521–31.

Hvenegaard, G.T., J.R. Butler, and D.G. Kristofak. 1989. 'Economic values of bird watching at Point Pelee National Park, Canada', *Wildlife Society Bulletin* 17: 526–53.

Jackson, R.L., and J.E. Leahy. 2004. 'Social aspects of coastal tourism', in M.J. Manfredo, J.J. Vaske, B.L. Bruyere, D.R. Field, and P.J. Brown, eds, *Society and Natural Resources: A Summary of Knowledge.* Jefferson, Mo.: Modern Litho, 175–86.

Keogh, B. 1989. 'Social impacts', in G. Wall, ed., *Outdoor Recreation in Canada.* Toronto: Wiley.

Kretchman, J.A., and P.F.J. Eagles. 1990. 'An analysis of the motivations of ecotourists in comparison to the general Canadian population', *Loisir & Société* 13, 2: 385–97.

Lankford, S.V., and D.R. Howard. 1994. 'Developing a tourism impact attitude scale', *Annals of Tourism Research* 21: 121–39.

Lemelin, R.H., and B. Smale. 2006. 'Effect of environmental context on the experience of polar bear viewers in Churchill, Manitoba', *Journal of Ecotourism,* 5: 176–91.

Lindberg, K. 2001. 'Economic impacts', in D.B. Weaver, ed., *The Encyclopedia of Ecotourism.* New York: CABI, 363–77.

Malcolm, C.D. 2003. 'The current state and future prospects of whale-watching management, with special emphasis on whale-watching in British Columbia, Canada', Ph.D. thesis, University of Victoria.

Manning, R.E. 2007. *Parks and Carrying Capacity: Commons without Tragedy.* Washington: Island Press.

Meadows, D., and D. Meadows. 1972. *Limits to Growth.* New York: Universe.

Murphy, P.E. 1994. 'Tourism and sustainable development', in W. Theobald, ed., *Global Tourism: The Next Decade.* Oxford: Butterworth-Heinemann, 274–90.

Newsome, D., S. Moore, and R. Dowling. 2002. *Natural Area Tourism: Ecology, Impacts and Management.* Toronto: Channel View Publications.

———, R. Dowling, and S. Moore. 2005. *Wildlife Tourism.* Toronto: Channel View Publications.

Orams, M.B. 1995. 'Towards a more desirable form of ecotourism', *Tourism Management* 16, 1: 3–8.

Owen, T.M., and B.K. Gilbert. 2005. 'Measuring the cost of risk avoidance in brown bears: Further evidence of positive impacts of tourism', *Biological Conservation* 123: 453–60.

Parks Canada. 1994. *Guiding Principles and Operational Policies.* Ottawa: Parks Canada.

———. 2007. 'Visitors learn, grow and discover', *Experiences* 2, 1: 1.

Parsons, E.C.M., M. Luck, and J.K. Lewandowski. 2006. 'Recent advances in whale-watching research: 2005–2006', *Tourism in Marine Environments* 3: 179–89.

Randall, B.C. 2000. 'An examination of visitor management issues within the Broken Group Islands, Pacific Rim National Park Reserve', MA thesis, University of Victoria.

Reed, D.G., H. Mair, and J. Taylor. 2000. 'Community participation in rural tourism development', *World Leisure* 2: 20–7.

Rollins, R. 1997. 'Validation of TIAS as a tourism impact tool', *Annals of Tourism Research* 24, 2: 37–41.

Roman, G., P. Dearden, and R. Rollins. 2007. 'Application of zoning and "limits of acceptable change" to manage snorkeling tourism', *Environmental Management* 39: 819–30.

Ross, S., and G. Wall. 1999. 'Ecotourism: Towards congruence between theory and practice', *Tourism Management* 20: 123–32.

Saltspring Island Chamber of Commerce. 2004. Salt Spring Island Perceptions of Tourism Survey.

Scace, R.C. 1993. 'An ecotourism perspective', in J.G. Nelson, R. Butler, and G. Wall, eds, *Tourism and Sustainable Development: Monitoring, Planning, Managing.* Waterloo, Ont.: Department of Geography Publication Series, Heritage Resources Centre, University of Waterloo, 59–82.

Sorice, M.G., C.S. Shafer, and R.B. Ditton. 2006. 'Managing endangered species within the use-preservation paradox: The Florida manatee (*Trichechus manatus latirostris*) as a tourism attraction', *Environmental Management* 37: 69–83.

Theberge, M., and P. Dearden. 2006. 'Detecting a decline in whale shark (*Rhincodon typus*) sightings in the Andaman Sea, Thailand, using ecotourist operator-collected data', *Oryx* 40: 337–42.

Vaske, J.J., M.P. Donnelly, and D. Whittaker. 2000. 'Tourism, national parks and impact management', in R.W. Butler and S.W. Boyd, eds, *Tourism and National Parks.* Toronto: John Wiley, 203–22.

Vaugeois, N., and R. Rollins. 2007. 'Mobility into tourism: Refuge employer', *Annals of Tourism Research* 34, 3: 630–48.

Weaver, D.B. 2002. 'The evolving concept of ecotourism and its potential impacts', *International Journal of Sustainable Development* 5, 3: 251–64.

Wight, P. 1993. 'Ecotourism: Ethics or eco-sell?', *Journal of Travel Research* 32: 3–9.

World Commission on Environment and Development. 1987. *Our Common Future.* Oxford: Oxford University Press.

World Tourism Organization. 1994. *Recommendations on Tourism Statistics.* United Nations and World Tourism Organization, Department for Economic and Social Information and Policy Analysis, Statistical Division, Statistical papers, Series M, No. 83.

World Travel and Tourism Council. 2006. 'Travel and tourism scaling to new heights', 3–18. At: <www.wttc.org/2006TSA/pdf/Executive%20Summary%202006.pdf>.

Wright, P., and D. McVetty. 2000. 'Tourism planning in the Arctic Banks Island', *Tourism Recreational Research* 25, 2:15–26.

Yasué, M., and P. Dearden. 2006. 'The potential impact of tourism development on habitat availability and productivity of Malaysian plovers (*Charadrius peronii*)', *Journal of Applied Ecology* 43: 978–89.

KEY WORDS/CONCEPTS

alternative tourism
amenity migration
Butler model of tourism life cycle
demonstration effect
Doxey model of tourism development
Duffus-Dearden model of wildlife tourism.
economic impacts of tourism
ecotourism
environmental impacts of tourism
mass tourism
social impacts of tourism
sustainable tourism

STUDY QUESTIONS

1. Outline some of the economic benefits and non-benefits of tourism.
2. Outline some of the environmental benefits and non-benefits of tourism.
3. Outline some of the social benefits and non-benefits of tourism.
4. Select a nature tourism operator in your community and illustrate possible economic, social, and environmental impacts of the operation.
5. Provide an example of Butler's tourism life cycle, using a tourism operation familiar to you.
6. What are the key features of ecotourism, and how do these differ from mass tourism?
7. Examine a park familiar to you and profile one tourism operator using the park. Does this company appear to use ecotourism principles?
8. Compare the concept of ecotourism with ecosystem management (Chapter 13).

CHAPTER 13

Protected Areas and Ecosystem-based Management

D. Scott Slocombe and Philip Dearden

INTRODUCTION

Earlier chapters have highlighted the ways in which protected areas are linked to their surroundings physically, ecologically, economically, and culturally. During the past 35 years, but particularly the last 20, the interaction between protected areas and their surroundings has been increasingly recognized. Managing protected areas without recognizing these linkages can lead to many problems, such as population increases in some wildlife species and decreases in others; ignorance of pollution originating outside the protected area; lack of support from local residents and failure to learn from their experiences; and institutional conflict and competition. As the title of Dan Janzen's (1983) influential article observed, 'No park is an island'.

As managers and conservationists have come to understand the challenges facing protected areas, particularly those related to the reduction of undisturbed ecosystems surrounding them, the need for a more regional, multi-stakeholder, science-based, and collaborative approach has become widely recognized (Dearden et al., 2005). Threats originating outside parks (see, e.g., Machlis and Tichnell, 1987; Dearden and Doyle, 1997; Parks Canada, 1998), the application of gap analyses to identify critical unprotected areas (e.g., Morgan et al., 2003), the need for inter-agency co-operation to facilitate networks of protected areas for biodiversity (see, e.g., Swinnerton and Otway, 2003), and understanding of the significance of local support for the long-term effectiveness of protected areas (e.g., Halpenny et al., 2003) all have fostered the development of more regionally integrated approaches to protected areas management.

These concerns are not limited to the management of protected areas; they influence and are influenced by similar changes in resource and environmental management. Multiple-use management, integrated resource management, watershed management, and comprehensive regional land-use planning have converged during the past decades to become known as ecosystem, or ecosystem-based, management. Although various definitions for the particular approaches exist, most share the concept of a regional 'greater ecosystem' for management based on biophysical rather than on administrative boundaries. Common themes include stress on the role of scientific knowledge as a basis for management, a collaborative and participatory process, and explicit definition of such complex management goals as ecological integrity, biodiversity maintenance, and sustainability.

Such approaches have developed over a number of years in Canada. By the early 1970s there was much work on ecosystem approaches in the context of problems of the Great Lakes (Caldwell, 1970; Francis et al., 1985; Vallentyne and Beeton, 1988). The concept and implications of the ecosystem approach were ultimately reflected in the Great Lakes Water Quality Agreement and the Remedial Action Plan process for areas of concern in the Great Lakes Basin (e.g., Hartig et al., 1998; MacKenzie, 1996). These efforts had wide influence in the development of regional-scale environmental management over many years. The earliest parks-related interest in Canada probably was catalyzed by biosphere reserves, especially Waterton Lakes National Park Biosphere Reserve in Alberta. Parks Canada and the US National Parks Service co-sponsored a workshop in June 1982 on the theme 'Towards the Biosphere Reserve: Exploring Relationships between Parks and Adjacent Lands'. By the mid-1980s Parks Canada was exploring the 'regional integration' of parks. This had strong economic and political dimensions, but was also concerned with ecological issues, such as resource and tourism development around parks (Community and Municipal Affairs, 1985). Several case studies were completed in the later 1980s, including one on Waterton.

The first book to explicitly link ecosystem management and protected areas appeared in 1988 (Agee and Johnson, 1988). Its authors were American, as were its examples—including Wyoming's Yellowstone National Park, a recurring American example in the history of ecosystem management. The contributors to the book stressed many of the themes highlighted above: external threats; the need for interagency co-operation; new scientific understanding of diversity, disturbance, and ecosystems; and a better appreciation for the human dimension of management. By this time research in Canada was growing on similar topics, most notably on environmental monitoring and ecosystem integrity. A workshop in Waterloo, Ontario, focused on ecological integrity and ecosystem management in parks and resulted in several influential publications (e.g., Woodley, 1993). In 1988, the National Parks Act was amended to require maintenance of ecological integrity as the top priority (Chapter 2). Perhaps not coincidentally, Canadian approaches to ecosystem-based management have tended to be strongly based in science and ecological integrity, and are arguably less participatory and process-oriented than those developed in the US and elsewhere.

Today, ecosystem-based management approaches are widely endorsed and increasingly implemented. For example, the 1994 Parks Canada Policy (Parks Canada, 1994) recognized ecosystem management and that it must be broad in scope, and based on wide support in terms of regional co-operation and integration of the park into the surrounding region (section 2.1.7). A similar approach is implicit in the National Marine Conservation Areas system plan and Act (Chapter 15), as well as in the Oceans Act (Vandermeulen, 2003). More recently, the report of the Panel on the Ecological Integrity of Canada's national parks made management of whole ecosystems—using the best scientific knowledge and involving local people and their concerns—a key part of its recommendations (Parks Canada Agency, 2000). Recent analyses of the minimum critical area for large carnivores in the national parks (Landry et al., 2001) have shown clearly that most parks are too small to sustain minimum viable populations of these

FIGURE 13.1 International boundary between Canada and the US linking Waterton and Glacier national parks. Political boundaries, even international ones, hold little significance for ecological processes. *Photo: P. Dearden.*

species without the use of appropriate habitat outside the parks, and that more complex analyses of scale and landscape patterns are needed in designing protected areas systems (Wiersma and Urban, 2004; Wiersma et al., 2005). Ecosystem management for parks is now a necessity for their survival and not an optional management fad (see Chapter 4).

The remainder of this chapter introduces the principles, tools, and methods of ecosystem-based park planning, and provides examples of its application.

THE ELEMENTS OF ECOSYSTEM-BASED PARKS MANAGEMENT

Ecosystem-based management has diverse roots. It draws on social and natural science, theory and practice, and protected and non-protected area examples. It is a new approach but with a long history. It has supporters as well as critics. In this section we will highlight different perspectives on the central ideas and elements of ecosystem-based management, moving towards a synthesis for protected areas management.

Agee and Johnson (1988) stressed nine principles that remain important today:

1. Co-operation and open negotiation are fundamental to success.
2. Individual agencies and stakeholders have different mandates, objectives, and constituencies.
3. Success should be measured by results.
4. Threshold management goals are stated by the park and wilderness legislation.

5. Problems defined clearly have a better chance of being resolved.
6. Over the long term, ecosystem management must accommodate multiple uses at a regional scale and dominant or restricted uses at the unit or site scale.
7. High-quality information is necessary to identify trends and respond to them intelligently.
8. Social, political, and environmental issues must be viewed in a system context, not as individual issues.
9. All management is a long-term experiment and decisions are always made with incomplete information.

Agee and Johnson's perspective was rooted in that of the professional park manager, and is particularly concerned with agencies, information, land management units, and satisfying the often conflicting demands of legislation, the public, and conservation.

Much of the discussion of ecosystem management is focused on public, government-owned land: Crown land in Canada, federal and state lands in the US. Hence, ecosystem management literature often takes the perspective of government land managers. This perspective tends to stress inter-agency co-operation and legislative foundations for management, public consultation, especially for multiple uses (perhaps instead of real public participation), and ways to link science and public policy to produce better outcomes. In most of the US and in the more populated regions of Canada, incorporating privately owned lands into ecosystem-based management is a challenge for protected area managers. It is of particular concern when private lands surround the protected area (Box 13.1). This is part of the general challenge of including private lands in an integrated land and resource management framework. The significance of this challenge is underscored by the growing interest in a range of forms of private protected areas, the role of which in ecosystem-based management remains to be explored (see Langholz, 2003, and Chapter 16). Possible approaches to this problem include improved understanding of the linkages between public and private lands, clarifying the space and time dimensions needed to ensure ecological integrity and resilience, co-operative efforts to diversify the local economic base, and consensual development of law and policy that can be used to promote ecological and economic sustainability

In contrast to the perspective outlined above, other writers have approached the issue from a more ecological point of view. Grumbine (1994) traced the origins of ecosystem management from Aldo Leopold's research and writing on wildlife, land, and ecosystems (1920s to the 1940s), early research on wildlife in the US national parks by George Wright and others (1930s and 1940s), and the Craigheads' bear research in Yellowstone National Park beginning in the 1960s. Ecological work, along with a great deal of experience in national park and national forest management, has uncovered the complexity of natural ecosystems and the need for large, intact areas to maintain diverse wildlife populations (Box 13.2).

With analysis based on an extensive review of the literature, Grumbine (1994: 31) identified 10 themes for ecosystem management: hierarchical context, ecological boundaries, ecological integrity, data collection, monitoring, adaptive management, inter-agency co-operation, organizational change, humans embedded in nature, and values. His definition—'ecosystem management integrates scientific knowledge of ecological relationships within a complex sociopolitical and values framework toward the general

BOX 13.1 Protected Landscapes in Canada

IUCN Category V lands (see Chapter 1)—protected landscapes/seascapes—are commonly set aside in many countries, especially in Europe, but much less so in Canada. These protected areas include private lands, and are ideal for countries such as the UK, where most lands, even inside national parks, are private. They are predominantly aimed at 'lived-in landscapes that demonstrate the ongoing interaction between people and their means of livelihood that is primarily dependent on the basic resources (cultural and natural) of the area' (Swinnerton, 2004: 79). These areas are particularly useful in places such as Europe where human occupancy has had a dominant impact on the landscape, and where, in fact, there may be few, if any, examples of natural landscapes remaining. However, these types of protected areas are also useful in countries such as Canada in extending biodiversity conservation protection values onto private lands.

Swinnerton (2004) has summarized the status of Category V lands in Canada and provided several examples, ranging from the Cooking Lake–Blackfoot Grazing, Wildlife and Provincial Recreation Area in Alberta to the National Capital Greenbelt (Ottawa). He also points out the potential use of Category V lands as corridors between protected areas, and Whitelaw and Eagles (2007) have provided a useful example of planning for the use of private lands as corridors in Ontario. Swinnerton also points out the potential importance of Category V lands for Aboriginal communities, especially in the North. The 2007 signing of the agreement between Parks Canada and the Deline First Nation and Deline Land Corporation to permanently protect the 5,587 km^2 Sahoyúé-?ehdacho Aboriginal cultural area on Great Bear Lake would have seemed like a prime opportunity, but instead it will proceed as a national historic site.

This observation illustrates some of the challenges being faced by Category V protected areas. For example, Swinnerton reports, based on 1997 data, that almost 10 per cent of Canada's protected area was classified as Category V. By 2005 the figure was 0.5 per cent. This change reflects the uncertainties in classifying and establishing standards for Category V areas that are discussed in more detail by Locke and Dearden (2005). Category V protected areas have a lot of potential to be strong allies in the further development of ecosystem-based management in the future, once better standards are developed and enforced.

goal of protecting native ecosystem integrity over the long term'—is widely cited, as is his list of goals:

- maintain viable populations of all native species in situ;
- represent native ecosystem types across a natural range of variation;
- maintain evolutionary and ecological processes;
- manage over long enough periods of time to maintain the evolutionary potential of species and ecosystems;
- accommodate human use and occupancy within these constraints.

BOX 13.2 Woodland Caribou and Nahanni National Park Reserve

Often, parks are created with one major threat to a particular resource in mind, and boundaries are drawn to reflect this. Nahanni National Park Reserve, on the border between Yukon and Northwest Territories, is one such example. The boundaries of the reserve were delimited to protect the river from hydroelectric development and resulted in a linear corridor centred on the South Nahanni and Flat Rivers. Important mammal populations, including the South Nahanni Woodland Caribou Herd (SNH), composed of the COSEWIC vulnerable-listed, western woodland caribou (*Rangifer tarandus caribou*), are not well protected by this configuration.

In response, Parks Canada sponsored a three-year study in co-operation with territorial governments and First Nations to learn more about the range of the SNH and the degree of protection offered by the reserve. At the outset of the study virtually nothing was known of the herd size, composition, and seasonal distribution. Results indicate that only a small proportion of the caribou's 4,000-km range is protected within the reserve, with the remainder in mineral-rich lands accessible by the Nahanni Range Road. As a result, the NWT government has stepped forward to take a lead role in conducting further studies to protect the wide-ranging herd. The initiative illustrates the importance of ecosystem-based management principles—such as thinking beyond park administrative boundaries, undertaking solid scientific research, and building partnerships with local stakeholders (see Tate, 2003). The Canadian Parks and Wilderness Society (CPAWS) is spearheading a campaign to get the federal government to agree to protect the entire South Nahanni watershed. In August 2007 the Prime Minister announced addition of a further 5,400 km^2 of protected land to the park, an important step in protecting the entire watershed (see cpaws.org).

Grumbine presents an ecological view of ecosystem-based management. Many ecologists and other natural scientists, who may not even agree with the use of 'ecosystem' in such a broad context or who dislike the vagueness of 'ecosystem management', are relatively content with species conservation as the basis for ecosystem management, coupled with its scientific, rational character (e.g., Wilcove, 1995). This approach can be strengthened by placing greater emphasis on the contributions of conservation biology, landscape ecology, and systems ecology to land management, resulting in an approach based on integrated ecological assessment and adaptive management (Jensen et al., 1996). A scientific (ecological or otherwise) basis for ecosystem management is not only intellectually desirable, but is essential. In a litigious system, such as that of the US, such a foundation enables managers to defend decisions against those who do not approve. Clearly, ecosystem management has received much attention for both scientific and political reasons (Ecological Society of America, 1995).

Slocombe (1993a, 1993b) addressed ecosystem-based management with a greater degree of social context. He drew particularly on experience with large, wilderness, protected areas and their surrounding regions, on comprehensive regional planning, and on the Great Lakes Basin experience to develop a characterization of an ecosystem approach (Table 13.1). Based on this characterization, Slocombe highlighted three dimensions of ecosystem-based management: defining management units, developing understanding, and creating planning and management frameworks. Overall, this approach sought to balance ecological and scientific dimensions with social and process-oriented ones.

The Great Lakes Basin Remedial Action Plan (RAP) process has provided fruitful lessons developed from locally run programs for relatively small sites, albeit ones of considerable significance. After reviewing and analyzing the RAP experience, Hartig and others (1998) developed eight principles and elements for effective ecosystem management:

1. broad-based stakeholder involvement;
2. commitment of top leaders;
3. agreement on information needs and interpretation;
4. action planning within a strategic framework;
5. human resource development;
6. results and indicators to measure progress;
7. systematic review and feedback;
8. stakeholder satisfaction.

In addition, government agencies were asked to change their traditional, top-down regulatory structure to one that is more co-operative, values-driven, and supportive. The authors identified central concepts to help guide public and private efforts, many of which are familiar from other ecosystem management writings. These included, among others: the watershed/bioregion as the management unit, partnerships, long-term vision, definition of principles to guide the process, geographic information systems (GIS) and decision support systems, data compilation, incentives, and utilizing of market forces. More recently, Great Lakes ecosystem planning has extended to Lakewide Management Plans, and the last 5–10 years have seen an innovative process developed around the proposal for a new national marine conservation area (NMCA) along the north shore of Lake Superior.

TABLE 13.1 Characteristics of Ecosystem Management

- Describe parts, systems, environments, and their interactions.
- Be holistic, comprehensive, trans-disciplinary.
- Include people and their activities in the ecosystem.
- Describe system dynamics, e.g., with concepts of homeostasis, feedbacks, cause-and-effect relationships, self-organization, etc.
- Define the ecosystem naturally, i.e., bioregionally, instead of arbitrarily.
- Consider different levels/scales of system structure, process, and function.
- Recognize goals and take an active management orientation. Include actor-system dynamics and institutional factors in the analysis. Use an anticipatory, flexible research and planning process. Enact implicit or explicit ethics of quality, well-being, and integrity.
- Recognize systemic limits to action.
- Define and seek sustainability.

Source: After Slocombe (1993a, 1993b).

Clearly, many of these perspectives are related: some are complementary, some include or are included in others. Steven Yaffee (1999) has suggested that a continuum of ideas and concerns is related to management, and stressed three points of view or approaches for ecosystem management:

1. environmentally sensitive multiple use (anthropocentric; human use subject to constraints that are usually beyond those in traditional resource management);
2. ecosystem approach to resource management (biocentric; ecosystem implies holistic thinking, dynamics, scale, complexity and cross-boundary management, problem orientation);
3. ecoregional management (ecocentric; managing landscape ecosystems in specific places, emphasizing ecosystem processes rather than biota; an integrated spatial unit).

Though ecosystem management approaches and theory enclose a common core of concerns and concepts, the complexity and diversity of protected areas imply the need for a multiplicity of approaches, methods, and techniques. The next section reviews various tools and methods used in ecosystem management.

TOOLS AND METHODS

Ecosystem-based management involves the use of various tools and methods, including systems approaches, conservation biology and landscape ecology, reserve design, adaptive management, GIS and EIS (environmental information systems), and participatory and collaborative approaches. However, in an ecosystem approach it is not so much the specific tools used that are important, but rather the way they are combined and integrated into an overall process.

A systems approach is integral to ecosystem-based management and, indeed, many environmental analyses foster connections within and between systems, and emphasize the dynamic nature of systems. This approach may be more or less quantitative and provides a framework for the analysis, description, and integration of different kinds of information (Slocombe, 2001). When we think of systems, natural systems often come to mind. Many people, for example, are familiar with the connectivity of the hydrological cycle. Social systems are equally interconnected and complex. Within protected areas there may be different but related human populations, such as residents, employees, and visitors. At the bioregional scale, these relate to local communities and economies; at the national scale they relate to legislation, policies, the media, and so on; and at the international level, treaties and conventions between and among nations can be of central importance. Natural and social systems are intimately connected through all these different scales. And they all permeate what we, at one time, considered to be the inviolate boundaries of the protected area. Knight and Landres (1998) provide a useful collection of papers on the challenges of 'stewardship across boundaries', while Chester (2006) provides a fine summary and a couple of detailed case studies of transboundary, park-centred, ecosystem-based management.

A fundamental premise of ecosystem management is to turn protected area managers from 'boundary thinking', which dominated plans and actions for virtually the first century of the national parks movement, to an understanding of the spheres of influence that affect parks beyond the administrative boundary (Figure 13.2). The difficulty of effecting this change should not be underestimated in terms of either theory or practice. This new 'thought boundary' may be extremely large, and will vary in size according to the particular threat being faced or resource being protected. Ecosystem-based boundaries are often bound to natural units, particularly the home ranges of key species (Beazley, 1999), or to watersheds (see, e.g., Colville and Rozalska, 2003), or to a combination of factors (Ruel, 1998). However, other dimensions of park values are also important and may not be supported by imposing another rigid boundary. Aesthetic threats, for example, may be defined in terms of the 'viewshed' from the park (Dearden, 1988). Park managers may also be able to wield powers of persuasion with neighbouring land users where legislative powers do not exist. In other cases, where ecological values are being degraded, as,

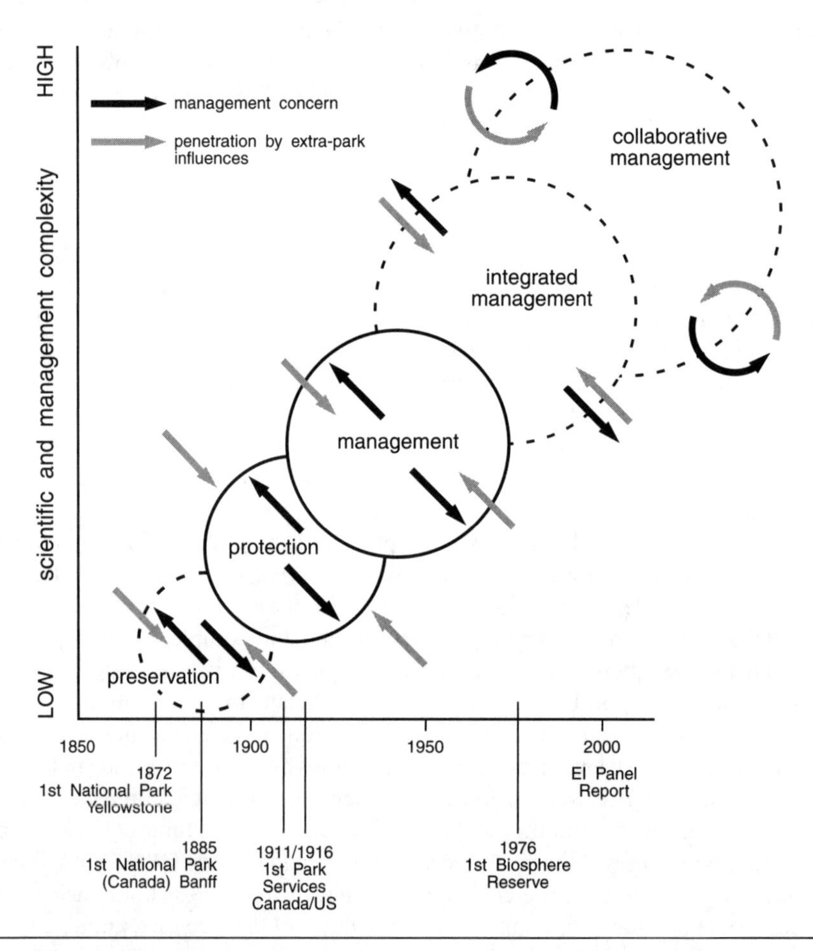

FIGURE 13.2 The evolving role of parks, from isolation to collaboration. The increasing size of the circles reflects the increasing amount of area protected. The dotted boundaries of the parks show more permeable boundaries in earlier and later years.

for example, by air pollution, determining the 'thought boundary' may be much more difficult for the manager who must address the problem (Welch, 2003).

Between these extremes of difficulty arise intermediate problems that park managers may be able to address in more innovative ways. Stresses on ecological integrity caused by visitor activities, where the social systems impact on the natural systems, provide numerous challenges. Since it has proven inadequate simply to deal with these social systems as 'visitors' when people are in the park, outreach beyond the park boundaries to relevant 'visitorsheds', or where the people originate, seems a solution. Such attempts raise awareness among potential visitors about the purposes of national parks, about appropriate and inappropriate activities, and about how visitors can reduce their impacts on the environment (Promaine, 1998).

Conservation biology and landscape ecology have been discussed in Chapters 4 and 5. Together with ideas such as Peterson and Parker's work on hierarchy and scale (1998), and with growing understanding of systems ecology and ecosystem function and processes (see Starzomski and Srivastava, 2003) the rationale for and the analytic tools of ecosystem-based management are developing well. This evolution can be traced in many papers in Samson and Knopff's *Ecosystem Management: Selected Readings* (1996). These theories underlie our perspectives on the effects of the size and spatial relationship of protected areas, and our decisions about which areas ought to be protected and how they should be connected. Schwartz (1999: 100) summarized the development over the previous 10 years of reserve choice criteria, from fine-filter (genes, species, populations) to coarse-filter (communities, habitats, ecosystems, landscapes). He suggests that embracing ecosystem-based management does not mean restricting conservation to large reserves; the key is 'focusing efforts on conserving interactions among species and processes within ecosystems'. However, holistic reserve design that integrates across scales is the desirable modus operandi.

Adaptive management is not a new idea (Holling, 1978), but it has been invigorated through ecosystem management initiatives and assessments of various resource management failures. Adaptive management is essentially management that adopts a systems approach, recognizes ubiquitous change, uncertainty, and complexity in natural and human systems, and accepts the necessity of learning by doing. It calls for the use of science and modelling in determining management actions, experimental management that is carefully monitored to provide feedback, and regular evaluation and revision of management based on what is learned (Chapter 5). It can be a challenge to people and institutions accustomed to certainty, who want to make a decision and move on without monitoring and re-evaluating. And it can be expensive. Still, its iterative process of establishing goals and objectives, developing hypotheses, making decisions and implementing them, and subsequently learning, monitoring, and evaluating is highly suited to ecosystem management and protected areas (Noble, 2004; Murray and Marmorek, 2003).

Geographic information systems have been a park management tool for the past two decades. Used co-operatively across disciplines, GIS can be a powerful means of organizing, presenting, and integrating information about a park and its surroundings. For example, Barnes and Ayles (1998) used GIS to develop a system for identifying and ranking ecosystem stressors at Prince Edward Island National Park. GIS was also used in a study to document landscape change at six of Canada's biosphere reserves (MAB, 2000).

Harshaw and Sheppard (2003) use GIS to assess the impacts of timber harvesting on recreation adjacent to a provincial park in BC.

If a GIS is too expensive to use, data are inadequate, or the process and results of its use are not made public, then GIS benefits will be few or potentially negative, especially if stakeholders feel that the technology is being used to hide information or support unpopular conclusions. Environmental information systems are more comprehensive systems of spatial and non-spatial information, often with a decision-support capability. Rothley et al. (2003), for example, describe the use of multi-objective integer programming to search for optimal solutions in reserve network design. Petersen (1998) provides an interesting example of a GIS/decision-support model for allocating land resources between grizzly bears and humans in Yoho National Park. Although there are relatively few examples of GIS application in protected areas management, initiatives around Yellowstone National Park and Waterton National Park have received some attention (see the comparison in Danby, 1998). Slocombe (2001) addresses information management and integration in more detail.

Participatory and collaborative approaches are increasingly stressed as the solution for making ecosystem-based management work. This may be especially true for protected area-centred and -led initiatives where the public and co-workers may be suspicious of the purposes of the leaders of the exercise. Local landowners, for example, may perceive collaboration as simple coercion by government agencies to gain greater control over their land. Participatory methods and strategies encompass a wide range, from simply informing the public to actively delegating decision-making to its members. Commonly used tools include public meetings, public submission of comments on draft plans, workshops to develop information products for a common base of understanding, facilitated discourse, and discussion. Drawing on the Yellowstone experience, Varley and Schullery (1996) highlight multiple dimensions and complexities of public involvement in ecosystem management, including identifying the public, producing credible information to support public understanding, and developing effective, targeted communication devices.

Daniels and Walker (1996) provide a detailed case study and evaluation of collaborative learning in the Oregon Dunes National Recreation Area (ODNRA) planning process. Collaborative learning emphasizes experiential learning, systemic improvement, and constructive discourse more than one would find in typical public participation programs. The process in ODNRA had three stages: (1) inform stakeholder groups and involve them in process design; (2) provide a common knowledge base about dunes issues, which involves identifying concerns about ODNRA management and generating suggested improvements; (3) organize improvements based on different strategic visions for the ODNRA and then debate improvement sets. Workshops involved four main activities: issue presentations, panel discussion, best and worst views and situation mapping, and individual and small group tasks. The program was quantitatively assessed and found to have improved dialogue among diverse communities, to have integrated scientific and public knowledge, and to have increased rapport, respect, and trust.

In some places collaboration and participation have moved beyond communicating information and participation in meetings. In much of Canada's North, park management is increasingly a matter of co-management among federal, Aboriginal, territorial, and sometimes other local interests, brokered through boards established directly or indirectly via comprehensive land claims settlements (Chapter 14). Few such boards have been in operation for more than a few years, so it is too early to attempt any conclusive assessment. They are, however, changing management and planning methods, particularly to include more consultation and local knowledge in many instances.

ECOSYSTEM-BASED MANAGEMENT IN PRACTICE

Ecosystem-based management programs have been implemented, if not always under that name, at many protected areas. It is arguable that the theory and practice have been more popular in the US than in Canada, but even if true in the past, this is changing. Here, we briefly discuss a number of ecosystem-based park management examples to illustrate the points made above, as well as the practical means and difficulties of implementing such programs. Ecosystem-based management is often much simpler in plan, and more complex in execution, than the theory would suggest.

Waterton Lakes National Park in Alberta and Glacier National Park in Montana (Figure 13.3), and the surrounding areas in Alberta, British Columbia, Montana, and Idaho (often referred to as the Crown of the Continent as it is the meeting place of the Atlantic, Pacific, and Arctic drainages), are among the most discussed examples of biosphere reserves, regional integration, or ecosystem management. The two national parks

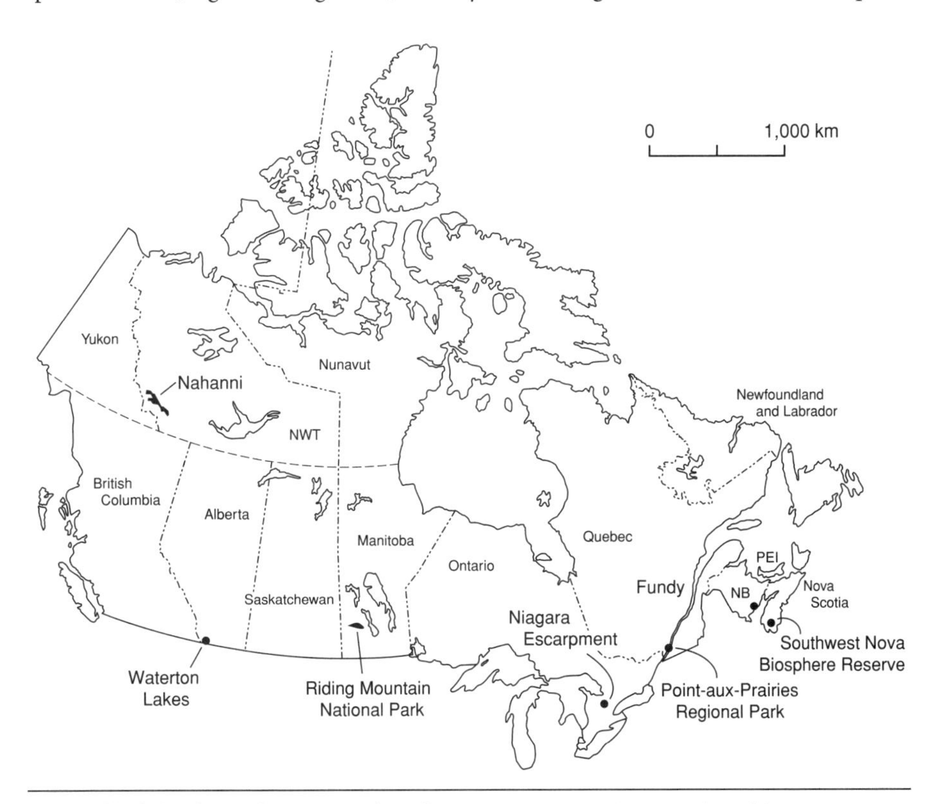

FIGURE 13.3 Sites of ecosystem-based management initiatives mentioned in text.

were jointly commemorated in 1932 as the Waterton–Glacier International Peace Park and received World Heritage designation in 1995. The area boasts an exceptional diversity of flora and fauna due to its varied topography where the prairie meets the mountains. Both geologically and ecologically, it is distinct from the larger mountain park block (Banff, Yoho, Kootenay, and Jasper) to the north. There is a strong influence from Pacific maritime weather systems, as this is the narrowest point of the Rockies, and elements of biota from the Pacific Northwest mingle with those from the interior plains and montane regions.

The complex mix of public and private ownerships and resource development activities surrounding the park matches this biophysical diversity. Alberta and British Columbia manage surrounding lands for multiple use, including extensive gas, oil, forestry, and ranching use. To the north and east of the park, much land is in private ownership, predominantly used for ranching, although small recreational holdings are increasing rapidly. There are also important multiple-use lands administered by the Blood Nation. The international border across the middle allows for interesting comparisons between American and Canadian experiences. Waterton and Glacier were designated as a biosphere reserve by UNESCO in 1979. Although undefined, the zone of co-operation extends about 20 km to the east and north of the national park. This region is mostly privately owned or is leased provincial Crown land. The Waterton Biosphere Association is composed of local ranchers, business people, and three park staff. Membership is open to anyone interested. An annual contribution from Parks Canada of $5,000 further helps to support general administrative costs.

Zinkan (1992) describes the characteristics and success of the Waterton National Park Biosphere Reserve and the origins of the Crown of the Continent initiative during the 1980s. Critical factors included early development of a vision statement and guidelines and identification of elements that would lead to a sustainable strategy. These included local initiative, involvement of all sectors, regional integration rather than reaction to threats, strategies based on sound vision and guidelines, educational activities, information made readily available in an integrated form, a clear research strategy and support, and realigning traditional government organization and scientific specialization. It was felt that, in the long run, political leadership in the form of legislation would be required, and that sustainable development models would become increasingly complex. While the latter is certainly true, no legislation or regulations govern overall management of the zone of co-operation. Waterton was identified (Parks Canada Agency, 2000) as among the best examples of co-operation among diverse stakeholders in national parks: hunters, ranchers, environmentalists, conservationists, and park staff. NGOs with such varied mandates as the Nature Conservancy of Canada and the Rocky Mountain Elk Foundation have worked with park staff and nearby landowners to address conservation, ranching, land development, and other issues. Throughout the 1980s and 1990s, Waterton illustrated the role of particular people and personalities in helping different groups work together. Despite these indicators of success, studies of landscape change in the greater Waterton ecosystem graphically illustrate the scale of the ongoing challenge in maintaining ecological integrity (Stewart et al., 2000). More recently a Crown of the Continent Managers' Partnership has focused on data collection and organization and development of GIS tools to facilitate management.

FIGURE 13.4 Waterton Lakes Biosphere Reserve. Some people might question whether an interpretive sign entitled 'A National Park' should contain as its only picture a modern industrial complex, even if the sign was paid for by Shell. *Photo: P. Dearden.*

Dolan and Frith (2003) have outlined what they feel to be the most important lessons learned from this oldest biosphere reserve in Canada:

- The biosphere concept is realized through the collective efforts of many individuals and organizations, often working collaboratively and independently towards the goals of the Man and the Biosphere program (MAB). It is rarely achieved through any one administrative structure but may be influenced through the leadership of specific individuals, government, or non-government institutions.
- The current model of biosphere reserve (i.e., protected core, buffer, and transition zones) does not promote an ecosystem-based approach. Rather, the model is often built on the concept of protected areas, which ultimately leads to an insular approach to natural ecosystems.
- Promoting the biosphere reserve as a 'jurisdiction' or 'administrative entity' is generally problematic and ineffective. The existing land base is already managed by different jurisdictions and any approach must support collaboration among jurisdictions rather than 'threaten' a particular administrative mandate or jurisdiction.
- An administrative structure for a biosphere reserve is an important element in implementing the Man and the Biosphere program. However, the focus of an administrative structure is best directed at supporting the community in the use of science, education/awareness, and best practices within the local communities.
- To maintain a successful administrative biosphere structure, a substantive and consistent level of fiscal resources is essential and community participation must ensure a diversity of interests.

The Southwest Nova Biosphere Reserve is an example of a more recent biosphere reserve in a different environment from Waterton. It crosses five counties in southwestern Nova Scotia and covers 13,770 km^2, with a population of about 99,500. Kejimkujik National Park and Nova Scotia's Tobeatic Wilderness Area form the core of the biosphere reserve, which also includes the Shelburne River, a Canadian heritage river. Initial academic and scientific interest in a biosphere reserve in this part of Nova Scotia was unsuccessful during the 1980s. Renewed interest in the 1990s, with a broader community and government base, led to establishment of the Southwest Nova Biosphere Reserve Association in 1998, a successful nomination in 2001, and final support from all levels of government in 2003. Scientific research had long been active in the region and an even stronger network emerged from, and played a role in, the designation. This has now been formalized in the Mersey Tobeatic Research Institute (MTRI), which has a base just outside Kejimkujik National Park. This is a fine example of the mutually collaborative and catalytic role of scientific research and monitoring in ecosystem-based planning and management of a protected area (see SWNBRA, 2007; Rehmann, 2007).

Another protected area that has stressed monitoring programs, and built on its biosphere reserve status, is the Niagara Escarpment Planning Area in southern Ontario. Its approach to improving management in the greater ecosystem has stressed a monitoring program tied to regional planning along the escarpment, developing stewardship, trust, and research programs, and creating new organizational initiatives involving communities and stakeholders (Ramsay and Whitelaw, 1998). Whitelaw and Hamilton (2003) provide a description of how governance processes and actors have changed as governance has evolved more towards a multi-party, community-based structure.

While many ecosystem-based management programs are for large wilderness parks and regions, this need not be the case. Urban protected areas are gaining much atten-

BOX 13.3 Biosphere Reserves

The Biosphere Reserve Program was started by the United Nations Educational, Scientific and Cultural Organization (UNESCO) as part of the Man and the Biosphere (MAB) program in 1968 as an aid to achieving MAB objective of striking a balance among conserving biodiversity, encouraging economic and social development, and preserving cultural values. Biosphere reserves are areas of terrestrial and coastal/marine ecosystems where, through appropriate zoning patterns and management mechanisms, the conservation of ecosystems and their biodiversity is ensured. Each biosphere reserve has a conservation function: to contribute to the conservation of landscapes, ecosystems, species and genetic variation; a development function: to foster economic and human development that is socially and ecologically sustainable; and a logistic function: to provide support for research, monitoring, education, and information exchange related to local, national, and global issues of conservation and development.

Typically, reserves should be divided into three zones:

- a *core zone* of strictly protected areas, with very little human influence, that serve as conservation areas for biodiversity and where natural changes in representative ecosystems can be monitored;
- a *buffer zone* surrounding the core zone where only low-impact activities are allowed, such as research, environmental education, and recreation; and
- a *transition zone* where sustainable use of resources by local communities is encouraged and these impacts can be compared to zones of greater protection.

How these zones are delimited is a national affair and the UNESCO designation does not add any legislative requirements, although some countries, such as Mexico, have enacted their own legislation to try to make the reserves more effective. There are now over 530 designated biosphere reserves in the world. However, it is a challenging task to find one that successfully implements all aspects of the program. In Canada there are 13 reserves, located at Clayoquot Sound and Mount Arrowsmith, BC; Waterton, Alberta; Redberry Hills, Saskatchewan; Riding Mountain, Manitoba; Niagara Escarpment, Long Point, Thousand Islands–Frontenac Arch, and Georgian Bay Littoral, Ontario; Charlevoix, Lac Saint-Pierre, and Mont St Hilaire, Quebec; and Southwest Nova, Nova Scotia. The Canadian Biosphere Reserve Association (CBRA) and Canadian Biosphere Research Network (CBRN) have recently been revitalized and are doing much to generate more attention for biosphere reserves (Francis and Whitelaw, 2004).

tion for ecological and socio-economic reasons (e.g., Trzyna, 2005) and deserve more from an ecosystem-based management perspective. Lajeunesse et al. (1995) present an ecosystem management approach for small, protected natural areas subject to intense user pressures, in both urban and suburban locations. Stages in the approach include initial ecological evaluation, followed by management interventions, which build on a sensitivity map created during the ecological evaluation stage, definition of ecological and social objectives and consultation, largely related to vegetation succession, and a follow-up monitoring scheme. They have applied the approach to Pointe-aux-Prairies Regional Park in Montreal, Quebec. This is an active, science-based management approach that, with consultation and monitoring, could lead to adaptive management. Lajeunesse et al. (2003) also address ecosystem-based management in PEI National Park, a small, intensively used national park with associated use conflicts.

A strong scientific background is essential to provide accurate advice for ecosystem-based initiatives. The Greater Fundy Ecosystem Group includes over 30 researchers from several universities, government departments, and industry who work closely with the Fundy Model Forest (FMF) and seek to provide the ecological background for sustainable forestry practices. This is especially relevant as the FMF is adjacent to Fundy National Park and a main goal is to complement the biodiversity conservation objectives of the park (Woodley and Forbes, 1995). In particular, research aims to:

FIGURE 13.5 Ecosystem-based management: the reality. Logging on the borders of Pacific Rim National Park Reserve. The reality is that managers often have little leverage to influence activities outside the administrative boundaries of the protected area. *Photo: P. Dearden.*

- identify strategies to maintain viable populations of native species;
- quantify species–habitat relationships;
- examine ecological stressors;
- identify operational management options.

The Group has produced a series of forest management guidelines for protecting native biodiversity. In so doing they used a combined top-down/coarse-filter approach and a bottom-up/fine-filter approach. The coarse-filter approach facilitates planning of larger arrangements of communities, taking into account factors such as composition, size, adjacency, and age-class distribution. The needs of specific species and species groups of interest can be taken into account in more detail by the operation of the fine-filter approach. This combination has resulted in some very specific management prescriptions that should aid considerably in biodiversity conservation in the greater Fundy ecosystem (Forbes et al., 1999)—a good example of the application of a science-based approach to ecosystem-based management to further protected area goals. However, as shown in Box 13.4, good social science is also necessary in order to understand local viewpoints and actually implement many of the practices required.

CONCLUSIONS

The late 1990s saw several efforts to learn from the previous decade's experience with ecosystem management. Some of this reflection has addressed the fundamental ideas and principles of ecosystem management. Stanley (1995), for example, has argued that ecosystem management symbolizes the arrogance of humanism and the doctrine of final causes, i.e., that nature is there primarily for human benefit. This is one reason why some authors have been careful to use the term 'ecosystem-based management' to make the point that we are managing human activities, not the natural environment. Stanley's other point is that ecosystem management cannot deliver what it promises due to issues of complexity, uncertainty, and control. This is where most authors see the need for adaptive, precautionary management.

Other authors have argued that ecosystem management is not a panacea for successful conservation, that scientific information is just one part, and perhaps not the most important part, of what is needed for good land and resource management. Terms like sustainability and integrity are variable and must be well defined, and management cannot escape being location-specific. The lesson here is that consultation, participation, and consideration of local people and their interests must also be a major part of management (Lackey, 1998). In a similar vein, Brunner and Clark (1997) sought to address the problem of ecosystem management principles being unsatisfactory for practical purposes. Three major approaches were evaluated: clearer ecosystem management goals, a better scientific foundation for management decisions, and comparative appraisal of current practices. They argued the first two were not enough as the problem is not primarily technical. Instead, they advocate and describe a practice-based approach, drawing on adaptive management, prototyping, and a process for inventorying and appraising and learning from decisions nationally.

BOX 13.4 Conflicts, Corridors, and Collaboration: Elk and Wolves between Duck Mountain Provincial Park and Riding Mountain National Park in Manitoba

For many years, conservation biologists assumed that corridors of continuous native vegetation were necessary for the movements of animals through the hostile matrix of human-dominated landscapes. It is now becoming evident that many large mammals can move through the matrix effectively and may even thrive in these human-modified environments (Nixon et al., 2007). However, wildlife survival is often based more on the attitudes and actions of people living in the matrix than on the habitat itself, where local residents largely determine the quality of the habitat.

Riding Mountain National Park (RMNP) in southwestern Manitoba has often been viewed as an extreme example of an isolated protected area completely surrounded by agriculture, despite a short 27-km gap between the park and Duck Mountain Provincial Forest and Provincial Park (DMPP). Of the land base between these protected areas, 88 per cent is privately owned farmland. Efforts in Manitoba to facilitate connections between RMNP and DMPP have included 'the Mountain to Mountain initiative' by the Canadian Parks and Wilderness Society and a conservation easement program of the Nature Conservancy of Canada that pays landowners to conserve natural habitats. While conservationists view these connections as a win-win situation, this is not always the case for the people living in the corridor who must also live with the impacts of wildlife that inevitably occur.

Elk are highly valued by local hunters as a food source and for their trophy value, while visitors come from throughout the province to harvest Riding Mountain and Duck Mountain elk. Tourists also come from around the world to view elk, which are the most commonly used symbol of RMNP. Farmers living within the corridor face important impacts from elk, including direct economic damage to crops that typically total more than $30,000 each year. This area also has the highest rate of bovine tuberculosis disease infection in cattle herds within Canada over the last decade, likely due to contacts between the infected elk population in and around RMNP and livestock. From the farmers' perspective, much of the blame for disease and wildlife damage is placed on RMNP and DMPP, which are seen as the primary source of elk and wolves in the region (Brook and McLachlan, 2006).

Wolves are the primary predators of elk, deer, and moose in the region and elk are their preferred food in RMNP (Carbyn, 1980). RMNP and DMPP play a critical role in conserving the predator–prey relationship between wolves and elk. Concern regarding disease transmission between elk and cattle means that the importance of the predator–prey relationship extends beyond the protected area boundaries (Stronen et al., 2007). Livestock predation is a direct expenditure to farmers in the region, but the risk of disease transmission also represents a cost. Carnivores such as wolves may influence the number and distribution of ungulates, and could play an important role in reducing the spread of diseases such as bovine TB between ungulates and livestock. Farmers, therefore, may benefit to some extent from having wolves as predators of elk on private lands. This creates a complex interrelationship among humans, livestock, wild ungulates, and predators in the areas between RMNP and DMPP.

The wolf population in RMNP has been estimated at approximately 75 animals in recent years (Parks Canada, unpublished data) and is vulnerable to factors such as disease and human-caused mortality. The genetic makeup of RMNP wolves differs from those in the Duck Mountain park (A. Stronen, unpublished data), suggesting low gene flow between the two areas. Wolves in both locations have high genetic variation but there is a risk of isolation and inbreeding over the long term without interchange between the two protected areas. The more variation small populations are able to maintain in their gene pool, the better the chances of surviving environmental changes such as outbreaks of canine distemper virus, which has caused wolf mortality in RMNP and is found in dogs adjacent to it. Almost every wolf pack in the park has territories including the RMNP border, which increases chances of contact with coyotes and dogs and transmission of disease.

Elk and wolves should both be conserved and managed as part of a functioning ecosystem and for their intrinsic value as components of the local fauna. In addition, the Riding Mountain–Duck Mountain situation may provide an opportunity to reconsider the importance of maintaining long-term functional predator–prey systems, and remind us that these relationships do not come to an abrupt halt upon meeting the boundary of a protected area. By supporting a healthy predator–prey relationship between wolves and elk both inside and, as far as possible, outside the federal and provincial protected areas in the region, we may at the same time reduce impacts of bovine tuberculosis and other infectious diseases on local farm operations. Thus, the development and maintenance of the 'corridor' required between RMNP and DMPP may largely depend on a willingness to accept long-term thinking about human–wildlife relationships.

Contributed by Ryan K. Brook, Environmental Conservation Lab, Clayton H. Riddell Faculty of Environment, Earth, and Resources, University of Manitoba, and Astrid V. Stronen, Department of Biology, University of New Brunswick

Slocombe (1998a) sought to draw practical lessons from experience with ecosystem management and related programs in Canada and Australia, revisiting his 1993 articles (Table 13.2). Building on that article, Slocombe (1998b) presented a theoretical discussion of the nature of goals for ecosystem-based management, both substantive (what we seek to achieve) and procedural (how we seek to get there). In practice, most planners and managers of ecosystems and economies continue to pursue traditional goals and targets, which miss many desirable characteristics of ecosystem-based management goals. Substantive goals can be grouped according to their relationship to system structure, organization, and process/dynamics, and their disciplinary or subsystemic breadth.

Procedural goals include four high-level goals: develop consensus; develop understanding of the system; implement a framework for planning and management; and do specific things that make a difference. A parallel, linked system of substantive and procedural goals at different levels of complexity and disciplinarity ideally would be needed to facilitate ecosystem-based management.

BOX 13.5 Ecosystem-based Management: The Reality

Ecosystem-based management makes so much sense it is difficult to find anyone who would argue against it as a means to improve the protection of parks and similar reserves. If this is the case, why are there so few examples of comprehensive ecosystem-based management actually in operation? One of the main impediments, especially in Canada, is that national parks, a federal jurisdiction, usually have limited influence on the provincial land bases surrounding the parks. And provincial parks, generally with a much weaker protection mandate and more limited resources, are rarely in a position to undertake ecosystem-management exercises. A major challenge for protected area managers, therefore, is to gain greater influence over areas over which they have little, if any, jurisdiction. There are several ways to extend influence. For example:

1. Protected area managers should understand the political decision-making system on lands outside their jurisdiction, so they will know where, when, and how they can formally intercede to protect park values. Parks Canada staff should routinely participate in local and regional planning initiatives with agencies that have jurisdiction in greater park ecosystems. Furthermore, it should not be forgotten that the federal government has jurisdiction over fisheries, endangered species, migratory birds, long-range air pollution, navigable waters, and related environmental impact assessments. All these existing powers should be mobilized to support ecosystem-based park management.
2. It is important to build formal and informal links with outside agencies, industry, local communities, First Nations, and NGOs. Increasingly, collaborative partnerships that bring together multiple sectors are playing a role in ecosystem-based management of parks and protected areas in Canada and elsewhere. This is part of improving governance for local conservation and development (Pollock, 2003).
3. Good communications skills are essential to inform other stakeholders of park concerns. Of special importance is the ongoing challenge of raising park literacy among the general public through interpretation programs so that people are aware of the roles that parks play in the landscape (Chapter 8).
4. Canadians love their parks. There is a huge resource of public goodwill directed towards parks. Sometimes this goodwill capital has to be drawn on when the formal and informal contacts mentioned above are inadequate to ensure ecological integrity through ecosystem-based management. Social capital (Sparkes, 2003) may be as important as ecological capital.
5. Information and communication technologies, and their products, are increasingly important for organizing and analyzing information, exploring alternatives, and presenting information and options to the public and decision-makers.

TABLE 13.2 Practical Lessons for Making Ecosystem-based Management Work

Defining Management Units	Developing Understanding	Creating Planning and Management Frameworks
Use meaningful units.	Describe and interpret many dimensions of the ecosystem.	Keep it simple; try not to layer new levels and organizations onto existing ones.
Be flexible; use multiple ways of defining units.	Make information available within and outside ecosystem.	Get top-level commitment and leadership.
Build on but don't be constrained by existing units.	Use local and traditional knowledge.	Implement close to the ground and ensure there are some immediate, visible benefits and products.
Ensure it is an operational unit in at least some way.	Be practical; when resources are limited, focus on understanding that would make a difference.	Focus on management processes, information flow, and planning and setting targets.
Maintain the interest of higher administrative levels in lower and newer units by communication, involvement.	When you've got information, use it: analyze, map, simulate, discuss.	Maintain flexibility, and ensure reviews to foster adaptation.

Source: After Slocombe (1998a).

Grumbine (1997), too, revisited his earlier article to address the question of why ecosystem management is still vague or difficult. He identified several issues: the politics of definition—who is involved in management and why; changing goals from resource extraction to ecosystem protection; the need for contextual thinking; and the lack of skills in problem definition. He found that his 10 themes, outlined above, were still accurate and relevant. In addition, Grumbine drew on the organizational change literature (e.g., Westley, 1995) to call attention to the sort of institutional changes needed to foster ecosystem management, something that other writers also are increasingly stressing: e.g., reorganization on ecosystem-based management themes, staff training in ecosystem-based management, teamwork, communication, consensus-building, employee incentives, staff training across disciplines, stewardship programs, independent audit and evaluation. He also placed greater emphasis on adaptive management topics, such as prototyping, monitoring, dealing with complex problems and multiple causes, and recognizing that there is no simple fix.

There has been relatively little systematic assessment of ecosystem management programs. An assessment of relatively new programs in Mount Revelstoke and Glacier national parks reached some initial conclusions (Feick, 2003). The programs' basis is a two-way flow of scientific information between the parks and surrounding land and resource management agencies. The ecosystem management programs have grown from trying to get a seat at the table for the parks to becoming invited partners in large, expensive, multi-agency programs. The results of the evaluation show the utility of scientific and technical information, but also the importance of other factors in ecosystem-based management decision-making.

Stephenson and Zorn (1997) reported on evaluation of the ecosystem management program in the Ontario region of Parks Canada. The report stressed how this program was successfully implemented through restructuring in-park conservation programs and a bottom-up change in philosophies and approaches that complemented top-down Parks Canada policies and programs (see also Zorn et al., 2001).

Quinn and Theberge (2003) reported on a national survey to assess ecosystem-based management in Canada. Canadian EBM was found to have fewer grassroots initiatives and formal institutional provisions than in the US. In Canada, the forestry sector has been an EBM leader, and although many agencies have incorporated some recognition of EBM approaches there are few consistencies of meaning, and formal incorporation into legislation has occurred, for the most part, only at the federal level. There are wide variations in adoption across the provinces, and a notable lack of human dimensions research compared to natural science. They identified six changes that would make a difference:

- more resources;
- more comprehensive participation;
- common vision/goals;
- shared land ethic;
- public involvement;
- greater influence and authority.

If one can conclude anything about ecosystem-based management, it might be that although science and information are critical foundations for good management, making ecosystem management effective and durable requires additional effort. Much greater attention to local and distant benefits, to inter- and intra-organizational issues, and to developing real collaboration with stakeholder groups is required. Collaboration is a rapidly developing area, and Halpenny et al. (2003) identify several priorities from a study of several northern Canadian parks and two American parks: defining roles and responsibilities; fostering communication; establishing mechanisms for collaborative decision-making, such as co-operative management boards or other structures; and development and implementation of projects to foster and demonstrate collaboration on key themes, such as ecological integrity or sustainability and local or traditional knowledge.

It is also true, of course, that park and regional and resource management will continue to evolve, and that other approaches will influence and perhaps eventually subsume ecosystem-based management. One such trend may well be bioregional approaches, which emphasize even larger regions, longer time scales, complex protected area networks, and stronger socio-cultural content (Chapter 1).

There is rising interest in regional approaches to protected areas and regional integration. Some of this interest reflects renewed government interest in park visitors and their experience, as well as the effects of parks on people in and around them. Some of this emerges from the 'new paradigm' of parks and protected areas that calls for protected area management that is less protectionist and more inclusive of local people, contentious as that may be (cf. Phillips, 2003; Locke and Dearden, 2005). Ultimately, this interest in regional approaches, and in strengthening a range of dimensions of ecosystem-based management, will foster exploration of a useful range of new approaches that can only strengthen conservation in the long run (see, e.g., Hanna et al., 2007).

REFERENCES

Agee, J.K., and D.R. Johnson, eds. 1988. *Ecosystem Management for Parks and Wilderness.* Seattle: University of Washington Press.

Barnes, S., and P. Ayles. 1998. 'GIS as a decision-making tool in the ecosystem conservation planning process', in Munro and Willison (1998: 705–11).

Beazley, K. 1999. 'Permeable boundaries: Indicator species for trans-boundary monitoring at Kejimkujik National Park', in *Protected Areas and the Bottom Line,* Proceedings of the 1997 Conference of the Canadian Council on Ecological Areas. Information Report MX 205E/F. Ottawa: Canadian Forest Service, 119–35.

Brook, R.K., and S.M. McLachlan. 2006. 'Factors influencing farmers' concerns associated with bovine tuberculosis in wildlife and livestock around Riding Mountain National Park, Manitoba, Canada', *Journal of Environmental Management* 80: 156–66.

Brunner, R.D., and T.W. Clark. 1997. 'A practice-based approach to ecosystem management', *Conservation Biology* 11, 1: 48–58.

Caldwell, L.K. 1970. 'The ecosystem as a criterion for public land policy', *Natural Resources Journal* 10: 203–21.

Carbyn, L.N. 1980. *Ecology and Management of Wolves in Riding Mountain National Park, Manitoba.* Final Report, Large Mammal System Studies, Report No. 10, Sept. 1975–Mar. 1979. Edmonton: Canadian Wildlife Service.

Chester, C.R. 2006. *Conservation across Borders: Biodiversity in an Interdependent World.* Washington: Island Press.

Colville, D., and K. Rozalska. 2003. 'Analyzing landscapes for ecological monitoring in southwestern Nova Scotia', in Munro et al. (2003).

Community and Municipal Affairs, National Parks Branch. 1985. *Regional Integration of National Parks.* Ottawa, Dec.

Danby, R.K. 1998. 'Regional ecology of the St. Elias mountain parks: A synthesis with management implications', MES thesis, Wilfrid Laurier University.

Daniels, S.E., and G.B. Walker. 1996. 'Collaborative learning: Improving public deliberation in ecosystem-based management', *EIA Review* 16: 71–102.

Dearden, P. 1988. 'Protected areas and the boundary model: Meares Island and Pacific Rim National Park', *Canadian Geographer* 32: 256–65.

——— and S. Doyle. 1997. 'External threats to Pacific Rim National Park Reserve, BC', in C. Stadel, ed., *Themes and Issues of Canadian Geography II.* Salzburg, Austria: Salzburger Geographische Arbeiten, 121–36.

———, M. Bennett, and J. Johnston. 2005. 'Trends in global protected area governance, 1992–2002', *Environmental Management* 36, 1: 89–100.

Dolan, B., and L. Frith. 2003. 'The Waterton Biosphere Reserve—Fact or fiction?', in Munro et al. (2003).

Ecological Society of America. 1995. *The Scientific Basis for Ecosystem Management: An Assessment.* Washington: ESA.

Feick, J. 2003. 'Does "good" science lead to "better" land use decisions?', in Munro et al. (2003).

Forbes, G., S. Woodley, and B. Freedman. 1999. 'Making ecosystem-based science into guidelines for ecosystem-based management: The Greater Fundy ecosystem experience', *Environments* 27, 3: 15–23.

Francis, G.R., A.P.L. Grima, H.A. Regier, and T.H. Whillans. 1985. *A Prospectus for the Management of the Long Point Ecosystem.* Ann Arbor, Mich.: Great Lakes Fishery Commission Technical Report 43.

——— and G.S. Whitelaw, eds. 2004. Special Issue: 'Biosphere reserves in Canada', *Environments: A Journal of Interdisciplinary Studies* 32, 3.

Grumbine, R.E. 1994. 'What is ecosystem management?', *Conservation Biology* 8, 1: 27–38.

———. 1997. 'Reflections on "What is ecosystem management?"', *Conservation Biology* 11, 1: 41–7.

Halpenny, E., M.E. Bowman, D. Aubrey and P.F.J. Eagles. 2003. 'Co-operative management in national parks', in Munro et al. (2003).

Hanna, K.S., D. Clark, and D.S. Slocombe, eds. 2007. *Transforming Parks and Protected Areas: Management and Governance in a Changing World.* London: Routledge.

Harshaw, H.W., and S.R.J. Sheppard. 2003. 'Assessing timber harvesting impacts to recreation in areas adjacent to parks and protected areas: An example from British Columbia', in Munro et al. (2003).

Hartig, J.H., M.A. Zarull, T.M. Heidtke, and H. Shah. 1998. 'Implementing ecosystem-based management: Lessons from the Great Lakes', *Journal of Environmental Planning & Management* 41, 1: 45–75.

Holling, C.S., ed. 1978. *Adaptive Environmental Assessment and Management.* Chichester: Wiley.

Janzen, D.H. 1983. 'No park is an island: Increase in interference from outside as park size decreases', *Oikos* 41: 402–10.

Jensen, M.E., P. Bourgeron, R. Everett, and I. Goodman. 1996. 'Ecosystem management: A landscape ecology perspective', *Water Resources Bulletin* 32, 2: 1–14.

Knight, R.L., and P.B. Landres. 1998. *Stewardship across Boundaries.* Washington: Island Press.

Lackey, R.T. 1998. 'Seven pillars of ecosystem management', *Landscape and Urban Planning* 40: 21–30.

Lajeunesse, D., G. Domon, P. Drapeau, A. Cogliastro, and A. Bouchard. 1995. 'Development and application of an ecosystem management approach for protected natural areas', *Environmental Management* 19, 4: 481–95.

———, R. Hawkins, L. Thomas, P. Ayles, and P. McCabe. 2003. 'Coastal ecosystem management: A battle of conflicting elements', in Munro et al. (2003).

Landry, M., V.G. Thomas, and T.D. Nudds. 2001. 'Sizes of Canadian national parks and the viability of large mammal populations: Policy implications', *George Wright Forum*: 18: 13–23.

Langholz, J.A. 2003. 'Privatising conservation', in S.R. Brechin, P.R. Wilshusen, C.L. Fortwangler, and P.C. West, eds. 2003. *Contested Nature: Promoting International Biodiversity with Social Justice in the Twenty-first Century.* Albany: State University of New York Press, 117–37.

Lessard, G. 1998. 'An adaptive approach to planning and decision-making', *Landscape and Urban Planning* 40: 81–7.

Locke, H., and P. Dearden. 1987. 'Economic development and threats to national parks: A preliminary analysis', *Environmental Conservation* 15: 151–6.

——— and ———. 2005. 'Rethinking protected area categories and the new paradigm', *Environmental Conservation* 32: 1–10.

MacKenzie, S.H. 1996. *Integrated Resource Planning and Management: The Ecosystem Approach in the Great Lakes Basin.* Washington: Island Press.

Man and the Biosphere Committee (MAB). 2000. *Landscape Changes at Canada's Biosphere Reserves.* Toronto: Environment Canada.

Morgan, L., P. Etnoyer, T. Wilkinson, H. Hermann, F. Tsao, and S. Maxwell. 2003. 'Identifying priority conservation areas from Baja California to the Bering Sea', in Munro et al. (2003).

Munro, N.W.P., P. Dearden, T.B. Herman, K. Beazley, and S. Bondrup-Nielsen, eds. 2003. *Making Ecosystem-based Management Work: Connecting Managers and Researchers.* Wolfville, NS: SAMPAA. At: <www.sampaa.org>.

——— and J.H.M. Willison. 1998. *Linking Protected Areas with Working Landscapes*, Proceedings of the Third International Conference on Science and Management of Protected Areas, Wolfville, NS.

Murray, C., and D. Marmorek. 2003. 'Adaptive management: A science-based approach to managing ecosystems in the face of uncertainty', in Munro et al. (2003).

Nixon, C.M., et al. 2007. 'White-tailed deer dispersal in an agricultural environment', *American Midland Naturalist* 157: 212–20.

Noble, B.F. 2004. 'Applying adaptive environmental management', in B. Mitchell, ed., *Resource and Environmental Management in Canada: Addressing Conflict and Uncertainty*, 3rd edn. Toronto: Oxford University Press, 442–66.

Parks Canada. 1994. *Guiding Principles and Operational Policies.* Ottawa: Supply and Services Canada.

———. 1998. *State of the Parks 1997.* Ottawa: Minister of Public Works and Government Services Canada.

Parks Canada Agency. 2000. *Unimpaired for Future Generations? Protecting Ecological Integrity with Canada's National Parks*, vol. 1: *A Call to Action*; vol. 2: *Setting a New Direction for Canada's National Parks.* Report of the Panel on the Ecological Integrity of Canada's National Parks. Ottawa.

Petersen, D. 1998. 'Allocation of land resources between competing species—Humans and grizzly bears in the Lake O'Hara area of Yoho National Park', in Munro and Willison (1998: 478–91).

Peterson, D.L., and V.T. Parker, eds. 1998. *Ecological Scale: Theory and Applications.* New York: Columbia University Press.

Phillips, A. 2003. 'Turning ideas on their head: The new paradigm for protected areas', *George Wright Forum* 20: 8–32.

Ponech, C. 1997. 'Attitudes of area residents and various interest groups towards the Riding Mountain National Park wolf population', Master's thesis, University of Manitoba.

Promaine, R.H. 1998. 'Applying ecosystem management principles to public education: A case study of Pukaskwa National Park', in Munro and Willison (1998: 633–42).

Quinn, M.S., and J.C. Theberge. 2003. 'Ecosystem-based management in Canada: Trends from a national survey and relevance to protected areas', in Munro et al. (2003).

Ramsay, D., and G. Whitelaw. 1998. 'Biosphere reserves and ecological monitoring as part of working landscapes: The Niagara Escarpment Biosphere Reserve experience', in Munro and Willison (1998: 295–307).

Ripple, W.J., and R.L. Beschta. 2004. 'Wolves and the ecology of fear: Can predation risk structure ecosystems?', *BioScience* 54, 8: 755–66.

Rehmann, Sami. 2007. 'Examining place-based governance principles in two Atlantic Canada protected areas', MES thesis, University of Waterloo.

Rothley, K.D., C.N. Berger, C. Gonzalez, E.M. Webster, and D.I. Rubenstein. 2003. 'Decision-support tools for identifying ecosystem reserves', in Munro et al. (2003).

Ruel, M. 1998. 'The Greater Kouchibouguac Ecosystem Project', in *Protected Areas and the Bottom Line*. Proceedings of the 1997 Conference of the Canadian Council on Ecological Areas. Information Report MX 205E/F. Ottawa: Canadian Forest Service, 136–9.

Samson, F.B., and F.L. Knopf, eds. 1996. *Ecosystem Management: Selected Readings*. New York: Springer-Verlag.

Schwartz, M.W. 1999. 'Choosing the appropriate scale of reserves for conservation', *Annual Review of Ecological Systems* 30: 83–108.

Slocombe, D.S. 1993a. 'Environmental planning, ecosystem science, and ecosystem approaches for integrating environment and development', *Environmental Management* 17, 3: 289–303.

———. 1993b. 'Implementing ecosystem-based management', *BioScience* 43, 9: 612–22.

———. 1998a. 'Lessons from experience with ecosystem management', *Landscape and Urban Planning* 40, 1–3: 31–9.

———. 1998b. 'Defining goals and criteria for ecosystem-based management', *Environmental Management* 22, 4: 483–93.

———. 2001. 'Integration of biological, physical, and socio-economic information', in M. Jensen and P. Bourgeron, eds, *A Guidebook for Integrated Ecological Assessment Protocols*. New York: Springer-Verlag, 119–32.

Southwest Nova Biosphere Reserve Association. 2007. The Southwest Nova Biosphere Reserve (SNBR). At: <www.snbra.ca/snbr.htm>.

Sparkes, J. 2003. 'Social capital as a dimension of ecosystem-based management', in Munro et al. (2003).

Stanley, T.R., Jr. 1995. 'Ecosystem management and the arrogance of humanism', *Conservation Biology* 9, 2: 255–62.

Starzomski, B.M., and D.S. Srivastava. 2003. 'What science tells us about ecosystems for ecosystem-based management', in Munro et al. (2003).

Stephenson, W.R., and P. Zorn. 1997. 'Assessing the ecosystem management program of St. Lawrence Islands National Park, Ontario, Canada', *George Wright Forum* 14, 4: 51–64.

Stewart, A., A. Harries, and C. Stewart. 2000. 'Waterton Biosphere Reserve landscape change study', in MAB (2000: 13–20).

Stronen, A., R.K. Brook, P.C. Paquet, and S.M. McLachlan. 2007. 'Farmer attitudes toward wolves: Implications for the role of predators in managing disease', *Biological Conservation* 135: 1–10.

Swinnerton, G.S., and S.G. Otway. 2003. 'Collaboration across boundaries—research and practice: Elk Island National Park and the Beaver Hills, Alberta', in Munro et al. (2003).

Tate, D. 2003. 'Expanding Nahanni National Park Reserve: The contributions of research, consultation and negotiation', in Munro et al. (2003).

Trzyna, T.C., ed. 2005. *The Urban Imperative: Urban Outreach Strategies for Protected Area Agencies*. Sacramento, Calif.: CIPA Publication 109.

Vallentyne, J.R., and A.M. Beeton. 1988. 'The "ecosystem" approach to managing human uses and abuses of natural resources in the Great Lakes Basin', *Environmental Conservation* 15, 1: 58–62.

Vandermeulen, H. 2003. 'Ecosystem-based management: The integrated management framework under Canada's Ocean Act', in Munro et al. (2003).

Varley, J.D., and P. Schullery. 1996. 'Reaching the real public in the public involvement process: Practical lessons in ecosystem management', *George Wright Forum* 13, 4: 68–75.

Welch, D. 2003. 'Atmospheric science and air issues in Canada's national parks, 2001', in Munro et al. (2003).

Westley, F. 1995. 'Governing design: The management of social systems and ecosystems management', in L.H. Gunderson et al., eds, *Barriers and Bridges to the Renewal of Ecosystems and Institutions.* New York: Columbia University Press, 391–427.

Whitelaw G.S., and P.F.J. Eagles. 2007. 'Planning for long, wide conservation corridors on private lands in the Oak Ridges Moraine, Ontario, Canada', *Conservation Biology* 21: 675–83.

——— and J. Hamilton. 2003. 'Evolution of Niagara Escarpment governance', in Munro et al. (2003).

Wiersma, Y.F., T.D. Nudds, and D.H. Rivard. 2004. 'Models to distinguish effects of landscape patterns and human population pressures associated with species loss in Canadian national parks', *Landscape Ecology* 19: 773–86.

——— and D.L. Urban. 2005. 'Beta-diversity and nature reserve system design: A case study from the Yukon', *Conservation Biology* 19: 1262–72.

Wilcove, D.S., and R.B. Blair. 1995. 'The ecosystem management bandwagon', *TREE* 10, 8: 345.

Woodley, S. 1993. 'Monitoring and measuring ecosystem integrity in Canada's national parks', in S. Woodley, J. Kay, and G. Francis, eds, *Ecological Integrity and Management of Ecosystems.* Boca Raton, Fla: St Lucie Press, 155–76.

——— and G. Forbes. 1995. 'Ecosystem management and protected areas', in T.B. Herman, S. Bondrup-Nielsen, J.H.M. Willison, and N.W.P. Munro, eds, *Ecosystem Monitoring and Protected Areas.* Amsterdam: Elsevier, 50–8.

Yaffee, S.L. 1999. 'Three faces of ecosystem management', *Conservation Biology* 13, 4: 713–25.

Zinkan, C. 1992. 'Waterton Lakes National Park moving towards ecosystem management', in J.H.M. Willison, ed., *Science and the Management of Protected Areas.* Amsterdam: Elsevier, 229–32.

Zorn, P., W. Stephenson, and P. Grigoriew. 2001. 'Ontario national parks ecosystem management program and assessment process for Ontario national parks', *Conservation Biology* 15, 2: 353–62.

KEY WORDS/CONCEPTS

adaptive management
biosphere reserve
boundary thinking
co-management
conservation biology
ecological integrity
ecosystem
ecosystem management
geographic information systems (GIS)
integrated resource management
landscape ecology
multiple use
regional land-use plan
systems approach
visitorshed
watershed
watershed management

STUDY QUESTIONS

1. Discuss each of Agee and Johnson's (1988) principles of ecosystem management. Speculate what difficulties may be encountered with the implementation of each principle.
2. Why is ecosystem-based management important?
3. Why are linkages between different management agencies important?
4. Why are linkages between parks and private landowners important?
5. What is significant about ecosystem-based management in Nahanni National Park?
6. What methods have been developed for participation and collaboration?
7. Examine a park in your area for evidence of 'boundary thinking', 'adaptive management', and co-operation with adjacent landowners.
8. Why is ecosystem-based management not more widely used?
9. What can park managers do to improve the use of ecosystem-based management?
10. What is the significance to ecosystem-based management of outreach programs?
11. Review a management plan from one of Canada's northern national parks for evidence of co-management.
12. Discuss how ecosystem-based management has been used in the Waterton Biosphere Reserve.

PART V

Thematic Issues

> May your trails be crooked, winding, lonesome, dangerous, leading to the most amazing view. May your mountains rise into and above the clouds.
>
> *Edward Abbey,* The Journey Home

This section expands on major challenges mentioned in previous chapters that deserve more in-depth consideration. Students are encouraged to identify other issues as they emerge or become more important. As mentioned in Chapter 1, one way to do this is to join an NGO such as CPAWS that produces regular national and regional newsletters on park issues.

The significance of adjacent communities has been raised within the context of ecosystem management, the management of Banff, and as one aspect of social considerations. Adjacent communities may benefit, but at times they are negatively impacted by parks and protected areas. Negative impacts may occur as a result of undesirable tourism developments or activities, or through the loss of access or use of an area when a new park is created. Another form of damage occurs when wildlife stray across park boundaries onto private land, such as the damage caused when elk, bears, and wolves enter private land from Riding Mountain National Park. Aboriginal communities are a particularly important component of this interaction between parks and adjacent communities. Aboriginal peoples have a stake in park development and management for a number of reasons. Parks contain important resources for many Aboriginal cultures—as places for traditional activities such as hunting, trapping, and fishing, as well as for the spiritual and cultural values Aboriginal cultures derive from park settings. The nature of Aboriginal involvement in the creation and management of parks and protected areas has been an important, but challenging and sometimes controversial process, as outlined in Chapter 14.

The second theme developed in this section deals with the frustrations surrounding attempts to create marine protected areas in Canada. Chapter 15 outlines the need to protect the biodiversity found in marine environments, and describes the role of marine protected areas as part of a marine conservation strategy. Difficulties in moving ahead with this initiative in Canada are described, as well as a number of issues related to the management of marine protected areas.

Stewardship is the third theme of this section. Here we examine conservation developments that take place outside traditional parks managed by government agencies. This includes private or private–public collaboration to achieve stewardship objectives. These initiatives are particularly important in light of the growing evidence that the conservation of biodiversity cannot be achieved entirely within park boundaries, but must involve conservation strategies in landscapes outside of parks. Hence, the work of such private non-governmental organizations as Ducks Unlimited and the Nature Conservancy of Canada has become very significant. Chapter 16 describes a number of conservation strategies, successes, and challenges experienced by these organizations. It also poses some questions about the role of private agencies in conservation initiatives.

The last chapter in this section is a new one that responds to feedback on earlier editions to include an international perspective. Chapter 17 points out that in many parts of the world, particularly in the tropics where biodiversity is the most threatened, protected areas are under considerable threat from the surrounding inhabitants trying to earn a living. It is important to ask what kinds of strategies can be used in these sorts of situations, and we must also ask ourselves whether these strategies can be applied in the Canadian context.

CHAPTER 14

Aboriginal Peoples and National Parks

Philip Dearden and Steve Langdon

INTRODUCTION

Modern concepts about protected areas and wilderness have evolved from ancient cultural and religious ideas related to spirituality and primeval nature. Although the notions of parks and wilderness are foreign to most Aboriginal languages, which reflect the place of humans as an integral part of nature, many cultures have embraced the idea of sacred places (Peepre and Jickling, 1994). However, national parks, as we understand them today, originated in the United States. Nash (1970) asserts that the origins of the national park idea can be traced to the year 1832. At that time, he observes, the artist-explorer George Catlin called for the creation of a 'nation's park' to protect the Indians and the wild animals of the American Plains. The institution proposed by Catlin differs very little from the essentials of the national park idea as it exists today. Perhaps ironically, the one significant difference was Catlin's proposal that Indian people be part of 'the [life] in the preserve' (Nash, 1970: 730).

On 1 March 1872 over 810,000 ha of northwestern Wyoming were designated as the world's first national park—Yellowstone. The park was set aside during an effort to subdue Plains Indian tribes, and the traditional inhabitants of the park moved to reservations or were forced out by the United States Army. In 1885, Banff National Park was established in Alberta, seven years after the Siksika (Blackfoot) and Nakoda (Stoney) tribes ceded much of southwestern Alberta to the Crown. The treaty allowed the tribes to continue hunting in the region, but the federal government decided these rights would not apply to Banff National Park (Morrison, 1995). Since the establishment of these early national parks, thousands of new protected areas have been created throughout the world. Many of these protected areas have been designated on lands traditionally used by Aboriginal peoples. Often, they have been established without the participation of Aboriginal peoples living in the regions affected. In many instances, the Aboriginal people have been forcibly removed from regions in which protected areas were established. Early establishment of national parks in Canada followed the same pattern, but in more recent times (post-1982) rights have been recognized and traditional activities have been allowed to continue.

The practice of establishing protected areas without regard for the needs of Aboriginal people has sometimes adversely affected both Aboriginal societies and protected area conservation initiatives. In effect, 'indigenous people have borne the costs of protecting natural areas, through the loss of access for hunting, trapping or other harvesting activities' (Morrison, 1995: 12). Displacement of Aboriginal people often disrupts traditional social and economic systems and results in serious social problems, such as malnutrition and loss of cultural identity (Dasmann, 1976). At the very least, such negative impacts may reduce popular support among Aboriginal peoples for protected areas. Consequently, the effectiveness of conservation in protected areas has been compromised because of poaching, clandestine exploitation of resources, or other forms of non-compliance with regulations governing protected areas.

In response to these problems, it became clear there was a need to understand Aboriginal perspectives on parks (Box 14.1), involve Aboriginal peoples in protected area planning and management, and further, to allow exploitation of resources in protected areas for subsistence purposes. The role of Aboriginal people in national parks has become an important area of concern globally for Aboriginal organizations, as well as for protected area managers and social scientists. In the last 30 years, the relationship between Aboriginal people and national parks in Canada has changed fundamentally, although these changes are uneven across the country. Aboriginal peoples in northern Canada have played a significant role in national park planning and development, while in southern Canada their role has varied from park to park. In general, national parks established since the Constitution Act, 1982, with its entrenchment of Aboriginal and treaty rights, have working relationships with Aboriginal people. Steady progress is being made in parks established before that date, as the value of strong working relationships is recognized by Parks Canada in its corporate plan. A number of northern national parks have been established in conjunction with Aboriginal land claim settlements, while park reserves await claims settlement before attaining full park status. Overall, more than 50 per cent of the land area in Canada's national park system has been protected as a result of Aboriginal peoples' support for conservation of their lands, and 17 formal co-operative management agreements exist in addition to numerous informal agreements. Dearden and Berg (1993) suggest that First Nations have emerged as the most dominant force influencing the establishment of national parks in Canada over the years since the 1982 Constitution Act, and the same is true up to the present. This chapter provides an overview of the past and present role of Aboriginal people in national park designation, planning, and management in Canada. An introduction to the changing social and legal status of Aboriginal people in Canada sets the historical context as it pertains to conservation and protected areas. This is followed by a discussion of Parks Canada's evolving policy, regulations, and legislation as it relates to Aboriginal peoples. Using examples from several parks in the Canadian North, the next section illustrates Aboriginal peoples' involvement in national park management where land claims agreements are in place. The contrasting situation in southern Canada is then outlined, with several examples of Aboriginal peoples' role in national park management. Finally, some recent trends and advances by Parks Canada, as well as future directions, are discussed.

BOX 14.1 A First Nations Perspective on Parks Management

Tribal peoples in Canada managed their lands for eons before the arrival of settler populations, often in a state that resembles the present lands now protected as parks. Many government land managers are, in fact, examining indigenous practices in their continued efforts to return lands to the conditions that settlers found, and which shaped their ideas of wilderness. At the same time, tribal peoples themselves are regaining jurisdiction over portions of their traditional territories and in some cases are co-managing some parks or protected areas.

It is important to note that territories called wilderness by settlers or modern park managers are thought of as homelands by First Nations people. These lands are full of evidence of long-standing continuous relationships between the tribe and the environment. A short walk from any beach on Haida or Nuu-chah-nulth territory, one encounters culturally modified trees, often centuries old. The Salish-Kootenai land still bears vegetative patterns reflective of centuries of controlled burns. In each case, their lands are far from untrammelled in tribal eyes, and humans certainly are not intruders into nature. The following discussion reflects the views of protected area management that have been developed by the Confederated Salish-Kootenai First Nations in British Columbia for the Mission Mountain Tribal Wilderness.

When tribal land managers speak of their stewardship role, a notion of both physical and spiritual protection emerges. While the physical protection of places is common to all land managers, spiritual protection is of special importance to tribal managers. Tribal societies have always believed that spiritual obligation to the land is as important as physical protection. This obligation may take the form of ritual observance on the land at sacred sites, of traditional practices associated with the hunting of game species, and of the return to the land of the remains of plant or animal harvest after human use. These centuries-old practices are considered vital by tribal communities for continued health of the land and of the people. A major factor in establishing the Mission Mountains Tribal Wilderness (MMTW) was the importance of the Mission Mountains to the spiritual well-being of the Salish-Kootenai people. The religious practices of the Salish-Kootenai people—conducting vision quests, hunting and gathering medicinal roots and herbs—continue today in the wilderness, and these practices are being passed on to the next generation.

Tribal land managers, often trained in Western resource management schools, also speak of the need to respect traditional land management and tenure systems. Many of these land tenure systems are organized around certain families, who have delegated certain responsibility to care for particular hunting areas or sacred sites. In most cases, their land management roles coexisted with their role as harvesters, unlike the Western system, which separates these functions. This integrated system, where hunters monitored their own areas, depended not on career managers but on family responsibility to the larger community.

The collective emphasis, as opposed to the individualistic emphasis of most non-tribal communities, also influences tribal land management. Tribal communities have always had decision mechanisms that focus on the collective, but this search for collective consent is increasingly difficult in a modern context. The unity of perspective gained by shared experiences of education, spiritual practice, and pursuits on the land is no longer evident. Communities now reflect some of the diversities that challenge decision-makers in the larger, dominant society, but communities show a continued desire to make the majority of decisions collectively, rather than leaving them to individuals.

Since many tribal communities are also impoverished ones, there is also considerable pressure on land managers to ensure that wilderness areas provide direct economic benefits to the community. Most tribal communities want to continue hunting, fishing, agriculture, and gathering on wilderness lands, even if they deny such opportunity to non-members of their community. In many Canadian tribal communities 'country food' continues to account for a majority of the people's diet.

Many communities also want a large stake in the tourist economy that often results from the designation of a park or protected area. Such economic spinoffs include gift shops, restaurants, lodging, and guiding. For example, some tribes have been attracted by the potential for sport hunting and fishing as a source of income from their traditional lands. However, some of these communities have serious ethical concerns about the very notion of hunting for sport, yet they recognize the growing impact of nature-based tourism. The issue for tribal land managers is how to accommodate this desire from the non-Aboriginal community without compromising either the needs of tribal members or the beliefs that underpin the tribal approach to land management.

Tribal land managers are also charged with cultural interpretation of both their lands and the people who live on them. Many non-tribal visitors to areas perceived as 'primitive' expect 'authentic' tribal culture to be part of that experience, and their notion of authenticity is usually rooted in settler reports of early contacts. Tribal communities are modern communities and do not wish to be held up to a standard of modernity that differs from other cultures. So the issue becomes one of how to portray relationship to the land in a way that does not make culture a commodity or portray it as a frozen artifact.

Another issue raised by tribal wilderness managers is the need to preserve knowledge about the land that is presently held by the elders of the community. To pass this knowledge on to the next generation, there is a need for younger tribal members to accompany elders onto the land. The elders, in turn, need to find a land that continues to resemble the one they know, so that they can pass on knowledge of animal behaviour or plant habitat. At the same time, as Western science and land management become more interested in traditional ecological knowledge, there is real concern in tribal communities about protection of the intellectual property rights of this community-held knowledge. Tribal land managers have to deal with who owns knowledge, and who can consent to its being shared, as well as identify who it will be passed on to and thus who they will consult in the future.

Source: Abstracted from MacDonald et al. (2000).

ABORIGINAL PEOPLES IN CANADA

Definition

In Canada almost one million people can claim at least partial Aboriginal ancestry according to the 2001 census. The Constitution Act, 1982 defines three categories of 'aboriginal peoples': Indian, Inuit, and Métis. However, these three categories are not homogeneous cultural groups, but contain a great variety of peoples with differing histories, languages, and cultures. Accordingly, the name 'First Nations' has been adopted by many Aboriginal peoples identified under the Constitution as 'Indian', to reflect their first arrival on this continent and a belief of their status as separate, and sovereign, entities.

Aboriginal Treaties in Canada

Southern Canada, with the exception of the Atlantic provinces, most of Quebec, and most of British Columbia, is covered by Indian treaties that lay out certain legal obligations of the Crown towards Aboriginal peoples. Between 1780 and 1850, a number of small treaties were negotiated with the Indians in what is now southern Ontario. These treaties usually involved small, lump-sum payments in return for extinguishing Aboriginal title. On rare occasions, fishing and hunting rights were guaranteed, and reserves were granted. In 1850, the Robinson-Superior and Robinson-Huron treaties, named after William Benjamin Robinson, a former fur trader who acted as the chief government negotiator, were negotiated with Indians of the upper Great Lakes region. In return for surrender of large areas of land, Indian people received lump-sum cash payments, annual payments to each person, and promises of continued hunting and fishing on unoccupied Crown land (Cumming and Mickenberg, 1972).

The Robinson treaties became the model for subsequent 'numbered' treaties that encompass much of Ontario, all of Manitoba, Saskatchewan, and Alberta, and portions of British Columbia, Yukon, and the Northwest Territories. These treaties, numbered 1 to 11, were completed between 1871 and 1929 by federal government representatives. The Williams treaties, completed in 1923 by the federal government, purportedly extinguished Aboriginal title to lands in southern Ontario. With some minor differences, treaties negotiated by the federal government are all similar. In return for cessation of their title, Aboriginal peoples received reserves, small cash payments, hunting and fishing gear, annual payments to each member of the signatory group, promises of continued hunting and fishing rights, and in some instances the promise of tools, seed, and livestock for farming.

About the same time as Robinson completed his treaties, James Douglas, as Chief Factor of the Hudson's Bay Company and later Governor of the Colony of Vancouver Island, began treaty-making with a number of Island tribes. Between 1850 and 1854, Douglas completed 14 treaties, extinguishing Aboriginal title to lands around Victoria, Nanaimo, and Fort Rupert (present-day Port Hardy). In return for surrender of their lands, Aboriginal people maintained possession of their village sites and fields, and were guaranteed the right 'to hunt on unoccupied lands, and to carry on [their] fisheries as formerly' (British Columbia, 1875: 5–11).

Aboriginal Rights in Canada

Most of Yukon, the Northwest Territories, British Columbia, Quebec, and the Atlantic provinces are free of the nineteenth-century treaties. Aboriginal peoples living in the areas where modern land claim agreements have not yet been settled have not ceded their Aboriginal title by treaty, nor have they been conquered by overt act of war. The Atlantic provinces are, however, covered by a series of 'peace and friendship' treaties that were put in place in the 1700s to facilitate European settlement. Nonetheless, they have been denied their Aboriginal rights and title, and they have been 'colonized' and marginalized over the last 100 years. In spite of this, Aboriginal people in Canada have never stopped pressing government for recognition of their Aboriginal rights (Frideres, 1988). Until recently, however, such efforts were largely unsuccessful. This situation is gradually changing following several important Supreme Court rulings and the addition to Canada's Constitution of the Constitution Act, 1982.

The Calder Case

In 1967, the Nisga'a initiated a suit before the Supreme Court of British Columbia, asking the court for a declaration that their Aboriginal title had never been extinguished (Sanders, 1973; Berger, 1982; Raunet, 1984). The trial, which came to be known as the *Calder* case, after Frank Calder, a Nisga'a chief and the founder of the Nisga'a Tribal Council, opened in the Supreme Court of British Columbia, which ruled against the Nisga'a, asserting that their Aboriginal rights were extinguished by overt acts of the Crown (*Calder v. Attorney General of British Columbia*, 1969). The Nisga'a appealed, but the British Columbia Court of Appeal (*Calder v. Attorney General of British Columbia*, 1970) upheld the lower court ruling. The Nisga'a then appealed to the Supreme Court of Canada (*Calder v. Attorney General of British Columbia*, 1973), and although they lost on a technicality, 4–3, six of the seven justices affirmed the existence of Aboriginal title under English law.

The *Calder* case had important repercussions for Aboriginal policy and law. For the first time, the Supreme Court of Canada recognized that Aboriginal title existed at the time of colonization as a legal right derived from the Aboriginal peoples' historical occupation and possession of the land, independent of any proclamation, legislative act, or treaty. Following this judgement, the federal government was willing to negotiate comprehensive land claim settlements in British Columbia, Quebec, and the two northern territories with Aboriginal groups that had never signed treaties (Canada, 1981; Sanders, 1983; Task Force to Review Comprehensive Claims Policy, 1985; Canada, 1987).

First Nations under treaty, at the same time, are able to file specific claims to address issues of government failure or wrongful action in regard to their treaty rights, when treaty lands, for example, have been taken over by government, sublet, or sold without full compensation and agreement on the part of the affected Aboriginal people. The negotiation process is drawn out and can be costly, and as it extends to years and decades the unlawful use of Aboriginal lands may continue unchecked, with unforeseen consequences. This happened in regard to Ipperwash Provincial Park in southwestern Ontario, which previously had been a military base, when a Native protestor of the Kettle and Stony Point First Nation was killed in 1995 by provincial police. Over a decade later, this ancient Chippewa burial ground was finally returned to the Aboriginal people.

Constitution Act, 1982

After intense lobbying by Aboriginal groups, the Constitution Act, 1982 was enacted containing two sections protecting Aboriginal rights. Section 25 protects Aboriginal, treaty, or other rights from infringement by other guarantees in the Charter. Section 35, entitled 'Rights of the Aboriginal Peoples of Canada', entrenches Aboriginal rights in the Constitution and also adopts and confirms the large body of common law, which has come to be known as the 'common law doctrine of Aboriginal rights'. This doctrine holds that the property rights, customary laws, and governmental institutions of Aboriginal peoples were assumed to survive the Crown's acquisition of North American territories.

The Constitution Act 'set the consideration of native law in a new context' (Elias, 1989: 4), which appears to be more favourable to the aspirations of Aboriginal people in Canada. This 'new context' is evident in the Supreme Court of Canada ruling in *Sparrow v. The Queen et al.* (1990), described below.

A new era in the relationship between Aboriginal peoples and the government of Canada was ushered in during the 1990s. The federal policy on land claims recognizes the inherent right to self-government of Aboriginal peoples, a policy that has had significant ramifications for protected areas and conservation in Canada.

The Sparrow Case

In May 1990, the Supreme Court of Canada handed down its landmark judgement in *Sparrow*. Ronald Edward Sparrow, a Musqueam Indian from British Columbia, was charged in 1984 under the Fisheries Act with using a drift net longer than that permitted

FIGURE 14.1 Vuntut National Park Reserve in northwest Yukon was created as part of the Inuvialuit Final Agreement. *Photo: P. Dearden.*

by the terms of his band's Indian food fishing licence. Sparrow admitted that the Crown's allegations were correct, but he defended his actions on the basis that he was exercising an existing Aboriginal right to fish, protected under s. 35(1) of the Constitution Act.

The Provincial Court held that the Musqueam did not have an Aboriginal right to fish. Sparrow appealed to County Court (*Sparrow v. The Queen*, 1986a) and the case was dismissed for similar reasons. The case was then appealed to the British Columbia Court of Appeal (*Sparrow v. The Queen*, 1986b), which held that the lower courts had erred in ruling that the Musqueam had no Aboriginal fishing rights. The Appeal Court also ruled that the Aboriginal right to fish existed at the time of enactment of the Constitution Act, and was therefore a constitutionally protected right that could no longer be extinguished by unilateral action of the Crown. The Court also held, however, that the trial judge's findings of facts were insufficient to lead to an acquittal. The ruling was appealed by Sparrow and argued before the Supreme Court of Canada in 1987.

In a unanimous ruling, the Supreme Court of Canada held that there was insufficient evidence on which to decide guilt or innocence. More importantly, the Court affirmed that the Musqueam people have an unextinguished Aboriginal right to fish. It also set forth a framework for defining the existence and scope of Aboriginal rights in Canada. In this regard, the Court (*Sparrow*, 1990: 16) held that prior to 1982 Aboriginal rights continued to exist unless they had been extinguished by an action of the Crown that was clearly intended to do so. Therefore, contrary to arguments made by the government of British Columbia, legislative action that is merely inconsistent with the concept of Aboriginal title cannot be construed as extinguishing such title. Following enactment of the Constitution Act, Aboriginal rights could no longer be extinguished by the Crown. The Supreme Court (*Sparrow*, 1990: 26) further defined the nature of constitutional protection of Aboriginal rights:

> the constitutional recognition afforded by the provision [s. 35(1)] therefore gives a measure of control over government conduct and a strong check on legislative power. While it does not promise immunity from government regulation in a society that, in the twentieth century, is increasingly more complex, interdependent, and sophisticated, and where exhaustible resources need protection and management, it does hold the Crown to a substantial promise. The government is required to bear the burden of justifying any legislation that has some negative effect on any Aboriginal right protected under s. 35(1).

In the eyes of the Court, both 'conservation' and 'resource management' constitute justifiable grounds for legislation that may have a negative effect on Aboriginal rights. However, even when such measures must be implemented, the Court held that Aboriginal people must be consulted so as to mitigate any impact upon their rights.

The *Sparrow* case directed the government to include Aboriginal people in co-operative management of natural resources. With respect to parks, it is clear that the ruling reinforced Aboriginal beliefs that they deserve special recognition with respect to management when their traditional territories coincide with park lands.

Delgamuukw

In *Delgamuukw v. British Columbia* (1997), seven years after the *Sparrow* decision, the Supreme Court ruled on Aboriginal title. In this case, the Gitksan and Wet'suwet'en claimed the right to their traditional lands in northern British Columbia. The lower court in BC had rejected the claim outright because it was based on oral tradition among the Aboriginal people, but the Supreme Court of Canada ruled otherwise, affirming the validity of oral historical accounts in land claim cases. Since Aboriginal title is an interest in land within the British common-law system, the relevant date for a court to examine whether Aboriginal title exists is the date of the assertion of British Crown sovereignty in an area. The case also affirmed that both Canadian law and the laws of the Aboriginal nations involved must be considered in providing definition to Aboriginal rights and title (Parks Canada Agency, 2000). This ruling could affect government resource dispositions on traditional lands as well as Aboriginal claims for title inside some national parks. In 1999, *Marshall v. The Queen*, which involved Mi'kmaq fishing rights based on an eighteenth-century friendship treaty, confirmed that oral tradition that can provide a context for a transaction is admissible to help a court interpret a treaty (Parks Canada Agency, 2000).

Haida and Taku

On 18 November 2004 the Supreme Court of Canada released its decisions in the cases of *Ringstad & B.C. Minister of Environment et al. v. Taku River Tlingit First Nation* and *B.C. Minister of Forests v. Haida Nation et al.* These decisions are unique in that they affect almost every department across government.

Both decisions were unanimous. Both cases involved lands subject to claims of Aboriginal rights and title. The Court ruled that the Crown has a legal duty to consult and, where indicated, to accommodate the concerns of Aboriginal groups when the Crown has knowledge of the potential existence of an Aboriginal right or title and contemplates conduct that might adversely affect it. The duty to consult is grounded in the honour of the Crown. The scope of consultation, and if appropriate, accommodation, will vary depending on the circumstances. There is no duty on third parties, such as private industry, to consult with First Nations as to potential Aboriginal rights. The Court appears to be telling governments that reasonable good-faith efforts to reconcile with Aboriginal peoples will be respected if Aboriginal peoples are appropriately engaged (Parks Canada Agency, 2006).

Sappier and Polchies

In 2004 the Court of Appeal of New Brunswick acquitted the respondents in these two treaty/Aboriginal rights appeals on charges of unlawful cutting and possession of timber taken from Crown lands, contrary to s. 67(1)(a)(c) of the provincial statute. The Court of Appeal held that the respondents have an Aboriginal right to harvest timber for personal use, including for the construction of housing and furniture and for fuel, and the Maliseet respondents, Sappier and Polchies, also have a treaty right to the same effect.

Comprehensive Land Claim Policy

Following the *Calder* case, the federal government released a policy statement announcing its willingness to negotiate the settlement of comprehensive land claims and the objectives to guide its involvement in such negotiations (Canada, 1981). After the recommendations of the 1985 Task Force to Review Comprehensive Claims Policy, a substantially modified land claim policy was unveiled by the federal government late in 1986 (Canada, 1987). Notwithstanding the Supreme Court judgement in *Sparrow*—and additional Aboriginal rights cases heard since then—the Minister of Indian Affairs and Northern Development has stated that the 'basic principles' of the current comprehensive land claim policy are not likely to change.

This policy requires Aboriginal peoples to surrender to the Crown their rights, interests, and 'Aboriginal' title in and to the land, water, and natural resources, in exchange for which they are to receive constitutionally protected rights, benefits, and privileges defined in land claim settlements. The comprehensive land claim policy does not deal explicitly with national parks or other forms of protected areas. This is not too surprising, for the intent of the policy and of the government's strategy in entertaining land claim negotiations is quite specific: that is, to clear the ill-defined Aboriginal title from the land in question. Whether national parks are included in the rights and benefits Aboriginal peoples obtain in return depends on the policy and strategy of Aboriginal peoples as well as the intent of government. It is clear, however, that comprehensive land claim settlements concluded under the existing (Canada, 1987) and the preceding policy (Canada, 1981) stress environmental conservation and protection of wildlife habitat. Moreover, many final agreements in the Territorial North deal with national parks.

PARKS CANADA POLICY, REGULATIONS, AND LEGISLATION RELATING TO ABORIGINAL PEOPLES

Many of Canada's national parks were designated at a time when both the federal and provincial governments did not acknowledge Aboriginal rights and title. Aboriginal peoples utilizing traditional lands or occupying reserves encompassed by newly designated parks were given little, if any, input in park planning and management. Indeed, when Riding Mountain National Park was established in 1933, the Keeseekoowenen band was evicted and their houses burned (Morrison, 1995). Aboriginal people with reserves in proposed park areas were encouraged by Parks Canada to sell or trade their reserves for lands outside proposed parks, and were prevented from hunting and trapping within them. There was little appreciation within government that parks could be used to support and maintain the land uses of Aboriginal peoples, and hence protect their land-based cultures. Instead, Parks Canada stressed the need for the parks system to represent biophysically defined natural areas. Adherence to the natural areas framework may have contributed to the estrangement of parks from Aboriginal peoples. Parks identified through the framework might have been excellent choices to represent natural areas, but were sometimes irrelevant to protecting vital wildlife habitat, the element of parks legislation that interested many Aboriginal groups. To Aboriginal peoples dependent on hunting, fishing, and trapping, the location of a park was the key to its utility and political acceptability.

The attitude of Parks Canada began to change in the 1970s as the aspirations of Aboriginal peoples became better known and appreciated by the Canadian public and political leaders. This process was aided by public hearings into oil and gas megaprojects, which brought representatives of Aboriginal peoples and environmental and other groups into the same camp. The Berger Inquiry of 1974–7 into a proposed gas pipeline from the Mackenzie Delta and northern Alaska, for example, noted the need for parks and conservation areas to be planned simultaneously with non-renewable resource development (Berger, 1977). In addition, Justice Thomas Berger, who earlier had acted as chief counsel to the Nisga'a, proposed a new type of park, a 'wilderness park', to preserve wildlife, wildlife habitat, and natural landscapes in northern Yukon, and to underpin the still-vibrant renewable resource economy of the Inuvialuit and Dene. This recommendation is now an acknowledged milestone in the debate that connects Aboriginal peoples with national parks.

The 1979 *Parks Canada Policy* tried to respond to Aboriginal issues and Justice Berger's report. The policy contained a number of sections that defined a new relationship between local people and potential national parks. In this regard, section 1.3.5 of the policy (Parks Canada, 1979: 39) stated that Parks Canada 'will contribute toward the cost of special provisions to reduce the impact of park establishment on occupants or other users of lands acquired for a national park.' While not directed specifically at Aboriginal people, this section indicated a willingness on the part of Parks Canada to be more sensitive to impacts on local people, including Aboriginal peoples, when establishing national parks. Consistent with the federal government's 1973 policy stating its intent to negotiate land claims in Quebec, British Columbia, and the territories, the 1979 parks policy also recognized the potential existence of certain Aboriginal rights in section 1.3.13 (Parks Canada, 1979: 40). Thus, the 1979 *Parks Canada Policy* embraced the concept of joint management by government and Aboriginal people eight years before this same concept was endorsed and adopted in the land claims policy.

In 1994, Parks Canada revised its policies, with a new and strong emphasis on ecological integrity, improved regional integration through co-operation with other jurisdictions, and a more comprehensive approach to working with Aboriginal peoples (Parks Canada, 1994). The 1994 *Guiding Principles and Operational Policies* sets out several policies with respect to Aboriginal interests:

- negotiation of comprehensive claims based on traditional uses and occupancy of land;
- rights and benefits in relation to wildlife management and the use of water and land, and the opportunity for participation on advisory or public government bodies;
- respect for the principles set out in court decisions, such as *Sparrow v. The Queen*, where existing Aboriginal or treaty rights occur within protected areas;
- at the time of new park establishment, respect for Canada's legal and policy framework regarding Aboriginal rights as affirmed by section 35 of the Constitution Act, 1982, and consultation with affected Aboriginal communities.

Of particular note, the policy also recognizes 'local knowledge' as valuable to management of heritage areas, although there is no explicit reference to Aboriginal ecological knowledge (ibid., 18). The *Guiding Principles* makes reference to collaboration and co-operation with 'Aboriginal interests to achieve mutually compatible goals and objectives. These relationships support regional integration, partnerships, co-operative arrangements, formal agreements, and open dialogue with other interested parties, including adjacent or surrounding districts and communities' (ibid., 19). This policy sets out a new approach directing park managers to work with a broad range of partners both inside and outside the national parks. Finally, in reference to national park agreements, the policy has provisions for 'continuation of renewable resource harvesting activities, and the nature and extent of Aboriginal peoples' involvement in park planning and management', and 'establishment pursuant to agreements with Aboriginal organizations' (ibid., 28, 51).

The 1994 *Guiding Principles and Operational Policies* made significant progress in redefining the relationship between Parks Canada and Aboriginal people in the national parks and national historic sites managed by Parks Canada. These principles have formed the foundation of the current approach to co-operative management with Aboriginal people found in numerous national parks. Today, however, the principles are somewhat dated in relation to working with Aboriginal people, given the succession of Supreme Court rulings in subsequent years. In addition, in 2000, Parks Canada became an agency (see Chapter 9), and one of the primary tools used to guide priority setting and accountability to the public is the corporate plan, which is updated annually (see Parks Canada website). The corporate plan clearly identifies the priority that Parks Canada Agency places on working in a co-operative fashion with Aboriginal people. The current version of the plan calls for priority in five areas: employment and development; economic opportunities; presentation of Aboriginal themes; commemoration of Aboriginal history and culture; and building relationships.

While Parks Canada still faces many legal situations related to the interpretation of legal rights, the organization is recognized as a national and international leader in co-operative management of protected areas.

The most recent *National Parks System Plan* reflects implementation of Parks Canada's policies with respect to Aboriginal people by endorsing 'a new type of national park where traditional subsistence resource harvesting by Aboriginal people . . . continues and where co-operative management approaches are designed to reflect Aboriginal rights and regional circumstances' (Parks Canada, 1997b).

Amendments to the National Parks Act in 1988 and 2000 also recognize the importance of traditional resource harvesting to Aboriginal peoples. The 1988 amendments allowed specific Aboriginal groups to carry out such harvesting in certain parks. It also extended, at the minister's discretion, traditional renewable resource harvesting rights in wilderness areas of national parks to Aboriginal peoples with land claim settlements, and allowed for regulation of traditional renewable resource harvesting in national parks by Order-in-Council.

The Canada National Parks Act of 2000 extended harvesting rights to a larger number of parks, including all those established by agreement. Section 10(1) of the Act also supports co-operative agreements with a wide range of organizations, including Aboriginal governments, for carrying out the purposes of the Act. The Act specifies that the federal cabinet may 'make regulations respecting the exercise of traditional renewable resource harvesting activities' in Wood Buffalo, Wapusk, and Gros Morne national parks, any park established in the District of Thunder Bay in the province of Ontario, and 'any park established in an area where the continuation of such activities is provided for by an agreement between the Government of Canada and the government of a province respecting the establishment of the park' (s. 17[1]). For the first time, the National Parks Act also provides for the removal of non-renewable resources in the form of carving stone in order to support traditional economies. Current harvesting activities are summarized in Box 14.2; however, no trend data are currently available.

The new National Parks Act does not guarantee co-operative management for Aboriginal peoples whose traditional lands fall within national parks; however, on a policy basis, Parks Canada has been very active in developing not only a formalized consultative process but also co-operative management arrangements. The Gulf Islands National Park Reserve is a good example of how Parks Canada, in advance of treaty settlement, has developed three co-operative management arrangements with Aboriginal groups to ensure consultation and input into major park decisions that affect them. In the case of Gwaii Haanas, the National Parks Act specifies in section 41(1) that 'the Governor in Council may authorize the Minister to enter into an agreement with the Council of the Haida Nation respecting the management and operation of Gwaii Haanas National Park Reserve of Canada.' Section 41(2) further allows for 'regulations, applicable in the Gwaii Haanas National Park Reserve of Canada, respecting the continuance of traditional renewable resource harvesting activities and Haida cultural activities by people of the Haida Nation to whom subsection 35(1) of the *Constitution Act*, 1982 applies.'

BOX 14.2 Summary of Limitations on Aboriginal Harvesting in National Parks

Limits to Aboriginal harvesting are specified by regulations under the Canada National Parks Act. These limitations:

- specify what are traditional renewable resource harvesting activities;
- designate classes of persons authorized to engage in those activities and prescribe the conditions under which they may engage in them;
- prohibit the use of renewable resources harvested in parks for other than traditional purposes;
- control traditional renewable resource harvesting activities;
- authorize the superintendent of a park to close areas of the park to traditional renewable resource harvesting activities for purposes of park management, public safety, or the conservation of natural resources;
- authorize the superintendent of a park to establish limits on the renewable resources that may be harvested in any period, or to vary any such limits established by the regulations, for purposes of conservation;
- authorize the superintendent of a park to prohibit or restrict the use of equipment in the park for the purpose of protecting natural resources.

Aboriginal people are permitted to harvest plants and animals in many national parks based on land claim settlements or on other policy-related arrangements. All national parks in the North have harvesting regimes, as do several in southern Canada. Some examples of the latter are Pukaskwa in Ontario and Gulf Islands, Gwaii Haanas, and Pacific Rim in BC. Animals that are harvested range from large mammals such as moose to fur bearers such as mink. Harvesting activities happen primarily in the northern parks, although exact records of harvest activities are not kept. While rights or granted access may exist, the level of harvest is relatively low. This may reflect the Aboriginal perspective that these jointly protected areas are special and, therefore, that harvesting should be minimized. It is also known that, in some situations, access to resources is easier in areas outside of national parks.

Perhaps the most important accommodation the National Parks Act makes to Aboriginal peoples lies in the term 'national park reserve', introduced through amendment to the statute in 1972. This designation applies, for example, to Nahanni and Pacific Rim, which are to become full national parks upon settlement of comprehensive land claims. The 'reserve' designation allowed Parks Canada to treat and manage the areas in question as national parkland, but did not extinguish any Aboriginal rights or title to the areas. Importantly, this designation does not prejudice the ability of Aboriginal peoples to select parkland in the course of land claim negotiations.

Canadian legislators seem to have chosen an ad hoc approach to accommodating the needs of Aboriginal peoples in national parks. Wood Buffalo National Park and Auyuittuq National Park Reserve provide examples of this ad hoc approach. The area around Wood Buffalo National Park was a favoured hunting ground of Aboriginal people for many years prior to its establishment as a park in 1922 (Lothian, 1976). When the park was established, Aboriginal people who had previously hunted and trapped in the area continued these activities under permit. In 1949, special district game regulations for Wood Buffalo were instituted, which superseded the National Parks Game Regulations and allowed for traditional hunting, trapping, and fishing by Aboriginal people (ibid.). The National Parks Act also enables the appointment of a Wildlife Advisory Board for the traditional hunting grounds of Wood Buffalo National Park, and this Board has a role, for example, in bison management and hunting, fishing, and trapping regulations (Canada National Parks Act, s. 37). Auyuittuq National Park Reserve, located on Baffin Island, was established in 1972 long before the Nunavut Agreement. Public park planning meetings in the early 1970s resolved that the Inuit, who had inhabited the region for almost 4,000 years, would retain traditional resource extraction rights within the park. When Auyuittuq National Park Reserve was established there was provision for a 'park advisory committee'. This represented one of the first co-operative efforts by Parks Canada. All of the national parks found in Nunavut currently have park advisory committees, and they have evolved into an effective means for the Inuit to participate in the planning and management of the national parks found in their traditional territories.

The interaction between Aboriginal peoples and national parks in Canada is not as clear as might be suggested by national park policy and legislation. Land claim settlements, in addition to Parks Canada policy and legislation and legal precedent, determine the role of Aboriginal peoples in planning for, and managing, national parks. This has given rise to subtly different kinds of parks in northern and southern Canada, for a significant park planning and management role is accorded Aboriginal peoples in northern Canada where parks are tied to settlement of land claims. Land claims in northern Canada will likely be completed sooner than many of those, for example, in British Columbia. In addition, many Aboriginal peoples in the south must look to treaties, legal precedent, and the National Parks Act and Parks Canada's *Guiding Principles and Operating Policies*, rather than to comprehensive land claim settlements, to protect their interests.

In response to this variation in approaches to working with Aboriginal organizations, the Panel on the Ecological Integrity of Canada's National Parks recommended that 'Parks Canada adopt clear policies to encourage and support the development and maintenance of genuine partnerships with Aboriginal peoples in Canada' (Parks Canada Agency, 2000: 7–8). The Panel also outlined key steps to foster trust and respect between Parks Canada and Aboriginal peoples, such as initiating a process of healing; providing adequate resources to maintain genuine partnerships; integrating Aboriginal culture, knowledge, and experience into education and interpretation programs; and ensuring protection of cultural sites, sacred areas, and artifacts. In an effort to move beyond the constraints of strict government legal positions, Parks Canada established an Aboriginal Secretariat in 1999, with the task of improving relationships with Aboriginal organizations throughout the national park system. Under the direction of the chief executive officer of the day, the Parks Canada corporate plan was also modified to better reflect the need for strong relationships with Aboriginal people and the important role they play in delivery of the Parks Canada mandate of protection, memorable experiences, and educational opportunities.

The following case studies describe current approaches to involvement of Aboriginal peoples in national park designation, planning, and management in both northern and southern Canada.

THE ROLE OF ABORIGINAL PEOPLES IN NATIONAL PARKS: NORTHERN CANADA LAND CLAIM AGREEMENTS

The Inuvialuit Final Agreement

The Inuvialuit of the Beaufort Sea region began land claim negotiations with the federal government in the mid-1970s and reached a Final Agreement in 1984 (Canada, 1984). Legislation to approve the final agreement and to amend the National Parks Act in consequence was passed in 1984. The Inuvialuit Final Agreement (IFA) requires the federal government to establish the western portion of northern Yukon as a national park, subsequently named Ivvavik (Figure 14.2). In line with Justice Berger's recommendations (Berger, 1977), the IFA characterizes Ivvavik as 'wilderness oriented', and requires that the planning for the park: 'Maintain its present undeveloped state to the greatest extent possible' (Canada, 1984: 18). Moreover, a central aim of the park is 'to protect and manage the wildlife populations and the wildlife habitat within the area'.

This objective reflects the national and international importance of the calving grounds of the Porcupine caribou herd, which are partially within the park. Not only did the IFA commit government to establish the park, define its boundaries, and specify its purposes and objectives, it mandated a Wildlife Management Advisory Council, composed of an equal number of government and Aboriginal members, to 'recommend a management plan for the National Park' (ibid.).

The 1978 Agreement-in-Principle promised that all of the Yukon north slope from the Alaskan border in the west to the NWT border in the east would be established as a national park. Due to the objections of the Yukon government and the oil and gas industry, the park boundaries defined in the final agreement were altered to divide the north slope into two zones (Fenge et al., 1986). The western portion, west of the Babbage River, was confirmed as a national park, but the eastern portion was excised from the proposed park to allow for the development of a transportation corridor. Nevertheless, this area was still to be subject to a 'special conservation regime whose dominant purpose is the conservation of wildlife, habitat and traditional native use' (Canada, 1984: 18). The Wildlife Management Advisory Council has responsibility for both the eastern and western portions of the north slope. This arrangement and the operations of the Porcupine Caribou Management Board link the national park to broader regional wildlife management and environmental conservation objectives. The IFA makes it clear that the Inuvialuit have the right to harvest wildlife for subsistence purposes throughout the north slope, and an exclusive right to do so in the national park. The door was left open, however, for Aboriginal peoples represented by the Council for Yukon First Nations to acquire harvesting rights in the park through their own land claim settlement.

In the case of Ivvavik National Park, the land claim settlement, rather than the intragovernmental work of Parks Canada, resulted in the establishment of this national park. Two additional national parks have been established in the Inuvialuit settlement area since 1984, Aulavik on Banks Island and Tuktut Nogait on the Arctic coast (Figure 14.2).

Yukon First Nations

The Umbrella Final Agreement (UFA) with the majority of Yukon's First Nations was ratified in 1994, enabling a wide range of co-operative conservation initiatives in the territory, including national parks. Chapter 10 of the UFA sets out the conditions for special management areas (SMA), a unique tool that allows First Nations to negotiate government-to-government arrangements for habitat protection, watershed protection, national wildlife management areas, parks, or other types of management agreements.

Vuntut National Park was established in 1993 through the Vuntut Gwitchin Final Agreement and was linked to a regional conservation package that included the Old Crow Flats SMA. Lands within Vuntut National Park are managed according to the National Parks Act and the Final Agreement, while the Final Agreement and a locally produced plan guide management of the SMA. Of interest were the different, and complementary, perspectives on park establishment held by Vuntut Gwitchin and Parks Canada. Gwitchin primarily viewed the park as a way to protect their traditional lands for harvesting and other cultural values, while Parks Canada saw the park as improving representation of the natural region (Njootli, 1994; Johnson, 1994). The objectives of the

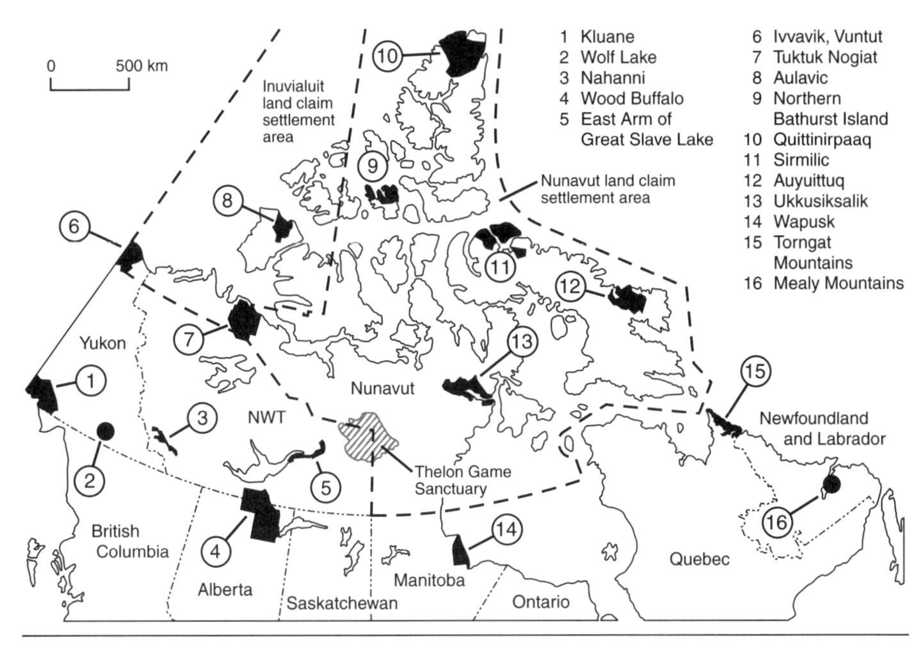

FIGURE 14.2 Parks Canada in the North.

park agreement include recognition and protection of the traditional and current use of the park by Vuntut Gwitchin in the development and management of the park (Canada, 1993: Schedule A, 105). The agreement provides economic and employment opportunities, recognizes oral history as a valid form of research, and gives Vuntut Gwitchin the exclusive right to harvest for subsistence at all times and for all species. Commercial trapping is also supported. Park management is based on a co-operative working arrangement with a local Renewable Resources Council that may make recommendations to the minister on all matters pertaining to the management of the park, and Parks Canada is directed to implement the recommendations of the Council that are accepted by the minister. The Vuntut Gwitchin First Nation also maintains a central role in park management, as shown by the 'Interim Management Guidelines' (Parks Canada, 1999). These guidelines were created in partnership with the Gwitchin, Parks Canada, and the Renewable Resources Council, whereby the park vision statement strongly reflects First Nation values. A complementary co-operation agreement was signed in 1998, which sets out clear roles and responsibilities in the park. Parks Canada retains overall responsibility for management and operations of the park.

Kluane National Park Reserve was established in 1972, followed two decades later by a new negotiated management arrangement through the Champagne and Aishihik First Nations Final Agreement for the southern part of the park. The Kluane First Nation, with traditional lands in the northern part of the park, settled their Final Agreement in 2004 and have also entered into a management agreement with Parks Canada. Unlike Vuntut National Park, the Kluane agreement establishes a management board consisting of First Nations and community representatives and an ex officio member from Parks Canada. Although the agreement has enhanced local co-operation, and the board provides advice to the minister through the park management plan, the agreement does not appear to meet the test of a true joint management regime where decisions and resources are shared equally.

In 1999, the Champagne and Aishihik First Nations proposed six goals for management of the park: renewing cultural ties to the park, learning and teaching cultural heritage; keeping plants and animals healthy for the future; creating training and employment opportunities; participating in tourism; and sharing responsibility for the park by working towards First Nation members becoming full co-managers of the park (Parks Canada, 2000).

The Inuit Land Claim

A key step in the political and social evolution of the eastern Arctic took place in 1993, with the signing of the agreement between the Inuit of the Nunavut Settlement Area and Canada. This agreement commits Parks Canada to work with regional Inuit communities and designated Inuit organizations in the development of co-operative structures for the parks as well as in the broader implementation of the Nunavut Land Claim Agreement (Parks Canada, 1999).

During the last two decades, Inuit generally have viewed national parks as 'friendly' land-use designations. In 1983, for example, the Nunavut Constitutional Forum (NCF), which represented Inuit in constitutional discussions within the NWT, proposed that approximately 25 per cent of the Arctic be set aside for park purposes (Doering, 1983). This proposal was not viewed favourably, hence the final agreement contains less dramatic and less far-reaching provisions on parks. The Inuit strategy during formal negotiations was later quite clear: parks were to be used to protect key wildlife habitat, allowing negotiators to concentrate landownership selections elsewhere.

While the IFA dealt with one park, the Nunavut Agreement did not commit government to establish any specified areas as national parks. Since the agreement was signed, significant progress has been made on new national park designations, including the establishment of Quittinirpaaq and Sirmilik national parks.

The park management provisions in the Nunavut Agreement are similar to those in the IFA. For each park, Inuit and government are to negotiate an Inuit Impact and Benefits Agreement (IIBA) to channel economic and social benefits from the park to local Inuit. As part of an IIBA, a joint Inuit/government parks planning and management committee can be set up to advise 'on all matters related to park management' (Tungavik Federation of Nunavut, 1990: 119). Management plans developed by Parks Canada will be based on the recommendations of the committee. Such plans have to be approved by the minister responsible for national parks. In conducting negotiations, TFN tried to persuade the federal government to adopt the term 'joint management regime' as used in the 1979 *Parks Canada Policy*, and to have this term enshrined in the final agreement. The federal government refused, saying that this concept was vague and open to misinterpretation.

The Nunavut Agreement also followed the lead of the IFA in characterizing national parks in the Arctic as 'wilderness oriented'. The Agreement stipulates that: 'each National

BOX 14.3 The Nunavut Agreement

The Nunavut Agreement is, in effect, a huge land claim agreement. It forms a modern treaty between the Inuit of the Nunavut Settlement Area, who were represented in negotiations by the Tungavik Federation of Nunavut (TFN), and the federal government. The agreement codifies an exchange between Inuit and the federal government. The Inuit agreed to give to the government their Aboriginal title to land, water, and the offshore. In exchange, the Inuit are to enjoy the rights and benefits set out in the agreement. These rights and benefits, which are 'guaranteed' under Canada's Constitution, include:

- title to approximately 350,000 km^2 of land, of which approximately 36,000 km^2 will include mineral rights;
- the right to harvest wildlife on lands and waters throughout the Nunavut Settlement Area;
- a guarantee of the establishment of (at least) three national parks in the Nunavut Settlement Area;
- equal membership with the federal government on new institutions of public government (established through the agreement) to manage the land, water, offshore, and wildlife of the Nunavut Settlement Area and to assess and evaluate the impact of development projects on the environment. These public institutions include the Nunavut Wildlife Management Board (NWMB), the Nunavut Water Board (NWB), the Nunavut Impact Review Board (NIRB), and the Nunavut Planning Commission (NPC).

Source: Canadian Arctic Resources Committee (1993).

Park in the Nunavut Settlement Area shall contain a predominant proportion of Zones I and II, as such zones are defined in the Parks Canada Policy.' Inuit also hope national parks will be tools for economic development. Very few 'outsiders' currently visit national parks in the Arctic; for example, less than 1,000 people visit the remote Auyuittuq National Park Reserve every year. Nevertheless, Inuit hope to use money from the land claim settlement to provide tourist and recreational facilities in communities adjacent to parks in order to attract more visitors to these places. It remains the intent of Parks Canada to work with the Inuit to examine means of maximizing tourism and economic opportunities while protecting the ecological values of the parks. (Figure 14.3)

THE ROLE OF ABORIGINAL PEOPLES IN NATIONAL PARKS: EXAMPLES FROM SOUTHERN CANADA

In southern Canada, Parks Canada has a wide spectrum of arrangements with Aboriginal peoples in park designation, planning, and management. Both Pacific Rim and Gwaii Haanas, for example, are designated as national park reserves, pending

FIGURE 14.3 Campsite at Sirmilik. *Photo: © Parks Canada/Wayne Lynch/3.01.07.04 (02).*

settlement of Aboriginal land claims encompassing the park areas. They differ, however, in the level of Aboriginal involvement in park designation, management, and planning; in the past Aboriginal people have had little involvement at Pacific Rim, although this has changed in recent years, while they enjoy significant involvement at Gwaii Haanas. This contrasting situation reflects the variability of many other national parks across the country, such as Pukaskwa in Ontario compared to Fundy in New Brunswick, or Riding Mountain compared to Wapusk in Manitoba. The continuum of Aboriginal involvement extends from limited direct involvement to co-operative management. Overall, however, the level of communication, consultation, and co-operation between Parks Canada and Aboriginal peoples has increased dramatically during the last 20 years. While Parks Canada still has a long way to go with respect to greater inclusion of Aboriginal peoples, the organization is often recognized for its leadership in this area.

Pacific Rim National Park Reserve

Pacific Rim National Park Reserve is located on the west coast of Vancouver Island and is divided into three distinct geographic units. The southerly West Coast Trail traverses reserve lands of the Nuu-chah-nulth; these people also have enclave reserves contained in the Long Beach and Broken Group Islands units of the park (Figure 14.4). In total, 28 Indian reserves belonging to seven different bands are either adjacent to the park or enclosed within its boundaries. There are also 289 recorded archaeological sites 'that relate to the native history within Pacific Rim National Park' (Inglis and Haggarty, 1986: 256). The park reserve is part of a larger area occupied by the Nuu-chah-nulth people for approximately 5,000–6,000 years (McMillan and St Clair, 2005). The Nuu-chah-nulth have never been conquered by Europeans, nor have they ceded this territory by treaty. As a result, they are now part of the British Columbia Treaty Process, and negotiations cover a large portion of the west coast of Vancouver Island, fully encompassing the national park reserve.

In spite of the fact that the Nuu-chah-nulth have a significant interest in the park area, a comprehensive land claim under negotiation that fully encompasses the park, and unextinguished Aboriginal rights to hunt and fish in the park region, in the past they had little say in the designation, planning, or management of the park. That, however, has changed significantly in recent years. The Ma-nulth Treaty, representing four Nuu-chah-nulth nations, was ratified in the fall of 2007 and formalized the arrangements for a co-operative management structure as well as for traditional activities within the park. Even before the treaty was in place, the Nuu-chah-nulth had enjoyed some special privileges within the park reserve, as they were allowed to continue harvesting park resources for subsistence purposes. Subsistence gathering of foods from the sea and other ocean resources are managed by the Department of Fisheries and Oceans (DFO), which maintained management jurisdiction over fishery resources within the

BOX 14.4 Pacific Rim National Park Reserve First Nation Initiatives

Pacific Rim has introduced numerous initiatives to improve relationships with surrounding First Nations peoples. Some of these include:

- Allowance of traditional activities, such as the harvesting of medicinal and sacred plants and the documentation and protection of important harvesting sites.
- Development of common conservation goals and the incorporation of traditional knowledge into management programs.
- Enhanced levels of First Nation employment in a variety of occupations.
- Development of training and mentoring programs.
- Promotion of economic opportunities for Aboriginal-operated tourism businesses, including guiding and outfitting.
- Establishment and funding of an agreement (Quu'aas Agreement) for enhanced First Nations presence on the West Coast Trail to provide for cultural resource protection, interpretation, and maintenance.
- Development of cultural interpretation programs.
- Establishment of joint archaeological projects.
- Respect for and use of First Nation place names in signs and brochures.
- Participation in the BC Treaty Process leading towards the establishment of a formal process for Nuu-chah-nulth involvement in the planning and operation of the Parks Canada Agreement with the Esowista First Nation for an expansion of the reserve on park lands to alleviate cramped housing conditions.
- Involvement in a wide variety of joint ecological, cultural, and visitor experience ventures.

park before it was gazetted under the National Parks Act. The Nuu-chah-nulth have special agreements in place that govern these activities. They also actively participate in the Aquatic Management Board (AMB), which provides advice to Fisheries and Oceans on fisheries and other matters pertaining to the ocean.

When Pacific Rim was first established, communication and co-operation with the First Nations who lived both in and around the park was not good and tension existed. For the past couple of decades both Parks Canada and the Nuu-chah-nulth have steadily improved relationships. There have been difficult times, but through consistent efforts at building relationships the situation is greatly improved to the point that other First Nations visit Pacific Rim to discuss how co-operation is achieved outside of a formalized agreement. Box 14.4 outlines some examples of joint initiatives that have helped to develop positive working relationships.

Gwaii Haanas National Park Reserve

Gwaii Haanas is located in the southern portion of the Haida Gwaii archipelago, 170 km offshore from Prince Rupert. In 1985, under the authority of the Haida Constitution, Gwaii Haanas was designated a Haida Heritage Site. On 11 July 1987, a Memorandum of Understanding to negotiate a national park reserve and national marine park on South Moresby Island was signed by the Prime Minister of Canada and the Premier of British Columbia (Sewell et al., 1989). This was followed, in 1988, by the South Moresby Agreement, formally designating the area as a national park reserve. In 1993 the Gwaii Haanas Agreement was signed, setting out the terms of co-management between the Haida Nation and the government of Canada and stipulating that:

> Gwaii Haanas will be maintained and made use of so as to leave it unimpaired for the benefit, education and enjoyment of future generations. More specifically, all actions related to the planning, operation and management of Gwaii Haanas will respect the protection and preservation of the environment, the Haida culture, and the maintenance of a benchmark for science and understanding.

Numerous Aboriginal heritage sites, including the Haida village of Ninstints—a UNESCO World Heritage Site—are situated within the park reserve. The Haida Nation is not actively engaged in the BC Treaty Process. They currently have a case before the Supreme Court of Canada seeking title to their traditional territories. If the Haida are successful, it will be necessary for Parks Canada and the Council of the Haida Nation to revisit the nature of their current relationship.

The level of Aboriginal involvement in park management at Gwaii Haanas is greater than that at Pacific Rim. The Canadian government and the Haida Nation negotiated an Archipelago Management Board (AMB) comprised of two representatives each from the Haida Nation and the government of Canada (Canada, 1990). The AMB is responsible for reviewing all aspects of park operation and management, including park management plans and annual work plans. The Haida are guaranteed continued access to Gwaii Haanas for a host of traditional activities, including: gathering traditional foods; gathering plants for medicinal or ceremonial purposes; cutting trees for ceremonial or

BOX 14.5 Haida Gwaii Watchmen

The Haida people, in recognizing that natural and cultural elements cannot be separated and that the protection of Gwaii Haanas is essential to sustaining Haida culture, initiated the Watchmen Program to protect culturally significant sites in the South Moresby region, now known as Gwaii Haanas. Since the co-operative management of Gwaii Haanas began, key elements of the Watchmen Program remain unchanged, with Watchmen posted at all the previously determined sites with the exception of Burnaby Narrows. The mandate of the program continues to be, first and foremost, the safeguarding of Gwaii Haanas. The presence of the Watchmen plays an important role in the protection of the sensitive sites, accomplished largely by educating visitors about the natural and cultural heritage of Gwaii Haanas and ensuring that visitors know how to travel without leaving a trace of their passage. General information about safety and the latest marine forecasts that come in by radio are also provided.

Source: Parks Canada (1999).

FIGURE 14.4 Totem poles at Ninstints, the World Heritage Site on South Moresby/Gwaii Haanas National Park Reserve. *Photo: P. Dearden.*

artistic purposes; hunting, fishing, and trapping; conducting ceremonies of traditional, spiritual, or religious significance; and seeking cultural and spiritual inspiration.

The AMB is empowered to examine the scope and intent of all Haida subsistence and traditional activities in the park reserve and to ensure that such activities are not contrary to national park purposes (Canada, 1990). Finally, the government of Canada provides training to assist Haida people to qualify for park employment, and the Haida participate in the selection of park employees. Currently, 50 per cent of the staff are Haida, including the field unit superintendent, the most senior position in the park and a member of the Parks Canada executive group.

Southern and Eastern Canadian National Parks

The role of Aboriginal peoples in national park management in the rest of Canada varies according to the status of treaties, the date of park establishment, the historic relationship between Parks Canada and local First Nations, the size of the park, and the proximity of Aboriginal communities to the park. In the mountain parks, Aboriginal people historically have had a very limited role in park management, although in recent years increased communication and co-operation is the trend. In Waterton Lakes National Park, for example, the Blood Nation is co-operating with Parks Canada on a 'good neighbour' basis within their traditional territory, even though they do not have direct jurisdiction in the park.

After a long history of exclusion of Aboriginal people at Riding Mountain, communications have increased in recent years, even though local Aboriginal people are still absent from a formal management role. A specific land claim has been settled with the Keeseekowenin First Nation and a parcel of land on the western shores of Clear Lake has been returned to the First Nation. A 'Senior Officials Forum' with First Nations participation has been established to discuss park management issues, and a historic use study has been completed in co-operation with Aboriginal people. Many Aboriginal people remain frustrated over their exclusion from hunting or ceremonial activities in the park. At St Lawrence Islands a working group on historic site management has been established with the Akwesasne. At Fundy National Park, historically there has been little communication between the Mi'kmaq and Parks Canada, although the First Nation is now a partner in the Fundy Model Forest.

Pukaskwa National Park has a progressive co-operative arrangement with the Ojibways of Pic River. In fact, the Pukaskwa situation is unique among those national parks in treaty areas, since harvesting activities described in the National Parks Act are allowed (Morrison, 1995: 19). According to Parks Canada, 'the Robinson-Superior Treaty Group (RSTG) will play a significant role in helping to achieve this [the park's] mission since the park is located within that portion of Ontario subject to the Robinson-Superior Treaty of 1850' (Parks Canada, 1999). The park management plan advocates strengthening the relationship with First Nations towards a common vision for the park, and further indicates that a joint wildlife management strategy for renewable resource harvesting will be developed in conjunction with representatives of the Robinson-Superior Treaty Group, in order to meet the commitment to allow First Nation harvesting and traditional use activities within the park.

In sum, the progress towards co-operative management with Aboriginal people in southern parks has been steady and widespread, although not yet on a par with the constitutionally mandated arrangements with First Nations where land claims have been settled, or where formal interim arrangements have been negotiated, such as at Gwaii Haanas. However, Parks Canada also must face challenges from other stakeholder groups that it has gone too far in accommodating Aboriginal interests, as discussed in Box 14.6.

BOX 14.6 Parks Canada, Natives Agree to 'Partnership'

Parks Canada and the two First Nation communities in Bruce County are entering into a 'partnership' agreement that gives Aboriginal people a much larger role in the operations of the two national parks in the Tobermory area.

The National Heritage and Resource Management Partnership agreement has been quietly in the works since last March, when Parks Canada officials met with representatives of the Chippewas of Nawash and Chippewas of Saugeen First Nations, jointly known as the Saugeen Ojibwa, 'to discuss areas of mutual concern and explore mutually beneficial interests and opportunities', Northern Bruce Peninsula council was told Monday.

The terms of reference for the new agreement envision 'five main areas of partnership activities' including an increased emphasis on Aboriginal heritage themes and presentation and 'improved resource management strategies and partnerships' for the Bruce Peninsula National Park and Fathom Five National Marine Park.

Many more national park jobs and park-related economic opportunities in the form of goods and services supply work for Nawash and Saugeen band members are also a focus of the agreement which is already being implemented.

Ivan Smith, superintendent of the parks, told council the aim is to have 30 per cent of the jobs in the two parks filled by Nawash and Saugeen band members within two years. One 'sole service' contract, for work being done at the Cyprus Lake campground, has been given to a Nawash contractor from Cape Croker and another such contract is in the works, he said.

That aspect of the partnership has already run into some local opposition. Rob Rouse, a Tobermory resident who also appeared before council, said those actions violate the federal–provincial agreement that created the two national parks in 1987.

That agreement contained numerous conditions the council of the former St Edmunds Township demanded as the price of the local community's support. Rouse recalled the national park was a hotly debated issue for several years after Parks Canada announced in the early 1980s it wanted to create one on the upper peninsula.

The federal agency asked St Edmunds and the former Lindsay Township to set up a local committee to study the proposal. The committee, which did not have any Aboriginal representatives, held numerous public meetings. Lindsay decided to opt out after a municipal election unseated a pro-park council. Meanwhile, St Edmunds drove a hard bargain, insisting on a condition that 75 per cent of the park jobs should go to local people. At the time that was defined as anyone who lived in St Edmunds or was eligible to vote in local elections, Rouse noted.

'It did not include the people living at Cape Croker or Saugeen communities, 50 and 70 miles from St Edmunds', he said. There was 'clear evidence from the beginning there were to be separate local and Aboriginal hiring programs', he added.

Rouse said all surviving St Edmunds councillors from the park debate days and all surviving members of the local park study committee have confirmed the original intent of the employment condition.

He said Parks Canada violated another condition in the 1987 federal–provincial agreement by not consulting the local community about the new Aboriginal partnership. He said he first heard about it when he was approached last fall by national park employees worried about their jobs.

Smith said Parks Canada has long considered people from the Nawash and Saugeen communities to be 'local' within the meaning of the condition cited by Rouse. The First Nations have a large hunting ground south of Tobermory and 'as landowners in the municipality, we accept them as local residents as well.'

Smith said local, non-Aboriginal residents now account for about 65 per cent of national park jobs—74 per cent if Aboriginal people are included. Parks Canada has 'done a lot for Tobermory, but had not come close for Aboriginal people. . . . We want to increase that.'

'Parks Canada is confident that the presence and active involvement of First Nations people at the (national parks) will enrich our decisions, stories and the experiences that we offer Canadians and international visitors, and so have agreed to the terms of reference', said a report handed out to members of council.

'These are the traditional lands of the Saugeen Ojibwa. We welcome and actively encourage their stories, traditions and practices', it said.

The Municipality of Northern Bruce Peninsula was created in 1999 when St Edmunds, Lindsay and Eastnor townships, and the former Village of Lion's Head were amalgamated. Rouse said only the local municipal council has the right to interpret the meaning of the 'local' employment condition in the 1987 park agreement. He asked council to make a 'clear and precise declaration of what "local employee" means'.

Mayor Milt McIver told Rouse council would respond to his request 'at a later date'.

Source: Phil McNichol, 'Parks Canada, natives agree to "partnership"', *Owen Sound Sun Times*, 14 Feb. 2007, A1.

DISCUSSION AND CONCLUSION

Prior to the late 1970s, Canadian national parks were designated with little consideration for Aboriginal peoples. However, following Justice Thomas Berger's (1977) groundbreaking northern pipeline inquiry, Parks Canada became more sensitive to the concerns of Aboriginal people. In fact, the 1979 *Parks Canada Policy* embraced the concept of joint management of parks by government and Aboriginal people fully eight years before this same concept was adopted in Canada's land claims policy.

Land claim settlements, Supreme Court decisions, and Aboriginal treaty rights now play as great a part as national park policy or legislation in determining the role of Aboriginal peoples in planning for and managing national parks. Diverse approaches to settlement of land claims, as well as varying treaty rights, result in differing relationships between Aboriginal people and national parks throughout Canada. For example, a significant park planning and management role is accorded Aboriginal peoples in northern Canada where parks are tied to settlement of land claims. Similar situations have evolved in southern Canada, now that Aboriginal and treaty rights have been given greater recognition by the courts. However, significant disparities still exist in the relations that various Aboriginal groups have with Parks Canada.

Parks Canada may have exacerbated this situation through its ad hoc approach to relations with Aboriginal peoples. On the one hand, such an approach affords a level of flexibility, allowing park managers to respond to the exigencies of individual situations. On the other hand, without clearly defined parameters for relations with Aboriginal people, other challenges are likely to emerge as all Aboriginal groups strive to achieve the highest possible standard of participation in park management. Other complex issues are likely to arise as Aboriginal and treaty rights continue to be defined. The question of Aboriginal and treaty rights to hunting and fishing has yet to be definitively dealt with by Parks Canada, although the courts are giving increasing recognition to an Aboriginal right to hunt and fish. These court rulings have a direct impact on the management of national parks situated in regions traditionally used by Aboriginal people, and for the future planning of parks proposed for such regions.

Similar issues emerge concerning treaty rights to hunt and fish. As part of most treaties in Canada, Aboriginal groups were promised continued rights to hunt and fish on 'unoccupied Crown land'. The current position of Parks Canada is that national parks are considered occupied Crown lands, and this distinction has yet to be fully resolved, either in the courts or in federal policy. However, Parks Canada has clearly changed its approach to managing existing national parks and to planning for, and designating, new ones that fall within lands traditionally used by Aboriginal peoples (see, e.g., Tate, 2003, for a detailed discussion of the interactions with Aboriginal peoples in the expansion of Nahanni National Park Reserve). A majority of national parks, and as yet unrepresented national park terrestrial natural regions, fall within territory traditionally used by Aboriginal peoples. Thus, Aboriginal and treaty rights are likely to have a continuing significant impact on national park management in Canada. Parks Canada is evolving to meet the demands of this changing situation, but it must give greater attention to the question of how Aboriginal and treaty rights are to be incorporated with park management practices if it is to adequately manage our national parks.

Since the 1988 amendment to the National Parks Act, Parks Canada's mandate has focused more closely on the maintenance of ecological integrity. This reaffirmation of the Agency's conservation mandate was further underscored by the Panel on the Ecological Integrity of Canada's National Parks (Parks Canada Agency, 2000). In recent years, Parks Canada Agency has focused attention on the three integrated elements of the mandate: protection, memorable visitor experiences, and education. The scientific definitions and objectives related to managing parks for ecological integrity are new to many Aboriginal peoples, as they are to the general public. Nonetheless, the Panel found widespread support for the concept of ecological integrity, provided that it encompasses human use as part of the ecosystem. Some parks, such as Kluane and Nahanni, have already prepared ecological integrity statements in co-operation with First Nations, where cultural landscapes and traditional activities are recognized and incorporated into ecological integrity considerations. Maintenance of ecological integrity appears to be consistent with the holistic world view of many Aboriginal peoples and may facilitate greater co-operation in the future management of national parks.

REFERENCES

Berger, T.R. 1977. *Northern Frontier, Northern Homeland*, 2 vols. Toronto: James Lorimer.

———. 1982. 'The Nishga Indians and Aboriginal Rights', in T.R. Berger, ed., *Fragile Freedoms: Human Rights and Dissent in Canada*. Toronto: Irwin, 219–54.

British Columbia. 1875, reprinted 1987. *Papers Connected With the Indian Land Question, 1850–1875*. Victoria, BC: Richard Wolfenden.

Calder v. Attorney General of British Columbia, 1969. 8 DLR (3rd) 59, 71 WWR 81 (Supreme Court of British Columbia).

Calder v. Attorney General of British Columbia, 1970. 13 DLR (3rd) 64, 74 WWR 481 (British Columbia Court of Appeal).

Calder v. Attorney General of British Columbia, 1973. SCR 313, 34 DLR (3rd) 145, [1973] 4 WWR 1 (SCC).

Canada. 1981. *In All Fairness: A Native Claims Policy*. Ottawa: Minister of Supply and Services.

———. 1984. *Inuvialuit Final Agreement*. Ottawa: Minister of Indian Affairs and Northern Development.

———. 1987. *Comprehensive Land Claims Policy*. Ottawa: Minister of Supply and Services.

———. 1990. Gwaii Haanas/South Moresby Agreement between the Government of Canada and the Council of the Haida Nation. Available from the Office of the Minister of Environment.

———. 1993. *Vuntut Gwitchin First Nation Final Agreement*. Ottawa: Department of Indian Affairs and Northern Development.

———. 2000. Bill C-27, an Act respecting the National Parks of Canada.

Canadian Arctic Resources Committee. 1993. 'Nunavut land claim agreement', *Northern Perspectives* 21: 3.

Cumming, P.A., and N.H. Mickenberg, eds. 1972. *Native Rights in Canada*, 2nd edn. Toronto: Indian Eskimo Association of Canada in association with General Publishing.

Dasmann, R.F. 1976. 'National parks, nature conservation and future primitive', *Ecologist* 6: 164–7.

Dearden, P., and L. Berg. 1993. 'Canadian national parks: A model of administrative penetration', *Canadian Geographer* 37: 194–211.

Delgamuukw v. British Columbia, 1997.

Dewhirst, J. 1978. 'Nootka Sound: A 4,000 year perspective', *Sound Heritage* 7: 1–30. (On file at Public Archives of BC.)

Doering, R.L. 1983. *Nunavut: Options for a Public Lands Regime*. Working Paper No. 3. Ottawa: Nunavut Constitutional Forum.

Elias, P.D. 1989. 'Aboriginal rights and litigation: History and future of court decisions in Canada', *Polar Record* 25: 1–8.

Fenge, T., I. Fox, B. Sadler, and S. Washington. 1986. 'A proposed port on the north slope of Yukon', in B. Sadler, ed., *Environmental Protection and Resource Development: Convergence for Today*. Calgary: University of Calgary Press, 127–78.

Fisher, R. 1977. *Contact and Conflict: Indian–European Relations in British Columbia 1774–1890*. Vancouver: University of British Columbia Press.

Frideres, J.S. 1988. *Native People in Canada: Contemporary Conflicts*, 3rd edn. Scarborough, Ont.: Prentice-Hall.

Inglis, R.I., and J.C. Haggarty. 1986. 'Pacific Rim National Park ethnographic history', unpublished manuscript on file at Royal British Columbia Museum.

Johnson, J. 1994. 'A Parks Canada perspective on Vuntut National Park', in Peepre and Jickling (1994).

Lothian, W.F. 1976. *A History of Canada's National Parks*, vol. 1. Ottawa: Parks Canada.

MacDonald, D., T. MacDonald, and L. McAvoy. 2000. 'Tribal wilderness research needs and issues in the United States and Canada', in S.F. McCool, D.N. Cole, W.T. Borrie, and J. O'Loughlin, comps, *Wilderness Science in a Time of Change Conference*, vol. 2, *Wilderness within the Context of Larger Systems*. Fort Collins: Colorado State University, 290–4.

McMillan, A.D. 1988. *Native Peoples and Cultures of Canada: An Anthropological Overview*. Vancouver: Douglas & McIntyre.

——— and D. St Claire. 2005. 'Ts'ishaa: Archaeology and ethnography of a Nuu-chal-nulth original site in Barkley Sound'. Report to Parks Canada.

Morrison, J. 1995. 'Aboriginal interests', in M. Hummel, ed., *Protecting Canada's Endangered Spaces: An Owner's Manual*. Toronto: Key Porter.

Nash, R. 1970. 'The American invention of national parks', *American Quarterly* 22: 726–35.

Njootli, S. 1994. 'Two perspectives—one park: Vuntut National Park', in Peepre and Jickling (1994).

Parks Canada. 1979. *Parks Canada Policy*. Ottawa: Department of Indian and Northern Affairs.

———. 1994. *Guiding Principles and Operational Policies*. Ottawa: Department of Canadian Heritage.

———. 1997a. *State of the Parks 1997 Report*. Ottawa: Department of Canadian Heritage.

———. 1997b. *National Parks System Plan*, 3rd edn. Ottawa: Department of Canadian Heritage

———. 1999. 'Vuntut National Park: Interim Management Guidelines', unpublished.

———. 1999. Website for individual national parks: <parkscanada.pch.gc.ca/parks/alphap2e.htm>.

Parks Canada Agency. 2000. *Unimpaired for Future Generations? Protecting Ecological Integrity with Canada's National Parks*, vol. 2, *Setting a New Direction for Canada's National Parks*. Report of the Panel on the Ecological Integrity of Canada's National Parks. Ottawa.

———. 2006. *A Handbook for Parks Canada Employees on Consulting with Aboriginal Peoples*. Ottawa: Department of Canadian Heritage.

Peepre, J., and B. Jickling, eds. 1994. *Northern Protected Areas and Wilderness*. Proceedings of a Forum, 1993, Canadian Parks and Wilderness Society and Yukon College. Whitehorse, Yukon.

Raunet, D. 1984. *Without Surrender, Without Consent: A History of the Nishga Land Claims*. Vancouver: Douglas & McIntyre.

Sanders, D. 1973. 'The Nishga case', *BC Studies* 19: 3–20.

———. 1983. 'The rights of the Aboriginal peoples of Canada', *Canadian Bar Review* 61: 314–38.

Sewell, W.R.D., P. Dearden, and J. Dumbrell. 1989. 'Wilderness decision-making and the role of environmental interest groups: A comparison of the Franklin Dam, Tasmania and South Moresby, British Columbia cases', *Natural Resources Journal* 29: 147–69.

Sparrow v. The Queen. 1986a. County Court, [1986] BCWLD 599.

Sparrow v. The Queen. 1986b. 9 BCLR (2nd) 300, 36 DLR (4th) 246, [1987] 2 WWR 577.

Sparrow v. The Queen et al. 1990. Supreme Court of Canada. Chief Justice Dickson, and Justices McIntyre, La Forest, Lamer, Wilson, L'Heureux-Dubé, and Sopinka. 31 May, *QuickLaw Reports*, file 20311.

Task Force to Review Comprehensive Claims Policy. 1985. *Living Treaties, Lasting Agreements: Report of the Task Force to Review Comprehensive Claims Policy*. Ottawa: Department of Indian Affairs and Northern Development.

Tate, D.R. 2003. 'Expanding Nahanni National Park Reserve: The contribution of research, consultation and negotiation', in N.W.P. Munro, P. Dearden, T.B. Herman, K. Beazley, and S. Bondrup-Nielsen, eds, *Making Ecosystem-based Management Work*. Proceedings of the Fifth International Conference on Science and Management of Protected Areas, Victoria, BC, May 2003. Wolfville, NS: SAMPAA. At: <www.sampaa.org/publications.htm>.

Tungavik Federation of Nunavut. 1990. Agreement-in-Principle between the Inuit of the Nunavut Settlement Area and Her Majesty in Right of Canada, mimeo.

KEY WORDS/CONCEPTS

Aboriginal Secretariat (1999)
ad hoc approach to Aboriginal involvement in parks
Berger Inquiry
Calder case
Catlin's view of national parks
comprehensive land claims
Constitution Act, 1982
co-operative management
Delgamuukw
Ecological Integrity Panel
Guiding Principles and Operational Policies (1997)
Haida Gwaii Watchmen Program
Inuit
Inuit land claims
Inuvialuit Final Agreement (1984)
local knowledge
national park reserve
National Parks Act (1988, 2000)
National Parks System Plan (1997)
Nunavut Agreement joint management
Parks Canada Policy (1979)
Robinson treaties
Sparrow case
Umbrella Final Agreement (1994)
Williams treaties

STUDY QUESTIONS

1. Discuss how Catlin's view of national parks differs from more traditional perspectives.
2. Outline the significance of each of the following for Aboriginal involvement with parks in Canada: *Calder*, Constitution Act, *Sparrow*, *Delgamuukw*, comprehensive land claim policy, Berger Inquiry.
3. Comment on the statement: 'Adherence to the natural areas framework may have contributed to the estrangement of parks from Aboriginal people.'
4. Outline the significance of each of the following for Aboriginal involvement with parks in Canada: 1994 *Parks Canada Policy*; 1997 *Systems Plan*; 1988 Amended National Parks Act; 2000 Canada National Parks Act.
5. Comment on the significance of national park reserves.
6. Of what benefit are national parks to Aboriginal people?
7. Of what benefit are Aboriginal people to national parks?
8. Discuss the variation in Aboriginal involvement with national parks between northern Canada and southern Canada.
9. Discuss the variation of Aboriginal involvement in national parks across southern Canada.
10. Examine a park management plan for evidence of Aboriginal involvement.
11. Outline some of the main improvements that have been made in how Aboriginal peoples are involved with park establishment and management in Canada.

CHAPTER 15

Marine Protected Areas

Philip Dearden and Rosaline Canessa

INTRODUCTION

As the global population continues to grow, the population is becoming increasingly littoral. Over half the world's inhabitants live within 200 km of the coast (Hinrichsen, 1998). More people mean more waste products and more people depending on the ocean for their livelihood. The United Nations Food and Agricultural Organization (FAO) estimates that 75 per cent of the world's marine fisheries are overexploited or fully exploited (FAO, 2007). For example, in 2001 the Fisheries Resource Conservation Council recommended that the annual catch of cod from the north and east coasts of Newfoundland be limited to 5,600 tonnes, a far cry from the 800,000 tonnes of the 1960s. The stock is acknowledged to be at its lowest level in recorded history despite a 15-year moratorium. And this is not an isolated case, as witnessed by similar concerns regarding fisheries on the Pacific coast. We are also reaching further and further down the food chain in our efforts to feed ourselves (Pauly et al., 1998), and marine habitats are becoming increasingly degraded (UNEP, 2006). The first major international survey of coral reefs found that 69 per cent of reefs were seriously degraded (*People and the Planet*, 1998), and over half the world's salt marshes, mangroves, and coastal wetlands have been destroyed. Recent surveys of the world's richest coral reef systems in the Indo-Pacific region show annual losses of over 3,000 km^2 per year, and a rate of destruction more than five times that of tropical rain forests (Bruno and Selig, 2007).

Yet, the oceans and their well-being are integral to sustaining life on this planet. They are key components in global cycles and energy flows. They are home to a vast array of organisms and display greater diversity of taxonomic groups than their terrestrial counterparts. Not only do these organisms help feed us, they are also the source of many valuable medicinal products. As well, we use the seas to dump our waste products and carry most of our goods around the world. The oceans enrich our cultures as individuals and nations draw strength and inspiration from their links to this vital and awesome life force.

Due to increasing degradation, many nations are seeking to improve marine conservation activities. One way to do this is through the development of international regulatory agreements that cover a wide range of human interactions with the oceans. Initiatives in this regard have included the London Convention on the Prevention of Marine Pollution by Dumping of Wastes and Other Matter (1972), the Convention for the

Prevention of Pollution from Ships (the Marpol Convention, 1973, 1978), the Montreal Guidelines for the Protection of the Marine Environment against Pollution from Land-based Sources (1985), and the environmental sections of the Law of the Sea Convention. A second way, used mainly to address conservation of commercial species, has been through regulations to limit fisheries catches. The inadequacy of this approach by itself can be clearly seen on the Pacific and Atlantic coasts of Canada. Overharvesting is also the main threat to marine species listed as endangered in Canada (Venter et al., 2007). A third way, and the focus of this chapter, is the protection of specific areas through designation as marine parks or reserves. In reality, all these approaches are necessary components of a comprehensive strategy to address marine conservation. It is the development of marine protected areas (MPAs), however, that has had the least attention, and currently only 0.6 per cent of the world's oceans are designated as MPAs (Wood, 2006).

One reason for this has been the difficulty in translating our well-established approaches for areal protection in terrestrial environments into the marine context. Setting a few boundaries, after all, will not prevent life or wastes from flowing in and out of a marine sanctuary. However, over time it has become apparent that these problems are not unique to the marine environment—many of our terrestrial parks suffer from trans-boundary effects (see Chapter 13). Further knowledge of marine environments has also indicated the values of designating specific areas for the protection of habitat and biodiversity, both for their own value and as an aid to maintaining viable fisheries (e.g., Shears et al., 2006).

Internationally, marine protected areas are defined as 'any area of intertidal or subtidal terrain, together with its overlying waters and associated flora, fauna, historical and cultural features, which has been reserved by legislation or other effective means to protect part or all of the enclosed environment' (Kelleher and Rechia, 1998). The potential contributions of MPAs include:

- protection of marine biodiversity, representative ecosystems, and special natural features (Sobel, 1993);
- support in rebuilding depleted fish stocks, particularly groundfish, by protecting spawning and nursery grounds (e.g., see Wallace et al., 1998);
- insurance against current inadequate management of marine resources;
- provision of benchmark sites against which to evaluate human impacts elsewhere and undertake scientific research;
- recognition of cultural links of coastal communities to biodiversity;
- provision of opportunities for recreation and education.

The effectiveness of marine reserves in providing for these benefits is largely conditional on size, how they are linked to other reserves, the kinds of activities allowed within the borders, the co-operation of local communities in supporting the reserve, the vulnerability to polluting influences from outside, and the ability to mitigate these negative impacts.

THE CANADIAN CONTEXT

Canada has the longest coastline of any country in the world along three oceans and the second largest area of continental shelf. For over 30 years Canada has sought to gradually extend its jurisdiction over these waters and was a significant contributor to the UN Law of the Sea Convention (UNCLOS) between 1974 and 1982. Nevertheless, it was not until 2003 that Canada ratified UNCLOS. Ratification allows Canada to extend its sovereign rights over the seabed to the edge of the continental shelf, where it extends beyond the 200 nautical mile (370.4 km) exclusive economic zone (EEZ), thus adding 1,750,000 km^2 (an area equivalent to the size of Canada's three prairie provinces). Once declared, Canada's jurisdiction will cover approximately 7 million km^2, some 2 per cent of the world's oceans. This vast area and diversity of habitats give rise to a spectacular marine life wherein all major groups of marine organisms are represented. There are some 1,100 species of fish and globally important populations of many marine mammals, including grey, bowhead, right, beluga, minke, humpback, and killer whales. Unfortunately, several of these species are also on Canada's endangered and threatened lists, including northern abalone, leatherback sea turtle, some salmon populations, Atlantic cod, and the beluga, bowhead, right, and Georgia Strait killer whales (COSEWIC, 2007).

These pressures on marine species, at least in part, are a result of the high proportion of the Canadian population living in the coastal zone. Manson (2005) suggests that almost 40 per cent of Canada's population lives within 20 km of a coast (including the Great Lakes) and this will increase by 5 million people by 2015.

Canada's biophysical diversity is paralleled by its jurisdictional complexity. The federal government, for example, is empowered to deal with navigation, fisheries, and general law-making and has a raft of legislation (e.g., the Fisheries Act, the Canada Shipping Act, the Canadian Environmental Protection Act, the Coastal Fisheries Protection Act) to enable it to do so. However, coastal provinces can also have considerable influence and have jurisdiction over activities such as aquaculture, fish processing and marketing, and ocean bed mining and drilling. In offshore areas the seabed is under federal jurisdiction, but in 'internal' waters, it is provincial. A 'federal' grey whale can enter a provincial marine park, and while still federal in the water column it may disturb the provincial substrate and eat a provincially protected amphipod that spent a short time as a free-swimming federally controlled larva, that subsists on federally supplied detritus from the water column. In subsequent years, the grey whale can enter a biosphere reserve in Mexico, a port authority in California, cross federal and state marine protected areas, become a target for Makah traditional fisheries, or even become a member of the International Whaling Committee sanctioned hunt in the Chuckchi Sea. Or wash up in a municipality on southern Vancouver Island and confound authorities as to who exactly is responsible for disposing of this once great and wonderful animal.

MPA PROGRAMS IN CANADA

At the federal level three agencies have a mandate to establish MPAs. In addition, many of the 11 sub-federal jurisdictions that have coastal waters also have multiple applying legislation, and Quebec, British Columbia, Newfoundland and Labrador, and Prince Edward Island have designated marine areas within their protected areas system (see Chapter 3). Estimates suggest that Canada has over 560 protected areas with a marine component, covering approximately 3 million hectares or 0.4 per cent of Canada's territorial and exclusive economic zone waters (Table 15.1) (Wood, 2007; Environment Canada, 2006). This section outlines some of the federal and provincial MPA programs. MPAs discussed in the following sections are shown in Figure 15.1.

Federal Initiatives

At the federal level, Fisheries and Oceans Canada (DFO), Environment Canada, and Parks Canada have authority to designate some form of MPA. Table 15.2 gives a summary of the various agencies and the legislative tools at their disposal. Fisheries and Oceans

FIGURE 15.1 Marine sites mentioned in text.

Table 15.1 Marine Protected Areas in Canada

Administrator	Type of Marine Protected Area	# of Marine Areas	Marine Area Protected (ha)	% Protected
Parks Canada	National Marine Conservation Area	1	11,500	
Parks Canada	National Park (Marine Portion)	11	938,000	32.2%
Parks Canada & Quebec	Saguenay–St Lawrence Marine Park	1	113,800	
Environment Canada	National Wildlife Area (Marine Portion)	13	152,317	47.6%
Environment Canada	Migratory Bird Sanctuary (Marine Portion)	51	1,417,155	
Fisheries and Oceans	Marine Protected Area	6	254,320	7.7%
Newfoundland and Labrador	Ecological Reserve (Marine Portion)	6	15,200	0.5%
Prince Edward Island	Terrestrial Protected Area (Marine Portion)	1	87	0.003%
Quebec	Terrestrial Protected Area (Marine Portion)	353	195,333	5.9%
British Columbia	Terrestrial Protected Area (Marine Portion)	120	201,129	6.1%
Total		**563**	**3,298,841**	**100%**

Sources: Environment Canada (2006); Canessa and Lunn (2005).

Canada, through section 29 of the Oceans Act (1997), is the lead agency in developing and implementing Canadian strategy for marine and coastal conservation, protection, and management. Recognizing the need for co-ordinated efforts among federal agencies, the government published *Working Together for Marine Protected Areas* in 1998, followed in 1999 by DFO's *Marine Protected Areas Policy* and the *National Framework for Establishing and Managing Marine Protected Areas.* Despite these reports, federal co-ordination on MPAs has remained inconsistent and opportunistic. In response, under the leadership of DFO the three federal agencies jointly developed a *Federal Marine Protected Areas Strategy* to co-ordinate their efforts to establish a network of marine protected areas of ecologically significant and representative areas within an integrated oceans management framework. The goal of this initiative is to contribute to the health of Canada's oceans and marine environments (Canada, 2005).

Parks Canada

Parks Canada differs from the other federal agencies in that there is an explicit statement regarding the encouragement of public understanding, appreciation, and enjoyment, whereas DFO and Environment Canada are concerned particularly with conservation purposes. Furthermore, Parks Canada has a specific goal of establishing a system of national marine conservation areas within each of the 29 defined marine natural regions (Figure 15.2) represented by one or more protected areas (Mercier and Mondor, 1995). Unlike national parks, national marine conservation areas (NMCAs) are managed for 'sustainable use' rather than strict protection of ecological integrity. They are to be managed on a partnership basis with local stakeholders, allowing most existing extractive uses to continue. Furthermore, unlike terrestrial parks, other agencies have jurisdiction

Table 15.2 Federal Statutory Powers for Protecting Marine Areas

Agency	Legislative Tools	Designations	Mandate
Fisheries and Oceans Canada	Oceans Act	Marine Protected Areas	To protect and conserve: • fisheries resources, including marine mammals and their habitats; • endangered or threatened species and their habitats; • unique habitats; • areas of high biodiversity or biological productivity; • areas for scientific and research purposes.
	Fisheries Act	Fisheries Closures	Conservation mandate to manage and regulate fisheries, including closing areas to fishing.
Environment Canada	Canada Wildlife Act	National Wildlife Areas Marine Wildlife Areas	To protect and conserve nationally significant habitats for migratory birds and wildlife for research, conservation, and interpretation.
	Migratory Birds	Migratory Bird Convention Act Sanctuaries	To protect coastal and marine habitats that are heavily used by migratory birds for breeding, feeding, and overwintering while providing for educational and interpretive purposes.
Parks Canada	National Parks Act	National Parks	To maintain the ecological integrity of natural environments that are representative of the 39 natural regions of Canada while providing opportunities for public understanding, appreciation, and enjoyment.
	National Marine Conservation Areas Act	National Marine Conservation Areas	To protect and conserve marine areas that are representative of Canada's ocean environments and Great Lakes, to provide a model for sustainable use of marine species and ecosystems, and to encourage public understanding, appreciation, and enjoyment of marine heritage.

Source: Fisheries and Oceans Canada (2005). *Canada's Federal Marine Protected Areas Strategy.* Reproduced with the permission of Her Majesty the Queen in Right of Canada, 2008.

within the NMCAs for managing renewable marine resources and navigation and shipping, although the federal government has clear title to all coastal and submerged lands within an NMCA. This requirement is currently one of the main factors delaying MPA formation in the southern Gulf Islands on the Pacific coast, as BC is unwilling to transfer provincial seabed to federal jurisdiction.

Within these multiple-use areas are zones where various levels of protection may be achieved. A minimum of two zones is required in an NMCA: at least one zone fully protects special features or sensitive elements of the ecosystems, where renewable resource harvesting and permanent facilities will not be allowed, nor is visitor use under normal circumstances; and another zone fosters and encourages ecologically sustainable use of marine resources. Other types of zones may also be included. Most traditional

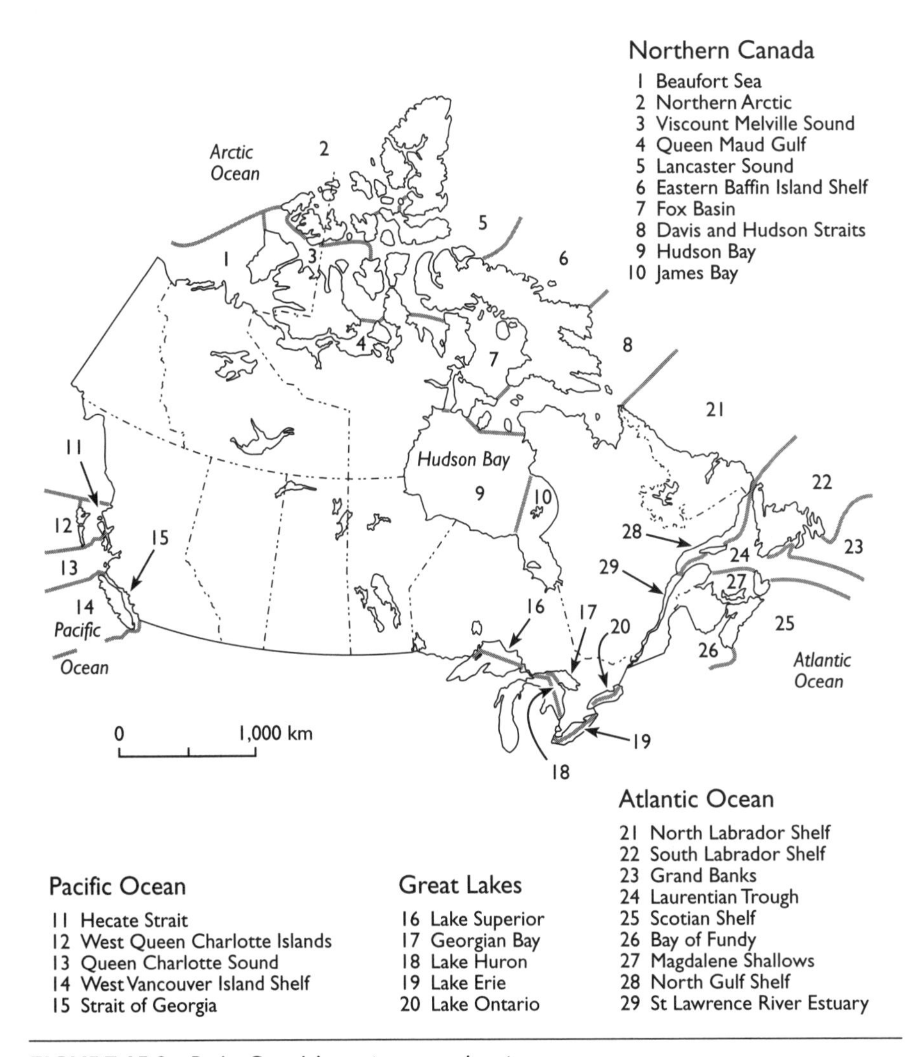

FIGURE 15.2 Parks Canada's marine natural regions.

fisheries will be allowed to continue, although Parks Canada will negotiate fisheries management plans with DFO in and around NMCAs. Only ocean disposal, seabed mining, and oil and gas extraction will be totally prohibited.

Zoning and other regulations are to be specified in NMCA management plans, which are to be produced within five years of establishment, and reviewed and tabled in Parliament every five years subsequently. These plans will specify provisions for ecosystem management, visitor use, protection, and zoning, with ecosystem management and the precautionary principle being the main foundations. An advisory committee will also be struck for each NMCA to advise the minister on the management plan.

The Act to establish NMCAs was passed in 2002. Although an agreement has been reached between the federal government and Ontario to establish the Lake Superior NMCA, at the moment no areas are protected under this legislation. Studies are currently underway to establish NMCAs in a number of locations, including Haida Gwaii (Queen Charlotte Islands), Isle de la Madeleine, and the southern Strait of Georgia in British Columbia. In addition, significant marine components are included within 11 established national parks.

Although currently there are no legally established NMCAs, Parks Canada is counting two existing sites as fulfilling a similar purpose. Canada's first national marine park was established on 1 December 1987 through an agreement between Canada and the province of Ontario. The site, Fathom Five, was formerly a provincial park that was transferred to federal jurisdiction with the establishment of the Bruce Peninsula National Park on adjacent lands. This site has long been recognized as important for its cultural significance in terms of shipwrecks and recreational diving. The Saguenay–St Lawrence Marine Park is an MPA with its own unique legislation, the Saguenay–St Lawrence Marine Park Act. Special legislation was required to bypass the necessity of transferring jurisdiction over the seabed to the federal government, which Quebec was not prepared to do. This site had only a limited commercial fishery, but there is no doubt as to its international significance as a major area of upwelling that is especially important for marine mammals. These facts were important in easing the establishment of the area, as well as in the collaborative nature of the partnership among federal, provincial, and Aboriginal governments and various stakeholder groups.

Fisheries and Oceans Canada

Fisheries and Oceans has adopted many of the same approaches as Parks Canada, in that marine protected areas (MPAs) will be oriented towards sustainable use and established through partnerships. Unlike Parks Canada, there is no predetermined goal of representing different regions in DFO's MPA system. Instead, areas of interest (AOIs) are nominated by regional groups, supplemented by regional overviews undertaken by technical interdisciplinary teams. These nominations are then subject to more intensive ecological, technical, and socio-economic evaluations to choose among various candidate areas. One proposal, for example, is to establish an MPA in the St Lawrence Estuary contiguous with the Saguenay–St Lawrence Marine Park discussed above. Such a proposal would improve conservation in the area by including summer distributions of beluga whales, haul-outs for harbour seals where they rest on rocks and shoreline areas, and the most important feeding sites for blue whales. This complementarity between the different types of MPAs may well be a harbinger of the structure of a more mature national network of MPAs that has yet to be defined.

Similar to the Parks Canada approach, it is envisioned that there will be internal zoning within DFO MPAs, allowing various levels of resource extraction. The Oceans Act provides for regulations to manage MPAs but does not specify what these are, leaving a wide degree of interpretation and regional flexibility. Management plans for individual MPAs will provide details on protection standards, regulations, permissible activities, and enforcement. A management plan is required before an area can be designated under the Oceans Act. The lack of national standards raises the quintessentially Canadian debate about whether the program is in fact 'national' at all, or might just as well be left to regional jurisdiction.

In the 10 years since passage of the Oceans Act six sites have been established. On the Pacific coast the only designated site is Endeavour Hydrothermal Vents, 250 km west of Victoria, which was designated in 2003. The hot vents are important sites for marine biodiversity. Of 236 species that have been collected at the vents, 223 were previously unknown to science, representing at least 22 new families and 100 new genera. The food chains around the vents rely not on the sun for energy supply, as do most other species, but on bacteria that make carbon compounds from the sulphur-rich emissions. Bowie Seamount, 180 km west of the Queen Charlotte Islands, is also actively being considered for designation. Bowie Seamount rises to within 25 m of the surface from depths of over 3,300 m. This is an important location for marine biodiversity, providing feeding sites of great importance to migratory seabirds, marine mammals, and distinctive rockfish communities.

Five Atlantic MPAs have been designated: Sable Gully near Sable Island, designated in 2004; Basin Head in the Gulf of St Lawrence, Eastport in Bonavista Bay, and Gilbert Bay in the Labrador Sea, all designated in 2005; and Musquash Estuary in the Bay of Fundy, designated in 2006. DFO has committed to establish nine MPAs between 2005 and 2010. To date, four have been designated within this time period and there are currently five proposals under consideration: Manicougan (Quebec), St Lawrence Estuary (Quebec), Beaufort Sea (NWT), Bowie Seamount (BC), and Race Rocks (BC). By 2012, DFO is projecting that 16–17 sites will have been established (Table 15.3).

Environment Canada

The main focus of the third federal component, Environment Canada (Table 15.2), is wildlife, particularly migratory bird species (Zurbrigg, 1996). The programs are administered by the Canadian Wildlife Service (CWS) and include designations for marine areas as national wildlife areas (NWAs) and marine wildlife areas (MWAs) created and managed pursuant to wildlife area regulations made under the Canada Wildlife Act and migratory bird sanctuaries (MBSs) under the Migratory Birds Convention Act of 1917 and 1994. NWAs must be owned by, or leased to, the federal government and might include a wide range of wildlife habitat types, not just for migratory birds, from uplands through to marine areas up to 12 nautical miles (22.29 km) from shore. MWAs extend this jurisdiction to the 200-mile (370.4 km) limit (CWS, 1999). MBSs differ in that federal ownership is not required. Habitats are mainly coastal, including marine waters surrounding islands.

BOX 15.1 Sable Gully

Sable Gully is the largest underwater canyon in eastern Canada. The Gully is located approximately 200 km off the coast of Nova Scotia, at the edge of the Scotian Shelf where the sea floor suddenly drops to over 2 km in depth. Over 70 km long and 20 km wide, this area is home to many interesting and unusual species. The Gully is a productive ecosystem that supports a diversity of marine organisms. The world's deepest diving whale, the bottlenose whale, is a 'vulnerable' species that lives in the Gully year-round. Fin whales and Northwest Atlantic blue whales, both also classified as 'vulnerable' by COSEWIC, make use of the Gully throughout the year. Deep sea corals are a significant feature of the benthic fauna in the area, and nine species are confirmed to live in the Gully. The occurrence of corals in such conditions provides exceptional opportunities for the study of these animals.

Sable Gully is a unique ecological site that has attracted the attention of a wide range of government agencies, researchers, area resource industries, and conservationists. In December 1998, Sable Gully was announced as an area of interest under the Marine Protected Area program of DFO and was designated in May 2004. Typically, MPAs in Canada and throughout the world have been located in coastal areas. Sable Gully is offshore, which presents unique challenges for management, research, and monitoring. The MPA regulations prohibit the disturbance, damage, destruction, or removal of any living marine organism or habitat within the Gully. The MPA contains three management zones, each providing varying levels of protection based on conservation objectives and ecological sensitivities. An ecosystem approach has been applied in the design of the MPA wherein human activities are assessed against the ecosystem features being protected in the Gully. The regulations also control human activities in areas around the Gully that could cause harmful effects within the MPA boundary.

Sources: Sable Gully CD; Fisheries and Oceans Canada, at: <www.dfo-mpo.gc.ca/oceans-habitat/oceans/mpa-zpm/factsheets-feuillets/sable_e.asp>.

Although in the future these designations are not expected to play a major role in the extent of marine area protected (Table 15.3), they currently account for almost 50 per cent of protected marine area (see Table 15.1). There are 13 (out of 51) NWAs with marine components, and 51 (out of 92) MBSs, adding up to a total of over 1.5 million ha protected. Unfortunately, over 95 per cent of this is concentrated in the Arctic, and as of yet no MWAs have been designated, although the Scott Islands, off the northwest tip of Vancouver Island, is anticipated to be the first MWA. Candidate MWAs are selected based on a minimum Canadian population or appreciable assemblage of migratory birds or species at risk, or critical habitat for a species-at-risk population. No minimum national set of regulations applies to extractive activities, as regulations are developed on a site-by-site basis.

TABLE 15.3 Projected Progress in MPA Designation by 2012

Federal Agency	Current Status	By 2012
Fisheries and Oceans Canada	6 MPAs 4–5 proposed MPAs in progress 6 new sites to be identified	16–17 MPAs
Environment Canada	No MWA 64 'marine' NWAs and MBSs	1 MWA 64 NWAs/MBSs + 4 new 'marine' NWAs
Parks Canada	2 NMCAs 11 national parks with marine components 4 proposed NMCAs in progress 1–3 NMCA proposals pending	6-9 NMCAs 11 national parks with marine components

Source: Landry et al. (2007).

BOX 15.2 Nirjutiqawik National Wildlife Area

Nirjutiqawik National Wildlife Area is located approximately 20 km off the southeastern tip of Ellesmere Island. This NWA includes Coburg Island, the Princess Charlotte Monument, and all the water within a 10-km radius. The total area of Nirjutiqawik is 1,650 km^2, 78 per cent of which is marine.

Approximately 65 per cent of Coburg Island is covered in glaciers and icefields, and the remaining terrain is rugged, mountainous highlands with peaks reaching more than 800 m above sea level. While extensive coastal glaciation has marked the western and northern sides of the island, the eastern and southern borders are covered in steep coastal cliffs. These cliffs, which range from 150 to 300 m, offer ideal habitat for the estimated 385,000 seabirds that nest in this NWA.

Local Inuit residents have long valued Coburg Island and the surrounding marine area because of its importance to their seal and polar bear hunts. The Canadian Wildlife Service began seabird research on Coburg Island in 1972, and recognized the biological importance of the region in 1975 when it was designated as an International Biological Program Site. Coburg Island was declared a key migratory bird habitat site by the CWS in 1984.

The local residents wished to see some form of protection for the area's seabird colonies, marine mammal populations, and surrounding waters. It was suggested in 1989 that the CWS and local community work together to develop a strategy for protection, and subsequently, the local Hunters and Trappers Association recommended that Coburg Island and surrounding environment be protected as a national wildlife area. Regional and national Inuit associations, various government departments, and the private sector have since supported the initiative, and the Nirjutiqawik NWA was officially created in 1995. A local management committee for the area has been formed, and negotiations with local Inuit on an impacts and benefits agreement have taken place. To ensure effective management, a key component of the planning process for this NWA has involved the compilation of Inuit history of the area.

Source: Environment Canada, at: <www.pnr-rpn.ec.gc.ca/nature/whp/nwa/coburg/df07s03.en.html>.

Provincial Initiatives

In addition to this complex federal scene, provincial and territorial jurisdictions also have been active in trying to create MPAs. Only Saskatchewan and Alberta are without sea coasts, and of the 11 jurisdictions with coastal components, all cannot be reviewed here. Our focus will be on British Columbia. For the most part, provincial initiatives have tended to be fairly small and often attached to coastal components of coastal provincial park systems. Nevertheless, such protected area systems in Quebec and British Columbia contribute over 10 per cent of the protected marine area in Canada (Table 15.1). Table 15.4 shows the various provincial statutes that have been used for areal protection of the marine environment in British Columbia. Among these, the Ecological Reserves Act and Parks Act are most significant. The Ecological Reserves program is oriented strictly towards a preservation mandate with no special orientation towards the marine environment. Nonetheless, a substantial number of the reserves (20) have a marine-oriented component and some also include subtidal components. They are, however, quite small. The provincial park system, similarly, has parks that

BOX 15.3 Scott Islands Proposed Marine Wildlife Area

The Scott Islands, off the northwest tip of Vancouver Island, are anticipated to be the first MWA designated under the Canadian Wildlife Act. Currently a provincial ecological reserve, the Scott Islands sustain the largest concentration of breeding seabirds (over 2 million) in the eastern North Pacific south of Alaska and are the most important seabird colony in British Columbia. The MWA will help to protect rare and endangered species assemblages and enhance protection of seabird forage areas for such internationally, nationally, and provincially significant species as Cassin's auklet (with as much as 55 per cent of the global and 70 per cent of the national population), the rhinoceros auklet, and the tufted puffin. The study area from which the final boundaries will be determined encompasses approximately 6 per cent of Canada's Pacific coast marine area. The majority of the study area covers the deep water west of the Pacific continental shelf and slope. The process to designate the Scott Islands as an MWA has been ongoing since the late 1990s. Although federal commitment was confirmed in the 2007 budget speech, designation is still likely several years away owing to federal–provincial–First Nations negotiations and public consultation.

Source: Environment Canada, at: <www.pyr.ec.gc.ca/scottislands/mwa_e.htm>.

include marine components (94), as do six wildlife management areas, but no separate marine legislation exists at the provincial level for marine designations and many of the marine parks are managed for their recreational rather than ecological values.

In 1994, an intergovernmental Marine Protected Areas Working Group and a senior Steering Committee were formed to develop a more integrated approach to protected area planning on the Pacific coast. In addition to Parks Canada, DFO, and BC Parks, the BC

TABLE 15.4 Provincial Statutory Powers for Protecting Marine Areas in BC

Agency	Legislative Tools	Designations	Mandate
Ministry of Environment, Lands and Parks	Ecological Reserve Act	Ecological Reserves	To protect: • representative examples of BC's marine environment; • rare, endangered, or sensitive species or habitats; • unique, outstanding, or special features; and, • areas for scientific research and marine awareness.
	Parks Act	Provincial Parks	To protect: • representative examples of marine diversity, recreational, and cultural heritage; and • special natural, cultural heritage and recreational features. To serve a variety of outdoor recreation functions including: • enhancing major tourism travel routes; and • providing attractions for outdoor holiday destinations.
	Wildlife Act	Wildlife Management	To conserve and manage areas of of importance to fish and wildlife and to protect endangered or threatened species and their habitats, whether resident or migratory, of regional, national, or global significance.
	Environment and Land Use Act	'Protected Areas'	To protect: • representative examples of marine diversity, recreational and cultural heritage; and • special natural, cultural heritage, and recreational features.

Sources: Canada and British Columbia (1998). Copyright © Province of British Columbia. All rights reserved. Reprinted with permission of the Province of British Columbia. www.ipp.gov.ca. Fisheries and Oceans Canada (1997). Reproduced with the permission of Her Majesty the Queen in Right of Canada.

Land Use Coordination Office (LUCO) and the BC Ministry of Agriculture, Fisheries and Food were invited to join the Working Group. The group organized multi-stakeholder forums to gather feedback regarding the nature of MPAs and the process by which they should be established. The main output of the Working Group's deliberations was a discussion paper, *Marine Protected Areas: A Strategy for Canada's Pacific Coast* (Canada and BC, 1998) outlining the joint governmental approach to creating a system of MPAs by the year 2010. Three important elements were emphasized: a joint federal–provincial approach, shared decision-making with the public, and building a comprehensive system of marine protected areas by 2010. The marine protected areas so created would:

- be defined in law by one or more of the statutes shown in Tables 15.2 and 15.4;
- protect some (but not necessarily all) elements of the marine environment in the MPA;
- ensure minimum protection standards prohibiting ocean dumping, dredging, and exploration for and development of non-renewable resources. Above these minimums, levels of protection would vary from area to area and also within areas.

The MPA system will be delivered as part of a comprehensive coastal planning process aimed at ensuring ecological, social, and economic sustainability. Six planning regions were identified, and the first step in establishing MPAs would be the nomination of key areas within each planning region by stakeholders or technical committees for evaluation.

While the discussion paper was an important step in co-ordinating efforts between provincial and federal agencies, there were several shortcomings. For example, there needs to be greater clarification of the relative priorities of the objectives listed above, emphasizing that the primary reason for the creation of marine protected areas is to protect biodiversity and ecosystem processes. The minimum standards to ensure that this occurs are also not very ambitious. There is clear evidence, for example, that bottom trawling is highly disruptive of marine ecosystems (see, e.g., Watling and Norse, 1998), and yet it is not excluded. The same claim can be made for other activities, such as finfish aquaculture. The process for establishment also lacks clarity.

These shortcomings will hopefully be addressed through the Memorandum of Understanding (MoU) to advance the implementation of Canada's Oceans Strategy signed by the governments of British Columbia and Canada (Canada and BC, 2004). One of the provisions in the MoU is the development of a subsidiary agreement to co-ordinate the establishment of MPAs for the Pacific coast. The agreement will outline appropriate mechanisms, processes, and structures to co-ordinate the review and establishment of new MPAs, and will include an assessment of existing federal and provincial marine protected areas (ibid.). The Department of Fisheries and Oceans, Parks Canada Agency, Department of Environment, Department of Natural Resources, BC Ministry of Agriculture, Food and Fisheries, BC Ministry of Sustainable Resource Management, Land and Water British Columbia Inc., BC Ministry of Water, Land and Air Protection, and the British Columbia Offshore Oil and Gas Team are all party to the MoU. The MoU and subsidiary agreements have been negotiated over the last three years and it is expected that they will be signed off in 2008.

Over the last seven years the governments of Canada and British Columbia and non-governmental organizations have sponsored scientific studies and public consultation on the technological, socio-economic, and environmental considerations for lifting the offshore oil and gas moratorium on Canada's Pacific coast. Although the moratorium remains in place, these activities have heightened awareness and concerns for establishing 'no-go' areas to protect sensitive habitats and species, as well as to protect industry's opportunities. The interest in lifting the moratorium, in particular on the part of the BC government, has also raised awareness of the need to address coastal environmental and socio-economic issues within integrated planning.

The establishment of MPAs will be embedded within a broader coastal zone planning process undertaken at a regional level through land and resource management plans (LRMP) and at the sub-regional level with coastal plans. The goal of the LRMP process is to develop a consensus-based integrated land-use plan that will meet the present and future needs of the area. A common planning approach involving issue and value identification, general management objective specification, determination of strategic-level zones and sub-zones, and specific management prescriptions is being followed. The Central Coast and the North Coast regions have explicitly attempted to integrate coastal nearshore areas in the planning process with little success, as in both cases the terrestrial and marine areas were eventually treated separately and in parallel rather than integrated. While numerous recommended MPAs emanated from the Central Coast LRMP, virtually none survived the complex negotiations. In the North Coast LRMP process, marine components of coastal terrestrial protected areas are limited to a 200 m stretch along the foreshore—essentially a marine buffer that supports administrative control and regulation of uses on the foreshore of terrestrial parks (R. Paynter, personal communication). At the sub-regional scale, coastal plans are developed to provide guidance for nearshore tenure applications. The coastal plans include areas recommended for conservation-focused management and notations of interest for marine protected areas. The notations of interest, along with those emanating from the North Coast LRMP, will be evaluated in the course of the federal–provincial MPA network system development process that emerges from the MoU. While integrating MPAs in integrated coastal planning in theory is good, in practice it leads to long delays in MPA establishment and, thus, should not eliminate the need for a separate process for the creation of MPAs that can proceed even in the absence of such broader processes.

ISSUES

Speed of Establishment

The most fundamental problem besetting MPAs in Canada is that, given such a vast and rich marine area within its jurisdiction, only 0.4 per cent of this area is designated for marine protection. The federal government has made commitments at both the international level (at the World Summit on Sustainable Development in Johannesburg in 2002 and under the Convention on Biodiversity in 2004) and at the national level (in the throne speech of October 2004) to complete a network of MPAs by 2012. Clearly, we will be a long way from meeting these commitments (Table 15.3). It is over 20 years since the MPA initiative began at the federal level. Although progress is occurring, the quality of the marine environment continues to deteriorate. It was more than 15 years ago when the Canadian Council of Ministers of the Environment, Canadian Parks Ministers Council of Canada, and Wildlife Ministers' Council of Canada signed the Tri-Council Statement of Commitment to 'accelerate the protection of areas representative of Canada's marine natural regions'. Fisheries and Oceans Canada, Parks Canada, and Environment Canada all have struggled since the mid-1990s to designate protected areas on the water. The Commissioner of the Environment and Sustainable Development was critical of the speed of progress by DFO in establishing MPAs. DFO did not meet its commitment to establish five MPAs by 2002. In fact, not one MPA had been designated by that time. The Commissioner's report noted that the MPA evaluation process alone was taking five to seven years (Office of the Auditor General of Canada, 2005). The report questioned the effectiveness of interdepartmental committees and the challenges DFO faces in shifting its focus from fisheries management to oceans management as mandated by the Oceans Act. It also cited the lack of performance expectations and accountability for the Oceans Action Plan and an anticipated higher level of funding by DFO to meet its goals. DFO is not alone. Parks Canada's proposed NMCAs and Environment Canada's proposed MWAs have similarly long gestation periods. This lack of performance puts into question the political will to create effective MPAs. Unfortunately, the second phase of the Oceans Action Plan once again received minimal support in the 2007 budget, emphasizing the lack of political support.

Level of Protection

Current levels of protection in existing MPAs are inadequate to ensure ecological integrity. MPAs are only effective in protecting ecological integrity insofar as the species and habitats are protected from destructive and disturbing activities. Large-scale habitat disturbance such as that caused by dredging, mining, oil or gas drilling, dumping, bottom trawling, dragging, finfish aquaculture, or other extractive activities have to be excluded if MPAs are to achieve minimum standards of protection. Watling and Norse (1998), for example, compare the impacts of bottom trawling with that of clear-cutting and conclude that it is both more destructive and more widespread.

It is hard to generalize on the level of protection afforded to MPAs since this depends on individual regulations and site-specific planning for each MPA. Most of the sites held either provincially or by the CWS have little legal protection from destructive activities. The Canadian and BC governments, as part of the *Marine Protected Areas Strategy* for the Pacific coast of Canada, have set minimum protection standards for all sites within their jurisdictions, including the prohibition of ocean dumping, oil and gas exploration and development, and dredging. This strategy does not, however, have any legal basis and has remained under discussion since it was prepared in 1998. These restrictions are also applied to Parks Canada's NMCAs. Removal of living marine organisms and mineral resources in DFO's MPAs are only permitted as part of a formal research plan to better conserve, protect, and understand the area. Prohibiting commercial fishing is a particularly contentious issue for MPAs and is usually accommodated with the trend towards zoned multiple-used MPAs. For example, the regulations for the Sable Gully MPA permit some level of commercial fishing in zones other than the most protected zone. There are now many studies on the impacts on marine life of different kinds of protective designation, particularly ones that support the need for no-take zones (e.g., Lubchenco et al., 2003; Wallace, 1999; Shears et al., 2006). Informed design of the location and size of no-take zones in the multiple-use MPAs allowed under federal legislation will be key in determining how effective the MPAs will be in the future in meeting their stated major objective of biodiversity conservation. Morgan et al. (2003) have also assessed the relative impacts of different types of fishing techniques and suggest how this might be used in MPA zoning to limit the collateral damage of commercial fishing.

Aboriginal Interests

Chapter 14 outlines in some detail the obligations of the government of Canada in taking into account the needs and aspirations of Aboriginal peoples. This must be particularly emphasized with respect to coastal peoples whose cultures and sustenance have depended on the sea for thousands of years. In many cases governments have failed to consult meaningfully with Aboriginal people or have altered agreements after consultations had been held. The proposed Race Rocks MPA was to be announced as Canada's first under the Oceans Act, but was withdrawn after it was revealed First Nations no longer supported the proposal following changes that had been made after they had given their initial approval (Leroy et al., 2003).

Aboriginal peoples often have different perspectives on conservation principles and mechanisms (Ayers, 2005), and it is essential that governments are open to addressing these different world views through meaningful consultation. However, there are also examples where Aboriginal interests seem to have been incorporated in a genuinely collaborative fashion. The Montagnais Essipit band, for example, is a strong supporter of the Saguenay–St Lawrence Marine Park and a member of the park co-ordinating committee. In Gwaii Haanas National Park Reserve, there is successful co-management between the Haida and the federal government, and it is anticipated that a similar structure will be put in place for the proposed NMCA. Aboriginal peoples are more than another stakeholder group, and present a major challenge to governments in trying to establish MPAs. However, without their consent any MPAs that do result would be functionally meaningless.

Stakeholder Involvement

The level of protection is often a function of local wishes. Globally, protected area agencies are falling over backwards to 'include the local community'. It has become, and deservedly so, the mantra of the new century. Many parks were created in the days of Big Government, when local populations had little say in their establishment and often contributed subsequently to making the park as ineffective as possible (Kessler, 2004). Those times are gone. Today, stakeholder involvement is viewed as a necessary condition for a successful MPA, as was learned in the Florida Keys National Marine Sanctuary (Delaney, 2003) and the Great Barrier Reef Marine Park (Thompson et al., 2004). There is a difference, however, in paying due heed to local stakeholders and in compromising the fundamental goals of protected area establishment as a commons resource for the good of all of society both now and in the future. In their efforts to appease local stakeholders, protected agencies are now reluctant to emphasize their responsibilities to a broader range of stakeholders. As a result, often MPAs are stymied, such as the attempts by Parks Canada to establish them in the West Isles in New Brunswick and also in Bonavista/Notre

Dame Bay in Newfoundland, due to objections from local resource extractors. On this basis it is doubtful whether scarcely a terrestrial park would have been created anywhere in the country, as some segment of society is almost always involved with resource extraction in the area. Indeed, Yellowstone, the world's first national park, would never have been established had the government of the US listened to the local stakeholders who were there to slaughter the last of the mighty plains bison.

Even where MPA establishment might run the gauntlet of local opposition, all legislation allows such flexibility in terms of regulations that actual protection might be nominal. When the Memorandum of Understanding was signed to establish Pacific Rim National Park Reserve, for example, crab fishing was allowed to continue, and has now grown to an extractive industry of over five times its original size. This kind of situation might in fact be more damaging than not having a park established. When people see a green space on the map, they assume that the resources are protected from extraction. Their conscience is assuaged. With the compromises permissible under existing MPA legislation, little such protection might be in place, yet the public is being misled into thinking that this is the case.

The foregoing is not to argue against local input by resource extractors into MPA designation and management. It is to point out that there are also broader societal responsibilities of protected area agencies and that the latter should not be uniformly sacrificed to the former, as is currently the case. This situation is not unique to Canada, nor MPAs. As Ray and McCormick-Ray (1995: 37) point out: 'Unfortunately, there is an expedient tendency to speak to the lowest common denominator in proposing MEPAs (marine

FIGURE 15.3 Bonavista Bay, Newfoundland. Difficulties have been encountered in establishing national parks and implementing conservation measures wherever local communities have a strong fishing tradition. *Photo: P. Dearden.*

BOX 15.4 Proposed NMCA at Newfoundland's Bonavista–Notre Dame Bays

The Bonavista and Notre Dame Bays marine area was selected as the representative site of the Newfoundland Shelf region for consideration as a possible national marine conservation area. In February 1997 the federal and provincial governments launched a feasibility study for a proposed NMCA of the two bays. An advisory committee comprised of professional fishermen, representatives of the aquaculture industry, fish processors, members of economic development boards, and residents of both bays was formed to assess the feasibility of the NMCA undertaking.

In March 1999, this committee expressed concerns on behalf of residents of the local communities, and voted against the establishment of the NMCA. The government made the decision to discontinue the feasibility study as there was insufficient support to proceed.

The local communities felt that once the NMCA was passed into legislation, fisheries might be curtailed by Parks Canada and this would threaten their livelihoods and negatively impact on their families and communities. Aquaculturists believed the NMCA to be incompatible with the objectives of their industry. The overall response was that the establishment of an NMCA in the region would pose a threat to the long-term sustainability of the aquaculture and fishery industries in Newfoundland. Local fishermen suggested that because they are already aware of the need for conservation and are involved in various conservation initiatives, such as lobster enhancement and conservation harvesting practices, the creation of an NMCA in the area was not necessary.

The failure of the Bonavista Bay initiative has been analyzed in more detail by Lien (1999). The analysis suggests that many factors concerned with the human dimensions of park establishment came together, acted in a synergistic way, and made further development of the initiative difficult. This experience is not unique to Canada and emphasizes the need to have a well-resourced and designed consultative process if support is to be forthcoming from local communities (Kelleher, 1999).

Presently, the Newfoundland Shelf area is not represented in the marine conservation area system. Following the failure of this project, Parks Canada reported that other options for representing this marine region will be investigated in due course. Current attention in Newfoundland is now directed at the south coast area.

Source: *St John's Telegram*, 11 Mar. 1999, 4; Lien (1999).

and estuarine protected areas) and their management, resulting from consensus-based participatory processes. This is self-defeating in the end, perhaps sooner than later.' However, it should be emphasized that the optimal situation is to establish MPAs that are ecologically viable and yet enjoy strong support from local communities. A case-sensitive approach is called for, but one that holds strong to conservation principles, one that invests in conservation education and actively tries to develop sustainable alternative futures for communities whose economic livelihoods are threatened.

System Plan and Connectivity

A key question for MPAs is site selection. This should hinge around the more fundamental question of what we are trying to protect and why. Typically, site selection favours locations about which more is known or those that might emphasize benefits to local fisheries rather than ecological criteria, that is, sites that benefit commercial species rather than non-commercial species, or, simply, sites that are easy to establish rather than those that require extensive consultation and negotiation with various stakeholders. Thinking must progress beyond a piecemeal approach of establishing individual reserves to the creation of systematic networks of reserves. A more explicit and comprehensive systems plan needs to be established on sound ecological criteria that all agencies with MPA interests can adopt. One of the challenges lies in the need to co-ordinate efforts among the myriad federal and provincial agencies that can designate MPAs according to their different objectives, and, in some cases, with internally competing objectives, such as Fisheries and Oceans Canada.

Neither DFO nor CWS has implemented such a systems plan. Parks Canada has adopted a systems framework using regional representation as a main tool to guide site selection. There are, however, questions about whether this is the most effective classification system on which to base representation, with the World Wildlife Fund proposing an alternative system (Day and Roff, 2000) and more detailed frameworks being suggested at the provincial level (e.g., see Zacharias et al., 1998). The need for a systematic and ecologically based approach to MPA siting has been highlighted in Canada's *Federal Marine Protected Areas Strategy*, in which a comprehensive analysis and identification of ecologically and biologically significant areas are proposed for identifying candidate MPAs within an MPA network. In the federal MPA strategy, a network is defined as 'a set of complementary and ecologically linked marine protected areas, consisting of a broad spectrum of marine protected areas, established and managed within a sustainable ocean management planning framework and linked to transboundary, global and terrestrial protected area networks' (Canada, 2005: 8). How that gets implemented remains to be seen, but policy-makers in Canada are looking towards reserve siting algorithms such as Marxan to develop such a systems approach (Leslie et al., 2003).

Integral to the ecological function of marine reserve networks is the concept of connectivity. Marine populations frequently have dispersive phases with little connection between reproduction and recruitment. It is therefore important to consider these kinds of movements, their temporal and spatial scale, and the kinds of connectivity required to maintain links between different habitats. Robinson et al. (2005) examined the contribution of the proposed Gwaii Haanas NMCA to a network of MPAs by simulating particle dispersion to and from the proposed NMCA. They concluded that Gwaii Haanas is both a source and sink, and therefore is an important piece of the MPA network puzzle. Examples such as this encourage marine park planners to favour several small reserves over one large reserve, if a choice has to be made (Done, 1998). The North American Marine Protected Areas Network (NAMPAN), a collaborative effort among Canada, the United States, and Mexico, is extending the concepts of connectivity and MPA networks internationally by creating functional linkages among protected critical marine habitats, thereby strengthening conservation of marine biodiversity throughout North America.

BOX 15.5 Connected Thinking: The 'Baja to Bering' Initiative

The high connectivity of the marine environment requires that conservation initiatives occur over large spaces through linked initiatives. This is the converse of what has happened in the past, where MPAs have been small and isolated, with no consideration of the links between them. One of the most ambitious of connectivity schemes has been suggested by the Canadian Parks and Wilderness Society (CPAWS), which is promoting the concept of a series of linked marine reserves extending from Mexico to the Arctic, the so-called 'Baja to Bering' initiative (Jessen and Lerch, 1999). This distance, over 20,000 km, includes some of the most productive and diverse marine habitats in the world and is also the home range of the Pacific grey whale that migrates along this route every year. The 'Baja to Bering' initiative has been adopted within the North American Marine Protected Areas Network (NAMPAN) co-ordinated by Canada, the US, and Mexico through the Commission for Environmental Co-operation. Co-operation in the Baja to Bering region focuses on four important aspects:

1. recognizing the pieces of the ecological puzzle that make up this diverse seascape;
2. identifying the critical habitats of this region;
3. enhancing support for conservation and management;
4. protecting and restoring the flagship migratory species that epitomize the plight of our shared ocean realm.

FIGURE 15.4 A grey whale migrating through Pacific Rim National Park Reserve.

Source: Commission for Environmental Co-operation, at: <www.cec.org/programs_projects/conserv_biodiv/project/index.cfm?projectID=19&varlan=english>.

Permeability

Although transboundary effects have become increasingly acknowledged in terrestrial parks, there is no doubt that the permeability of MPA boundaries is far higher. This creates great management challenges as species are more mobile and the reserves often cannot protect all life phases of any given organism. Indeed, for highly migratory and mobile species, MPAs may afford little realistic protection (e.g., see Duffus and Dearden, 1995; Clay, 1999). MPAs are also more vulnerable to transboundary flows from outside the reserve. Thus, when an oil spill of 875,000 litres occurred off the coast of the state of Washington, the beaches of Pacific Rim National Park Reserve were soiled and 46,000 birds were killed. Similarly, the Saguenay–St Lawrence MPA is downstream from some of the most polluted waters in North America, and MPAs can only be as healthy as the surrounding waters to which they are connected.

The porous nature of MPA boundaries puts an even greater emphasis on ecosystem-based management (Chapter 13) than in terrestrial parks. In practice this means, for example, that attention must be devoted to establishing large, highly protected areas and that these must be buffered by areas with less stringent protection. Sobel (1995) argues that one of the most common mistakes is to establish core areas that are too small and then to wonder why they are not effective in meeting their goals.

Management Experience

Parks Canada is the oldest parks service in the world (see Chapter 2) and has a great deal of experience in managing terrestrial parks. Such is not the case in the marine environment, nor is it the case with the other involved agencies. There is a greater wealth of experience available in other jurisdictions, such as Australia and New Zealand, and Canada can profitably draw on this experience. This lack of experience raises challenges particularly

FIGURE 15.5 Killer whale at Robson Bight Ecological Reserve, BC. Can marine protected areas help fulfill conservation goals for wide-ranging species such as whales?

for DFO and the fundamental shift in management responsibility embedded in the Oceans Act. As the Commissioner on the Environment and Sustainable Development asks, 'Can a department that has historically dedicated most of its resource to managing one of the key ocean sector industries—the fishery—transform itself to represent and integrate a broader oceans interest?' including biodiversity conservation (Canada, 2005: 10).

One implication of this lack of management experience in the marine environment is the need to take an adaptive approach to management (Chapter 5). Adaptive management is particularly appropriate where levels of uncertainty are high. Management actions are undertaken as experiments from which we learn and modify our actions as a result of learning from them. The Great Barrier Reef Marine Authority, for example, found that the original zoning system was inadequate to protect the range of biodiversity on the Reef and has had to expand the no-take areas from 4.5 per cent to over 33 per cent to adjust to these inadequacies (Fernandes et al., 2005). Canada can build on these experiences as we start our own adaptive management program with full attention to the precautionary principle (Lauck et al., 1998). Both adaptive management and the precautionary principle are guiding principles in Canada's *Federal Marine Protected Areas Strategy*. However, this should be more than just a licence to 'learn by doing'; it requires a more systematic experimental strategy to build certainty in establishing and managing a network of MPAs.

Lack of Information and Monitoring Systems

MPAs suffer even more than their terrestrial counterparts from a lack of knowledge of their ecosystem functioning. Lack of information is often used as an excuse to delay establishing MPAs. Research that can assist in greater understanding of these ecosystems as well as help guide management activities and implement an adaptive management approach needs to be promoted. In particular, managers need to know (Kelleher, 1999):

- What is the state of the ecosystem, and particularly its dominant biota, rare and endangered species, ecological processes (such as sedimentation, absorption of nutrients and toxic substances), and ecological states (such as water temperature and quality)?
- What are the pressures on the system, whether natural (e.g., El Niño, severe storms) or human caused (e.g., habitat destruction, pollution)?
- What is the range of management responses and what are the implications of each?
- Are management activities having the desired response, and, if not, how should management be modified?
- Is management meeting objectives?

Understanding marine ecosystems in general and MPAs in particular is compounded by the effects of climate change on species populations and communities (Soto, 2002). As Hannah et al. (2002: 264) note, 'The notion of conserving communities and ecosystems as they presently exist may be obsolete' due to inevitably shifting patterns of biodiversity. Therefore, when selecting MPAs one must also consider:

- How will climate change affect representative habitats in the future?
- Will the range of representivity shift with climate change?
- Will areas selected to protect unique or critical habitats still be relevant in future climate change scenarios?

- What new species will be at risk or flourish due to climate change?
- How can MPA establishment and management systems remain flexible to respond to climate change in the future, yet respond to today's conservation needs?

To answer these kinds of questions a mixture of monitoring, resource assessment, and research activities are required. Few agencies are equipped in terms of either knowledge base or infrastructure to answer all these questions, and this deficiency, of necessity, requires building links with the scientific community. Parks Canada has now formed an NMCA Science Network on the Pacific coast, composed of both social and natural scientists, to address this need. Agencies should also not overlook the involvement of local community groups in these activities. Not only can they offer valuable local knowledge, but often they have resources such as boats and volunteers that can be made available. Since local communities are frequently affected by management regulations that emanate from research activities, understanding can be increased by including these local stakeholders at the earliest stages.

Lack of Resources

Despite Canada's pride regarding the extent of our coastline and continental shelf, despite Canada's development of an MPA strategy, despite Parks Canada's global reputation as the oldest and one of the best national park services in the world, despite public awareness of the plight of our fisheries on both coasts and the need for more effective conservation, the level of resources, human and financial, accorded to MPA programs is minimal. Notwithstanding the best intentions in the world, agencies concerned with the establishment and management of MPAs cannot perform the task satisfactorily if they are chronically underfinanced. Only public pressure on political masters is likely to solve this problem.

CONCLUSIONS

The fundamental conclusion from the foregoing is that while there is a plethora of principles, strategies, frameworks, and agreements regarding MPAs in Canada, we are not doing very well at translating these into establishing and managing marine protected areas in the seas that border the country on the east, west, and north. Protection of the marine environment through areal designations is one of the weakest, if not the weakest, aspect of Canada's protected area programs. In Canada, a grand total of 0.4 per cent of the marine area is in some form of protective designation, and many would dispute whether even these areas are afforded any real protection. After more than 20 years of trying to establish a national MPA program, momentum seems to be growing—six of the seven federal MPAs have been designated in the last four years. However, we cannot rest on numbers. It is critical that these MPAs be more than mere 'paper' parks but rather perform their fundamental role in the protection of marine biodiversity.

Despite some progress, Canada's MPA program remains tenuous because agencies seem embroiled in process and the resources committed are inadequate (Guenette and Alder, 2007). Public apathy is at least partially responsible for this lack of progress. This stems from the still mistaken perception of the vastness of the oceans.

As Rachel Carson put forth almost 50 years ago in the Preface of her classic work, *The Sea Around Us*:

> Although man's record as a steward of the natural resources of the earth has been a discouraging one, there has long been a certain comfort in the belief that the sea, at least, was inviolate, beyond man's ability to change and despoil. But, this belief, unfortunately, has proved to be naive. (Carson, 1961: xi)

Not only do we need to establish MPAs and ensure that they have strong protection, but we also need to ensure that they play a major educational role in raising public awareness of the complexities of the seas and our dependence on their environmental processes. Given the publicity accorded to the Parks Canada Ecological Integrity Panel and its report (Parks Canada Agency, 2000), it may well be that a marine equivalent is required to maintain the momentum in developing Canada's MPA program.

REFERENCES

Agardy, T., ed. 1995. *The Science of Conservation in the Coastal Zone*. Gland, Switzerland: IUCN.

Ayers, C. 2005. 'Marine conservation from a First Nations' perspective: A case study of the principles of the Hul'qumi'num of Vancouver Island, BC', Master's thesis, University of Victoria.

Bruno, J.F., and E.R. Selig. 2007. 'Regional decline of coral cover in the Indo-Pacific: Timing, extent, and subregional comparisons', *PLoS ONE* 2: 1–8. At: <www.plosone.org/doi/pone.0000711>.

Canada and British Columbia. 1998. *Marine Protected Areas: A Strategy for Canada's Pacific Coast*. Ottawa and Victoria.

——— and ———. 2004. *Memorandum of Understanding Respecting the Implementation of Canada's Oceans Strategy on the Pacific Coast of Canada*. Ottawa and Victoria.

Canadian Wildlife Service. 1999. *Marine Protected Areas—Opportunities and Options for the Canadian Wildlife Service*. Discussion paper prepared for the Marine Protected Areas Working Group. Ottawa: Environment Canada, Jan.

Canessa, R., and K. Lunn. 2005. 'Developing and evaluating indicators to measure the effectiveness of coastal and marine protected areas on Canada's Pacific coast', paper presented at North American Marine Protected Area Network Symposium and Workshop, Loreto, Mexico, Mar.

Carson, R. 1961. *The Sea Around Us*. Oxford: Oxford University Press.

Clay, D. 1999. 'Marine protected areas in Canada: An inadequate strategy for bluefin tuna (*Thunnus thynnus thynnus* L)', in *Protected Areas and the Bottom Line*. Proceedings of the 1997 Conference of the Canadian Council on Ecological Areas, Fredricton, NB. Information report MX 205E/F. Ottawa: Natural Resources Canada, 198–208.

Committee on the Status of Endangered Species in Canada (COSEWIC). 2007. *Canadian Species at Risk*. Ottawa: COSEWIC.

Day, J.C., and J.C. Roff. 2000. *Planning for Representative Marine Protected Areas: A Framework for Canada's Oceans*. Toronto: WWF.

Delaney, J. 2003. 'Community capacity building in the designation of the Tortugas Ecological Reserve', *Gulf and Caribbean Research* 15, 2: 163–9.

Done, T. 1998. 'Science for management of the Great Barrier Reef', *Nature and Resources* 34: 16–29.

Duffus, D., and P. Dearden. 1995. 'Whales, science and protected area management in British Columbia, Canada', in Agardy (1995: 53–64).

Environment Canada. 2006. *Canadian Protected Areas Status Report, 2000–2005*. Warsaw, Ont.: Canadian Parks Council.

Faucher, A., and H. Whitehead. 1995. 'Importance of habitat protection for the northern bottlenose whale in the Gully, Nova Scotia', in N.L. Shackell and J.H.M. Willison, eds, *Marine Protected Areas and Sustainable Fisheries*. Wolfville, NS: SAMPAA, 99–102.

Fernandes, L., et al. 2005. 'Establishing representative no-take areas in the Great Barrier Reef: Large-scale implementation of theory on marine protected areas', *Conservation Biology* 19: 1733–44.

Fisheries and Oceans Canada. 1997. *An Approach to the Establishment and Management of Marine Protected Areas under the Oceans Act*. Ottawa: DFO.

———. 2005. *Canada's Federal Marine Protected Areas Strategy*. Ottawa: Fisheries and Oceans Canada.

Food and Agriculture Organization. 2007. *The State of the World Fisheries and Aquaculture 2006*. Rome: FAO.

Guenette, S., and J. Alder. 2007. 'Lessons from marine protected areas and integrated coastal management in Canada', *Coastal Management* 35: 51–78.

Hinrichsen, D. 1998. *Coastal Waters of the World: Trends, Threats and Strategies*. Washington: Island Press.

Jessen, S., and N. Lerch. 1999. 'Baja to the Bering Sea—A North American marine conservation initiative', *Environments* 27: 67–89.

Kelleher, G. 1999. *Guidelines for Marine Protected Areas*. Gland, Switzerland: IUCN.

——— and C. Rechia. 1998. 'Lessons from marine protected areas around the world', *Parks* 8: 1–4.

Kessler, B. 2004. *Stakeholder Participation: A Synthesis of Current Literature*. Silver Spring, Md: National Oceanic and Atmospheric Administration and National Marine Protected Areas Center.

Landry, M., A. McCormack, and D. Yurick. 2007. 'Planning for Canada's federal MPA network', paper presented at SAMPAA Marine Workshop, SAMPAA VIth International Conference, Wolfville, NS, May.

Lauck, T., C.W. Clark, M. Mangel, and G.R. Munro. 1998. 'Implementing the precautionary principle in fisheries management through marine reserves', *Ecological Applications* 8: S72–8.

LeRoy, S., R. Dobell, T. Dorcey, and J. Tansey. 2003. 'Consensus vs implementation: Toward the creation of an MPA at Race Rocks, BC', in Munro et al. (2003).

Leslie, H., M. Ruckelshaus, I.R. Ball, S. Andelman, and H. Possingham. 2003. 'Using siting algorithms in the design of marine reserve networks', *Ecological Applications* 13, 1 (Suppl.): S185–98.

Lien, J. 1999 'When marine conservation efforts sink: What can be learned from the abandoned effort to examine the feasibility of a national marine conservation area on the NE coast of Newfoundland?', paper presented at 16th Annual Conference of Canadian Council for Ecological Areas (CCEA), Ottawa, 4–6 Oct.

Lubchenco, J., S. Palumbi, S. Gaines, and S. Andelman. 2003. 'Plugging a hole in the ocean: The emerging science of marine reserves', *Ecological Applications* 13, 1 (Suppl.): S3–7.

Manson G.K. 2005. 'On the coastal populations of Canada and the world', *Proceedings of the Canadian Coastal Conference 2005*. Ottawa: Canadian Coastal Science and Engineering Association.

Mercier, F., and C. Mondor. 1995. *Sea to Sea: Canada's National Marine Conservation System Plan*. Ottawa: Parks Canada.

Miller, C.A., M.M. Ravindra, and J.H. Willison. 1999. 'Towards a Scotian Coastal Plain biosphere reserve for southwestern Nova Scotia', in *Protected Areas and the Bottom Line*. Proceedings of the 1997 Conference of the Canadian Council on Ecological Areas, Fredericton, NB. Information report MX 205E/F. Ottawa: Natural Resources Canada, 177–97.

Morgan, L.E., R. Chuenpagdee, S. Maxwell, and E.A. Norse. 2003. 'MPAs as a tool for addressing the collateral impacts of fishing gears', in Munro et al. (2003).

Munro, N.W.P., P. Dearden, T.B. Herman, K. Beazley, and S. Bondrup-Nielsen, eds. 2003. *Making Ecosystem-based Management Work.* Proceedings of the Fifth International Conference on Science and Management of Protected Areas, Victoria, BC, May. Wolfville, NS: SAMPAA. At: <www.sampaa.org/publications.htm>.

Office of the Auditor General of Canada. 2005. *Report of the Commissioner of the Environment and Sustainable Development to the House of Commons*, Chapter 1: 'Fisheries and Oceans Canada—Canada's Oceans Management Strategy'. Ottawa: Office of the Auditor General of Canada.

Parks Canada Agency. 2000. *Unimpaired for Future Generations? Protecting Ecological Integrity with Canada's National Parks*, vol. 2: *Setting a New Direction for Canada's National Parks.* Report of the Panel on the Ecological Integrity of Canada's National Parks. Ottawa.

Pauly, D., V. Christensen, J. Dalsgaard, R. Froese, and F. Torres Jr. 1998. 'Fishing down marine food webs', *Science* 279: 860–3.

People and the Planet. 1998. 'Reef check', 7: 5.

Ray, C.G., and M.G. McCormick-Ray. 1995. 'Critical habitats and representative systems in marine environments: concepts and procedures', in Agardy (1995: 23–40).

Robinson, C.L.K., J. Morrison, and M.G.G. Foreman. 2005. 'Oceanographic connectivity among marine protected areas on the north coast of British Columbia, Canada', *Canadian Journal of Fisheries and Aquatic Sciences* 62: 1350–62.

Shears, N.T., R.V. Grace, N.R. Usmar, V. Kerr, and R.C. Babcock. 2006. 'Long-term trends in lobster populations in a partially protected vs no-take marine park', *Biological Conservation* 132: 222–31.

Sobel, J. 1993. 'Conserving biodiversity through marine protected areas: A global challenge', *Oceanus* 36: 19–26.

———. 1995. 'Application of core and buffer zone approach to marine protected areas', in Agardy (1995: 47–52).

Soto, C. 2002. 'The potential impacts of global climate change on marine protected areas', *Reviews in Fish Biology and Fisheries* 11: 181–95.

Thompson, L., B. Jago, L. Fernandes, and J. Day. 2004. 'Barriers to communication—how these critical aspects were addressed during the public participation for the rezoning of the Great Barrier Reef Marine Park'. Great Barrier Reef Marine Park Authority.

United Nations Environment Program (UNEP). 2006. *Marine and Coastal Ecosystems and Human Well-being: A Synthesis Report Based on the Findings of the Millennium Ecosystem Assessment.* Nairobi: UNEP.

Venter, O., N.N. Brodeur, L. Nemiroff, B. Belland, I.J. Dolinsek, and J.W.A. Grant. 2006. 'Threats to endangered species in Canada', *Bioscience* 56: 903–10.

Wallace, S.S. 1999. 'Evaluating the effects of three forms of marine reserve on northern abalone populations in British Columbia, Canada', *Conservation Biology* 13: 882–7.

———, J.B. Marliave, and S.J.M. Martell. 1998. 'The role of marine protected areas in the conservation of rocky reef fishes in British Columbia: The use of lingcod (*Ophiodon elongatus*) as an indicator', in N.W.P. Munro and J.H.M. Willison, eds, *Linking Protected Areas with Working Landscapes.* Proceedings of the Third International Conference on Science and Management of Protected Areas, Wolfville, NS: SAMPAA, 206–13.

Watling, L., and E.A. Norse. 1998. 'Disturbance of the seabed by mobile fishing gear: A comparison to forest clearcutting', *Conservation Biology* 12: 1180–97.

Wood, L. 2006. *Summary Report of the Current Status of the Global Marine Protected Area Network and of Progress Monitoring Capabilities.* UNEP/CBD/COP/8/INF/4.

———. 2007. 'MPA Global: A database of the world's marine protected areas'. Sea Around Us Project. UNEP-WCMC and WWF. At: <www.mpaglobal.org>.

Zacharias, M., and D. Howes. 1998. 'An analysis of marine protected areas in British Columbia, Canada, using a marine ecological classification', *Natural Areas Journal* 18: 4–13.

———, ———, J.R. Harper, and P. Wainwright. 1998. 'The British Columbia marine ecosystem classification: Rationale, development and verification', *Coastal Management* 26: 105–24.

Zurbrigg, E. 1996. *Towards an Environment Canada Strategy for Coastal and Marine Protected Areas*. Hull, Que.: Canadian Wildlife Service.

KEY WORDS/CONCEPTS

adaptive management
aquaculture
bottom trawling
connectivity
Department of Fisheries and Oceans Canada
Ecological Reserves Act
ecosystem management
Environment Canada
exclusive economic zone (EEZ)
fisheries management
fish farms
gap analysis
integrated management
international agreements
jurisdictional complexity
lack of funding (resources)
lack of information
land and resources management plan
level of protection
management experience
marine protected areas
migratory bird sanctuary
monitoring
multiple-use areas
national wildlife areas (NWAs)
Oceans Act (1994)
Parks Canada
permeability
precautionary principle
protected marine areas (PMAs)
regional flexibility
speed of establishment
stakeholder involvement
stakeholders
strict protection
sustainable use
systems plan
Tri-Council Statement of Commitment
UN Law of the Sea Convention
zoning system

STUDY QUESTIONS

1. What values or benefits flow to humans from marine environments?
2. Provide reasons to account for the slow progress in creating marine parks.
3. Outline the major threats to the viability of marine protected areas.
4. Compare 'sustainable use' with 'strict protection'.
5. Compare the marine protection strategies for each of the following agencies: Parks Canada; Fisheries and Oceans Canada; and Environment Canada.
6. Comment on the failure of the proposed Bonavista and Notre Dame Bay National Marine Conservation Area.
7. What is significant about each of the following marine protected areas: Sable Gully; Race Rocks; Nirjutiqawik?
8. Discuss the challenges of stakeholder involvement in MPA planning.
9. List major issues involved in the management of marine protected areas.
10. Why is the 'Baja to Bering' initiative so important?
11. Review the Parks Canada systems plan for marine protected areas. Identify any one natural area and conduct research to determine what efforts have been made to establish an MPA. What obstacles have been encountered? What suggestions can you make?

CHAPTER 16

Stewardship: Expanding Ecosystem Protection

Jessica Dempsey and Philip Dearden

INTRODUCTION

Historically, Canadian protected areas have been located in the public domain, established and controlled by federal, provincial, and local governments. This approach has worked to secure large pieces of public land, like Banff and the Tatshenshini-Alsek Provincial Park in British Columbia, for conservation and is the subject of most of this book. However, progress has been slow. Some Canadians and Canadian environmental organizations are increasingly dissatisfied with the 'top-down' protected areas approach to ecosystem protection, and have called for a more 'community-based' approach that also embraces private lands. Furthermore, shifts in park theory, based largely on ecosystem science, as emphasized in Chapters 4, 5, and 13, have altered traditional conservation practices, which largely concentrated on activities *inside* park boundaries. From the 1960s to the 1980s, areas outside the park were mapped as blank and generally ignored in the planning documents. Ecosystem-based theory shows that isolated 'island' parks in a sea of unsustainable land use fail to protect ecosystems even within individual parks.

Expanding protected areas through land acquisition *and* reducing external impacts on established protected areas are important tasks for protecting ecological integrity. This chapter discusses what stewardship is and explores its emerging importance. It discusses the 'why' of stewardship broadly and within the Canadian context. It surveys a 'steward's toolbox' to determine how it often is pursued, and looks at the major players involved in stewardship. Finally, the chapter examines the challenges and limitations of stewardship in the Canadian landscape, raising some critical questions and avenues for future research.

WHAT IS STEWARDSHIP?

Stewardship is a difficult concept to define—it is widely used, but has different meanings, in different contexts, to different people. For Roach et al. (2006: 47), 'This lack of uniform definition reflects the present wide use of the concept—if not its overuse—

FIGURE 16.1 Parks such as Waterton Lakes depend heavily on co-operation from surrounding landowners to sustain wildlife populations. *Photo: P. Dearden.*

in the realms of public policy, philosophical debate, and religious application.' Indeed, stewardship is a concept embraced even by President George W. Bush (not known for his 'environmental' policy-making by any means). On Earth Day 2007, he called for 'good stewardship of land and oceans' (see press release, at: www.whitehouse.gov/news/releases/2007/04/20070420-8.html). Corporations often describe themselves as 'good corporate stewards', which may actually reflect advances in ecological behaviour—or may be 'greenwash' (Colchester and Rose, 2004). ('Greenwash' is 'green' public relations material and images backed by little actual change in environmental behaviour by a corporation or business.) Others argue that, when done right, stewardship initiatives are the ultimate environmentalist act—changing and increasing the value that individuals, communities, and corporations place on non-human species (Roach et al., 2006).

Stewardship can be described generally as 'people taking care of the earth' (Brown and Mitchell, 1998: 8). This wide definition encompasses such actions as recycling, composting, walking instead of driving, reducing consumption, or donating land to a trust. This chapter focuses on actions contributing *directly* to land or water ecosystem protection, in other words, on people, collectively or individually, '*caring for or being responsible for a local area or resource*' (Roach et al., 2006: 48; emphasis in original). This reflects the notion of stewardship as defined by Canadian advocates, such as the Alberta Stewardship Network, which defines it as: 'an ethic whereby citizens participate in the careful and responsible management of air, land, water and biodiversity to ensure we have healthy ecosystems for present and future generations'.

In practice, stewardship takes many forms, encompassing many different governance arrangements and approaches 'to create, nurture and enable responsibility in users and owners to manage and protect land and natural resources' (Brown and Mitchell, 1998: 8). It includes landowners who voluntarily restrict damaging land use, plant native species over exotic, and place protective covenants on their land. It includes community members contributing to wildlife monitoring programs, doing passive education for tourists and visitors, and participating in collective restoration. Stewardship is demonstrated by park visitors who voluntarily choose not to hike a sensitive trail or decide to participate in park host programs (see, e.g., Cardiff, 2003). Corporations can also practise stewardship through sustainable land practices that reduce damage to wildlife habitat. In many cases stewardship restricts or reduces destructive activities, while still maintaining a working landscape. With all these potential approaches under the stewardship umbrella, one can see why the concept generates political controversy.

Stewardship deviates from regulated protection as found in most park legislation or other types of regulated environmental management, such as endangered species legislation. One commonality found within stewardship, as Roach et al. (2006) emphasize, is its voluntary nature. To those promoting them, stewardship types of approaches to ecosystem or biodiversity conservation are often touted as more effective and efficient

FIGURE 16.2 Creston Valley Wildlife Centre nestled between the Selkirk and the Purcell mountains in southeastern British Columbia is a stewardship initiative with a substantial interpretive component. Although created originally with government funding, the Centre now relies extensively on private contributions to maintain this important wildlife area. *Photo: P. Dearden.*

means of protecting species or the environment than government-imposed regulations, often referred to as 'command and control'. But while activities falling under stewardship are 'voluntary', stewardship is certainly not de-linked from governments, which often play a critical role in providing incentives for conservation or by creating the supportive climate for private organizations (Brown and Mitchell, 1998).

EMERGING IMPORTANCE OF STEWARDSHIP

Scope

Stewardship is claiming more space in the geography of protected areas. The Nature Conservancy of Canada (NCC) alone, one of many conservation organizations, has protected approximately 1.9 million ha since 1962 (Nature Conservancy Canada). In 2006, Ducks Unlimited Canada secured over 46,000 ha of wetland (DUC, 2006). According to its website, Ducks Unlimited is responsible for 'positively influencing' over 10 million hectares of wetlands in Canada since 1938. For a benchmark, Parks Canada, established in 1885, is responsible for over 27 million ha. While this is not a direct comparison of conservation value (national parks offer stronger ecological protection, and many other stewardship organizations are protecting land besides Ducks Unlimited and the Nature Conservancy), it does demonstrate the significance of stewardship as a force for conservation.

In the last decade or so, land trusts have grown enormously; for example, in Ontario they increased from 13 to 36, in British Columbia from 24 to 37, and in Quebec from 1 to 10 (Campbell and Rubec, 2006: 2), and there are over 200 land trusts now active in Canada (Canada, 2006). In Nova Scotia, where there is a high proportion of private landownership (over 70 per cent), the Nova Scotia Wilderness Areas Protection Act requires the government to promote the designation of private land as wilderness. Various programs exist to encourage private conservation and to acquire and secure lands in ecologically significant natural areas for conservation purposes. The programs include designation of private land under the Act, donation of private land to a recognized conservation agency, conservation easements that restrict development on private land, designation of ecologically sensitive land under the federal Ecogifts Program, as well as partial donation and split-receipting (Nova Scotia, 2005). Each program involves different landownership regimes and financial incentives. The implementation of the private land conservation initiative is augmented by land trusts that receive government financial support, purchase private lands, and then transfer ownership to the provincial government. However, to date less than 0.1 per cent of Nova Scotia has been formally protected on private land (ibid.).

Stewardship also is of interest to the federal government. In 2002, *Canada's Stewardship Agenda* emerged from a federal–provincial–territorial initiative. The *Agenda* is 'a plan for collaboration that proposes a national vision and operating principles for stewardship' that emerged from nationwide consultation (Federal–Provincial–Territorial Stewardship Working Group, 2002: 3). A national web portal for stewardship now exists (www.stewardshipcanada.ca), and is supported by various government agencies.

Numerous funding programs have been created for community stewardship, such as the Canadian Wildlife Service's Habitat Stewardship Program funding to 'stewards' for implementing activities that protect or conserve habitats for species designated by the Committee on the Status of Endangered Wildlife in Canada (COSEWIC) as nationally 'at risk' (endangered, threatened, or of special concern) (see www.cws-scf.ec.gc.ca/hsp-pih/default.asp?lang=En&n=59BF488F-1). The British Columbia government and the government of Canada jointly produced the 'Stewardship Series', a set of guides and handbooks for BC residents and local governments (see www.stewardshipcentre.bc.ca/publications/default.asp). Federal legislative changes amenable to stewardship have also been passed, resulting in increased incentives and encouragement for ecological gifts and donations. Over $170 million in federal funds were leveraged by ENGOs over the period 1986–2003 to secure 1.8 million ha of lands worth $3.2 billion (Canada, 2006).

Government interest in promoting stewardship was reinforced in the spring 2007 federal budget when the government pledged $225 million in matching grants to land-acquiring stewardship organizations such as the Nature Conservancy and Ducks Unlimited. With this money the government aims to protect a further 200,000 ha in southern Canada that have high ecological values and yet are threatened by development. Indeed, such is the scale of private conservation lands now in Canada that many provinces officially include them in their protected area networks.

Rationale for Increase

This increased presence of stewardship in the Canadian landscape emerges from a number of forces, of which three will be discussed here. First, the increased role of stewardship is part of a larger policy shift in the late twentieth century, a policy shift sometimes characterized as 'neo-liberalism'. While definitions and debates about neo-liberalism abound, generally it refers to a shift away from centralized governance towards the self-regulating market (McCarthy and Prudham, 2004). In the contemporary neo-liberal policy environment, 'State functions aimed at curbing socially and environmentally destructive efforts are "rolled back" . . . and "restructured" in a variety of ways' (ibid., 276), including budgetary restraints (Shultis, 2003), privatization (Dearden et al., 2005), and a move from binding to voluntary or 'results-based' regulations (e.g., a new results-based forestry code in BC; see Hoberg and Paulsen, 2004) and towards approaches that may include 'public–private co-operation, self-regulation, and greater participation from citizen coalitions, all with varying degrees of capacity and accountability' (McCarthy and Prudham, 2004: 276). Shultis (2003) has explored the implications of this change in political climate for park agencies in Canada in more detail.

In this policy environment, command-and-control approaches, such as protected area designation or endangered species legislation (binding regulations), may be deemed inefficient, too expensive, or simply a failure (Holling and Meffe, 1996). Further, Merenlender et al. (2004: 66) argue that landowners and even provincial and local governments are threatened by top-down regulations and, as such, 'obviously favor incentive-based, voluntary conservation for private-land resources'.

While space prohibits a full exposition of these arguments, stewardship proponents argue that voluntary approaches are more flexible, responsive, and adaptive. Roach et al. (2006: 48) write, 'The advantages of stewardship over the strategies of land acquisition and government regulation include greater community support (often stemming from the fact that the program is voluntary), lower costs, a longer-term commitment, and an emphasis on changing people's behavior through efforts to change attitudes and values.' Where public protected areas require acquisition of funds difficult to get, public consultations take time, and intergovernmental negotiations can take decades, conservation easements and stewardship approaches 'can often react more quickly, flexibly and cost efficiently than government agencies to land conservation opportunities' (Brown and Mitchell, 1998: 14). So, proponents argue, when sensitive ecosystems become available for acquisition or are under threat, private organizations can act and build support quickly. Of course, critical questions should be posed regarding the long-term effectiveness and cost-efficiency of these 'fast-acting' initiatives for conserving biodiversity. Concerns related to accountability and equity are raised later in this chapter.

A second reason behind the growth of stewardship initiatives is the recognition of increased environmental pressures. Biodiversity loss in our generation has been unprecedented in human history (Millennium Ecosystem Assessment, 2005a). Although protected areas are a main tool to address this degradation, that conservation efforts need to have a much wider footprint on the landscape is widely agreed (e.g., Wiersma and Nudds, 2003). In the Canadian context, habitat loss is the greatest threat to endangered species and is caused primarily by agricultural and urban land conversion (Venter et al., 2006). Much of the habitat used by these species is on private land (Barla et al., 2000; Kerr and Deguise, 2004). Conservation science and planning now emphasize the need for connected networks of government and private lands (e.g., Beazley, 2003), in other words, a broadening of thinking to larger ecosystem scales (as discussed in Chapter 13). As Brown and Mitchell (1998:: 8) write, 'The stewardship approach . . . offers a means of extending conservation practices beyond boundaries of conventional protected areas, to address needs on the "land between".' An outstanding example is the Yellowstone to Yukon corridor (see Chapter 1).

Finally, with increased public recognition of environmental issues, especially with recent attention on global warming, communities more often demand to be involved in the stewardship of their local environments. But while some stewardship projects and policies respond to this increasing demand, the purpose of many stewardship projects and programs continues to be directed towards changing people's values and actions, to 'encourage local people's concern for the natural environment of their community and to harness this concern in order to protect the environment' (Roach et al., 2006: 49). Roach et al. (ibid., 50) further argue that stewardship works by moving people 'from a non-environmental to a pro-environmental perspective and makes unnecessary the threat of legislated sanctions or expensive land acquisitions'. Stewardship projects and policies are part of a larger project of creating 'environmental citizens' within a political environment that tends to shy away from direct government action and regulation.

Canadian Context

In Canada, stewardship is playing a larger role in ecosystem conservation and is being promoted by federal, provincial, and territorial governments as well as by a wide range of organizations and foundations. In areas of high private ownership and landscape modification (for example, near urban areas), private or smaller-scale conservation initiatives are used to protect rare ecosystems or species. For example, the Garry Oak Ecosystems Recovery Team, a partnership of governmental and non-governmental agencies and organizations on southern Vancouver Island, is working to improve, restore, and protect this ecosystem, as less than 5 per cent of Garry oak ecosystems remain. According to the website, more than 100 species of plants, mammals, reptiles, birds, butterflies, and other insects are currently officially listed as 'at risk of extinction' in Garry oak and associated ecosystems, and several species have already been extirpated. A key objective is '[t]o motivate public and private protection and stewardship activities by supplying critical information to the appropriate audiences' (see www.goert.ca/about/goals.htm).

A considerable amount of critical endangered species habitat (like the Garry oak ecosystems above), estimated at around 60 per cent, lies in private hands (Barla et al., 2000: 95). But Canada's endangered species legislation (SARA) provides inadequate protection for terrestrial species on private lands (less than the US Endangered Species Act)—only habitats on *federal* lands are strictly protected (Boyd, 2003; Dearden and Mitchell, 2005). Due to these legislative weaknesses, stewardship in some cases has ensured habitat is maintained or, in other cases, restored. To encourage private landowners to protect endangered species habitat, provincial governments and non-governmental organizations are using a variety of techniques. Nature Saskatchewan's Operation Burrowing Owl, established in 1987, aims to protect burrowing owl grassland habitat from cultivation, to monitor population changes, and to increase awareness of the owl (Box 16.1). A broader approach to protect grassland birds overall that links government science to a wide range of stewardship initiatives in Saskatchewan is described by Davis et al. (2003). However, very little research examines how these voluntary initiatives are working to conserve biodiversity effectively (e.g., Jones et al., 2003). Indeed, the Ontario government has recently (2007) brought in endangered species legislation, hailed by some environmental groups as the best in the country, that may be an important precedent for other provinces and also a *necessary* complement to these voluntary stewardship initiatives.

Finally, and as noted above and in Chapters 5 and 13, conservation science is emphasizing the need for protected areas to be a part of the larger landscape. Even large parks like Banff cannot survive alone; in order to conserve connecting habitats, they must be embedded in a greater park ecosystem practising sustainable land management. Smaller parks rely on surrounding lands to an even greater extent. As such, park neighbours (public and private) are increasingly being brought into stewardship initiatives (see Box 16.2).

STEWARDSHIP'S TOOLBOX

Various stewardship tools and methods are in use. For Brown (1998: 10), 'Specific tools vary according to social, legal, ecological, and institutional constraints, but all operate to encourage, enable, or formalize responsible management.' The following description draws largely from Brown and Mitchell (1997, 1998), who provide an organized and concise description of tools and techniques for stewardship. Although the tools are shown as separate in Table 16.1, often they are used together, in pursuit of stronger and longer-term land conservation.

BOX 16.1 How Is Saskatchewan's Operation Burrowing Owl Working?

Within Operation Burrowing Owl (OBO), landowners (farmers and ranchers, predominately) voluntarily agree to conserve grassland habitat for endangered burrowing owls and other prairie wildlife. In return for involvement, participants receive a gate sign, an annual newsletter, and educational literature. These are non-binding agreements that can be terminated at any time.

In their research assessing the value of these verbal agreements, Warnock and Skeel (2004) found that the program was effective for retaining grassland habitat, with grassland retention significantly higher at OBO sites than at random locations during an era of accelerated loss of grassland (1987–93). However, despite this success, the OBO sites did not significantly affect overall grassland loss, nor the decline in burrowing owl populations, as the OBO sites comprise only 0.7 per cent of the remaining grasslands in the study area. Indeed, between 1987 and 2001, the burrowing owl population in Saskatchewan sharply declined. Skeel et al. (2001) found a 95 per cent decline in the owl population between 1988 and 2000. Warnock and Skeel conclude that the stewardship program is 'working', but the small amount of grassland set aside by landowners participating in the program is simply too small to affect population change, and that 'other factors such as habitat improvements that enhance productivity (more prey, less predation), reduced pesticide use, and habitat conservation measures on wintering and migration stop sites . . . are also needed for this species to recover' (Warnock and Skeel, 2004: 315). The authors suggest economic incentives to increase participation in the program, but we add the possibility of stronger endangered species legislation and/or regulations on pesticide use.

Education

Landowners, whether private, community, government, or corporate, may be unaware of existing important or sensitive habitats on their property. Education about and communication of these features comprise one of the most powerful stewardship tools. Education types and formats are limitless and variable to the subject, area, and audience. For example, on the west coast, a joint initiative by the provincial government and the federal government identified sensitive ecosystems for the southern Gulf Islands and eastern Vancouver Island. Efforts have been made to educate people who participate in land-use and planning decisions (local government staff, councillors and committee members, conservation organization representatives, staff from federal and provincial agencies, First Nations) about these sensitive ecosystems and management needs. One-day workshops were organized, which included field trips to sensitive areas. Passive education methods, such as informative pamphlets and websites, also are used to convey information. Similar techniques can be used to foster public concern for particular species and ecosystems. Dubois et al. (2003), for example, describe a project in southwestern British Columbia targeted at raising public awareness of the threats to bighorn sheep. This educational program led to action on community monitoring, restoration projects, and further research and studies on the economic returns to local communities from the sheep.

BOX 16.2 Keeping Gros Morne Park Connected to the Wider Ecosystem

Gros Morne National Park is located on the west coast of Newfoundland. A UNESCO World Heritage Site, it encompasses over 1,800 km^2 and is home to many species, such as the woodland caribou, Arctic hare, pine marten, and moose. While it is a relatively large protected area, most of the land to the south and east of the park is under the control of Corner Brook Pulp and Paper, for timber harvesting. Concerned about the threat to ecological integrity of the park stemming from timber harvesting, in 2001 the provincial government, the national park, and Corner Brook Pulp and Paper signed a framework agreement for 'Maintaining the Connectivity of Gros Morne National Park within the Greater Ecosystem'. The main goal of this agreement is to 'contribute to maintaining the ecological integrity of Gros Morne National Park by developing scientifically-based solutions to maintaining connectivitiy between Gros Morne National Park and the larger landscape for agreed-upon indicator species.' The chosen indicator species are the Newfoundland marten (an endangered species), woodland caribou (a wide-ranging species that moves in and out of the park over the year), resident birds, and lynx.

The work has been science-based, with strong participation from Acadia University in Nova Scotia. Field studies are being conducted to better understand where martens live in the park and within the greater ecosystem and to define the occurrence and abundance of a variety of resident and migratory birds. Since the agreement was signed, tangible changes have been made in policy related to the harvesting of forest and to regional connectivity. For example, Corner Brook Pulp and Paper ceased all clear-cutting operations in an important watershed (Main River) and is implementing harvesting techniques on the ground as part of a co-operatively designed adaptive management approach.

Source: Based on Anderson and VanDusen (2003).

TABLE 16.1 Toolbox Summary

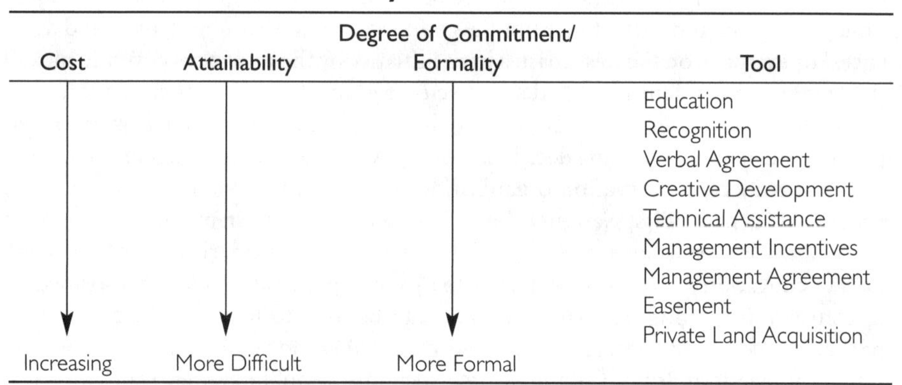

Cost	Attainability	Degree of Commitment/ Formality	Tool
			Education
			Recognition
			Verbal Agreement
			Creative Development
			Technical Assistance
			Management Incentives
			Management Agreement
			Easement
			Private Land Acquisition
Increasing	More Difficult	More Formal	

Source: After Brown and Mitchell (1997: 105).

Recognition

Education can be very effective in reaching landowners and changing land-use practices. Public recognition of outstanding work 'rewards' land stewards. This can be done in different ways—from formal awards ceremonies to plaques or signs showing participation. Operation Burrowing Owl in Saskatchewan (see Box 16.1) provides a sign for participants, recognizing them as partners. Wildlife Habitat Canada (a national, non-profit conservation organization) manages a national award program for both forest and agricultural stewardship, designed to honour exemplary stewardship efforts of Canadians (see www.whc.org/EN/stewardship/stewardship_awards_CSC.htm).

Verbal Agreements

For those landowners unwilling to enter a written, legally binding agreement, but who still want to participate in 'taking care of the land', a verbal agreement may be used. Although this technique depends on the goodwill of the landowner to carry out the agreement (thus, it is susceptible to 'greenwash'), it can be effective in creating a sense of responsibility and sustained stewardship. For example, in the southern Ontario Carolinian zone, the Ontario Heritage Foundation has facilitated over 1,000 verbal agreements with private landowners who now conserve over 6,000 hectares of rare and threatened ecosystems. Awards were given to participating landowners to recognize their efforts (Ontario Heritage Foundation—Carolinian Canada).

Technical Assistance

A landowner may wish to participate in stewardship restorative activities but lack the technical skills and equipment. In this case, a stewardship organization or the government might provide assistance for restoring habitat on the land in exchange for a verbal agreement or perhaps something more formal. For example, the Ontario Wetland Habitat Fund provided financial and technical assistance to landowners to improve the ecological integrity of wetland habitats on their property. Through this program, over 20,000 ha of habitat were enhanced for the benefit of waterfowl and wildlife on private land (the program was discontinued in 2007, demonstrating the fragility or tenuousness of stewardship approaches).

Another example of technical assistance is the computer-based decision support tool that assists landowners and community groups on how best to deal with invasive species in the endangered Garry oak ecosystem of southern Vancouver Island. The tool assists in deciding whether management is necessary and, if so, what kinds of techniques might be used and how best to apply them (Murray, 2003). A particularly interesting feature of this approach is that users are provided with feedback forms, which record their decision regarding management need and approaches, their results, and what they learned. Incorporating stewardship learning is fundamental to adaptive management and improved outcomes. The tool development was co-sponsored by a stewardship organization, the Nature Conservancy of Canada, and the Habitat Stewardship Program of the government of Canada.

Creative and Eco-Friendly Development

In many parts of Canada, development is one of the greatest risks to sensitive or rare ecosystems. Creative development encourages careful planning of site location and building design to maintain all or some habitat. For example, urban developers might choose to save important trees, by changing development plans and building layout. 'Smart growth' urban development, which restricts low-density suburban sprawl from expanding into important surrounding ecosystems, also protects habitat and is a form of stewardship practised by local governments and developers alike. One example is the development of Vancouver's False Creek lands, formerly industrial land close to Vancouver's downtown core, into a 'sustainable community'. The plan calls for some restoration of this waterfront neighbourhood, including an island for bird habitat and a wetland. This neighbourhood will also be home to the Vancouver Olympic Village in 2010 (see www.city.vancouver.bc.ca/commsvcs/southeast/). One issue with 'smart growth' strategies (conservation of urban lands) is that it can impact, and usually reduce, the amount of land available for housing stock, which can have a negative effect on affordability.

Management Incentives

Landowners who choose to become stewards by maintaining or changing land practices/management are sometimes eligible for financial incentives offered by a stewardship organization or, more likely, by the government. These might include incentives for keeping land forested or for changing sewage treatment practices to protect marine ecosystems. For example, in Ontario the Conservation Land Tax Incentive Program encourages and supports long-term stewardship on private lands by providing tax relief to those who participate. Lands classified as 'conservation lands', which include the most environmentally important habitats, are completely exempt from property tax. Managed forests (with approved management plan) and farmlands are assessed at 25 per cent of residential property tax rates. Increasingly, Canadian farmers are being encouraged to change their management practices through payments for ecological goods and services (see Box 16.3).

Management Agreement

While management incentives work for more passive land management, sometimes active management is required to maintain or restore declining ecosystems (Chapter 5). Management agreements are used when landowners are willing to let another organization manage parts or all of their land actively for conservation purposes, while retaining ownership. This technique is being used heavily in Prince Edward Island, where the provincial government has negotiated over 300 voluntary agreements allowing active wetland conservation (Brown and Mitchell, 1998). Incentives, rent payments, or formal leases (providing income to the landowner) might be added to this agreement to encourage and sustain stewardship over the long term. One critical drawback, however, is that management agreements are not binding on a new owner.

In Newfoundland, 11 management agreements have been signed between communities and the Eastern Habitat Joint Venture, a public–private partnership established to increase the waterfowl populations of eastern Canada by protecting and enhancing both

BOX 16.3 Alternative Land-Use Services and Ecological Goods and Services

A brainchild of Ian Wishard, a Manitoba farmer, Alternative Land Use Services (ALUS) is a pilot project that pays farmers to adopt environmentally beneficial land-use practices, practices that yield public 'ecological goods and services'. Ecological goods and services (often shortened to ecosystem services) are defined by the Millennium Ecosystem Assessment (2005b: 27) as 'the benefits people obtain from ecosystems'. These include: provisioning services such as food and water; regulating services such as regulation of floods, drought, land degradation, and disease; supporting services such as soil formation and nutrient cycling; and cultural services such as recreational, spiritual, religious, and other non-material benefits. On the Canadian Prairies, protecting ecosystem services might mean that a farmer leaves a wetland rather than draining it (wetlands are critical bird habitat and are important for water and pollutant filtration) and not taking the first cut of forage until after nesting time. The argument is that these small changes in farming practices yield public 'ecosystem goods and services' that have a price, and as such, these farmers should be compensated for providing them. The ALUS program is promoted as 'by farmers, for farmers', but one key supporter has been Delta Waterfowl, a US conservation and hunting lobby with a keen interest in conserving ducks for its members' recreational hunting.

Ecological goods and services are becoming more commonplace as a policy tool, and reflect a larger trend of protecting the environment by pricing nature's services, assigning clear property rights to them, and then trading these services in a market. This approach raises many questions, though, such as who bears the burden of conserving ecological goods and services, who decides a service is worth paying for, and how do we determine the 'value' of that service? Geographers such as McAffee (1999) have warned that this type of approach is 'selling nature to save it' and can exacerbate unequal relations. Others, such as Robertson (2004), focus on the difficulty of creating the unit of trade in an ecological service, the abstract measure that can be generalized as necessary for trade. For example, trade in wetlands often substitutes an 'acre' for the services it provides, but clearly the one acre of a wetland in one place does not necessarily correspond to another acre elsewhere.

the abundance and quality of wetlands in eastern Canada. In these wetland stewardship agreements, landowners, managers, and municipalities all pledge to conserve specific wetlands and associated uplands within their jurisdiction (Roach et al., 2006).

Conservation Easements

While the above tools and techniques are subject to the goodwill of landowners and the renegotiation of agreements, verbal or written, a conservation easement is a far-reaching agreement between a landowner and an organization that restricts certain land uses and management. These restrictions are built into the land deed in perpetuity. Easements can take many forms and can be made flexible to suit the landowners' and the organizations' needs and desires. For example, easements can apply to all or to only a portion of the property, can set management restrictions in sensitive forest areas, or can allow for trail access across private land.

In Canada, all provinces and territories have some kind of easement legislation, but exact names for this tool vary throughout Canada: in BC they are called conservation covenants and in Manitoba they are conservation agreements. Sometimes the provincial legislation includes modest incentives, like tax relief, to encourage landowners to participate and to ensure long-term land protection. Some organizations 'buy' easements, providing the landowner with some financial compensation. In most cases, though, the easement is a voluntary donation. Conservation easements are becoming commonplace all over the country, in part due to changes in legislation and to the boom in land trust establishment. They sometimes are purchased using a blend of private and public funds. While easements are more secure than verbal agreements, they do not necessarily mean biodiversity is being conserved—many easements are not monitored for compliance, and they also raise questions about social equity (Box 16.4).

Private Land Acquisition

The strictest form of ecosystem protection is complete land acquisition, where a conservation organization acquires all rights to the property by donation or purchase. This is a costly stewardship technique due to land expense and ongoing land management costs. However, creative arrangements reduce some of these barriers. A landowner might donate land to a conservation organization for a one-time tax receipt (an ecological gift). He or she might sell the land at full price but then donate some back for management. A person could donate the property as part of a will, or might donate funds to the conservation organization for land acquisition. Some land trusts work out mortgage agreements with lending institutions to spread payments over a longer term or collaborate with government agencies to reduce management costs. A piece of land might be owned by a land trust, but managed by the government. This way, land organizations can reduce their ongoing costs and free up resources for further land acquisition.

Kirkland and Rose Islands, located off Vancouver, were purchased by the Nature Trust of BC in 1989 to protect sensitive bird habitat. The islands were sold to the Nature Trust at market value for about $3 million by a group of wealthy hunters. The hunters then donated $1.5 million back to the Nature Trust, with a provision that the group continue to have *exclusive* hunting rights on the islands from 1 September to 31 January each year. In addition, the hunters pay the property taxes, take out liability insurance, and finance management of the property for the benefit of waterfowl habitat, costs ranging into the tens of thousands each year (*Vancouver Sun*, 27 Nov. 2000). As this example demonstrates, land acquisitions by private NGOs are flexible arrangements, catering to the needs of the current owner as well as meeting *some* conservation goals. This acquisition, however, also raises critical question about who is benefiting from these arrangements—in this case, the hunters, who while maintaining access to their properties (and financing the ongoing management of the property) also gained $1.5 million, and probably significant tax deductions for this 'ecological gift'.

BOX 16.4 Critical Questions about Conservation Easements

Merenlender et al. (2004: 67) warn conservation biologists to be aware of the 'scientific and policy assumptions that underline easement specifications'. Examining the US context, the authors found little systematic data on the location and types of resources conserved, and to what degree they were conserved. The assumptions held about conservation easements do not necessarily reflect the reality: 'Frequently', they write, 'easements are assumed to be good for conservation because they at least abate the risks for the land being subdivided or developed to its highest economic use, and this is considered a benefit for all forever. The real story is much more likely to be that, with the conservation easement in place, where there is currently one house there will be two or three houses, with the easement protecting an unknown quantity of open space of unidentified ecological integrity for an undetermined amount of time' (ibid., 70). The authors found that monitoring is a major challenge, especially when ownership of the title changes over time. Land trusts, they argue, focus most of their attention on acquiring easements and not enough on ensuring they actually are upheld. Finally, the authors raise critical issues of social equity. Little is understood about how these private transactions, which often blend private and public funds, affect equitable access to the benefits of conserved resources (ibid., 72).

Private land acquisition is not without problems, particularly related to social equity and accountability. In Canada, one critical question is what responsibility private land acquisition organizations have towards First Nations people, and the extent to which these trusts could also promote decolonization and reconciliation through greater participation of First Nations and perhaps even by transferring title or ownership. This is particularly the case in British Columbia, where many comprehensive land claims have not been settled (Chapter 14). In much of the land trust material reviewed for this chapter, the discussion around First Nations seemed to focus on how can we get them to help us, rather than how can we use our resources to help them.

STEWARDSHIP PLAYERS

Although stewardship is associated with private, individual actions 'caring for the land', many different actors contribute to and encourage stewardship. Non-governmental organizations (NGOs) are very active participants. National organizations like the Nature Conservancy and Ducks Unlimited play important roles, as do a plethora of local land trusts and conservation organizations. These private organizations acquire land and negotiate conservation easements (Brown and Mitchell, 1998), and act as monitors over these protected areas. Some organizations provide technical assistance to landowners who wish to improve wildlife habitat, or at least direct them to the people and resources. Other conservation organizations play an educative role, informing the public about endangered species and spaces. For example, the Canadian Nature Federation does not own land, but rather educates Canadians about important spaces and species, giving them the opportunity to volunteer on conservation projects and programs.

BOX 16.5 Private Lands and Accountability

While there is no doubt that stewardship initiatives have protected significant habitat, the question of *accountability* in a time of increased private environmental management looms large. In a case study of six different Canadian private protected areas, Hannah (2006) found that, although their governance processes were generally sound, all lacked performance monitoring and had only modest engagement with local stakeholders. In another recent study near Redberry Lake, Saskatchewan, Reed (2007) found that private organizations like the Nature Conservancy of Canada (NCC) and Ducks Unlimited Canada were helping to usher in an increasingly private, and unaccountable, environmental management regime. NCC, she found, was negotiating land donations, easements, and agreements, of which the details often were kept confidential and private. These organizations also were undertaking strategic, scientific planning for the area without engaging local residents or their local knowledge (ibid., 333). Their efforts, writes Reed, 'focus first on engaging with government and academic scientists to identify key conservation criteria and geographic areas and then they subject their plan to "expert review" once a plan is near completion' (ibid.), failing to engage local residents in the key planning and decision-making phases. Locals, Reed found, have identified both NCC and Ducks Unlimited as 'groups that dictate to locals how to protect their land and reduce the collective commitment within the reserve area' (ibid.).

The situation in Redberry Lake does not necessarily transfer to other places where land trusts operate in more publicly engaged ways. Reed explains differences or unevenness in accountability and governance by suggesting that 'where public and civic sectors are weaker, ENGOs can operate more like private organizations that negotiate private and confidential contracts rather than as public or civic organizations operating in the public domain' (ibid., 335). She argues that with strong public and civil sectors, the situation can be quite different, as in Clayoquot Sound on the west coast of BC.

This said, important questions should be asked regarding the accountability of conservation organizations. For example, in 2003, the *Washington Post* (Ottaway and Stephens, 2003a, 2003b, 2003c) presented the results of a two-year investigation into the US Nature Conservancy, reporting that the land trust arm of the organization was involved in some shady land dealings—particularly in permitting ill-advised (if not illegal) land transactions that benefited individual board members, trustees, and donors, who then reaped tax savings. While this was in the US and should not be seen as illustrative of land trust organizations, it demonstrates how these organizations are not infallible, and should be scrutinized and regulated.

In Canada, the public sector also contributes to the success of stewardship in a number of ways. By sponsoring education programs, governments disseminate important information about harmful activities and positive changes. For example, some government ministries sponsor stewardship websites that act as clearinghouses and one-stop information centres for interested citizens (see Alberta's at www.landstewardship.org). Through funding programs like EcoAction 2000, the federal government provides resources for community organizations to participate in environmental stewardship.

BOX 16.6 The Island Nature Trust—Prince Edward Island

The Island Nature Trust is a non-government, not-for-profit organization dedicated to protection and management of natural areas on Prince Edward Island. It acquires lands to be held in trust for future generations, manages these lands as an example of appropriate and sustained use, and helps private owners voluntarily protect their lands. Its vision is to work with government and private landowners to create a true natural areas network on Prince Edward Island, consisting of core protected areas connected by corridors.

In 1979, Island Nature Trust was incorporated as the first private, provincially based nature trust in Canada. The Trust is governed by a board of directors that includes 12 members of the Trust and one representative from each of the four founding organizations: PEI Museum and Heritage Foundation, PEI Wildlife Federation, Natural History Society of PEI, and the Biology Department of the University of Prince Edward Island.

The Trust has protected nearly 2,800 acres of land through acquisition. All of these properties have been permanently protected under the Natural Areas Protection Act. The properties are managed to maintain, protect, and enhance the significance of the sites. Many of the sites serve as demonstrations for landowners having properties with similar features.

Because so much of PEI is privately owned, another activity of the Trust is to provide landowners of significant ecological areas with information on how the land should be managed to best protect and enhance the site. Private stewardship is of great importance as it allows for the protection of property without the need to expend funds for acquisition. Some property owners have permanently protected their significant areas by designating them under the Natural Areas Protection Act.

The Trust has developed educational presentations for classrooms, youth groups, adult groups, and seniors. In addition, the Trust is involved with the protection of species at risk. A great effort has been placed on monitoring and protecting piping plovers nesting on beaches outside PEI National Park. The protection involves identifying nesting areas, erecting signage and ropes to warn people to stay away from nests to minimize disturbance, and, in cases where predators have been a problem, installing predator enclosures. Where nests are present, beach users are frequently contacted to explain the problems facing plovers and the need to remain near the water edge and to keep pets on a leash. The Island Nature Trust protects species and habitats both on land and sea.

The Nature Trust also is involved with initiatives to provide protection for the diminishing number of common terns on PEI, and in 2007 was seeking funds to help protect a remnant stand of old-growth hardwoods that otherwise was slated for harvesting.

Legislative changes and incentive programs (as discussed above) also enable and encourage stewardship. Since 1995, donations of ecologically sensitive land, or easements, covenants, and servitudes on such land, have been eligible for special tax assistance. These so-called 'ecological gifts' can be made to environmental charities approved by Environment Canada, as well as to any level of government in Canada. Environment Canada is responsible for certifying that the land is ecologically sensitive, approves the recipient, and certifies the fair market value of the gift. In 2006, this deal was sweetened when ecological gifts were exempted from any capital gains tax accruing to the property (see www.cws-scf.ec.gc.ca/egp-pde/default.asp?lang=En&n=522AB5A3-1). According to the Canadian Wildlife Service website, over 560 ecogifts valued at over $182 million have been donated across Canada, protecting 48,000 hectares of wildlife habitat. To encourage the continued protection of these lands, lands taken out of protection are subject to a substantial tax penalty equal to 50 per cent of the value of the land at the time of disposition.

The final stewards to consider are individuals who participate as landowners or citizens in stewardship initiatives. Although financial 'carrots' are important, incentive for participation in these activities also comes from personal satisfaction and a 'conviction of individual responsibility for the health of the land' (Leopold, 1949: 240). This is what Leopold has called an ecologically necessary land ethic, which 'enlarges the boundaries of the community to include soils, waters, plants and animals, or collectively: the land' (ibid., 239).

However, most stewardship occurs when the players act *together* for conservation goals in partnerships. For example, Ontario Parks annually funds the Nature Conservancy of Canada, which in turn uses the funding to invest in land acquisitions and protection. The NCC, it is argued, delivers many times the amount of protected areas with these funds than the government could have with the same amount (Parks Canada Agency, 2000: 9–12). In Victoria, The Land Conservancy (TLC) acquired a significant acreage of Garry oak ecosystem, in part due to a landmark arrangement where the local municipality has agreed to advance TLC a three-year, interest-free loan of almost $400,000 to acquire the property. On a smaller scale, a single conservation easement is usually the result of many different players: the individual who owns the land, the local land trust with the advice and expertise, the government providing the legislation and financial incentive, and science (government and research institutions) providing direction and inventory of critical habitat.

CHALLENGES AND LIMITATIONS OF STEWARDSHIP

Stewardship is playing a larger role in Canadian protected areas every year and is helping to sustain ecosystems within and outside of existing parks. Four challenges or limitations to the approach, however, need to be explored.

First, there are serious questions concerning the extent to which these approaches contribute to biodiversity conservation. Lands restored or protected via stewardship are subject to many of the same challenges facing national and provincial parks. As the Panel for the Ecological Integrity of Canada's National Parks acknowledged (Parks Canada, 2000), protecting ecological integrity is a difficult task that requires active management

(Chapter 5). Like the national parks, private organizations may lack the science and training necessary to manage lands or to gauge priorities for land acquisition. Easements and private protected areas are plagued by a lack of monitoring (Hannah, 2006) and, as Operation Burrowing Owl (Box 16.1) exemplifies, voluntary initiatives can only go so far in actually stemming species losses. While conservation organizations are able to respond quickly to development pressures or opportunities, acting fast and quickly also has downfalls—many resources are required to generate support, and the area may not fit into an overall plan that protects biodiversity. As Brown (1998: 6) notes, 'There is an essential tension between responding to opportunities that arise and taking a strategic approach', and thus a real need exists to 'strike a balance between opportunism and strategy'. Stewardship cannot replace continued investment in strong federal and provincial protected areas, which have powerful legislated mandates for ecological integrity. And a vital need persists for endangered species legislation to protect critical habitats on a systematic, networked basis. Ontario's new legislation will pave the way, hopefully, for other provinces to improve their own legislation. Furthermore, other natural resource legislation and policy, such as environmental impact assessment and forest harvesting codes, must be strengthened (not deregulated) in concert with stewardship initiatives.

Second, the rise of stewardship and of private or public–private conservation has brought new decision-makers into environmental governance. The government remains an important presence and plays a critical role in providing incentives, but within stewardship regimes governments do not select lands for conservation or determine the terms. This leads to questions of accountability and governance. As Reed's research in Saskatchewan demonstrates (Box 16.5), and as Raymond and Fairfax (2003: 636) write, 'The interplay—or lack thereof—between land trust activities and the more public, participatory, and accountable public planning process is a critical issue for practitioners. These private groups and individuals, for most purposes, are significantly removed from public scrutiny, public accountability, and public participation.' The multiplication of land trusts 'constitutes a major problem with public scrutiny' (ibid., 628). Reed argues that this can lead to 'uneven environmental management', where some communities are left out of major land-use changes, and where some areas are governed by increasingly private and unaccountable environmental management regimes.

Third, as Raymond and Fairfax (2003: 637) state, 'All of this has . . . consequences for the equitable distribution of environmental goods and services.' Although there are many cases where private land acquisitions can be used and accessed publicly, sometimes properties or easements are gained without access. Indeed, with the amount of wetlands conserved, one could argue that hunters are some of the biggest benefactors (Roach et al., 2006). In addition, while the rationale for tax breaks is to compensate landowners for restricting development and thus maintaining a reduced land value, some research has found that these easements actually increase property values, thereby further benefiting landowners (Pfeffer and Lapping, 1994; Mills, 1984). As Raymond and Fairfax (2002: 638) note, 'we fear that the conservation community may be allowing the social justice side of the equation to whither in favor of an assertion of ecological "rights".' Conservation organizations should be challenged to bring in equity issues related to Aboriginal restitution and other issues like affordable housing, which is an issue sometimes directly related to urban land-use containment strategies such as 'smart growth'.

Fourth, some might question the narrow use of the term 'stewardship' by many conservation organizations to activities that 'care for the land' directly. In doing so, this narrow definition detracts from driving forces of biodiversity loss, such as climate change from increased CO_2 emissions and the seemingly ceaseless expansion of production and consumption (and profits). Stewardship, or caring for the Earth, must necessarily involve more than voluntary agreements or land acquisitions here and there. A 'land ethic' should not only encourage landowners to adopt environmentally friendly practices, but also should encourage people to reduce their ecological footprint and to address questions of uneven growth and development—the more systemic aspects of biological diversity and habitat loss. Stewardship, to be meaningful, cannot mean simply setting land aside and then proceeding to profit and/or shop as usual!

Notions of protected areas in Canada are expanding from predominately top-down, centralized, and government-led approaches to more wide-ranging efforts by governments, NGOs, corporations, and individuals at the local, regional, provincial, national, and international scales. National parks and other agencies, both private and public, are paying much more attention to the development of conservation strategies and networks, building frameworks for improved ecosystem protection and human well-being. Stewardship is a fast-growing part of this emerging protected area paradigm because it can be flexible, responsive, and partnership-based, and may cultivate conservation values outside park boundaries, within society at large. However, as with other approaches to biodiversity conservation, stewardship is not a panacea. Indeed, with the explosive growth, increased research attention, and analysis of stewardship initiatives, it is necessary to gauge how well they are working not only in terms of species conservation and ecosystem protection, but also in relation to governance and social equity.

ACKNOWLEDGEMENTS

Jessica Dempsey would like to thank the Trudeau Foundation and the Social Sciences and Humanities Research Council (SSHRC) for their support.

REFERENCES

Alberta Stewardship Network. n.d. At: <www.ab.stewardshipcanada.ca/stewardshipcanada/home/scnABIndex.asp>.

Anderson, J., and G. VanDusen. 2003. 'Keeping Gros Morne National Park connected with the broader landscape', in Munro et al. (2003).

Barla, P., J.A. Doucet, and J.D.M.S. Green. 2000. 'Protecting habitats of endangered species on private lands: Analysis of the instruments and Canadian policy', *Canadian Public Policy* 26: 95–110.

Beazley, K. 2003. 'Systems planning and transboundary protected areas management: An example from Nova Scotia, Canada', in Munro et al. (2003).

Boyd, D. 2003. *Unnatural Law: Rethinking Canadian Environmental Law and Policy*. Vancouver: University of British Columbia Press.

Brown, J.L. 1998. 'Stewardship: An international perspective', *Environments* 26: 1–7.

——— and B. Mitchell. 1997. 'Extending the reach of national parks and protected areas: Local stewardship initiatives', in J.G. Nelson and R. Serafin, eds, *National Parks and Protected Areas.* New York: Springer, 103–16.

——— and ———. 1998. 'Stewardship: A working definition', *Environments* 26: 8–17.

Campbell, L., and C.D.A. Rubec. 2006. *Land Trusts in Canada: Building Momentum for the Future.* Ottawa: Wildlife Habitat Canada.

Canada. 2006. *Canadian Protected Areas Status Report 2000–2005.* Ottawa: Environment Canada.

Cardiff, S. 2003. 'Trail stewards: Involving users in human use management in Jasper National Park of Canada', in Munro et al. (2003).

Coalition on the Niagara Escarpment. 1998. *Protecting the Niagara Escarpment: A Citizen's Guide.* Acton, Ont.: Coalition on the Niagara Escarpment.

Colchester, M., and M. Rose. 2004. 'Green corporate partnerships—are they an essential tool in achieving the conservationist mission, or just a ruse for covering up ecological crimes?', *The Ecologist* (July–Aug.).

Davis, S., G. McMaster, D. MacDonald, S. Wiles, J. Lohmeyer, and J. Hall. 2003. 'Application of an ecosystem-based stewardship approach to the conservation of grassland bird habitat in Saskatchewan', in Munro et al. (2003).

Dearden, P., and B. Mitchell. 2005. *Environmental Change and Challenge: A Canadian Perspective,* 2nd edn. Toronto: Oxford University Press.

———, M. Bennett, and J. Johnstone. 2005. 'Trends in global protected area governance', *Environmental Management* 36: 89–100.

Dubois, J., B. Swan, and A.D. Dibb. 2003. 'Research in our backyard: A co-operative ecosystem-based education, research, and management project', in Munro et al. (2003).

Ducks Unlimited Canada (DUC). 2006. *DUC Annual Report.* At: <www.ducks.ca/aboutduc/news/annual_report/index.html>.

———. n.d. 'Our wetland and wildlife progress'. At: <www.ducks.ca/aboutduc/progress/index.html>.

Federal–Provincial–Territorial Stewardship Working Group. 2002. *Canada's Stewardship Agenda: Naturally Connecting Canadians.* Ottawa: Canadian Wildlife Service.

Hannah, L. 2006. 'Governance of Private Protected Areas in Canada: Advancing the Public Interest?', Ph.D. dissertation, University of Victoria.

Hoberg, G., and A. Paulsen. 2004. 'The BC Liberals' "New Era of Sustainable Forestry": A progress report', Issue Brief. At: <www.policy.forestry.ubc.ca/newera.html>.

Holling, C.S., and G. Meffe. 1996. 'Command and control and the pathology of natural resource management', *Conservation Biology* 10: 328–37.

Jones, R., N.A. Sloan, and B. DeFreitas. 2005. 'Prospects for the northern abalone (*Haliotis kamtschatkana*) recovery in Haida Gwaii through community stewardship', in Munro et al. (2003).

Kerr, J.T., and I. Deguise. 2004. 'Habitat loss and the limits to endangered species recovery', *Ecology Letters*: 1163–9.

Leopold, A. 1949. *Sand County Almanac.* New York: Oxford University Press.

McAffe, K. 1999. 'Selling nature to save it? Biodiversity and green developmentalism', *Society and Space* 17: 133–54.

McCarthy, J., and S. Prudham. 2004. 'Neoliberal nature and the nature of neoliberalism', *Geoforum* 35: 275–83.

McGinnis, M.V., ed. 1999. *Bioregionalism.* New York: Routledge.

Merenlender, A.M., L. Huntsinger, G. Guthey, and S.K. Fairfax. 2004. 'Land trusts and conservation easements: Who is conserving what for whom?', *Conservation Biology* 18: 65–76.

Millennium Ecosystem Assessment. 2005a. *Ecosystems and Human Well-being : Current State and Trends: Findings of the Condition and Trends Working Group*. Washington: World Resources Institute.

———. 2005b. *Ecosystems and Human Well-being: Biodiversity Synthesis*. Washington: World Resources Institute.

Mills, J. 1984. 'Conservation easements in Oregon: Abuses and solutions', *Environmental Law* 14: 555–83.

Munro, N.W.P, P. Dearden, T.B. Herman, K. Beazley, and S. Bondrup-Nielsen, eds. 2003. *Making Ecosystem-based Management Work*. Proceedings of the Fifth International Conference on Science and Management of Protected Areas, Victoria, BC, May 2003. Wolfville, NS: SAMPAA. At: <www.sampaa.org/publications.htm>.

Murray, C. 2005. 'Decision support for managing invasive species in Garry oak ecosystems', in Munro et al. (2003).

Nature Conservancy Canada. 'About NCC'. At: <www.natureconservancy.ca/site/PageServer?pagename=ncc_about_index>.

Nova Scotia. 2005. *Protecting Nature on Your Property*. Halifax: Department of Natural Resources. At: <www.gov.ns.ca/enla/protectedareas/docs/PrivateLandBrochur.pdf>.

Ontario Heritage Foundation—Carolinian Canada. At: <www.heritagefdn.on.ca/userfiles/HTML/nts_1_2772_1.html>.

Ottoway, D., and J. Stephens. 2003a. 'Corporate ties enrich, entangle Nature Conservancy—$1.6 million home loan among president's perks', *Washington Post*, 3 May, A1.

——— and ———. 2003b. 'Nonprofit land bank amasses billions', *Washington Post*, 4 May, A1.

——— and ———. 2003c. 'On Eastern Shore, for-profit "flagship" hits shoals', *Washington Post*, 5 May, A11.

Parks Canada. 1998. *State of the Parks: 1997 Report*. Ottawa: Department of Canadian Heritage and Canadian Government Publishing Centre.

Parks Canada Agency. 2000. *Unimpaired for Future Generations? Protecting Ecological Integrity with Canada's National Parks*, vol. 2: *Setting a New Direction for Canada's National Parks*. Report of the Panel on the Ecological Integrity of Canada's National Parks. Ottawa.

Pfeffer, M.J., and M.B. Lapping. 1994. 'Farmland preservation, development rights and the theory of the growth machine: The views of planners', *Journal of Rural Studies* 10: 223–48.

Raymond, L.S., and S.K. Fairfax. 2003. 'The "shift to privatization" in land conservation: A cautionary essay', *Natural Resources Journal* 42: 599–640.

Reed, M. 2007. 'Uneven environmental management: A Canadian comparative political ecology', *Environment and Planning A* 39: 320–38.

Roach, C.M., T.I. Hollis, B.E. McLaren, and D.L.Y. Bavington. 2006. 'Ducks, bogs, and guns: A case study of stewardship ethics in Newfoundland', *Ethics and the Environment* 1: 43–70.

Robertson, M.M. 2004. 'The neoliberalization of ecosystem services: Wetland mitigation banking and problems in environmental governance', *Geoforum* 35: 361–73.

Skeel, M.A., J. Keith, and C.S. Palaschuk. 2001. 'A population decline recorded by Operation Burrowing Owl in Saskatchewan', *Journal of Raptor Research* 35: 399–407.

Shultis, J. 2003. 'Implications of neo-conservatism and fiscal conservatism on science and management in protected areas', in Munro et al. (2003).

Vancouver Sun. 2000. 'Hunters stalk waterfowl preserve: Four islands off Ladner are the playground of an exclusive club of affluent wildfowlers', 27 Nov.

Venter, O., N.N. Brodeur, L. Nemiroff, B. Belland, I.J. Dolinsek, and J.W.A. Grant. 2006. 'Threats to endangered species in Canada', *BioScience* 56: 903–10.

Warnock, R.G., and M.A. Skeel. 2004. 'Effectiveness of voluntary habitat stewardship in conserving grassland: Case of Operation Burrowing Owl in Saskatchewan', *Environmental Management* 33: 306–17.

Wiersma, Y.F., and T.D. Nudds. 2003. 'On the fraction of land needed for protected areas', in Munro et al. (2003).

KEY WORDS/CONCEPTS

accountability
Canadian Nature Federation
connectivity
creative development
Ducks Unlimited
easement
ecogift program
ecological goods and services
education
equity
governance
greater park ecosystem
greenwash
land trusts
management agreement
management incentive
monitoring
Nature Conservancy of Canada
neo-liberalism
non-governmental organization (NGO)
private land acquisition
recognition
stewardship
tax incentives
technical assistance
uneven environmental management
verbal agreement

STUDY QUESTIONS

1. Why is stewardship needed in Canada? Why is it growing in Canada?
2. List some examples of stewardship.
3. How does stewardship differ from the management of parks and protected areas?
4. What are ecological goods and services and how is this approach being used in Canada?
5. What can governments do to promote stewardship?
6. What can individuals do to promote stewardship?
7. Describe three pros and three cons of stewardship, using examples.
8. Select any environmental NGO and research the recent activities of the organization. Discuss the extent to which these activities complement the role of existing parks and protected areas.
9. How can stewardship lead to uneven environmental management or raise questions of accountability and governance? How could these issues be mitigated or avoided?

CHAPTER 17

International Perspectives

Philip Dearden

INTRODUCTION

There are few areas of such firm global consensus as the degradation of environmental quality. It is one factor that seems to span the awareness of the most poverty-stricken to the richest nations. The interest is not merely academic. Humanity depends on planetary life-support systems for survival. As these systems become degraded they become less able to support human activities. The UN's Millennium Ecosystem Assessment concluded that over the past 50 years there has been 'a substantial and largely irreversible loss in the diversity of life on Earth' (MEA, 2005: 18; see Box 17.1). The assessment

BOX 17.1 Millennium Ecosystem Assessment

Over the past 50 years, humans have changed ecosystems more rapidly and extensively than in any comparable period of time in human history, largely to meet rapidly growing demands for food, fresh water, timber, fibre, and fuel. This has resulted in a substantial and largely irreversible loss in the diversity of life on Earth.

The changes that have been made to ecosystems have contributed to substantial net gains in human well-being and economic development, but these gains have been achieved at growing costs in the form of the degradation of many ecosystem services, increased risks of nonlinear changes, and the exacerbation of poverty for some groups of people. These problems, unless addressed, will substantially diminish the benefits that future generations obtain from ecosystems. The degradation of ecosystem services could grow significantly worse during the first half of this century and is a barrier to achieving the Millennium Development Goals.

The challenge of reversing the degradation of ecosystems while meeting increasing demands for their services can be partially met under some scenarios the MA [Millennium Assessment] has considered, but these involve significant changes in policies, institutions and practices that are not currently under way. Many options exist to conserve or enhance specific ecosystem services in ways that reduce negative trade-offs or that provide positive synergies with other ecosystem services.

Source: Millennium Ecosystems Assessment (MEA) (2005).

suggests that human activities have increased species extinction rates by as much as 1,000 per cent over background rates and that up to 30 per cent of mammal, bird, and amphibian species are currently threatened with extinction.

These estimates represent the work of over 1,360 experts from 95 countries that were reviewed by 850 other experts. The results are further confirmed by work such as the Living Planet Index produced by WWF (2006), which is based on trends from 1970 to 2003 in over 3,600 populations of more than 1,300 vertebrate species from around the world. Over the 30-year period the index shows an overall population decline of around 30 per cent. Just as with global climate change, there is no scientific dispute regarding the scale and speed of environmental degradation and resulting loss in biodiversity now occurring across the globe. With global populations projected to increase from the 6.5 billion of 2006 to about 9 billion by 2050, there are going to be very significant challenges in conserving Earth's remaining biological heritage in the future.

As a result of this concern, increasing attention is being given to biodiversity at the international level. In particular, the Convention on Biodiversity was signed following the World Summit on Sustainable Development, held in Rio de Janeiro in 1992. Canada was the first industrial country to sign the Convention and the Secretariat is housed in Montreal. The Convention requires signatories to develop protected area systems, and the Seventh Conference of the Parties in 2004 produced a detailed work plan to aid in protected area system expansion and effective management.

Protected area systems have seen a remarkable growth over the last decade, expanding from roughly 4 per cent of the land surface to almost 12 per cent by 2003 (Chape, 2003; see Figure 17.1). Nonetheless, some biomes, such as grasslands, remain badly underrepresented, and under 0.5 per cent of the world's oceans are under reserve designation.

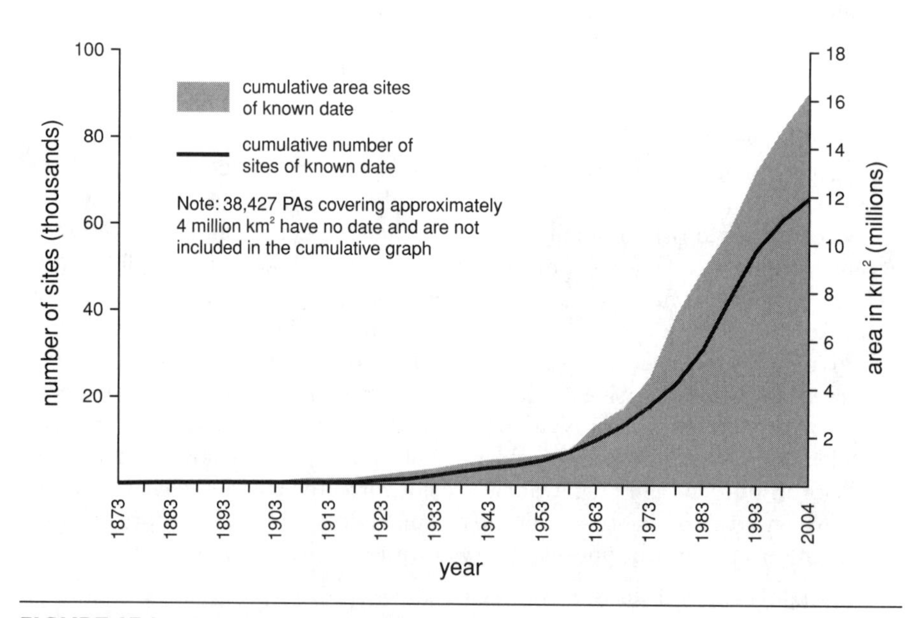

FIGURE 17.1 Historical growth in global protected areas.

Rodrigues et al. (2004) concluded that over 1,400 terrestrial locations alone are critically needed as additions to protected area systems in the future if biodiversity losses are to be slowed. Although many protected areas are under threat, the evidence suggests that, on the whole, they are one of the most effective means of protecting biodiversity (Bruner et al., 2001). Most biodiversity is in the tropics, and indices such as the Living Planet Index clearly show that most biodiversity erosion is occurring in the tropics. For that reason, many protected areas are being established in tropical countries, countries that are often the poorest on the planet. Tanzania, for example, is ranked as the country with the lowest purchasing power parity (PPP) in the world and has almost 40 per cent of its area under protection. The Congo, ranked third lowest in PPP, has 18 per cent, and Zambia, at seventh, has 41 per cent (Scherl et al., 2004). Thus, protected areas are not randomly distributed but tend to be concentrated in those countries with high poverty levels, and this relationship between parks and poverty is the main issue on the global protected areas agenda and was the focus of the Fifth World Parks Congress, 'Benefits Beyond Boundaries', held in Durban, South Africa, in 2003.

ENVIRONMENT, POVERTY, AND PARKS

Poverty means different things in different societies, but should be understood to mean more than just lack of income. Poverty often includes factors such as: lack of opportunities for self-improvement, vulnerability to disasters and ill health, exclusion from decision-making, and lack of capacity to defend community interests. When the United

FIGURE 17.2 Virtually all the possessions of this family in Botswana can be seen in the photograph. *Photo: P. Dearden.*

Nations articulated its Millennium Development Goals in 2000, the first one was to halve both the proportion of people whose income is less than $1 day and the people who suffer from hunger between 1990 and 2015. However, in many cases the creation of protected areas to conserve biodiversity causes the foreclosure of future land-use options, and this may create significant economic opportunity costs that act against programs intended to alleviate poverty. Often, the ones bearing these costs most heavily are the local people, who are also often the poorest people in society.

Part of the Problem: Parks May Exacerbate Poverty

Park establishment and management do not have a good history overall with local peoples. As seen in Chapters 2 and 14, historically, even in Canada, local peoples were forcibly displaced from their homes as a result of park establishment and many would argue that Aboriginal peoples are still being displaced from their lands and seas. Even when allowed to remain within park boundaries or settle on park peripheries, local peoples are generally forbidden to use park resources and may suffer further through crop-raiding animals or predators that kill their livestock or even family members. The result of these impacts is often debilitating poverty. West and Brockington (2006) provide an overview of some of the 'unexpected consequences' of protected areas on local peoples, and there are many examples of the costs borne by local peoples as a result of park establishment. In Madagascar, for example, Ferraro (2002) estimated the opportunity cost to communities that harvested wild resources at between $19 and $70 per

FIGURE 17.3 Elephants emerging from the parks in the dry season are a major problem for farmers in Sri Lanka, who can lose their whole crop during one night. *Photo: P. Dearden.*

household per year. Another Madagascar study, which also included the cost of abandoning slash-and-burn agriculture, put the cost at between $93 and $191 (Brand et al., 2002). These costs are likely about 10 per cent of annual household income, with a total opportunity cost of $3.37 million (Ferraro, 2002).

Hence, local people often harbour a justified sense of resentment towards parks, and out of necessity may resort to poaching to kill animals (Figure 17.4) or collect plant material, and to encroachment to create agricultural lands inside the park. In sub-Saharan Africa, Wilson and Wilson (2004) reported that the biggest threats to World Heritage Sites are poverty-related, with unsustainable resource extraction affecting 71 per cent of sites and encroachment for agriculture and grazing, 38 per cent. At Doi Inthanon National Park in northern Thailand, Dearden et al. (1996) found that 85 per cent of villagers illegally cut firewood in the park and 37 per cent were involved in illegal hunting. Obviously, these activities are inconsistent with biodiversity conservation, and all large mammals and birds had been extirpated from the park. Social surveys showed that almost 40 per cent of the local people were in debt and the forest remained their only source of income.

BOX 17.2 Sri Lanka—Parks and Poverty

Sri Lanka is arguably the richest country in the world in terms of biodiversity per unit area and is part of the Western Ghats Biodiversity Hotspot globally recognized by scientists (Mittermeier et al., 1999). In Sri Lanka, 28 per cent of flowering plants, 65 per cent of amphibians, 52 per cent of reptiles, 41 per cent of freshwater fishes, and 14 per cent of mammals are recognized as endemic species. At the same time, however, over 480 plant species and 75 per cent of the endemic vertebrates are recognized as threatened.

Sri Lanka has established an excellent system of protected areas covering some 14 per cent of the country. These PAs, however, are under threat. Although there are no communities within the parks, over 1.5 million people live in the zone surrounding the PAs and surveys suggest that over 90 per cent of the people depend on PAs for their livelihoods. Up to 75 per cent of the families receive poverty alleviation assistance from the Sri Lankan government. Sri Lanka is also a very densely populated country, with almost 20 million people in an area roughly the size of Vancouver Island.

As a result of these pressures the protected areas suffer from encroachment for agricultural land, extensive livestock grazing, poaching, firewood collection, and gem mining. In order to combat these problems the Sri Lankan government has embarked on a program to develop alternative income opportunities for communities in the vicinity of parks. Much of this program hinges on the very same biodiversity that is under threat, through aiming to attract more eco-tourists to the island. Not only is the wildlife diverse, it is also abundant and easily viewed compared to other Asian locations. It is probably the best place in the world to see leopards and the only place where large aggregations of Asian elephants can be seen (Figure 17.3). Unfortunately, these attempts to bolster tourism income are being critically hampered by the ongoing conflict between the government of Sri Lanka and the Tamil minorities in eastern and northern Sri Lanka.

FIGURE 17.4 This leopard was poached in Horton Plains National Park, Sri Lanka. *Photo: P. Dearden.*

Furthermore, at a nearby reserve where intensive agriculture had been developed outside the boundaries, Tungittiplakorn and Dearden (2002a) found that additional income from the farming reduced illegal activities in the park, and the few remaining poachers were also the poorest families in the community. There appears to be a strong inverse relationship between poverty and biodiversity conservation (Wells, 1992).

In sum, establishing protected areas may displace local peoples from their traditional lands and livelihoods, and also foreclose future economic opportunities for them from those lands. The result is a sense of antagonism towards the park and illegal use of park resources, often resulting in biodiversity declines. A large investment from the park in protection and guarding is often required in an effort to mitigate these declines.

Part of the Solution: Parks and Ecosystem Services

Sound environmental management is also a foundation for a more sustainable future. This was recognized in the Millennium Development Goals, where Goal 7 called for an 'integration of the principles of sustainable development into country policies and programmes . . . [to] reverse the loss of environmental resources'. An indicator for progress on this goal is the 'land area protected to maintain biodiversity'. Protected areas produce many ecosystem services that are valuable to society. The Millennium Assessment (MEA, 2005) recognizes four main categories. First are the provisioning services that produce natural products such as food, fresh water, fuel, and medicines that are useful to local communities. Water availability is often an overriding benefit recognized by local people. For marine protected areas, a lot of evidence shows that establishment of no-take zones has a beneficial impact on fishing outside the zones (Russ et al., 2004). Fish increase in abundance and size as a result of protection and are then caught by fishers as expanding populations spread beyond the reserve. There is also evidence that similar 'spillover effects' are evident in terrestrial ecosystems (Hart, 2001).

Other services include regulating services, such as climate regulation, watershed protection, coastal protection, carbon sequestration, water purification, and pollination; cultural services, such as cultural heritage, spiritual values, tourism, and education; and supporting services, such as soil formation, primary production, and nutrient cycling. Some of these are felt most by communities close to the reserve (e.g., water purification, pollination), but others, such as climate regulation, are global in nature. Under cultural services there should also be some accounting for option, existence, and bequest values. These values arise from retaining the possibility of use in the future (option values), knowing that an area still exists even if the beneficiary has no intention of actually visiting an area (existence values), and the value of being able to pass on things to future generations (bequest value).

Protected areas also provide economic opportunities for local populations by providing jobs such as rangers and guides. Tourism may provide significant income opportunities. The spillover effect described above may also give rise to sport hunting opportunities that otherwise may have not been available (Johnson, 1997). Furthermore, some countries have programs that return money to local communities from entrance fees paid to PAs. In Uganda, for example, 12 per cent of gross revenue goes back to communities (Worah, 2002).

ADDRESSING THE CHALLENGE

The discussion above indicates that there are both costs and benefits to the creation of protected areas. The main challenge is that both costs and benefits occur at different scales. Balmford and Whitten (2003) argue that the main benefits of PAs are in regulatory services and option, existence, and bequest values and that these benefits accrue mainly to the international community. However, most of the costs are born by the local communities. Therefore, the international community should be actively engaged in trying to mitigate the costs and disperse the benefits to local communities. The following discussion revolves around this challenge and the challenge of making PAs more effective at the global level.

Creating Parks Is Not the Goal

One way to address the challenge created by park establishment in areas of high poverty is to recognize that establishing PAs is not the goal for conservation. It is a means towards an end: the protection of biodiversity, ecosystem services, and the other values outlined in Chapter 1. Many indigenous cultures have conservation ethics and practices that predate modern conservation systems. These systems are based on protecting what is acknowledged, through traditional ecological knowledge (TEK), to be important at the local level, and is 'administered' by the local people. In contrast, modern protected area networks tend to be designed around scientific knowledge, recognizing sites of regional and national importance, and often are administered by national governments far removed from the scene (Figure 17.5). In many cases these state-driven systems have displaced traditional conservation mechanisms. In so doing they may have replaced functioning local systems with more formal, but often dysfunctional state systems. In Africa, for example, there were royal hunting preserves among the Zulu and Swazi, advanced land management systems among the people of Botswana, and resource management protocols among the pastoralists in East Africa and in the coastal forests of East Africa (Barrow and Fabricus, 2002). All these were swept away and replaced by colonially determined, state-run park systems, and these colonial actions fuelled antagonism towards the parks among local peoples that exists to this day (Fabricus et al., 2001).

The protected area literature has many debates as to which system is superior. In fact, both systems have advantages and disadvantages (Dearden, 2002), and one of the key tasks of national conservation strategies should be to determine what conservation systems are already in place and then to work out what kind of approach might work best in specific locations. This requires a thorough understanding of particular place context. For example, in a remote area of northeast Cambodia in a newly established national park, Baird and Dearden (2003) provide persuasive evidence suggesting that biodiversity conservation goals will be furthered if the people indigenous to that area are given jurisdiction over certain resources. In other words, if the goal is to conserve biodiversity, then a mix of the two models shown in Figure 17.5—rather than one or the other—will be required for optimal biodiversity conservation. In the future this is the kind of context-specific approach that will be needed instead of automatic subscription to a goal of creating one kind of protected area with one kind of management approach.

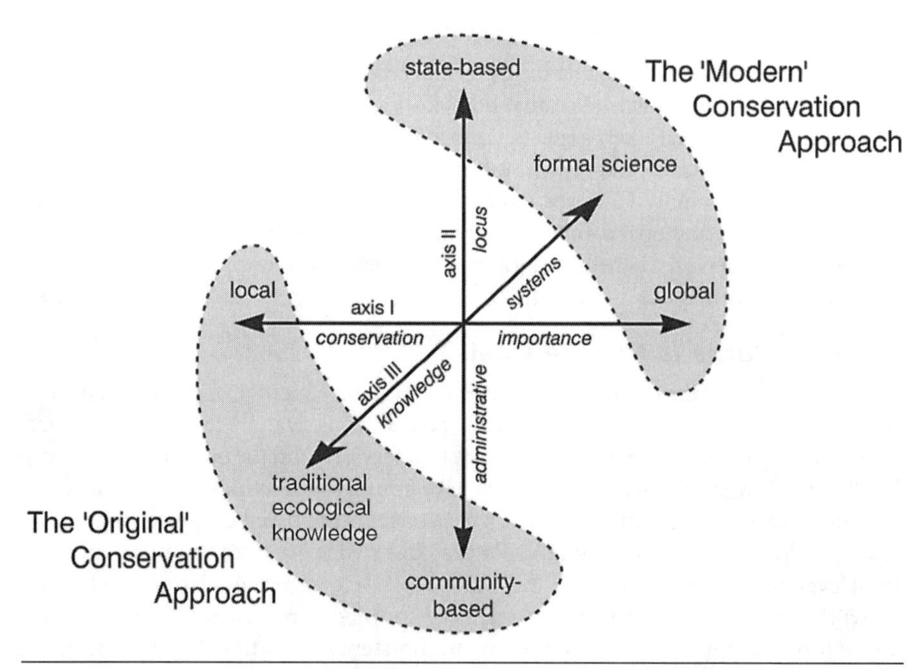

FIGURE 17.5 The 'original' and 'modern' approaches to conservation. These should not be seen as mutually exclusive, but simply as more (or less) appropriate in different circumstances.

The Categories Debate

The World Conservation Union (IUCN) has categorized protected areas to allow for the kind of flexibility mentioned above. The system is described in Chapter 1. Some countries, and even continents, such as Africa, have adopted the categories as the basis for planning and managing their own PA systems. The system is based on management objectives and recognizes that some PAs are more stringently protected than others. For example, Category VI PAs allow for sustainable resource use and do not prohibit mineral exploration and development, but at least two-thirds of the area should be maintained in 'predominantly unmodified natural systems' (Phillips, 2003). Despite the newness of this concept compared with the much longer history of more strictly protected areas, such as national parks, this category now amounts to almost one-quarter of the global protected area and is likely to grow, as governments, even in Canada, look increasingly towards designating areas as cat-

BOX 17.3 Community Conserved Areas

Community conserved areas (CCAs) are where local communities voluntarily protect significant biodiversity, ecological services, and cultural values through customary laws or other effective means. They cover a wide variety of initiatives but generally encompass the following attributes:

- The protectors place high value on the target area either for livelihood or for cultural or spiritual reasons.
- The management is effective in conserving habitats, species ecological services, and cultural values.
- Local peoples hold the power over local decision-making.

There are many different examples, ranging from community forests and reef areas through to sacred sites, and many have been very effective in their conservation efforts. Chernela et al. (2002) provide interesting examples using MPAs in Brazil, Australia, and Tanzania to illustrate the diversity and success in some of these kinds of initiatives.

Many CCAs also face management challenges, however. The new systems of conservation shown in Figure 17.5 often swept away the management systems that were in place. The very existence of the CCA may not be recognized by the authorities. The traditional knowledge base that used to bring continuity to management is often now no longer in existence, as more recent generations have become disconnected from the environment with the advent of modern education systems and the inroads of the mass media. They may also be vulnerable to political change in communities. Tungittiplakorn and Dearden (2002b) describe a situation in northern Thailand where a headman had placed a protective order on the last remaining herd of sambar deer in the area. For years the herd flourished. However, when a new headman took over, anxious to prove his power, he removed the protective order and the deer were soon extirpated.

egory VI lands. The designation is popular because of the very flexibility it was designed to give. It allows designation of lands, and even the downgrading of biodiversity conservation as the prime goal, yet these lands can be counted towards international goals of achieving targets for setting aside lands for biodiversity while still allowing extractive uses.

Although in theory this category would allow for improved conservation overall, as well as boost local livelihoods, there is also the danger that governments will opt to designate increasing numbers of areas as Category VI and neglect the higher categories that provide a greater degree of protection to biodiversity. In so doing they will also generate inflated figures for the amount of land devoted to biodiversity protection as a prime concern. In the United Nations Economic Commission for Europe Temperate and Boreal Forest Assessment, for example, the US included all its national forests as Category VI, including areas that were heavily logged and used for mining and oil and gas extraction. The result is that the US has almost 40 per cent of its forest area classified as 'protected'. These figures do not reflect reality and are being used by anti-conservation forces to argue the case that there is too much protected area in the US. Even Canada has sought to use this argument in trying to promote areas that have not been logged in the commercial forest as 'protected areas' (Neave and Neave, 2003). Unfortunately, there is no guarantee that these lands will not be logged in the future, nor is there protection from logging roads or oil and gas development.

Concern over these difficulties with Category VI and its increasing dominance and focus on human-dominated landscapes led Locke and Dearden (2005) to suggest that Categories V and VI be allocated into a new type of classification—'Sustainable Development Areas'—rather than masquerade as protected areas where protection of wild biodiversity is the main concern. A global summit called to examine the issue acknowledged that many difficulties had arisen with Category VI and is now seeking to tighten the criteria so that the category will not be abused in future. The debate over categories has served to highlight some of the challenges with trying to instill a degree of flexibility into PA management to accommodate social concerns while maintaining the goal of biodiversity conservation as the top priority.

The Transition to Social Fencing

The hardships created by park establishment have been recognized by park planners for several decades. The 1982 World Parks Congress agreed on the principle that the needs of local people should be systematically integrated into protected area planning, and much work has been undertaken on this front in the last 25 years. Critics of orthodox protected areas such as national parks often call them the 'fences and fines' approach to conservation. In other words, the dominant relationship between the PA and local people is one of antagonism. The PA is seen by local people as a threat to their livelihood and illegal use of park resources is rampant, leading to biodiversity declines. The management response to this situation is to exclude the people through boundary fences and by establishing penalties for transgression.

One transition that has to take place is the replacement of these 'hard fences' by social fencing. The concept behind social fencing is that the local people provide a social fence to the park—they protect it themselves because it is in their own best interests to do so. In this scenario, there is no need for the huge investment that park authorities must make in protection; the park is self-policed, and the relationship between the park and people is symbiotic, with both parties benefiting.

The challenge is how to create the conditions for this situation to develop. There have been many attempts over the last decade, some with greater success than others. One initiative has been the development of integrated conservation and development projects (ICDPs). Since the international funding agencies, such as the World Bank, made the connection between poverty, development, and the sustainable use of resources, they have been keen to fund projects to see how this might be achieved. This resulted in a flood of funding, and the rapid development of many projects aimed at providing alternative income sources that would allow people to benefit economically from conservation and reduce environmentally damaging practices. In theory the concept appears sound, and likely to result in 'win-win' situations for both conservation and development. However, the reality has proved much more elusive as the two goals often conflict and development has occurred at the expense of conservation.

McShane and Wells (2004) have summarized some of the main reasons behind the failure of many ICDPs (Box 17.4). Adams et al. (2004: 1147) characterize ICDPs as 'overambitious and underachieving', and warn that, 'in most cases, hard choices will be necessary between goals, with significant costs to one goal or the other.' Ferraro (2001) argues that it might be more effective, rather than designing and implementing complicated development schemes, to simply pay local people directly for ecosystem maintenance through a contracted system of performance payments.

Despite these sentiments, it should be pointed out that most ICDPs have not been in operation for any appreciable time period, and the challenges need long-term strategies, not short-term projects. Successes have occurred. For example, following the establishment of ICDPs surrounding two national parks in Uganda, pro-park attitudes among local people changed from 47 per cent in 1992 to 76 per cent in 2003 (Scherl et al., 2004). The World Wildlife Fund (n.d.) has analyzed six case studies of species conservation projects that have contributed significantly to improving local livelihoods and meeting four of the Millennium Development Goals, and a new generation of ICDPs is now trying to adopt the lessons outlined in Box 17.4.

Ecotourism

One kind of initiative where perhaps the goals of conservation and development are more compatible is in the development of ecotourism (also see Chapter 12). In theory, ecotourism is the perfect solution to reconciling the poverty-conservation nexus. Visitors are attracted to see elements of the natural environment. If these elements are conserved, rather than degraded by local communities, and visitation takes place, then a positive feedback loop develops between conservation and income generation. The natural element of the environment becomes worth more in its pristine state than it does if it is killed or degraded. This is a perfect example of incentive-based conservation (Hutton and Leader-Williams, 2003) that has the potential to form social fencing. The economic returns can be substantial. For example, it is estimated that in 2001 the total income value added generated from tourism in the Caprivi, Kueene, and Erongo regions in Namibia was approximately US $16 million. About $1.5 million of this was in the form of income captured at the local level, such as in the form of wages, communal

BOX 17.4 Shortcomings of the First Generation of ICDPs

- The flawed assumption that planning and money alone were sufficient to achieve win-win scenarios.
- Attempting to implement ICDPs within the framework of a time-bound 'project cycle' and failure to adapt to the pace of local communities by trying to meet externally imposed deadlines.
- Failure to identify, negotiate, and implement trade-offs between the interests and claims of multiple stakeholders.
- Lack of adaptive management and flexibility to respond to evolving scenarios.
- Failure to cede significant decision-making to local stakeholders so that ICDPs remained outside local systems, thereby reducing the likelihood that any gains they may have achieved would persist beyond the project life.
- Perceived or actual bias towards the interests of either the protected area management agency or an environmental NGO.
- A focus on activities (social programs and income creation through alternative livelihoods) rather than impacts (on biodiversity).
- Addressing local symptoms while ignoring underlying policy constraints or, conversely, dealing with macro-level issues while ignoring local realities.
- Regarding 'local communities' as homogeneous entities when the reality was a wide range of different stakeholders with different needs and aspirations.

income, and profits on community-owned enterprises (WWF, n.d.). This makes a substantial difference in an area recognized as being the most poverty-stricken in one of the poorest countries in the world. On the marine front, Troeng and Drews (2004) calculated a gross revenue of over $6 million from sea turtle tourism at Tortuguero National Park in Costa Rica, and the Great Barrier Reef generates over $1 billion annually in direct revenues (UNEP, 2006).

Clearly, there are significant economic benefits that can be derived from ecotourism, but there are also significant challenges.

- Ecotourism is not a panacea for reconciling conservation-poverty problems that will work in all locations. Following some early successes, many development projects tacked on an ecotourism component in the expectation that the success would be replicated. However, success is judged by the market, and tourists will not visit unattractive locations just to meet development demands. Ecotourism can work very well, but only in certain places. From a meta-analysis of 251 case studies, Kruger (2005) concludes that ecotourism is more sustainable where charismatic mammal and bird species are involved, for example.
- Successful ecotourism depends on having the right attraction at the right place at the right time, with the necessary infrastructure to access the target resources. In other words, successful ecotourism does not just happen but involves skilled planning and management (Wall, 1997).

- Planning and management must involve local communities from the outset and ensure that specific and significant benefits go to needy communities.
- Communities are not homogeneous and rarely do benefits get distributed evenly. Marginalized sectors of society often stay that way unless redistributive steps are taken (e.g., Bookbinder et al., 1998; Spiteri and Nepal, 2006).
- Ecotourists are usually wealthy and tend to be very discerning in terms of the tourism facilities they will patronize. Private-sector involvement is often required to ensure that facilities are managed up to an acceptable standard.
- Some private-sector operators have developed world-class, environmentally sensitive facilities that result in significant benefits to communities and conservation. Others are in it just for the money, and standards have to be set in order to ensure that ecotourism is not just mass tourism in disguise.
- Ecotourism is vulnerable to the same charges as ICDPs in that, although development may prosper, it is also necessary that conservation also benefit (Hvenegaard and Dearden, 1998). Kruger (2005) found that only 17 per cent of the 251 projects reviewed claimed an improvement in conservation values.
- Encouraging tourists presupposes that the PA has a carefully constructed visitor management component to the management plan. The fundamental goal of parks is to conserve biodiversity, and if excessive and ill-planned tourism occurs, then considerable damage can be sustained by park resources. An effective monitoring program must be in place to detect possible negative impacts of visitation, and the institutional capacity must exist to implement limits on tourism (Buckley, 2003; Roman et al., 2007).

FIGURE 17.6 The presence of large charismatic species is one characteristic that makes Kruger National Park in South Africa a major money-generator for conservation. *Photo: P. Dearden.*

Governance

Institutional capacity to implement limits to tourism in protected areas is part of the overall challenge of governance. Although there have been major increases in the number and area of PAs over the last decade, these expansions will only contribute to biodiversity conservation if the PAs are managed effectively. Governance refers to how decisions are made and, particularly, who is involved. A gradation of inclusivity can be recognized, ranging from total control by PA authorities through to total control by local people (Figure 17.7). In reality most management structures fall somewhere between these two extremes.

When areas are co-managed, authority, responsibility, and accountability are shared among two or more parties. The parties may include indigenous peoples, government agencies, NGOs, private interests, and local stakeholders. Joint management occurs in situations where none of these parties has ultimate authority in their own right, but must reach agreement with the other parties. Collaborative management is when authority is held by one party (often the government) but there is a requirement to collaborate with other interests.

There have been some dramatic changes in the balance among these options over the last 15 years. Dearden et al. (2005) undertook a global survey of governance trends between 1992 and 2002 for the 2003 World Parks Congress and reported a strong trend towards greater inclusivity. The amount and strength of stakeholder involvement had increased dramatically and participatory management was now required in more than half of the countries responding to the survey. In 1992, 42 per cent of countries reported that the government was the sole decision-making authority, compared with only 12 per cent a decade later. There were no joint management arrangements in 1992, but 15 per cent of countries reported them by 2002, with an increase from 12 per cent to 30 per cent of countries embracing co-operative decision-making. The survey results clearly demonstrate that park management recognizes the need to seriously engage other interests in park governance.

The survey also documented other changes in global protected area governance. Accountability, for example, reflects how accountable managers are to the public they represent and what measures are made of accountability. The survey found strong growth in the use of accountability tools, such as Canada's State of the Parks reports, annual reports, external audits, national advisory committees, and stakeholder round-

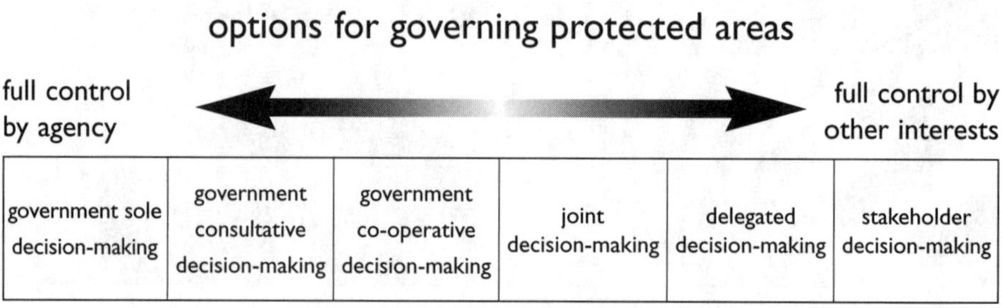

options for governing protected areas

full control by agency ⟵⟶ full control by other interests

government sole decision-making	government consultative decision-making	government co-operative decision-making	joint decision-making	delegated decision-making	stakeholder decision-making

FIGURE 17.7 The spectrum of power-sharing in protected areas.

tables. More than two-thirds of respondents judged that these measures had improved park management. One main means of judging the effectiveness of management is through the park management plan. The implementation of such a plan is now required in more than two-thirds of the countries responding to the survey, although public participation in the plan is required by fewer than half the agencies.

Effective management also requires resources, both financial and human. The survey reported overall increases in both dimensions, but also noted significant shortfalls. Financially, although there had been increases in funding, especially from non-traditional sources (e.g., NGOs, user fees), two-thirds of respondents felt that funding had not kept pace with management needs. They also recognized significant gaps in human resources and sought training in areas such as environmental education, community involvement, park planning and administration, enforcement, and conflict management. Overall, however, 90 per cent of managers reported that the governance of their protected area system had improved in effectiveness between 1992 and 2002. The challenge now is to target key areas for further improvements while consolidating, and learning from, the very rapid changes that took place during that time period.

CONCLUSION

Just as protected area legislation, policy, and management practices changed rapidly through the 1990s in Canada (Chapters 2 and 9), so did global trends. Many of the driving mechanisms are the same—increased public and political awareness of environmental degradation; increased pressure on parks as major depositories of biodiversity; growing awareness of the need to incorporate the views of local communities into management; and the need to extend PA values outside the boundaries into the rest of the landscape. Furthermore, the global discourse on the relationship between poverty and protected areas is not irrelevant to Canada. There are significant pockets of poverty in Canada, and they are concentrated largely among indigenous communities that often live in close proximity to existing or planned parks. Approaches to addressing these challenges at the global scale should be examined for their applicability to Canada.

REFERENCES

Adams, W.M, R. Aveling, D. Brockington, B. Dickson, J. Elliott, J. Hutton, D. Roe, B. Vira, and W. Wolmer. 2004. 'Biodiversity conservation and the eradication of poverty', *Science* 306: 1146–9.

Baird, I., and P. Dearden. 2003. 'Biodiversity conservation and resource tenure regimes: A case study from NE Cambodia', *Environmental Management* 32: 541–50

Bajracharya, S.B., P.A Furley, and A.C. Newton. 2006. 'Impacts of community-based conservation on local communities in the Annapurna Conservation Area, Nepal', *Biodiversity and Conservation* 15, 8: 2765–86.

Balmford, A., and T. Whitten. 2003. 'Who should pay for tropical conservation, and how should costs be met?', *Oryx* 37: 238–50.

Barrow, E., and C. Fabricus. 2002. 'Do rural people really benefit from protected areas—rhetoric or reality?', *Parks* 12: 67–79.

Bookbinder, M.P., E. Dinerstein, A. Rijal, H. Cauley, and A. Rajouria. 1998. 'Ecotourism's support of biodiversity conservation', *Conservation Biology* 12: 1399–1404.

Brand, J., T. Healy, A. Keck, B. Mintern, and C. Randrianarisoa. 2002. *Truths and Myths in Watershed Management: The Effects of Deforestation in the Uplands on Rice Productivity in the Lowlands.* Antananarivo, Madagascar, and Ithaca, NY: FOFOFA and ILO.

Bruner, A.G., R.E. Gullison, R.E. Rice, and G.A.B. Fonesca. 2001. 'Effectiveness of parks in protecting tropical biodiversity', *Science* 291: 125–8.

Buckley, R. 2003. 'Ecological indicators and tourist impacts in parks', *Journal of Ecotourism* 2: 54–66.

Chape, S., S. Blythe, L. Fish, P. Fox, and M. Spalding. 2003. *2003 List of UN Protected Areas.* Gland, Switzerland: IUCN, and Cambridge: UNEP-WCMC.

Chernela, J.M., A. Ahmad, F. Khalid, V. Sinnamon, and H. Jaireth. 2002. 'Innovative governance of fisheries and ecotourism in community-based protected areas', *Parks* 12: 28–41.

Dearden, P. 2002. '"Dern Sai Klang": Walking the middle path to biodiversity conservation in Thailand', in P. Dearden, ed., *Environmental Protection and Rural Development in Thailand.* Bangkok: White Lotus Press, 377–400.

———, M. Bennett, and J. Johnstone. 2005. 'Trends in global protected area governance', *Environmental Management* 36: 89–100.

———, S. Chettamart, D. Emphandu, and N. Tanakanjana. 1996. 'National parks and hill tribes in Northern Thailand: A case study of Doi Inthanon', *Society and Natural Resources* 9: 125–41.

Fabricus, C., E. Koch, and H. Magome. 2001. *Community Wildlife Management in Southern Africa: Challenging the Assumptions of Eden.* London. IIED.

Ferraro, P.J. 2001. 'Global habitat protection: Limitations of development and a role for conservation performance payments', *Conservation Biology* 15: 990–1000.

———. 2002. 'The local costs of establishing protected areas in low-income nations: Ranomafana National Park, Madagascar', *Ecological Economics* 43: 261–75.

Hart, J.A. 2001. 'The impact and sustainability of indigenous hunting in the Ituri Forest, Congo-Zaire: A comparison of hunted and non-hunted duiker populations', in J.G. Robinson and E.L. Bennett, eds, *Hunting for Sustainability in Tropical Forests.* New York: Columbia University Press, 106–53.

Hutton, J.M., and N. Leader-Williams. 2003. 'Sustainable use and incentive-driven conservation: Realigning human and conservation interests', *Oryx* 37: 215–26.

Hvenegaard, G., and P. Dearden. 1998. 'Linking ecotourism and biodiversity conservation: A case study of Doi Inthanon National Park, Thailand', *Singapore Journal of Tropical Geography* 19: 193–211.

Johnson, K.A. 1997. 'Trophy hunting as a conservation tool for Caprineae in Pakistan', in C.H. Freese, ed., *Harvesting Wild Species: Implications for Biodiversity Conservation.* Baltimore: Johns Hopkins University Press, 393–423.

Kruger, O. 2005. 'The role of ecotourism in conservation: Panacea or Pandora's box?', *Biodiversity and Conservation* 14: 579–600.

Locke, H., and P. Dearden. 2005. 'Rethinking protected area categories and the "new paradigm"', *Environmental Conservation* 32: 1–10.

McShane, T.O., and M.P. Wells. 2004. *Getting Biodiversity Projects to Work: Towards More Effective Conservation and Development.* New York: Columbia University Press.

Millennium Ecosystem Assessment (MEA). 2005. *Ecosystems and Human Well-Being: A Framework for Assessment.* Washington: Island Press.

Mittermeier, R.A., N. Myers, and C.G. Mittermeier, eds. 1999. *Hotspots: Earth's Biologically Richest and Most Endangered Terrestrial Ecoregions.* Mexico City: Conservation International.

Neave, D., and E. Neave. 2003. *The Web of Conservation Lands within Canada's Forested Landscapes.* Ottawa: Canadian Forest Service.

Phillips, A. 2003. 'Turning ideas on their heads: A new paradigm for protected areas', *George Wright Forum* 20: 8–32.

Rodrigues, A.S.L., et al. 2004. 'Effectiveness of the global protected area network in representing species diversity', *Nature* 428: 640–3.

Roman, G., P. Dearden, and R. Rollins. 2007. 'Application of zoning and "limits of acceptable change" to manage snorkeling tourism', *Environmental Management* 39: 819–30.

Russ, G.R., A.C. Alcala, A.P. Maypa, H.P. Calumpong, and A.T. White. 2004. 'Marine reserve benefits local fisheries', *Ecological Applications* 14: 597–606.

Scherl, L.M., A. Wilson, R. Wild, J. Blockus, P. Franks, J.A. McNeely, and T.O. McShane. 2004. *Can Protected Areas Contribute to Poverty Reduction?* Gland, Switzerland: IUCN.

Shears, N.T., R.V. Grace, N.R. Usmar, V. Kerr, and R.C. Babcock. 2006. 'Long-term trends in lobster populations in a partially protected vs no-take marine park', *Biological Conservation* 132: 222–31.

Spiteri, A., and S.K. Nepal. 2006. 'Incentive-based conservation programs in developing countries: A review of some key issues and suggestions for improvements', *Environmental Management* 37: 1–14.

Troeng, S., and C. Drews. 2004. *Money Talks: Economic Aspects of Marine Turtle Use and Conservation.* Gland, Switzerland: WWF International.

Tungittiplakorn, W., and P. Dearden. 2002a. 'Biodiversity conservation and cash crop development in northern Thailand', *Biodiversity and Conservation* 11: 2007–25.

——— and ———. 2002b. 'Hunting and wildlife use in some Hmong communities in northern Thailand', *Natural History Bulletin of the Siam Society* 50: 57–73.

United Nations Environment Program (UNEP). 2006. *Marine and Coastal Ecosystems and Human Well-Being: A Synthesis Report Based on the Findings of the Millennium Ecosystem Assessment.* Nairobi: UNEP.

Wall, G. 1997. 'Is ecotourism sustainable?', *Environmental Management* 21: 483–91.

Wells, M. 1992. 'Biodiversity conservation, affluence and poverty—mismatched costs and benefits and efforts to remedy them', *Ambio* 21: 237–43.

West, P., and D. Brockington. 2006. 'An anthropological perspective on some unexpected consequences of protected areas', 20: 609–16.

Wilson, A.C., and E.B. Wilson. 2004. *A Review of Threats to World Heritage Sites 1993–2002.* World Commission on Protected Areas. Gland, Switzerland: IUCN.

Worah, S. 2002. 'The challenge of community-based protected area management', *Parks* 12: 80–93.

World Wildlife Fund (WWF). n.d. *Species and People: Linked Futures.* Gland, Switzerland: WWF International.

———. 2004. *How Effective Are Protected Areas?* Gland, Switzerland: WWF International.

———. 2006. *Living Planet Report 2006.* Gland, Switzerland: WWF International.

KEY WORDS/CONCEPTS

Aboriginal peoples
community conserved areas
Convention on Biological Diversity
ecosystem services
ecotourism
governance
integrated conservation and development projects (ICDPs)
IUCN protected area categories
local communities
Millennium Ecosystem Assessment
Millennium Development Goals
opportunity costs
poverty

STUDY QUESTIONS

1. What is meant by the 'opportunity costs' of park establishment?
2. Describe some of the links between poverty and park establishment.
3. What types of ecosystem services may be provided by PAs?
4. Why is creating parks not the goal for conservation?
5. Why is there a debate over the role of IUCN protected area categories?
6. Discuss some of the shortcomings of the first generation of ICDPs. What would you do to address them?
7. Outline the steps to guide ecotourism initiatives to a more successful conclusion.
8. What is governance?
9. What were the main global governance trends between 1992 and 2002?

PART VI

Concluding Perspectives

The idea of wilderness needs no defence. It only needs more defenders.

Edward Abbey, The Journey Home

There is one chapter in this section. The final chapter summarizes some of the key concepts, drawing on a brief case study of Goldstream Provincial Park close to Victoria. However, in this concluding chapter we argue that being current is not sufficient, and we challenge the reader to think in concrete terms about the future. The management of parks requires an understanding of the historical forces that have influenced ecological integrity and visitor experiences, including, for example, population growth, changes in transportation systems, landscape changes near parks, pollution levels, new technologies, changing leisure patterns, and public support for parks. However, the future is seldom a simple extension of past trends, and a broader array of factors needs to be explored as we consider the challenges facing managers in the future. The chapter presents a way of structuring how we examine the future—in order to anticipate what scenarios are possible, and what ones are most likely.

CHAPTER 18

Challenges for the Future

Rick Rollins and Philip Dearden

In this chapter we reflect on some of the main themes revealed in the various sections of the book, while suggesting some challenges for the future. Many of these themes and issues can be illustrated in almost any park in Canada, or in parks found elsewhere in the world. We will confine our examples to a few parks close to our homes on Vancouver Island, and encourage readers to look for similar examples in parks near to them. In this chapter we also want to stress 'thinking about the future'. Protected area managers make decisions that influence the future integrity of a park, yet these decisions are usually based on current circumstances, without giving full attention to possible future scenarios and the forces that will likely influence the future (McNeely, 2005). This chapter moves from the present to a consideration of the future, an element of park management we feel requires more attention.

Goldstream Provincial Park, only 398 ha, is located just 15 minutes from Victoria. Easily accessed from the Trans-Canada Highway, the park receives high numbers of local visitors and tourists. Arriving in the park, one is struck by a stand of huge cedar trees comprising the remnants of the old-growth forest that was once prevalent in the region. In addition to the cedars, one of the most important features and visitor attractions is the annual salmon run that takes place in the park each fall. During this run, park visitors can easily view at close quarters a stream choked with large salmon, struggling slowly upstream against the current to find just the right gravel bed for spawning. Then they die. Watching this drama unfold is an emotionally charged experience, bringing visitors into a close personal contact with the mysteries of the natural world.

INTERPRETIVE SERVICES

Visitors who are moved by the magnificence of this salmon run often want to better understand the story of the salmon: How do the salmon know to find their way back to this very stream where they emerged from eggs several years previously? Where do they go in their journey away from the stream into the open ocean? How do they know when to return? To help visitors better appreciate the story of the salmon, and other features in the park, BC Parks provides a number of interpretive services. One form of interpretation is an interpretive trail, located along the pathway beside the stream. A number of interpretive signs present short informational vignettes related to the life cycle of the salmon. A second form of interpretation is in the form of talks provided by a professional interpreter in an outdoor theatre located in the park campground. A third form of interpretation is provided by a visitor centre, located near the estuary, where the stream spills into the ocean. The visitor centre has a number of exhibits illustrating the natural history of the park, as well as free informational pamphlets and natural history books offered for sale. A live video camera located in the estuary is linked to a large screen in the nature centre, allowing visitors to observe bald eagles and other features difficult to see from the visitor centre. The video camera is interactive, with a 'joy stick' that allows visitors to adjust the camera position to observe different features.

Because of these interpretive services, and the close proximity of the park to the city of Victoria, many schoolchildren are brought to the park as part of environmental education programs operated by local schools. These programs complement existing science programs offered in schools, but also provide a form of experiential education difficult to replicate in the classroom or in the urban environment. Outdoor educators have maintained that this type of experiential education is important for human development, for it makes people aware of the natural world and our role in sustaining natural processes. Research also shows that people who are provided with outdoor experiences in their youth are more likely to participate in nature-based activities throughout their life, and maintain this appreciation of nature (Iso-Ahola, 1980).

FIGURE 18.1 Visitors crowd the banks of the Goldstream River in the fall to watch the annual salmon migration. *Photo: P. Dearden.*

Interpretive programming at Goldstream also stresses the connection between First Nations cultures and park features, notably salmon and cedar. Salmon were the main staple food source for First Nations cultures in the area, and cedar was the principal building material, used for many essential items including homes, canoes, boxes, masks, and totem poles. First Nations near Goldstream Provincial Park retain certain rights to harvest salmon in the park. However, First Nations have little involvement in park management. In other parts of Canada, First Nations' involvement in the creation of new parks has been essential, usually as part of comprehensive land claim settlements, as described in Chapters 11 and 14. Often, this collaboration has led to forms of co-management, a very significant advance in the approach to decision-making in parks. One of the more significant challenges in the creation and management of new parks, particularly in the Canadian North, will hinge on the resolution of land claims and maintaining good working relations between park agencies and First Nations communities.

Visitation at Goldstream Park is consistently high, due in part to the interpretive services provided, but this is not the case with all provincial parks in BC, where attendance has declined in recent years (as has participation in a number of outdoor pursuits, notably fishing and hunting). The reasons for this decline in park use are not well understood, and new research is underway to better understand this phenomenon. However, many speculate that a major cause is 'nature deficit disorder', a term coined to express a growing disconnect across Canada between nature and society (as discussed in Chapter 1). For a number of years young people have had less experience with nature in their youth, and this may be influencing park visitation, particularly in the 18–30 age group. Reasons for this may be: concern for children's safety; lack of time parents have to nurture outdoor experiences; and the emergence of computer-based entertainment competing for the attention of children. What is not certain is the impact declining park visitation may have on support for parks and protected areas in the future. In the past 20 years, much of the impetus for the creation of new parks and protection of parks has been public support, as evidenced in public opinion polls, and the increasingly sophisticated strategies of environmental NGOs in mobilizing public opinion to influence conservation policies. Some park agencies are concerned about the impacts of declining park use and are implementing interpretive strategies delivered in the urban environment outside of parks (see Chapter 8). However, BC Parks has drastically reduced funding to interpretive programs in provincial parks, perhaps exacerbating the risk of nature deficit disorder. Better research is needed to determine what forms of interpretation are most effective in terms of influencing visitor understanding, attitudes, and behaviour. This research is needed in order to provide interpretive services in ways that are demonstrated to meet park objectives (see the discussion in Chapter 1 regarding the potential of interpretation to influence everyday, non-park behaviour).

Rather than using park staff, interpretive programs at Goldstream are provided through a contract tendered to the private sector. As well as interpretation, other services in the park, including maintenance, are delivered through contracts negotiated with private-sector companies. Interestingly, research conducted by BC Parks failed to reveal any decline in service quality, as perceived by parks visitors, when these services were offered through private contracts compared to being offered by parks staff. While many observers are concerned about the implications of private-sector involvement in park operations, the consequences of this approach require further investigation.

REVENUE GENERATION

In British Columbia, the 'Protected Area Campaign' of the 1990s, engineered by the BC Provincial Parks Department, more than doubled the total area of the province under protection, to over 12 per cent. At the same time, provincial debt led to a reduction in funding to parks, including cuts to interpretation and other services. The resulting shortfall in operational funding has compelled BC Parks and other park agencies to consider 'revenue streams', something unheard of in the past and very controversial today. One new form of revenue stream is the use of parking fees, introduced at Goldstream Park and other heavily used provincial parks across the province. This initiative has been met with strong resistance in many parks, sometimes expressed through vandalism of parking meters. This and other protests led the provincial government to remove parking meters from many parks. Nearby, at the Long Beach unit of Pacific Rim National Park, park user fees have increased dramatically to $6.95 per person per day, or $34.65 per person per year. The issue of revenue generation in parks needs more public debate, as this reflects possible shifts in public policy not unlike the challenges facing health care and education. It is interesting to observe, for example, the concern raised by government officials in BC over declining participation in hunting and fishing. This decline has had a significant impact on licence sales for fishing and hunting permits, which in turn has impacted on the operational budgets that these revenues supported. At a time when environmental groups and the general public have expressed concerns about declines in fish and game populations, one might have expected the agencies responsible to be more supportive of relaxed pressure on the resource. A similar issue has occurred on the West Coast Trail in nearby Pacific Rim National Park Reserve. Here, a quota system of 52 people per day was established in 1993 in response to high use levels and the concern of park staff and visitors about diminished wilderness experiences and possible threats to the park environment. To defray the high cost to maintain the trail, a user fee was also established, based in part on visitor research (see Chapter 6). Over the years, use levels on the trail declined somewhat, but mainly in the shoulder seasons (spring and fall). This led park management to increase the quota from 52 to 60 people per day in order to meet revenue targets. This is an example of revenue generation taking precedence over ecological integrity *and* quality of visitor experience. In terms of ecological integrity, no commitment was made to monitor natural conditions to determine impacts of higher use levels on park resources. In terms of visitor experience, survey research in 1994 and 2000 (prior to the quota increase) showed highest support for retaining current use levels or reducing use.

INNOVATION AND INVOLVEMENT

As park agencies struggle with revenue streams and the lack of public funding, it is apparent that protected area managers will need to be innovative in order to find resources to sustain adequate levels of park management. At Goldstream, for example, a well-developed volunteer system provides some of the interpretive services described above. On the West Coast Trail, in nearby Pacific Rim National Park Reserve, a Nuu-chah-nulth 'watchman' service provides an important supplement to an understaffed warden service. However, in the adjacent Broken Group Islands segment of the park, some challenges remain. Here, the park does not have the resources to communicate essential park messages to visitors, most of whom travel by kayak into and through the park. The park has taken the innovative approach of developing licensing agreements with kayak guiding companies whereby kayak guides are required to use sound environmental practices of minimal-impact camping when taking groups into the park. Surveys of kayak groups indicate that the guides are very effective in this responsibility; however, other aspects of messaging are often neglected. For example, many park visitors travelling with kayak guides are not aware they are in a national park, and many are disappointed at the lack of active interpretation provided by their kayak guide. Kayak guides in this example illustrate the potential role that the ecotourism industry can play in park management. While some critics would prefer that parks be resourced sufficiently to provide all important messaging and interpretation, there is also a compelling argument that private-sector involvement in the form of ecotourism can provide more innovative services at a lesser cost compared to what park agencies are able to provide.

The involvement of ecotourism in Canadian parks, so critical in the context of developing countries, will require greater efforts to develop properly (see Chapter 12). A recent study of rural tourism development in central British Columbia (Vaugeois et al., 2007) revealed that almost no communication existed between park managers and the local business community. It appeared that local business leaders failed to see the opportunity to partner with parks to promote a package of tourism attractions to the region. Park managers, for their part, seemed too busy with their day-to-day management tasks to find time to engage with local communities. When confronted, park managers acknowledged the need to support local tourism objectives and to solicit community involvement in addressing conservation issues related to the management of the greater ecosystem shared by the parks and local rural communities. At a time when rural communities are struggling to develop a sustainable economy, and parks struggle to effectively conserve natural processes and promote appropriate use, it is apparent that better collaboration, through ecotourism and other forms of partnering, is required.

PROTECTION AND CONSERVATION

Goldstream Park provides context for several other lessons in park management. Equally important to the unique visitor experience of observing the ritual of the salmon run is the protection the park provides for salmon spawning habitat. To protect salmon habitat in the park, visitor pathways are placed carefully, where good observation opportunities occur and without damaging the stream bank. However, salmon move in and out of park waters, living most of their lives outside of the park, where considerable pressure from overharvesting and a host of other factors have greatly reduced salmon abundance. Obviously, Goldstream Park does not provide sufficient protection to sustain salmon populations, although the park plays an important role. Protection of salmon and other marine life requires other conservation strategies, as outlined in

Chapter 16. For example, near Goldstream Park a national marine conservation area (NMCA) is being considered by Parks Canada in the Gulf Islands region between Victoria and Vancouver. If approved, this NMCA will help sustain salmon that spawn in Goldstream Park and other marine life found in the area. Yet, public support for various forms of marine protected areas is mixed, contributing to the slow development of the NMCA and related programs across the country. Marine conservation strategies require a high level of consensus within society, and particularly within local stakeholder groups, some of whom view marine conservation as a threat to their livelihood or lifestyle. Park agencies will need to engage with the public in a more significant way to develop support for the various marine protected area initiatives.

Salmon are a conspicuous component of the greater Goldstream ecosystem, but are part of an interconnected web of life only partly understood. Only recently has ecological research demonstrated the extent that salmon contribute to nutrient cycling in the forests surrounding salmon streams, facilitated often by bears who feed on salmon and leave the salmon carcasses on shore. Removal or depletion of any of the components of this food web has severe implications for all other components. Parks need to protect whole ecosystems, as far as possible, to ensure that at least some areas of the world still function in a natural manner (see Chapters 4, 5, and 13). These links illustrate not only the *hospital role* of protected areas but their importance in providing the opportunity for research and understanding of natural processes (Chapter 1).

Salmon at Goldstream Park also provide a food source for bald eagles. However, the abundance of bald eagles at the park has been relatively small until recent years, the

FIGURE 18.2 Salmon bring nutrients from distant oceanic sources back into the river and provide an excellent illustration of the need for ecosystem-based management. *Photo: P. Dearden.*

reasons for which were not understood until now. In 1994, a bear was reported feeding on berries found in the estuary region of the park where visitors often went to watch the salmon or view other features in the area. Out of concern for public safety, the estuary was closed to visitors for a number of weeks until the bear moved away from the area. During this closure, park staff noticed a sharp increase in the number of bald eagles in the area. Park staff did some library research that revealed that bald eagles are very sensitive to human interactions, often not returning to an area for hours after a human has passed through. These observations led park staff to undertake a wildlife experiment that involved keeping the estuary closed to the public and monitoring bald eagle sightings over a number of years. Over the life of this program, bald eagle sightings have been in excess of 200, compared to less than 20 before the closure of the estuary. This illustrates the potential contribution of adaptive management and monitoring in park settings (see Chapter 5). Without such a study it may have been difficult to convince the public and senior park officials of the need to keep the estuary closed to park visitation. To sustain public interest in this management action, a video camera was installed in the estuary (described earlier in this chapter), so that visitors in the nature house at the edge of the estuary could observe bald eagles and other wildlife in the estuary.

Goldstream contributes to the protection of salmon and other natural features characteristic of the region, including the endangered marbled murrelet, which is dependent on old-growth forests. The park is not large enough, however, to adequately protect all the natural features contained therein. Much of the more recent thinking about ecosystem management suggests that parks like Goldstream can become islands in a sea of development, with a resulting loss in biodiversity, unless consideration is given to connectivity between natural areas, as well as the overall management of contiguous

FIGURE 18.3 Bald eagles feeding. *Photo: P. Dearden.*

landscapes (Chapters 4 and 13). These considerations are evident in the Goldstream example, in that the park is part of a continuous network of natural areas known locally as the 'Sea to Sea Greenbelt'. The recent completion of this greenbelt required considerable innovation under the leadership of the Capital Regional Parks Department (CRD Parks). Ten years ago, CRD Parks realized that time was running out to acquire the remaining lands for the Sea to Sea Greenbelt, as well as to acquire other lands identified in the master plan, and that funding was not available for these purchases. In response, the agency conducted a public survey of residents in the regional district to determine public willingness to contribute a special fee through property tax (approximately $10 per household per year) that would be used only to acquire new park lands in the region. The survey demonstrated strong public support for this measure, and this was confirmed by a referendum held later in the year. With this demonstrated public support (at a time when tax increases were not generally popular), local politicians implemented the tax. As these tax funds accumulated, CRD Parks approached the Nature Conservancy, which partnered with CRD Parks, allowing for the purchase of a total of 2,888 hectares of park land, mainly in the Sea to Sea Greenbelt (Ward, 2007). This example illustrates the role of public opinion in shaping park policy, as well as the role of 'stewardship' (the Nature Conservancy) in augmenting the conservation strategies of park agencies (Chapter 16).

PROTECTED AREAS IN POPULATED REGIONS

The work of CRD Parks in the area around Goldstream Park also illustrates the challenge of creating new protected areas in the more densely populated regions of southern Canada, areas often poorly represented in protected area frameworks. Typically, these lands are under intense development pressures due to expanded industry or urban housing, and are already fragmented by existing development. This is similar to the scenario under which the new Gulf Islands National Park Reserve was established in 2003. The park is quite different from most national parks in that it is located in close proximity to two major population centres, Vancouver and Victoria, as well as several small towns and villages scattered throughout the Gulf Islands. The park consists of numerous parcels of land across several islands in the area. These lands were acquired by a variety of means: gifts from local landowners, purchases by Parks Canada, and as land transfers from the province (including many small provincial parks located in the region). This is a small park, fragmented across and between numerous islands that are significantly populated with rural communities, where farm land is mixed with forest and small human settlements characterized by beach cottages, marinas, resorts, and small local businesses. The park has a different feel from that of most national parks, as the various park segments are small and the human footprint on nearby lands is more conspicuous. The park, therefore, does not provide much in the way of 'wilderness experiences' but does provide ecological integrity at a different—but important—scale from that found in most other national parks. Essentially, the park helps protect a rural landscape in order to limit the scale and type of human development in the region. Management of Gulf Islands National Park Reserve illustrates the need for a high level of dialogue with a variety of interest groups, including community residents, First

Nations, tourism associations, and a number of government agencies. Consequently, Gulf Islands demonstrates the leadership role of park agencies in ecosystem management (see Chapter 13).

From a visitor perspective, Goldstream Park differs from the experience provided at other parks nearby. Goldstream is an example of a 'frontcountry experience' easily accessible on a day-trip basis by people living in or visiting Victoria. Arriving in the parking lot, visitors find many facilities and services, including picnic tables, fire pits, modern toilets, and running water, as well as the nearby campground and interpretive services described earlier. In contrast, we can consider the history of Banff, Canada's first national park. Banff was created to promote the early tourism industry in western Canada, and did this by providing a variety of tourism facilities that eventually resulted in the modern townsite evident today, as described in Chapters 2 and 10.

Another level of visitor service is illustrated on the West Coast Trail segment of Pacific Rim National Park, which provides a rugged wilderness experience that can only be accessed by people willing to backpack for 5–10 days and to be totally self-sufficient during their visit. Social science has revealed that market segments exist for a variety of types of park experience, usually characterized by variability in the nature of facilities and services suggested in the above examples. We argue here that ecological integrity is the most important priority of parks but that it is possible, up to a point, to cater to a variety of different visitor experiences, ranging from the frontcountry experiences available at Goldstream Park to the backcountry experiences provided on the West Coast Trail. We do not see a role for townsites within parks; however, gateway communities such as Canmore, Whistler, and Tofino, located near parks, can provide a type of visitor experience that serves to accommodate larger visitor numbers in more highly serviced facilities. These gateway communities provide infrastructure to support an important nature-based tourism industry, and may create more awareness and support for park protection. However, within parks, more subtle zoning systems and regulations should be used to allow for low-impact use in ways that respect ecological integrity, as outlined in Chapters 6 and 7. Social science and conservation science have informed this debate, as illustrated in the case study of Banff (Chapter 10) and elsewhere in the book. Nonetheless, we remain concerned about proposals for higher levels of tourism development proposed for some parks, such as the initiative of the BC government in 2006 to develop resorts within Cape Scott Park, on the northern tip of Vancouver Island, and in other provincial parks.

THE FUTURE OF PARKS AND PROTECTED AREAS

As we consider these and other developments affecting the management of parks in Canada, it is important to consider future scenarios and forces likely to influence parks in Canada and other parts of the world. Park managers in Goldstream, Pacific Rim, Banff, and elsewhere across the country make decisions that influence the character of parks into the future, sometimes far into the future. As we attempt to consider the future implications of park management decisions made today and train park staff who will manage these future scenarios, it is important to consider the forces likely to shape the future. Among these forces are the following (McNeely, 2005):

- *Uncertainty.* Uncertainty is present in a number of dimensions that will influence the future condition of parks: uncertainty about natural processes, uncertainty about the consequences of human activities and interventions, and uncertainty about future political and economic environments.
- *Values attached to protected areas.* Values will shape the types of visitor facilities provided, management of lands adjacent to parks, and development of other conservation strategies in society.
- *Demographics.* Population will likely increase. This may lead to more tourism and financial resources for parks, or it may lead to more pressure on the landscape, undermining the conservation goals that parks contribute to.
- *Social equity.* If parks are viewed as benefiting only the rich, lower-income groups or those who feel they bear the cost of parks (adjacent communities) may choose behaviours that undermine parks. In the developing world, the reduction of poverty, linked to pro-conservation behaviour, will be critical to park management.
- *Non-government agencies.* Groups like the Canadian Parks and Wilderness Society and the Nature Conservancy will continue to support parks and conservation objectives outside of parks.
- *The Internet and other advances in communications technology.* Improved communication systems provide the opportunity to improve knowledge transfer and awareness, thereby influencing the way people think about parks. However, this same technology can create a disconnect, reducing the amount of authentic first-hand experiences with nature.
- *Improved transportation.* Improved transportation allows more people to visit protected areas and so develop greater support for conservation goals. Alternatively, higher park visitation may overstress park features if appropriate visitor management strategies are not developed.
- *Improved technology.* In the past, improved technology has often served parks well, as in the case of composting toilets. On the other hand, technology has contributed to acid rain, climate change and ozone depletion, and other developments that threaten parks.
- *Climate change.* Climate change will not be reversed very quickly, so the future will see changes to many park environments as vegetation and wildlife regimes alter in response to changes in temperature and moisture.

The impact of these and other factors will vary from place to place and over time, but such factors should be considered by all park managers. We cannot assume that the future will be the same as today, and parks need to be resilient to contrary forces, and opportunistic in capitalizing on new opportunities as they occur. The IUCN has prepared three possible scenarios for the year 2023 (McNeely, 2005), as outlined in Box 18.1. These scenarios could relate to Canada, and we encourage students to debate the likelihood of each of these occurring, and why.

In considering different possible scenarios for the future and the factors likely to influence parks, it is evident that the future is not necessarily an extension of the past, and we should not be complacent in this respect. Fifteen years ago the first edition of this book did not fully anticipate the tremendous increase in parks and protected areas

BOX 18.1 IUCN Scenarios of Protected Areas in the Year 2023

Scenario 1: The Triple Bottom Line

By 2023, the world community has finally concluded that its self-interest will best be served through considering the planet to be one world. The Triple Bottom Line world treats economic growth, social well-being, and environmental sustainability as three intertwined goals. Governance follows the principle of subsidiarity, with decision-making as close as possible to the citizen. The 'Global Alliance', a tripartite international body of governments, the corporate sector, and civil society, has replaced the United Nations to become an international governance body, and the nation-state has become less important as a decision-maker. In the Triple Bottom Line world, protected areas are more financially sustainable, as their value for providing environmental and social services has become recognized and converted into policy. They still are constantly threatened by alternative land uses.

Scenario 2: The Rainbow

In the year 2023, the Rainbow world has gone through tumultuous changes that essentially reversed the move towards globalization that seemed inevitable back in 2003. One result was that protected areas are no longer seen as worldwide, or even national, concerns, but are managed for the benefit of local communities. Inevitably, some protected areas that had been imposed by national interests have been converted to agriculture, and communities have sprung up in arable locations within former national parks. But in many cases, the local communities saw it as in their enlightened self-interest to maintain the protected areas, with some even attaining a sacred status. In the Rainbow world, local interest dominates, with profound implications for protected areas, both positive and negative.

Scenario 3: Buy Your Eden

Economics is the dominant theme in the Buy Your Eden world, and the gap between the rich and the poor has widened in 2023. Many protected areas have been privatized, and new ecotourism multinationals are running the worldwide system of 'The World's Greatest Nature', appealing to the prosperous international tourism market. These fortunate few outstanding protected areas (which were called World Heritage Sites until they were purchased by the consortium of private tourism internationals) are very well managed for tourism objectives, which often include maintaining biodiversity, especially of the charismatic type. But the numerous other protected areas that are not deemed to be of sufficient profit potential are suffering from inadequate investment, and many fall prey to the growing number of desperate rural poor.

Source: McNeely (2005). Reprinted with permission.

that has occurred in Canada and elsewhere in the world; nor did we anticipate that park visitation might decline in the future, even as population increased along with improved communication and transportation. We have sought in this edition to capture current thinking about protected areas and their management, and hope this will provide a useful benchmark for students and other readers. At the same time, we encourage readers to stay current by consulting the many other sources of information about parks available on the Internet, in libraries, and in the popular media.

REFERENCES

Iso-Ahola, S.E. 1980. *The Social Psychology of Leisure and Recreation.* Dubuque, Iowa: W.C. Brown.

McNeely, J.A. 2005. 'Protected areas in 2023: Scenarios for an uncertain future', *George Wright Forum* 22, 10: 61–74.

Vaugeois, N., R. Rollins, and D. McDonald. 2007. *Rural Tourism Development in Central British Columbia: Observation Report from the Tourism Research Innovation Project (TRIP)*, June. At: <web.mala.bc.ca/vaugeois/default.html>.

Ward, J. 2007. Personal communication with CRD Parks planner, July.

KEY WORDS/CONCEPTS

adaptive management
declining park visitation
ecosystem management
factors shaping the future
financial resources for parks
First Nations
future scenarios
hospital role of protected areas
land claims
marine conservation
parks and rural community development
private-sector involvement in park management
revenue generation in parks
stewardship
tourism activity in protected areas

STUDY QUESTIONS

1. Review the scenarios presented in Box 18.1 and discuss the feasibility of each. Discuss which scenario seems most likely, and why.
2. Review the future scenario ('vision') developed for Canada's national parks by the Ecological Integrity Panel, and discuss what forces or conditions would need to occur to make this scenario feasible. Do you think this scenario is possible? Discuss.
3. Discuss why visitation in parks has declined in recent years. In what ways might this present a problem for park managers, or in what ways might this be a positive development.
4. Interview a park manager to solicit her/his views regarding ecological integrity and ecosystem management. Ask for concrete examples of how these concepts are being implemented. Do these measures seem adequate?
5. Examine the protected area systems plan for the province in which you live. Conduct research to determine how complete/incomplete the systems plan appears to be. Conduct interviews with park officials to determine what impediments exist. Discuss possible solutions.
6. Interview a representative from any environmental organization, such as the Canadian Parks and Wilderness Society or World Wildlife Fund Canada. Discuss the outstanding current issues in parks in your region and elsewhere, and what actions or strategies might be effective in resolving some of these issues.
7. Look for recent articles and reports about park issues appearing in newspapers and magazines and on television, radio, or the web. Outline the issue presented in the article and prepare a response to the issue, drawing on specific concepts found in this text.

APPENDIX

Recommended Websites

Interest in protected areas has grown rapidly over the last few years as their importance has become increasingly recognized. This means that new developments in the field are always occurring, ranging from changes in legislation and policy through to the designation of new areas and developments in specific parks. The best way to stay abreast of these changes is through the Internet. Most parks have their own websites that can be easily located with a web search. At the global level, the website of the World Parks Commission of IUCN is an excellent resource (www.iucn.org/themes/wcpa/), and at the national level it is useful to check the national (www.pc.gc.ca) and provincial park websites, as well as those of leading NGOs such as CPAWS (www.cpaws.org). Below are some suggestions of useful sites.

Alberta Land Stewardship: www.landstewardship.org
Alberta Stewardship Network: www.ab.stewardshipcanada.ca/stewardshipcanada/home/scnABIndex.asp
BC Parks and Protected Areas Research Forum: www.unbc.ca/bcparf/
British Columbia Marine Protected Areas: ilmbwww.gov.bc.ca/cis/coastal/mpa/index.html
British Columbia Parks: www.env.gov.bc.ca/bcparks/
British Columbia Stewardship Centre: www.stewardshipcentre.bc.ca/stewardshipcanada/home/scnBCIndex.asp
Canadian Boreal Initiative: www.borealcanada.ca
Canadian Council on Ecological Areas: www.ccea.org
Canadian Nature Federation: www.cnf.ca/
Canadian Parks and Wilderness Society: www.cpaws.org
Canadian Wildlife Service, Ecogifts Program: www.cws-scf.ec.gc.ca/egp-pde/default.asp?lang=En&n=EC1F7288-1
Canadian Wildlife Service, Habitat Stewardship Program: www.cws-scf.ec.gc.ca/hsp-pih/default.asp?lang=En&n=59BF488F-1
Children and Nature Network: www.cnaturenet.org
City of Vancouver, Southeast False Creek: www.city.vancouver.bc.ca/commsvcs/southeast/
Committee on the Status of Endangered Wildlife in Canada: www.cosewic.gc.ca/
Conservation International: www.conservation.org

Ducks Unlimited Canada: www.ducks.ca
Eastern Habitat Joint Venture Newfoundland: www.env.gov.nl.ca/env/wildlife/Wildlife/EasternHabitat.htm
Environment Canada, EcoAction 2000: www.ec.gc.ca/ecoaction/index_e.htm
Environment Canada, Marine Wildlife Areas: www.pyr.ec.gc.ca/scottislands/mwa_e.htm
Environment Canada, Protected Area Network: www.hww.ca/hww2.asp?pid=0&id=231&cid=4
Fauna & Flora International: www.fauna-flora.org
Fisheries and Oceans Canada, Marine Protected Areas: www.dfo-mpo.gc.ca/oceans-habitat/oceans/mpa-zpm/index_e.asp
Garry Oak Ecosystem Recovery Team: www.goert.ca/
Georgian Bay Trust Foundation: www.gblt.org/
Habitat Acquisition Trust Foundation (BC): www.hat.bc.ca
Islands Nature Trust (Prince Edward Island): www.islandnaturetrust.ca/
Land Conservancy of BC: www.conservancy.bc.ca
Millennium Ecosystem Assessment: www.millenniumassessment.org/en/index.aspx
Nature Child Reunion: www.naturechildreunion.ca
Nature Conservancy of Canada: www.natureconservancy.ca/
Nature Trust of BC: www.naturetrust.bc.ca/
Ontario conservation land tax incentive program: www.mnr.gov.on.ca/MNR/cltip/
Ontario Heritage Foundation, Carolinian Canada: www.heritagefdn.on.ca/userfiles/HTML/nts_1_2772_1.html
Ontario Wetland Habitat Fund: www.whc.org/wetlandfund/
Operation Burrowing Owl: www.naturesask.ca/stewardship_burrowingOwl.php
Parks and Protected Areas Research Forum of Manitoba: www.umanitoba.ca/outreach/pparfm/
Parks Canada, National Parks: www.pc.gc.ca/progs/amnc-nmca/system/index_e.asp
Parks Canada NMCA Program: www.pc.gc.ca/progs/amnc-nmca/index_E.asp
Parks Research Forum of Ontario: www.prfo.ca
Science and Management of Protected Areas Association: www.sampaa.org
Sensitive Ecosystems Inventories of British Columbia: www.env.gov.bc.ca/sei/
Smart Growth: www.smartgrowth.bc.ca/
Stewardship Canada: www.stewardshipcanada.ca/
United Nations Environment Programme, World Conservation Monitoring Centre: www.unep-wcmc.org
Wildlife Conservation Society: www.wcs.org
Wildlife Habitat Canada Stewardship Awards: www.whc.org/EN/stewardship/stewardship_awards_CSC.htm
World Commission on Protected Areas: www.iucn.org/themes/wcpa/
World Wildlife Fund: www.panda.org
Yellowstone to Yukon: www.y2y.net

Index